WASTES IN THE OCEAN

Volume 1: Industrial and Sewage Wastes in the Ocean
Iver W. Duedall, Bostwick H. Ketchum, P. Kilho Park, and Dana R. Kester, Editors

Volume 2: Dredged Material Disposal in the Ocean
Dana R. Kester, Bostwick H. Ketchum, Iver W. Duedall, and P. Kilho Park, Editors

Volume 3: Radioactive Wastes and the Ocean
P. Kilho Park, Dana R. Kester, Iver W. Duedall, and Bostwick H. Ketchum, Editors

Volume 4: Energy Wastes in the Ocean
Iver W. Duedall, Bostwick H. Ketchum, P. Kilho Park, and Dana R. Kester, Editors

Volume 5: Deep Sea Waste Disposal
Dana R. Kester, Bostwick H. Ketchum, Iver W. Duedall, and P. Kilho Park, Editors

Volume 6: Near-Shore Waste Disposal
Bostwick H. Ketchum, P. Kilho Park, Dana R. Kester, and Iver W. Duedall, Editors

LEAD AND LEAD POISONING IN ANTIQUITY
Jerome O. Nriagu

INTEGRATED MANAGEMENT OF INSECT PESTS OF POME AND STONE FRUITS
B. A. Croft and S. C. Hoyt, Editors

VIBRIOS IN THE ENVIRONMENT
Rita R. Colwell, Editor

WATER RESOURCES: Distribution, Use and Management
John R. Mather

COGENETICS: Genetic Variation in Susceptibility to Environmental Agents
Edward J. Calabrese

GROUNDWATER POLLUTION MICROBIOLOGY
Gabriel Bitton and Charles P. Gerba, Editors

CHEMISTRY AND ECOTOXICOLOGY OF POLLUTION
Des W. Connell and Gregory J. Miller

SALINITY TOLERANCE IN PLANTS: Strategies for Crop Improvement
Richard C. Staples and Gary H. Toenniessen, Editors

ECOLOGY, IMPACT ASSESSMENT, AND ENVIRONMENTAL PLANNING
Walter E. Westman

CHEMICAL PROCESSES IN LAKES
Werner Stumm, Editor

INTEGRATED PEST MANAGEMENT IN PINE-BARK BEETLE ECOSYSTEMS
William E. Waters, Ronald W. Stark, and David L. Wood, Editors

PALEOCLIMATE ANALYSIS AND MODELING
Alan D. Hecht, Editor

BLACK CARBON IN THE ENVIRONMENT: Properties and Distribution
E. D. Goldberg

GROUND WATER QUALITY
C. H. Ward, W. Giger, and P. L. McCarty, Editors

TOXIC SUSCEPTIBILITY: Male/Female Differences
Edward J. Calabrese

ENERGY AND RESOURCE QUALITY: The Ecology of the Economic Process
Charles A. S. Hall, Cutler J. Cleveland, and Robert Kaufmann

AGE AND SUSCEPTIBILITY TO TOXIC SUBSTANCES
Edward J. Calabrese

(*continued on back*)

PLANT STRESS–INSECT INTERACTIONS

PLANT STRESS–INSECT INTERACTIONS

EDITED BY

E. A. HEINRICHS

Department of Entomology
Louisiana State University
Baton Rouge, Louisiana

A WILEY-INTERSCIENCE PUBLICATION

JOHN WILEY & SONS

NEW YORK • CHICHESTER • BRISBANE • TORONTO • SINGAPORE

Library of Congress Cataloging in Publication Data:

Plant stress–insect interactions / edited by E. A. Heinrichs.
p. cm. — (Environmental science and technology, ISSN
0194-0287)

"A Wiley-Interscience publication."
Bibliography: p.
ISBN 0-471-82648-0
1. Insects—Host plants. 2. Plants, Effect of stress on.
3. Insect pests. 4. Plants—Disease and pest resistance.
5. Plants, Protection of. I. Heinrichs, E. A. II. Series.

SB931.P53 1988 88-255
632—dc19 CIP

Printed in the United States of America

10 9 8 7 6 5 4 3 2 1

CONTRIBUTORS

Miguel A. Altieri
Agricultural Experiment Station
Division of Biological Control
University of California, Berkeley
Albany, California

Thomas L. Archer
Texas A & M University System
Agricultural Research and Extension Center
Lubbock, Texas

J. H. Benedict
Texas A & M University
Corpus Christi, Texas

May Berenbaum
Department of Entomology
University of Illinois
Urbana, Illinois

Bruce C. Campbell
Plant Protection Research Unit
USDA-ARS
Albany, California

D. L. Dahlman
Department of Entomology
University of Kentucky
Lexington, Kentucky

D. Dale
Department of Entomology
Kerala Agricultural University
Kerala, India

Abner M. Hammond
Department of Entomology
Louisiana State University Agricultural Center
Baton Rouge, Louisiana

Tad N. Hardy
Department of Entomology
Louisiana State University Agricultural Center
Baton Rouge, Louisiana

J. L. Hatfield
Cropping Systems Research
USDA-ARS,
Lubbock, Texas

E. A. Heinrichs
Department of Entomology
Louisiana State University Agricultural Center
Baton Rouge, Louisiana

Thomas O. Holtzer
Department of Entomology
University of Nebraska
Lincoln, Nebraska

Patrick R. Hughes
Boyce Thompson Institute for Plant Research
Cornell University
Ithaca, New York

John M. Norman
Department of Agronomy and Center for Agricultural Meteorology and Climatology
University of Nebraska
Lincoln, Nebraska

Dale M. Norris
Department of Entomology
University of Wisconsin
Madison, Wisconsin

Thomas J. Riley
Department of Entomology
Louisiana State University Agricultural Center
Baton Rouge, Louisiana

Merle Shepard
Department of Entomology
International Rice Research Institute
Manila, Philippines

C. Michael Smith
Department of Entomology
Louisiana State University Agricultural Center
Baton Rouge, Louisiana

SERIES PREFACE

Environmental Science and Technology

The Environmental Science and Technology Series of Monographs, Textbooks, and Advances is devoted to the study of the quality of the environment and to the technology of its conservation. Environmental science therefore relates to the chemical, physical, and biological changes in the environment through contamination or modification, to the physical nature and biological behavior of air, water, soil, food, and waste as they are affected by agricultural, industrial, and social activities, and to the application of science and technology to the control and improvement of environmental quality.

The deterioration of environmental quality, which began when humans first collected into villages and utilized fire, has existed as a serious problem under the ever-increasing impacts of exponentially increasing population and of industrializing society. Environmental contamination of air, water, soil, and food has become a threat to the continued existence of many plant and animal communities of the ecosystem and may ultimately threaten the very survival of the human race.

It seems clear that if we are to preserve for future generations some semblance of the biological order of the world of the past and hope to improve on the deteriorating standards of urban public health, environmental science and technology must quickly come to play a dominant role in designing our social and industrial structure for tomorrow. Scientifically rigorous criteria of environmental quality must be developed. Based in part on these criteria, realistic standards must be established and our technological progress must be tailored to meet them. It is obvious that civilization will continue to require increasing amounts of fuel, transportation, industrial chemicals, fertilizers, pesticides, and countless other products, and that it will continue to produce waste products of all descriptions. What is urgently needed is a total systems approach to modern civilization through which the pooled talents of scientists and engineers, in cooperation with social scientists and the medical profession, can be focused on the development of order and equilibrium in the presently disparate segments of the human environment.

Most of the skills and tools that are needed are already in existence. We surely have a right to hope a technology that has created such manifold environment problems is also capable of solving them. It is our hope that this Series in Environmental Sciences and Technology will not only serve to make this challenge more explicit to the established professionals, but that it also will help to stimulate the student toward the career opportunities in this vital area.

ROBERT L. METCALF
WERNER STUMM

PREFACE

Despite the current food surplus in some countries, there is concern that the combined effects of continued increases in the human population, loss of forests and arable crop land, and the ever-present specter of plant stresses portend worldwide food, fiber, and lumber shortages in the future. The role of the agricultural scientist in averting these food shortages is to (1) modify the environment in which a plant grows by minimizing the abiotic and biotic stresses and (2) genetically modify plants so that they are able to tolerate the various stresses.

Recent knowledge has increased our awareness of the severe constraints that environmental stresses impose on plant growth and development. Some stresses, such as moisture extremes, caused by drought or excess, have limited crop production through the ages. Other stresses have only become important in recent years. Of considerable concern to crop specialists is the depletion of soil minerals, increased soil salinity, and a global rise in environmental pollution. The necessary widespread use of chemicals, including insecticides, fungicides, and growth regulators, is an unavoidable stress factor that has an ever intensifying effect on crop production. In addition, the biotic stresses—herbivorous insects, plant pathogens, and weeds—are constant threats and cause unmeasurable crop losses. Because of the myriad of interactions among the various abiotic and biotic stresses, it is difficult to quantify their combined effects. However, an understanding of the various plant stresses and their interactions is of extreme importance, and is a subject that, until recently, has received only limited research emphasis.

This book draws together the diverse literature on the effects of the various abiotic (physicochemical) and biotic stresses on plants, and describes how the physiological, chemical, and morphological responses of plants to stress may alter their suitability as hosts for herbivorous insects. Chapter 1 discusses the importance of plant stress on global food production and introduces the concept of plant stress and the general effects of stress on plants. Chapters 2–9 cover the responses of plants and their subsequent suitability as hosts when affected by the physicochemical stresses, soil minerals, moisture, temperature, light, insecticides, herbicides and growth regulators, air pollution, and mechanical damage. The biotic stress herbivorous insects is discussed in Chapter 10; the interactions of plant stress, insect pests, and natural enemies in Chapter 11; diseased plants as hosts for insects in Chapter 12; and the interrelationship between crops, weeds, and herbivorous insects in Chapter 13.

Many of the contributing authors have indicated a dearth of scientific data on the effect of stresses on host plant susceptibility to insects, and because of the extreme practical significance of these stress effects on plants, each chapter expresses the urgent need for accelerated research efforts. Environmental stresses such as pollution and salinity will become increasingly important in the next decade, and this book explains the role of plant protection scientists in dealing with plant stress problems.

Effective plant stress research teams should involve an interdisciplinary approach consisting of entomologists, plant pathologists, weed scientists, plant biochemists, plant physiologists, ecologists, soil scientists, and plant breeders. All agricultural scientists represented by these disciplines will find this book to be a useful resource for their research programs, and for students in these disciplines, the book will serve as a supplement to course texts.

I thank the authors for their contributions and extend my appreciation to B. M. Rheams and S. F. Kelley for their assistance in the preparation of the manuscript.

E. A. HEINRICHS

Baton Rouge, Louisiana
January 1988

CONTENTS

PLANT STRESS–INSECT INTERACTIONS

1

GLOBAL FOOD PRODUCTION AND PLANT STRESS

E. A. Heinrichs

Department of Entomology
Louisiana State University Agricultural Center
Baton Rouge, Louisiana

1. INTRODUCTION

The purpose of this chapter is to inform the reader of the importance of abiotic and biotic stresses as constraints to production of food and fiber and to provide an introduction to the following chapters. Despite the current grain surpluses in some regions of the world, especially North America and Europe, there are dramatic deficits in others. Plant stresses not only decrease productivity, but attempts to minimize stress effects increase the cost of production and minimize profits. Stresses thus have extreme economic significance in the current attempt to increase profitability of agricultural production.

In 1965 Lester Brown stated that "the less developed world is losing the capacity to feed itself" (Brown 1976). The gap between food production and consumption in many countries was wide. Famines in some countries had been averted because crop surpluses grown on North American farms were provided in the form of food aid. In 1965 food grain shipments to developing countries were sufficient to feed 100 million people (Brown 1970).

1.1 Causes of Food Shortages

1.1.1 Population

The "green revolution" in rice and wheat which consisted of improved cultivars, expanded irrigation, and increased use of chemical fertilizers and pesticides began in the 1960s and contributed significantly to narrowing the gap between food production and population (Brown 1981). From 1950 to 1980 world food production doubled, with the increases in developing countries exceeding that of the developed countries. However, populations in developing countries increased at twice the rate of the developed world. Increases in per capita food production in developing countries since the 1950s have only been one-third those of developed countries because increased production has been offset by the high population growth rate (Brown 1981). Although it appears that fertility is decreasing and that the population growth rate may have peaked, the current population structure will result in a population of two billion more people by the year 2000 (Toenniessen 1984), an increase of 70 to 90 million people per year. More than 90% of the population growth will occur in the developing countries (Table 1) where life expectancy has dramatically increased in the last 40 years. During all of history the world population in this century reached one billion, but from 1950 to 1975 it increased from two and one-half billion to four billion and is expected to reach six to seven billion by the year 2000 (Brown 1981). Food production will have to increase by about 100% by the year 2000 to keep pace with population growth (Ritchie 1980). Even if birth rates decrease and the world population does not reach the level predicted, there will be a significant pressure on a limited land resource, with the greatest

Table 1. Population Projections for World, Major Regions, and Selected Countries

	Population (millions)		Percent Increase by 2000	Average Annual Percent Increase	Percent of World Population in 2000
	1975	2000			
World	4090	6351	55	1.8	100
More developed regions	1131	1323	17	0.6	21
Less developed regions	2959	5028	70	2.1	79
Major Regions					
Africa	399	814	104	2.9	13
Asia and Oceania	2274	3630	60	1.9	57
Latin America	325	637	96	2.7	10
USSR & Eastern Europe	384	460	20	0.7	7
North America, Western Europe, Japan, Australia, and New Zealand	708	809	14	0.5	13
Selected Countries and Regions					
People's Republic of China	935	1329	14	1.4	21
India	618	1021	65	2.0	16
Indonesia	135	226	68	2.1	4
Bangladesh	79	159	100	2.8	2
Pakistan	71	149	111	3.0	2
Philippines	43	73	71	2.1	1
Thailand	42	75	77	2.3	1
South Korea	37	57	55	1.7	1
Egypt	37	65	77	2.3	1
Nigeria	63	135	114	3.0	2
Brazil	109	226	108	2.9	4
Mexico	60	131	119	3.1	2
United States	214	248	16	0.6	4
U.S.S.R.	254	309	21	0.8	5
Japan	112	133	19	0.7	2
Eastern Europe	130	152	17	0.6	2
Western Europe	344	378	10	0.4	6

From G. H. Toenniessen, *Salinity Tolerance in Plants,* R. C. Staples and G. H. Toenniessen, eds., © 1984; reprinted by permission of John Wiley & Sons, Inc.

pressure occurring in low- and middle-income countries where population growth is greatest and where 80% of the population will live in the year 2000 (Hopper 1981).

There are regional differences with respect to the food shortage problem. Hunger and malnutrition are problems in Latin America, but there they are

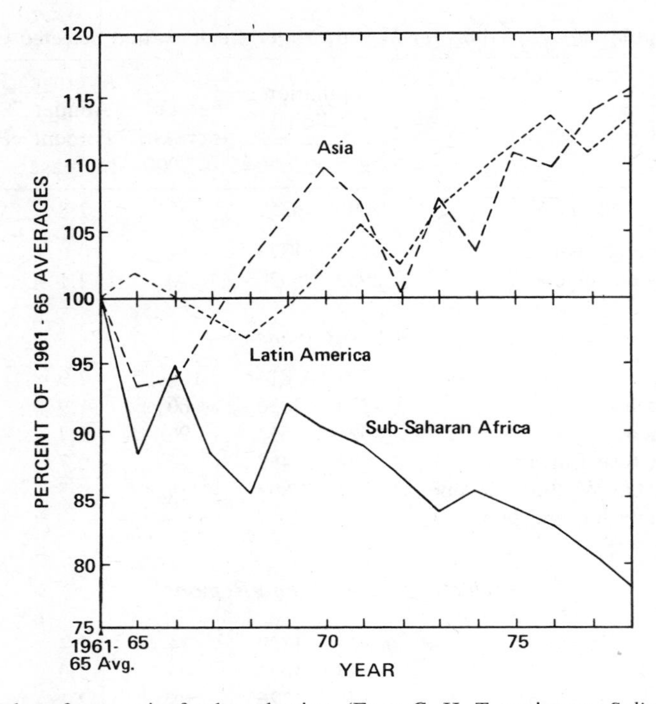

Figure 1. Index of per capita food production. (From G. H. Toenniessen, *Salinity Tolerance in Plants,* R. C. Staples and G. H. Toennissen, eds., © 1984; reprinted by permission of John Wiley & Sons, Inc.)

the result of inequalities in the distribution of income and food. The total number of undernourished and malnourished people is highest in Asia (Toenniessen 1984). Although investment in research in Asia is paying off and is expected to provide further gains in narrowing the food production–population gap, food production per capita in most of the 45 countries of sub-Saharan Africa has steadily decreased since 1965 (Fig. 1). It is in this area of the world that food production programs face their greatest challenge in keeping pace with population increases.

Population growth has generated a need for more land, not only for food production but also for urbanization, energy production, and transportation (Brown 1981). Growth and expansion of cities has been an important cause of cropland loss. The world's urban population increased from 29 to 39% from 1950 to 1975 and is expected to reach 49% by the year 2000. Unfortunately, much of the land lost to urbanization is fertile cropland.

Energy production has been a major factor in the loss of cropland to nonagricultural uses as increasing populations demand more energy. Hydroelectric dams have inundated vast expanses of fertile farmland. In Africa the Kariba dam has displaced 29,000 farmers; a portion of Ghana's rich farmland lies beneath the waters of the Volta dam; and in the Ivory Coast, prosperous cocoa and coffee farms have been submerged in the waters behind the Kossou dam (Brown 1981).

Automobile-based transportation systems are also significant consumers of cropland. In the United States millions of hectares of cropland are required to provide parking area for the approximately 150 million vehicles, but even more important as consumers of the nation's cropland are the streets, highways, and the service stations constructed to service the transportation industry (Brown 1981).

1.1.2 Affluence

Affluence is another major contributor to food shortages (Toenniessen 1984). Of the increased demand for food, 75% is estimated to be due to population increase and 25% to changes in food habits (Ritchie 1980). Increasing affluence in both developed and developing countries has caused an increase in meat consumption, and most animal products require about four pounds of grain for every pound of meat produced (Slater 1981). About 78% of the grain produced in the United States is fed to animals, much of it to pets. This is equivalent to 20 million tons of protein that could be eaten by protein-starved humans. In the United States alone, dogs and cats daily consume protein sufficient to sustain four million humans.

1.1.3 Environmental Stresses

Agricultural productivity is determined by a complex of interactions of plants with their abiotic and biotic environment. Climate is a major abiotic factor determining what crops will be grown and what production will be. The Irish potato famine in the 1840s is a paragon of a climate-related food disaster. Cool, damp weather contributed to the blight that reduced potato production (Schneider and Bach 1981). The potato famine reduced the Irish population by 50% due to death and migration. In the twentieth century concern with the impact of climate on food production was heightened in 1963 when the Russian crop failure ended a period of low-cost surplus grain (Slater 1981). Additional Russian droughts, and monsoon failures in India, led to expanded wheat production in export countries.

In the last two decades unfavorable weather patterns have had severe impacts on food production (Barr 1981, Schneider and Bach 1981). In 1974 adverse weather patterns in the United States brought a record cold winter, followed by a wet spring and an early fall freeze. The most severe drought of the century gripped the Soviet Union and Europe in 1975, and extended drought occurred in the United States in 1976 and 1977. In 1980 droughts occurred in Australia and the southern United States, while excessive moisture was a constraint to crop production in portions of China and the Soviet Union. A decade-long drought has recently taken its toll in the Sahel. In

1986 a severe drought in the southeastern United States drastically reduced crop production.

As world population has continued to increase, there has been intensified pressure on existing cropland, especially in developing countries. Until about 1950 an increase in cultivated area was able to meet the increasing demands for food. From 1950 to 1975 the world population increased at a more rapid pace than did the opening up of new land to cultivation. This resulted in a decrease of per capita area in cereal crops from 0.241 to 0.184 ha (Brown 1981). According to Slater (1981), the loss and deterioration of cropland is not just a current problem. The Tigris–Euphrates valley was once described as the Fertile Crescent and is believed to have supported many more people than it does today. North Africa was once considered the granary of the Roman Empire but has lost much of its productive capacity. However, it is significant to note that the current rate of cropland loss on a world basis is unprecedented and affects all of the countries of the world—both the rich and the poor.

Despite the need for additional cropland, the amount of cropland abandoned annually has increased. The major reason for the abandonment of cropland is the combination of economics, social and ecological forces causing erosion, desertification and waterlogging, and salinization of irrigated land (Brown 1981). It has been estimated that more than 100 tons of topsoil per hectare are lost to erosion annually as tropical hillsides are brought under cultivation (Slater 1981). At this rate approximately 15 cm of tropical topsoil will be lost by the year 2000. These marginal farming areas will then be barren as most have only 15 cm of topsoil. The deforestation and misuse of hillside areas by land hungry Javanese farmers in Indonesia has resulted in erosion at rates faster than reclamation programs can restore the soil (Slater 1981). In the Punjab area of Pakistan soil erosion is laying waste to more than 5000 ha of land yearly. Ethiopia has been described as ''literally going down the river'' because more than one billion tons of topsoil are lost from denuded highlands each year due to trees being cut for firewood (Brown 1981). These are catastrophic losses, but even moderate losses of topsoil can cause decreases in fertility that affect crop production. Pimentel et al. (1976) cite three studies conducted in the United States that indicate corn yields decline four bushels per acre for every 2.5 cm of topsoil lost, starting with a topsoil base of 30 cm or less. Loss of topsoil to erosion and the subsequent loss of arable cropland is a major concern because with the best of agricultural practices, the development of 2.5 cm of topsoil takes one century; if left to nature, it takes much longer (Brown 1981).

Desertification, the transformation of arable land to desert, is progressing at the rate of six million hectares per year (Hekstra 1981). The Sahara Desert which is expanding westward into Senegal and eastward into the Sudan is claiming vast areas. Between 1958 and 1975 this desert expanded southward about 100 km (Brown 1981).

Irrigated areas of the United States that were once lush are becoming

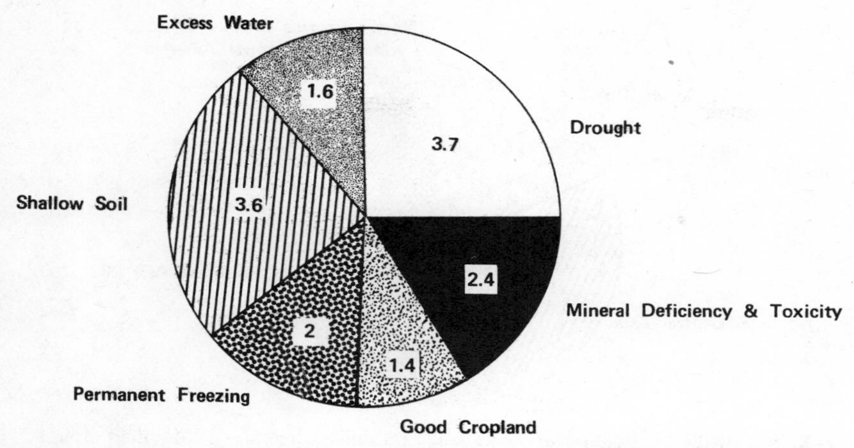

Figure 2. Area of marginal land in the world (in billion hectares). (Modified after Dudal 1976)

unproductive. In the arid western states water is being diverted from irrigation to assuage the thirst of the rapidly expanding cities and in the western Great Plains water is being diverted from farmland to satiate a thirsty population in Denver, Colorado. In Arizona and California cropland that was once irrigated and verdant has turned to dusty brown and has been abandoned (Brown 1981). In many areas the removal of groundwater for irrigation is causing the water table to drop faster than it can be replenished. Where groundwater is depleted, farmers will have to change to growing drought tolerant crops, and some cropland may be totally abandoned altogether.

Much of the world's land is not suitable for cultivation or is marginal for crop production. According to Dudal (1976), only 1.4 out of 14.7 billion hectares of land in the world is good cropland (Fig. 2). The remainder consists of marginal land, 2.4 billion hectares of which is limited in productivity by mineral deficiencies and toxicities, 3.7 million by drought, 1.6 million by excess water, 3.6 million by shallow soil, and 2.0 million by permanent freezing.

Saline soils alone cover 343 million hectares of the earth's surface (Fig. 3). Of these, 230 million hectares are marginal and have limited crop production potential. Of the moderately saline areas, 37 million are in Africa, and 56 million hectares are in the densely populated humid tropics of south and southeast Asia where food and arable land are scarce (Ponnamperuma 1984a).

In addition other irrigated land is being threatened by waterlogging and salinization. One-third of the 210 million hectares of irrigated land has deteriorated because of salinization (Ponnamperuma 1984b). Waterlogging and salinity problems occur where river water is diverted to land that has inadequate underground drainage. As the water table rises, water evaporates through the soil surface concentrating salt and minerals near the surface.

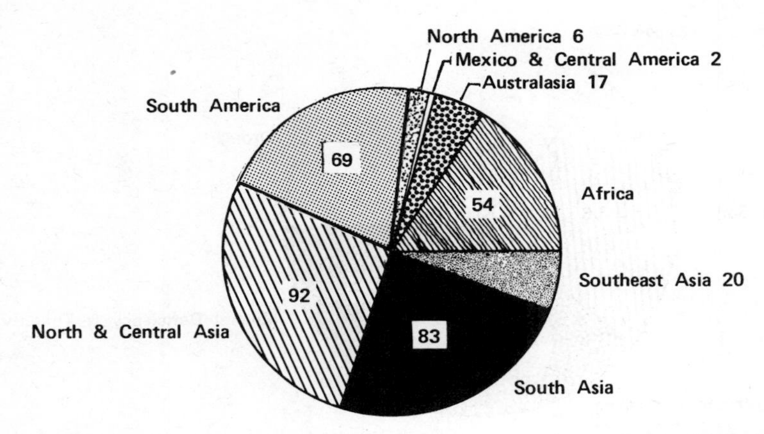

Figure 3. Worldwide distribution of saline lands (in million hectares). (Modified after F. N. Ponnamperuma, *Salinity Tolerance in Plants,* R. C. Staples and G. H. Toenniessen, eds., © 1984; reprinted by permission of John Wiley & Sons, Inc.)

The decline of some of the world's leading civilizations has been due to waterlogging and salinization of cropland (Brown 1981). Waterlogging and salinization are affecting irrigated areas in China, India, and Pakistan; the Aswan dam has caused salinization of land in Egypt. In Argentina two million hectares have been affected by salinization, and in Pakistan almost all of the irrigated land is being threatened by waterlogging and salinization.

Deforestation is another human activity that has indirectly caused severe stress on crop production. Deforestation causes soil erosion which results in flooding. About 200 million people depend on shifting cultivation for a livelihood (Hekstra 1981). Under this farming system the forests act to restore fertility after cultivation in a given area ceases. Logging operations have shortened the fallow period between cultivation periods; and because of the low soil fertility, farmers have moved further up mountain slopes where erosion is severe when the land is tilled. Shifting cultivation in the Ivory Coast removed 30% of the forested area from 1956 to 1966. In the Philippines, shifting cultivation denudes about 3500 km^2 per year. To balance timber harvests with reforestation in developing countries, 10 to 15 million hectares should be planted annually but only 5 million hectares are actually being planted (Thalwitz 1981).

It is evident that inadequate water, soil, and forest management have severely altered the quality of soil over vast areas of the world. As a result plants growing in these areas are stressed by the adverse soil conditions and suboptimal water quantities. Other physicochemical stresses of human origin (e.g., air pollution, and subsequent acid rain, and pesticides) combined with the stresses produced by the previously discussed human actions, play major roles in plant growth and agricultural productivity. These stresses are major concerns of all citizens as the quality of life of the present and future generations is threatened.

2. PLANT PRODUCTION AND GLOBAL FOOD REQUIREMENTS

Plants are the most important source of food for humans, and as population increases, plant products will continue to play an even increasing role in feeding people. Of the total food harvested, 98% comes from the land and 2% from the sea and inland waters (Borlaug 1981). Of the world food harvest, plant products contribute about 80%, and animal and marine products about 20%. Historically, more than 300 plant species have been utilized by people for food. That number has gradually decreased over the centuries. Today, the world's population depends on 23 crop species—primarily the cereal grains, rice, wheat, maize, and sorghum. Cereals occupy approximately 50% of the cultivated land in the world and contribute half of the calories and total protein intake (Borlaug 1981). It is expected that plants will continue to serve as the major source of food in the future. Despite the relative abundance of food in the 1980s, to feed the world's population in the year 2015, crop production will have to double what it was in 1975, an increase from 3.3 to 6.6 billion tons. The maximum genetic potential of many of the crop species is significantly above the average yields achieved today. However, grain yield is dependent on the interaction between the genetic makeup of a plant and the environment in which it is cultivated. Maximum yield can only be achieved in pest-free environments with optimum soil fertility and physics, moisture, temperature, wind, and sunlight. Normally, several of these conditions are suboptimal and yield reductions, varying from moderate to extreme, occur.

In the following section I will introduce the various environmental factors that stress plants and briefly discuss the effects that these stresses have on plants. The chapters that follow discuss in more detail the various environmental factors that cause plant stress, the effect on plants, and the interactions between stressed plants, herbivorous insects that utilize the stressed plants for food, and the natural enemies of those herbivorous insects.

3. ENVIRONMENTAL STRESS EFFECTS ON PLANTS

All plants are dependent on their environment for growth and development. "Environment includes all of the factors and forces prevailing internally and externally on, around, and in the plant" (Treshow 1970). Throughout the centuries people have selected plants that survive adverse environments. Biologists have used the term "stress" for "any environmental factor potentially unfavorable to living organisms" (Levitt 1972). According to Jaffe and Telewski (1984), "stress is an environmentally induced change in a plant which is potentially injurious.'' Grime (1981) has defined ''plant stress'' as ''external constraints which limit the rate of dry matter production.'' Phytochemists consider stress as an external constraint that limits the normal production of secondary metabolites in plants (Timmermann and Steelink 1984). In this book we consider stress as ''any abiotic or biotic factor of the environment that affects plant physiology, chemistry, growth, and/or development in such a way that plants perform below the average for a region.'' Christiansen (1979) has aptly quoted Bob Bergland, former U.S. secretary of agriculture, on the subject of plant stress: ''In twenty-eight years of farming, I have never seen a normal year. It has always been too hot or cold or too wet or dry.'' Indeed, environmental conditions that cause plant stresses are all too common and severely impact on food production in the United States and throughout the world.

Increasing world food production can be accomplished by (1) expanding agriculture into areas not presently suitable for crop production and by (2) increasing per hectare yields on land that is now under cultivation. In both cases it is necessary to develop an agricultural production system in which environmental stress is minimized by altering the environment to fit the plant's requirements or by altering the plant to fit the environment (Christiansen 1982). Throughout the process of domestication of crop plants, actions were taken to improve the adaptation of a crop to its environment or to develop agronomic practices designed to improve the crop environment, and thus attenuate the impact of environmental stress on the plant.

Stress injury can be divided into primary and secondary categories (Levitt 1972). A primary stress injures a plant by the strain that it produces. When a stress injures a plant, not by the strain that it produces, but by giving rise to a second stress, secondary stress injury occurs. Thus a given stress may not itself be injurious but produce another strain that is injurious to the plant. For example, a high temperature may not injure the plant but may produce a water deficit (secondary stress) that is injurious.

Environmental stress consists of two major types, physicochemical (abiotic) and biotic (Fig. 4). Physicochemical stresses consist primarily of (1) physical and chemical properties of soils, (2) moisture deficit and excess, (3) temperature extremes, (4) electromagnetic energy, (5) plant growth regulators and pesticides, (6) air pollution, and (7) mechanical damage. Major biotic stresses primarily consist of (1) insect herbivores, (2) plant competitors, and (3) plant pathogens. These stresses can reduce the performance of a plant; and when they do, the suitability of the plant as a host to herbivorous insects may be altered. In the following section the effects of the various physicochemical and biotic stresses on plants are discussed.

3.1 Physicochemical Stresses

3.1.1 Chemical Properties of Soils

Along with water and temperature stresses, mineral deficiencies and excesses occur throughout the world and are among the most important con-

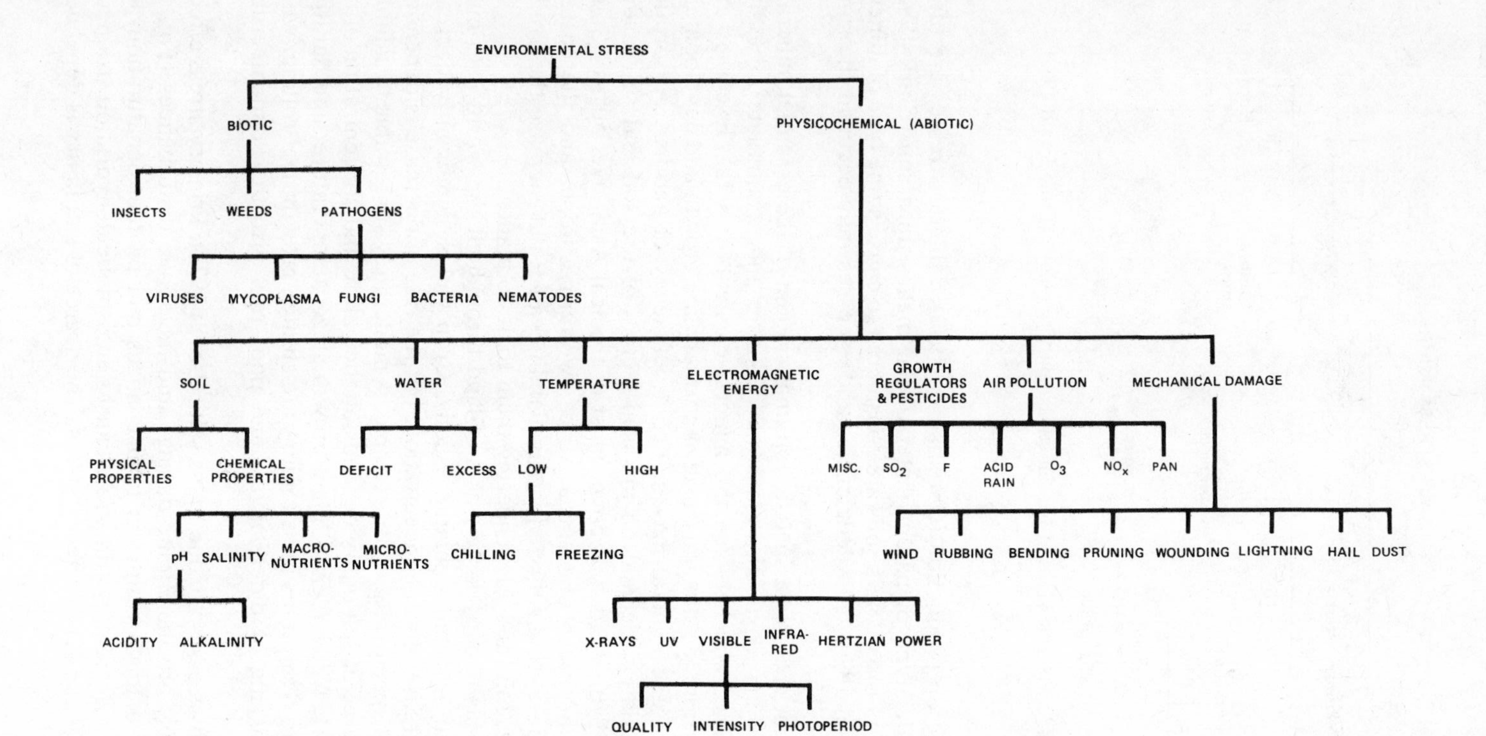

Figure 4. Types of environmental stresses to which a plant may be subjected. (Modified after Levitt 1972)

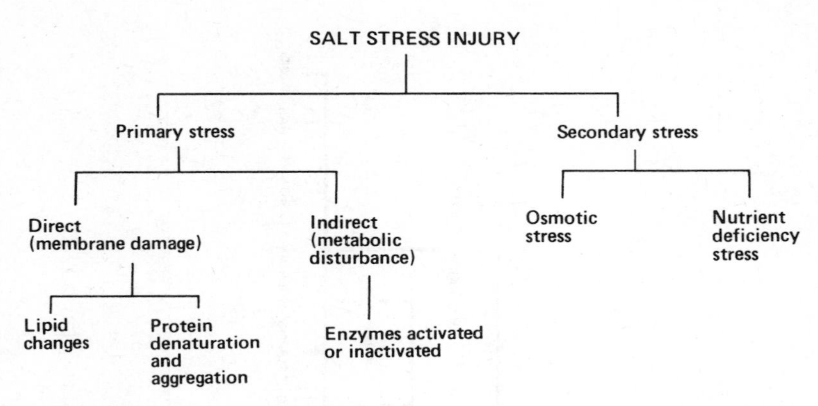

Figure 5. Possible kinds of Na salt stress injury to plants. (From Levitt 1972)

straints to crop production (Christiansen 1979). Mineral stresses are often related to the type of parent material that makes up a soil (Clark 1982). About 25% of the world's soils are considered to cause some type of mineral stress (Dudal 1976), and deficiencies and excesses may occur in the same soil (Clark 1982).

Each plant requires a specific, optimal amount of the essential minerals, but if a surplus exists, the plant may take up and accumulate excessive amounts. The ratio between the mineral elements is vital. Excesses and deficits disrupt the ratio and cause abnormal plant growth and development. Excesses of one mineral may induce a deficiency of another. Absorption and accumulation depend on physical factors of the soil and soil acidity (Treshow 1970). Acid soils cover most of tropical Asia, over 50% of South America, and much of Central Africa, western Europe, and the eastern United States. Acid soils reduce availabilities of P but increase the availability of Fe, Mn, and Al, rendering them toxic to plants.

Saline soils occupy large tracts of land throughout the world with most occurring in arid and semiarid regions. Crop yields significantly decrease with an increase in salt concentration, but the rate of decrease is dependent on the plant species. Rice, for example, is moderately susceptible to salinity, and the degree of salt injury depends on the nature and concentration of the salt, soil pH, water regime, plant growth stage, temperature, and duration of exposure. Symptoms of salt injury are stunted growth, leaf rolling, white leaf tips, drying of older leaves, and poor root growth (Ponnamperuma 1984a).

Most salt stresses are due to Na salts (Levitt 1972). Primary stress effects of salt injury are membrane damage and metabolic disturbances (Fig. 5). Membrane damage results in lipid changes and protein denaturation and aggregation. Secondary salt stress causes an osmotic dehydration strain and a nutrient (P and K) deficiency stress. The degree of salt injury is dependent on interactions with other environmental stresses such as temperature and light.

Macronutrients necessary for plant growth are N, P, K, Mg, Ca, and S. Nitrogen is the most commonly deficient element (Treshow 1970). It is a component of the amino acids, peptides, nucleic acids, nucleotides, and coenzymes. In addition to being a component of plant nutrients consumed by herbivorous insects, N is a component of some allelochemics that affect the feeding behavior of insects.

Fertilizer is a major component of agricultural and silvicultural systems. Because most soils are deficient in N, fertilizer is applied to increase crop yields. The high yield of the ''green revolution'' varieties is partially due to their response to N. Nitrogen is a limiting factor for many herbivores (Mattson 1980). Scriber (1984) suggests that heavy applications of N fertilizer may be responsible for the increased crop damage by insects during the past 30 years.

According to White (1984), deficiencies, of both the macronutrients N, P, and K and the various micronutrients, increase the proportions of soluble N in the plant. In addition the environmental stresses, acidity, salinity, rainfall, temperature extremes, radiation, mechanical damage, and the biotic stresses increase the proportion of soluble N in the tissues of the affected plants which is made available to herbivorous insects. It is important to note that the physiological changes, which occur in a plant as the result of stress, often do not produce visible symptoms. Despite the absence of visible symptoms, the altered plant chemistry can have profound effects on the suitability of plants as hosts for insects.

The effects of N fertilizer application on insect populations varies. Populations have been reported to increase, decrease, or show no change depending on the circumstances (van Emden et al. 1969, Smirnoff and Bernier 1973, Mitchell and Paul 1974). In a survey of the literature by Scriber (1984), there were 115 studies in which insect damage, growth, fecundity, or populations increased and 44 where they decreased under high N concentrations.

Mineral stresses affect primarily plant physiology (nutritional and allelochemical) but can also cause changes in plant morphology and phenology (see Chapter 2). These changes affect the suitability of plants as hosts for insects and plant pathogens. Insect natural enemies and biotic agents that interact with plant pathogens are also impacted. Likewise, the ability of plants to compete against weeds is affected (see Chapters 11, 12, and 13). The role of a plant as a host is altered by mineral stresses because stresses affect host selection, insect survival, growth, reproduction, and wing morphism (see Chapter 2). The level of fertilizer-induced physiological and morphological plant response effects on insects depends on the feeding guild to which an insect belongs—that is whether it is a phloem or xylem feeder, defoliator, stem borer, leaf miner, and so on (Scriber 1984). The activity of pollinators is related to floral characters which are also modified by mineral stress.

The literature abounds with examples of plant-mediated fertilizer effects on herbivores (see Chapter 2). It is well recognized that fertilizer plays an important role in the quality of a plant as a host for insects. However, the primary role of fertilizer in pest management systems is to provide the nutrition needed for the plant to compensate for insect damage (tolerance) and to rapidly pass the most susceptible growth stages so that it can escape insect attack (Scriber 1984). It must be noted that mineral stresses differentially affect the nutritional and allelochemic composition of various plant species and plant-mediated mineral stresses that are detrimental to one insect species may be beneficial to another.

3.1.2 Water

Water is necessary for every stage of plant development, from seed germination to maturation. Water is directly or indirectly required for all life processes and every chemical reaction. Water is the continuous phase between the substrate and the plant and between cells, and it is an essential part of plant tissues, the cells that make up the tissues, and the protoplasm that composes the cells. Water is the medium in which essential nutrients are carried from the soil solution to the cells (Treshow 1970). The rate and physiological activity of many metabolic reactions depends on the hydration level of enzymes. Water is also important as a major reagent in many biological reactions, including respiration and photosynthesis.

Water stress is the most important environmental factor affecting plant growth and development. Water stress may result from either a deficit or an excess of water (Fig. 4). In the literature, however, the term "water stress" usually refers to a deficit, rather than to an excess (Levitt 1972). Water is unevenly distributed over the earth's surface, and rarely is water available in the right amount, at the right place, and right time for optimal plant growth. Droughts and floods occur throughout most of the world, although the extent of their regularity and severity varies from one region to another (Lindh 1981). Much of the world's land is exposed to water deficits ranging from semiarid to extremely arid (Fig. 6) (Simpson 1981). About 50% of Africa and 90% of Australia experience arid conditions.

It is common for plants to be subjected to mild or severe drought at some time during their growth cycle. Numerous reviews have been written on the subject of drought stress (Kozlowski 1968, 1972; Levitt 1972, 1980). Drought has been defined as "any period during which plant and/or soil water deficiencies affect the growth and development of crop plants" (Quizenberry 1982). Moisture stress affects several plant processes, and stress-directed alteration of the chemical and physical composition of plant tissues also alters the quality of plants as food for insects.

Moisture deficits cause morphological and physiological changes in plants (Parsons 1982). Morphological responses to water deficits consist of leaf shedding, leaf rolling, leaf angle changes, and an increase in the root to shoot

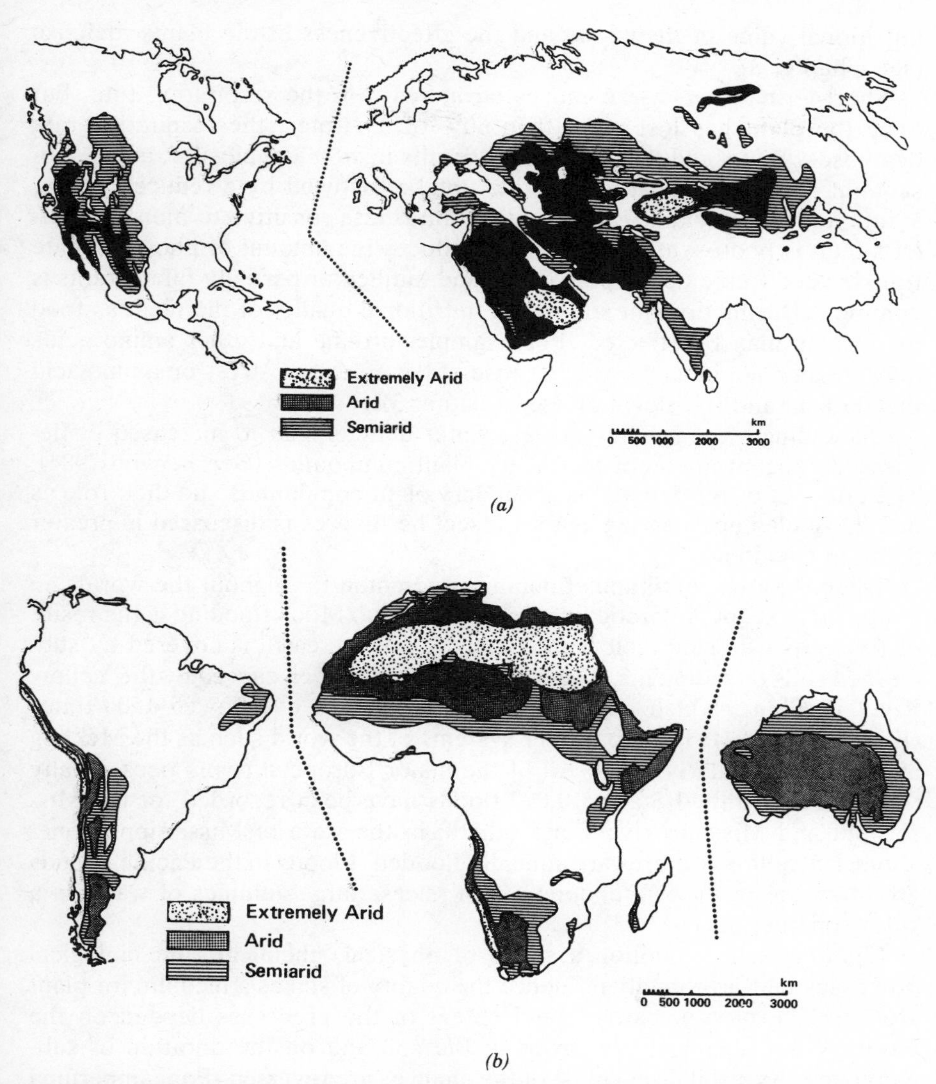

Figure 6. Arid lands of (*a*) North America and Asia; (*b*) South America, Africa, and Australia. (Modified after Simpson 1981, and Meigs 1953. From *Water Stress on Plants* by G. M. Simpson 1981. Reprinted by permission of Praeger Publishers.)

ratio. Physiological changes caused by water deficits include leaf cuticular wax thickness, transpiration, respiration, stomatal behavior, photosynthesis, translocation, mineral nutrition, and protein breakdown (Parsons 1982, Treshow 1970). These factors may impact on insects by altering their microenvironment and by altering such characteristics as the toughness and nutritional value of their food and the effectiveness of the plants' defense (see Chapter 3).

Short-term water stress causes an increase in the respiration rate, but once the plant has lost more than 60% of its water, the respiration rate decreases. Water reduction in leaves results in stomatal closure, and a subsequent reduction in carbon dioxide uptake followed by a reduced photosynthetic rate (Levitt 1972). Translocation is less sensitive to moisture deficits than is photosynthesis. Drought reduces the amount of photosynthate translocated to the developing grain, and unfilled or partially filled grains is common. During drought stress, the nutritional quality of the plant as food for insects may be affected. For example, proline and other amino acids may accumulate in leaf tissue because of the effects of stress on amino acid metabolism and translocation (see Chapter 3).

Depending on the plant species, water deficits lead to increased or decreased concentrations of secondary plant compounds (Gerzhenzon 1984). The effect of water deficits on secondary plant compounds and their role as defensive chemicals acting against insect herbivores is discussed in greater detail in Chapter 3.

Temporary or continuous flooding is common throughout the world; no continent, except Antarctica, is free of flooding. Much flooding is the result of rivers overflowing their banks, and 72% of the earth is covered by submerged soils or sediments. The worst floods have been caused by the Yellow River in China, which has been recorded to have overflowed 1500 times (Kozlowski 1984). The great river systems of the world such as the Mekong and Ganges regularly flood. All of the major European rivers occasionally flood. In the United States 10,000 floods have been recorded for the Mississippi and Missouri rivers and others; in the state of Mississippi alone, about 1.5 million hectares are annually flooded. Floods in the Pacific Islands are often the result of typhoons which release large volumes of water in a short time period.

Flooding sets in motion a series of physical, chemical, and biological processes that profoundly influence the quality of soil as a medium for plant growth. The nature, pattern, and extent of the processes depend on the physical and chemical properties of the soil and on the duration of submergence. As a soil dries, most of the changes are reversed (Ponnamperuma 1984b).

Flooding fills the pore spaces of the soil with water and severely limits gas exchange of the plant roots. At O_2 levels below 2%, leaves of cereal crops become chlorotic, growth ceases, and death occurs (Treshow 1970). In addition to oxygen depletion, flooding decreases transpiration and thus decreases water absorption and photosynthesis. According to Krizek (1984), hormonal metabolism and mineral uptake are affected by flooding. Gibberellins in the xylem are released, abscisic acid levels, indole acetic acid contents, and ethylene production are increased. Nutrient uptake is decreased, resulting in chlorosis. The mineral content of waterlogged soils is altered.

Salt accumulation accompanies waterlogging as frequent irrigation can lead to an accumulation of salts.

The plant-mediated effects of flooding on herbivorous insects are not covered in this book because of a dearth of literature on the subject. However, many of the environmental stresses have similar effects on plants, and the effect of flooding on plant physical and chemical processes (e.g., hormone content, mineral uptake, and ethylene production) has similar effects on insects as that of other stresses which are discussed in this book.

3.1.3 Temperature

Plant species have an optimum temperature for growth and development that is determined by a complex of genetic, developmental, and cultural factors (McDaniel 1982). Although some plant species can survive temperature extremes, most are confined to a temperature range of 10° to 40°C (Treshow 1970). Exposure to temperatures below and above this range stresses the metabolic activities of plants and leads to changes in plant nutrition and chemical and morphological defenses against insects (see Chapter 4).

Plant development is dependent on temperatures suitable for metabolic activity. Development is a function of biochemical reactions controlled by enzymes and the rate at which enzymes act is a function of temperature (Treshow 1970). Photosynthetic activity is generally low at temperatures below 10°C, increases to about 30°C, and again decreases above 30°C. Temperatures also determine the solubility of gases in plant cells and influence the uptake of soil minerals and water by the roots. The optimum temperature for metabolic activity varies among the plant species and among the various plant organs and plant growth stages within a species.

Levitt (1972) has classified temperature stress into low and high temperature categories and has further divided low temperature into chilling and freezing (Fig. 4). Chilling is considered as a low temperature above freezing or a ''temperature that is cool enough to produce injury but not cool enough to freeze the plant'' (Levitt 1980). DiCosmo and Towers (1984) list the phytochemical alterations caused by chilling injury in several plant species.

Responses of plants to high temperature are closely related to the water status of the plants (McDaniel 1982). Because high temperature stress is usually associated with drought conditions, it is difficult to separate the influence of these two stresses (see Chapter 4). Heat stress influences phytochemical synthesis including increased protein and free amino acid content and decreased alkaloid levels (DiCosmo and Towers 1984, White 1984).

It has been predicted that the increased concentration of CO_2 in the atmosphere will cause a ''greenhouse'' effect and that as a result temperatures will be increasing. It is projected that a doubling of atmospheric CO_2 by the year 2050 will increase temperatures in North America by 2°C, and rainfall in certain regions will decline by 10% (Pimentel 1981). These changes in temperature and moisture are predicted to reduce yields of corn and wheat by 10 to 25%. The altered climate would have a direct effect on insect populations and an indirect effect through changes in growth and chemistry of host plants.

3.1.4 Electromagnetic Energy

Electromagnetic energy effects on plants involve primarily the ultraviolet (40–390 nm), visible (390–780 nm), and infrared (780–2500 nm) regions of the spectrum (Fig. 4). The term "light" refers only to the visible portion of the electromagnetic radiation spectrum (Cathey and Campbell 1982). High intensity radiation in the short (UV) and long (infrared) wavelengths generally has a negligible effect on plants because it is absorbed by the ozone in the atmosphere. However, at high elevations where ozone is minimal, UV radiation may cause plant damage (Treshow 1970).

Light is the basic form of energy that directly or indirectly propels the life processes of most living organisms. Plant responses (growth, development, differentiation, and reproduction) are determined by the quality, intensity, and duration (photoperiod) of light. Light quality affects metabolic processes that determine the concentrations of various phytochemicals (DiCosmo and Towers 1984), including secondary metabolites that affect insect growth and development. Low and high light intensity adversely affect plant growth. Photoperiod regulates the development of many plants, and these plants may escape insect injury by reaching maturity and/or by a change in their value as a food source before insect populations reach damaging levels.

Sources of ionizing radiation most detrimental to plant physiology include x-rays, beta rays, and gamma rays. Ionizing radiation-induced plant injury is rare under natural conditions; however, because of contamination from artificial sources, quantities of ionizing radiation in nature have increased, and this stress is likely to become an increasing problem in the future (Levitt 1972).

Insect populations have been reported to increase on plants that were exposed to intermediate doses of gamma irradiation (White 1984). Irradiation of growing plants disrupts protein synthesis and increases protein fractions and amino acids in the plant tissues (Levitt 1980). These and other changes in plant nutrition resulting from exposure to irradiation have been associated with insect outbreaks (see Chapter 5).

In recent years there has been a great deal of interest in attempting to influence plant growth by exposing plants to electric fields, magnetism, and sound stresses. Electrical fields (see Chapter 5) have been reported to inhibit plant growth, affect growth orientation, and to increase respiration. Magnetic fields and sound also have been shown to affect plant growth and development (Levitt 1972), but they are of no importance in nature and their role in plant mediated effects on herbivores has received little attention by scientists.

3.1.5 Pesticides and Growth Regulators

Pests are major constraints to production in agricultural and forest ecosystems and annually cause severe losses by reducing yields and quality and by increasing production costs. Insects damage plants by feeding on all plant parts, from the seedling stage to maturity, and act as vectors of plant pathogens. Similarly, fungal, bacterial, and viral pathogens and nematodes derive nourishment from plants, and weeds compete with crop plants for space, light, soil nutrients, and soil moisture. The various pesticides and plant growth regulators used to manipulate agricultural systems have direct and indirect effects on insects, which thus influence their population levels.

Insecticides and fungicides have a direct effect on insects by inhibiting or stimulating feeding, growth, and reproduction and an indirect effect through destruction of natural enemies or through host plant-mediated effects (see Chapter 6). The extent of the direct effect of insecticides and fungicides on insects is dose dependent. They may be toxic at high dosages and stimulate insect growth and reproduction at low dosages (Luckey 1968). Although it is often difficult to separate direct effects of chemicals from plant-mediated effects on insects, there is an increasing array of literature on the latter (White 1984). White reports on the evidence that many insect outbreaks appear to result from a physiological alteration of host plants that changes their nutritional value, and thus affects the biology of herbivorous insects.

Herbicides and plant growth regulators may also have a direct or plant-mediated effect on insects, both of which are dose dependent. Although herbicides have occasionally been reported to have a direct effect on certain insects (Norris 1982), plant-mediated effects appear to be of most importance (White 1984). Modification of plant physiology by herbicides has been implicated in causing increased insect populations (Norris 1982). Herbicides and plant growth regulators have a multitude of phytochemical effects on plants, and the literature is replete with examples (see DiCosmo and Towers 1984) of phytochemical changes that would be expected to alter the suitability of plants as hosts for insects.

Herbicides and plant growth regulators also alter the physical and morphological characters of host plants (Campbell et al. 1984). These changes may involve an increase in physical barriers and early termination of host plant growth, both of which adversely affect the plant's suitability as a host for insects. By killing weeds, the habitat is altered, and the ecological change can impact on insect populations on crop plants (see Chapter 13). Growth regulators either stimulate or inhibit plant development processes, and thus they disturb the relationship between the host plant and the insect (Scheurer

1976). The role of herbicides and growth regulators as plant stress agents in plant–insect interactions is described in detail in Chapter 7.

3.1.6 Air Pollution

Air pollutants are defined as "aerial substances that have some adverse effects on plants, animals, or materials" (Treshow 1984). Although air pollution was a problem for the populace in and around European cities during the Middle Ages, it is only since the 1950s that it has been considered a stress that can injure plants and affect crop production. It is vividly apparent that air pollution is at present a severe problem and will continue to be one of the more important problems facing us for some time in the future (Mudd and Kozlowski 1975).

Although natural events such as volcanic eruptions contaminate the air, air pollution has been described as a social disease, anthropogenic in origin, which has been primarily generated by human activities (Gillette 1984). Air pollution has become a global problem, transcending natural and political boundaries from densely populated regions to remote areas. The recently perceived problem of acid rain, which was first observed in northern Europe and more recently in the United States and Canada, is a widespread problem with alarming implications that has caught public attention.

Air pollution is associated with both direct (primary) and indirect (secondary) effects on insect populations (Alstad and Edmunds 1982). Pollutants may act directly as insecticides or may impact on insect populations by affecting natural enemy populations and through host plant-mediated effects where the plant is rendered more or less suitable as a host for herbivorous insects. Our interest in this book is in the latter and is the subject of Chapter 8.

Phytotoxic air pollutants are classified by their physical and chemical properties. Physically they consist of gases and particles, and chemically they are separated into oxides of sulfur and nitrogen, ozone, halogens, hydrocarbons, heavy metals, and so on (Weinstein and McCune 1979).

Plants vary in their susceptibility to air pollutants, and the extent of their susceptibility is dependent on an array of factors, including the plant species involved, nature of the pollutant, exposure conditions, and numerous environmental factors (Fig. 7). These environmental factors may act as plant stresses and thus may modify the response of the plant to air pollutants. In turn, air pollution also interacts with other environmental conditions and may modify the plant's response to other environmental stresses.

The effects of pollutants on plants have been described by Weinstein and McCune (1979). The biochemical and physiological changes induced in plants by pollutants alter the quality of plants as hosts for insects. Insect feeding, growth, and reproduction may be affected (Hughes and Laurence 1984). Chemical and physical changes in the plant may alter the nutritional value of the plant and may also alter plant properties that influence insect

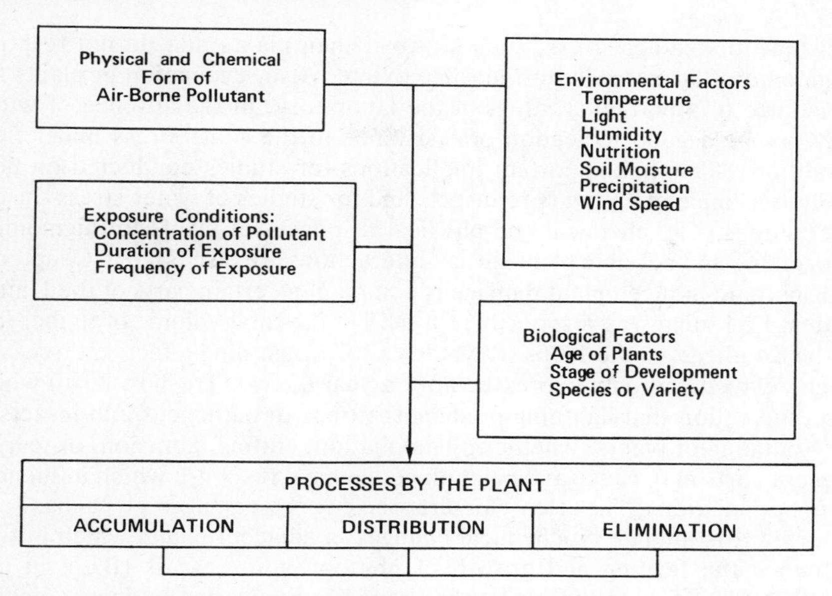

Figure 7. Factors affecting plant response to air pollutants. (From L. H. Weinstein and D. C. McCune, *Stress Physiology in Crop Plants,* H. Mussell and R. C. Staples, eds., © 1979; reprinted by permission of John Wiley & Sons, Inc.)

behavior or serve as defenses against insect attack (e.g., secondary plant compounds) (see discussion by White 1984). Chapter 8 reviews the impact of air pollutants on insects and the air pollutant–plant–insect interactions.

3.1.7 Mechanical Damage

Mechanical stress effects on plants have received relatively little attention in comparison to other plant stresses. Mechanical stresses can be divided into those involving natural environmental factors (climatic extremes), such as wind, dust, lightning, hail, sleet, and snow, and those stresses caused by people and animals. Mechanical stresses caused by climatic factors, human beings, and animals (including herbivorous insects) result from wounding of various plant parts and by mechanical perturbations such as shaking, rubbing, and bending (Fig. 4).

Thigmomorphogenesis is a plant growth and developmental response resulting from mechanical perturbations such as wind stress and rubbing, and it has been reported in 39 plant species (Jaffe and Biro 1979). Mechanically perturbed plants are less susceptible to other stresses than unperturbed plants (Jaffe and Telewski 1984). Jaffe and Biro (1979) have thus emphasized the importance of considering thigmomorphogenesis when integrating data from experiments dealing with stress conditions. For example, in field tests, plants that have undergone wind-induced thigmomorphogenesis are more resistant to weather stress, such as frost, than plants that do not respond thigmomorphogenetically to the same extent. Also, evaluation of plants for resistance to a particular stress in the laboratory, in the absence of wind, may not be a good indication of resistance to the same stress under field conditions. This has important implications for studies conducted on host plant resistance of cultivars to insects and for studies of water stress–insect interactions. The chemical and physical responses of plants to thigmomorphogenesis and possible plant–insect interactions are discussed in Chapter 9.

Lightning-induced plant damage is common in certain parts of the United States. Lightning was responsible for 50% of the total volume of timber lost in the Southeast in the 1930s (Reynolds 1940). Lightning effects on trees are often delayed and only expressed after a year or two (Treshow 1970) which is an indication that lightning predisposes trees to pathogens and insects.

Wounding of plants, whether by defoliation, cutting, abrasion, or tearing of plant parts may cause a distinct plant chemical response which influences insect populations. The release of airborne cues by damaged plants has been reported to stimulate biochemical changes in adjacent plants which in turn influence the feeding and growth of phytophagous insects (Baldwin and Schultz 1983). Generally, plants that have been wounded by insect feeding have an enhanced level of resistance to other stresses including mechanical and biotic stresses (e.g., other insects) as described in Chapter 10.

3.2 Biotic Stresses

The biotic environment includes all living things—animals, plants, and pathogens—and the interrelated actions and reactions that they directly or indirectly impose on each other. Plants are the fundamental units of the ecosystem on which other organisms directly or indirectly depend. There is increasing evidence that herbivores, plant pathogens, and other plants cause changes in plant defensive systems, and thus their suitability as hosts for insects (Rhoades 1983). Weeds that compete with cultivated plants stress them directly by competing for light, water, and nutrients. They may also interact with cultivated plants by serving as a host for insect pests and their natural enemies. Plant pathogens, fungi, bacteria, viruses, mycoplasma, and parasitic nematodes are biotic stresses that directly attack plants and alter their suitability to insects. Feeding by insects induces stress factors that cause a change in the response of plants to the insect itself, or to other insects that attack the plant.

3.2.1 Herbivorous Insects

Insects are biotic agents that can cause mechanical and chemical stress. Insects feed on plants by chewing on various plant parts, removing plant sap from the xylem or phloem with their piercing-sucking mouthparts, mining in leaves, or boring in stems, and cause gall formation. Chewing on leaves

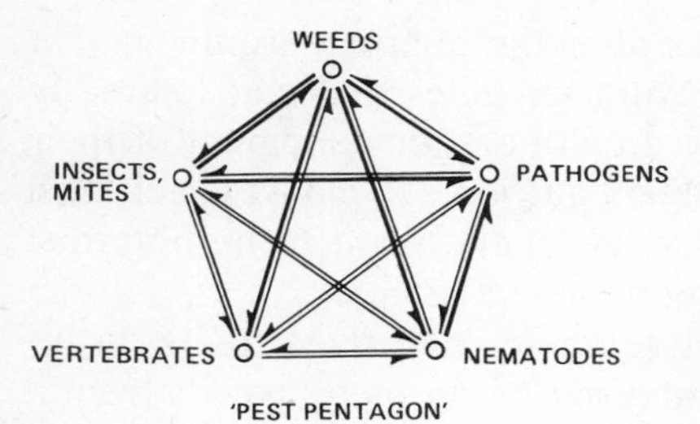

Figure 8. Potentials for interactions between different pest organisms. (After Norris 1982)

or other plant parts causes a mechanical stress that is in some respect similar to that of leaf removal or tearing by weather factors—wind, rain, hail, and sleet or by human damage (see Chapter 9). Mechanical damage to leaves causes alterations in plant growth and chemistry that affect plant suitability for insects (Mattson 1980, Rhoades 1983).

Herbivorous insects cause chemical stress to plants directly or indirectly through their feeding activity. Some sucking insects modify plant metabolism by secreting salivary juices directly into plant tissues (Mattson 1980). Insects may serve as a source of plant growth regulators through the secretion of salivary agents that promote plant growth (Dyer and Bokhari 1976). Certain heteropterous bugs, aphids, and psyllids have salivary secretions that contain growth regulators, for example, indoleacetic acid (Norris 1979). Insect damage may alter the sensitivity of plants to chemical, physical, and biotic stresses including other insects, plant pathogens, and weeds (see Chapters 9 and 10).

3.2.2 Plant Diseases

Inoculation of plants with fungi, bacteria, or viruses alters the levels of resistance to further attack by the same or different pathogens (Suzuki 1980, Goodman 1980, Hamilton 1980) (see Chapter 12). Induced plant defenses play an important role in controlling the spread of disease between and within plants (Rhoades 1983). According to Ryan (1983), attacks of certain pests cause substances to be released (or produced) called elicitors that induce synthesis of specific phytoalexins near the site of pathogen attack, supposedly to arrest pathogen activity and possibly deter insect invasion (see Fig. 1, Chapter 10).

3.2.3 Weeds

Weeds compete with crop plants for sunlight, moisture, and soil nutrients. Weeds also interact with crop plants by serving as alternate hosts for plant pathogens and herbivorous insects. Weeds may directly affect insects and other pests (Fig. 8), or they may indirectly affect them through their interaction with the host plant. Weeds are the only pests that are primary producers and serve as food sources for both insect pests and natural enemies that attack the pest (Norris 1982). Weeds alter the microenvironment that directly affects insect pests and their natural enemies and can indirectly affect herbivorous insects by altering the growth and development of their host plants. Weeds alter the availability of soil nutrients to plants which then modifies the appearance and nutritional value of the plant to herbivorous insects and the ability of the plant to tolerate insect attack.

Weeds have direct and indirect, positive and/or negative effects on insects. Chapter 13 explains the stresses imposed on crops by weeds from a viewpoint beyond direct competition by incorporating three trophic level system interactions.

4. PLANT-MEDIATED STRESS EFFECTS ON HERBIVORES

Until recently, insect outbreaks were attributed primarily to density-independent factors (e.g., weather) that had a direct effect on the insect or to density-dependent biotic factors (e.g., competition for food, parasitism, predation, and disease) that were presumed to vary with population density. According to Rhoades (1983), neither of these factors satisfactorily explains temporal fluctuations in abundance, and this suggests that quality as well as quantity of food and defensive characteristics of plants (host plant suitability) must also be taken into account in determining the causes of insect outbreaks.

Host plant suitability involves plant characteristics and insect herbivore adaptations as well as modification of both the plant and the insect by abiotic and biotic factors (Scriber 1984). These factors affect plant growth and development in various ways and may have a positive, negative, or neutral effect on an herbivorous insect. Stresses affect plant growth, and thus the amount of biomass that is available as food for insects. Plant morphological changes determine the degree of attractiveness that a plant has to a potential herbivore. The influences of stress on the chemical nature of plants has extremely important effects on their value as hosts. The content of primary plant compounds (nutrients) and secondary plant compounds (allelochemics) is affected by environmental stresses to which plants are exposed.

4.1 Primary Plant Compounds

The content of primary plant compounds (e.g., vitamins, nucleic acids, lipids, steroids, carbohydrates, amino acids, and water) may be altered by environmental stresses. Their presence determines the acceptance of a plant for insect feeding and determines the value of a plant as a food source for insect growth, survival, and reproduction (Scriber 1984). A compound that is a nutrient for one growth stage of a species may not be one for another stage: a compound that is a nutrient at one concentration may be toxic at a greater concentration (Reese and Schmidt 1986).

Environmental stresses have a profound influence on the nutritional quality and quantity of a plant (Rhoades 1983). Drought can lead to an enhanced nutritional quality (increase in total available carbohydrates and total foliage nitrogen content) of plants that results in an increased insect survival and fecundity. Soluble plant metabolites of potential nutritional importance in insects (e.g., proline, sugars, glycerol, malate, and shikimate) increase in concentration in response to drought, freezing, salinity, and root flooding. A relative shortage of nitrogenous food is one of the important nutritional factors that limit the abundance of many insect species (McClure 1983). Because of its role in metabolic processes, cellular structure, and genetic coding, N is a critical element in all organisms (Mattson 1980). The role of plant stresses in regulating the quantity of N in plants and the effect (positive and negative) of low N values on herbivorous insects are discussed in detail in the following chapters.

4.2 Secondary Plant Compounds

Secondary plant compounds have until recently been considered metabolic waste products with no definite function (Schoonhoven 1972). Their role in the ecology of plant species was recognized by Dethier (1947) and Fraenkel (1953), and their role in plant–insect relationships is described in Fraenkel's (1959) paper, "The Raison d'Être of Secondary Plant Substances." Kogan (1986) has described secondary plant compounds as the "most ecologically active" plant compounds, although the distinction between primary and secondary plant compounds is not absolute, chemically or ecologically.

Schoonhoven (1972) has defined secondary plant compounds as "those compounds that are not universally found in higher plants but are restricted to certain plant taxa, or occur in certain plant taxa at much higher concentrations than in others and are of no nutritional significance to insects." Secondary plant compounds that are produced by one organism and affect other organisms are referred to as allelochemicals (Kogan 1986). Allelochemicals are called allomones if their activity favors the producing organism and kairomones if their activity favors the receiving organism (Table 2). The numerous phytochemicals that act as allelochemicals in plant–insect interactions have been listed by Schoonhoven (1972), Rhoades (1979), Ryan (1983), and DiCosmo and Towers (1984). Various chapters in this book describe the effects of various physicochemical and biotic stresses on the content of allelochemicals in different plant species.

By definition, allelochemicals are not nutrients, but they have been shown to interact with nutrients (Reese and Schmidt 1986). Many of the deleterious metabolic effects of allelochemics may be due to these interactions. There is increasing evidence that secondary metabolic products exist in a dynamic equilibrium and are involved in cycles that include primary compounds such as sugars and amino acids (Seigler and Price 1976). Secondary plant compounds may possibly act as intermediaries in metabolic processes or as reg-

Table 2. Principal Classes of Chemical Plant Factors (allelochemics) and Corresponding Behavioral or Physiological Effects on Insects

Allelochemical Factors	Behavioral or Physiological Effects
Allomones	Give adaptive advantage to the producing organism
Antixenotics	Disrupt normal host selection behavior
Repellents	Orient insects away from plant
Locomotory excitants	Start or speed up movement
Suppressants	Inhibit biting or piercing
Deterrents	Prevent maintenance of feeding or oviposition
Antibiotics	Disrupt normal growth and development of larvae; reduce longevity and fecundity of adults
Kairomones	Give adaptive advantage to the receiving organism
Attractants	Orient insects toward host plant
Arrestants	Slow down or stop movement
Feeding or oviposition excitants	Elicit biting, piercing, or oviposition; promote continuation of feeding

From Kogan 1986.

ulators of biochemical processes. Secondary plant compounds may have a role in metabolic functions along with primary compounds, as well as an important role in plant defense.

4.3 Interactions between Nutrients and Defense Chemicals

Evidence indicates that allelochemicals can interact with nutrients in various ways, as described by Reese (1979). The determinant as to when an allelochemical is an allelochemical and when it is a nutrient depends on the situation and is not a characteristic of the molecule.

Stress acting on plants can cause an imbalance between the proximate nutrient content and defensive capabilities of plants, which results in an increased nutritional quality for herbivores. A model of this has been illustrated by Rhoades (1983) (Fig. 9). Insect population growth in the long term is checked by plant defense systems that decrease herbivore fecundity, increase mortality, and increase insect susceptibility to natural enemies (Fig. 9A). Through the influence of stress acting on the plants, there is an increase in the availability of highly nutritious food, which then results in decreased insect mortality, increased fecundity, and resistance to natural enemies (Fig. 9B). Plants respond to the high insect population by increasing their commitment to defense, and as a result the nutritional quality decreases below the equilibrium level (Fig. 9C). The insect population then collapses to a level below the equilibrium because of a decrease in the nutritional quality

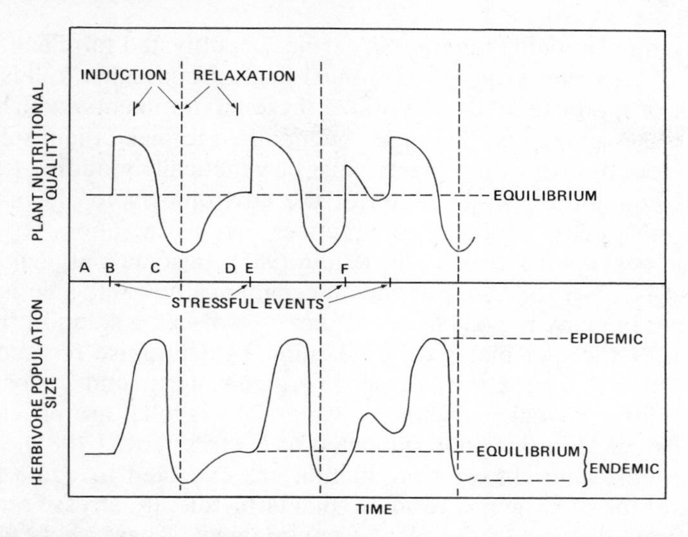

Figure 9. Model of herbivore population fluctuations based on changes in host plant nutritional quality as affected by physical stress and attack-induced defensive responses of plants. (After Rhoades 1983)

of the plants and a decrease in resistance to natural enemies whose populations increase during the collapse phase. The plants relax their defenses and in the absence of further stress, equilibrium plant nutritional quality and insect populations are reestablished (Fig. 9D). Additional plant stress (Fig. 9E) causes another increase in the insect population. When a factor causing stress closely follows a previous stressful event before plants have relaxed their defenses, the increase in the insect population is minor (Fig. 9F). Continued stress and high insect population levels result in the destruction of the plant and a subsequent collapse of the insect population because of a shortage of food.

5. SUMMARY

With the continued increases in population and with the resultant decrease in cropland area, deterioration of soils, depletion of ground water, increased air pollution problems, and various other factors that cause plant stress and food shortages are constant threats in developing countries and may someday return to haunt developed countries if production increases are not maintained. Unforeseen natural, economic, and political events can suddenly change the global food outlook, and the margin of error is slight. Without continued investment in science and education, sustained food productivity cannot be ensured. Certainly as Lee and Taylor (1986) have suggested, it is preferable to err on the side of surplus rather than food shortages.

Because of the limited amount of fertile, uncultivated farmland, most of the gains in agricultural production must come from higher yields and the utilization of marginal lands. To utilize these environments, which impose severe stresses on plants, they must be modified to make them suitable for crop production, or the crop plants must be genetically modified to tolerate the stress conditions. Modification of the environment to alleviate stress factors involves crop management practices such as irrigation, fertilization, tillage, and pest control technology (Blum 1985). In some marginal lands the addition of N, P, K, and Zn provide an economically viable solution to low fertility, but in many regions farms cannot afford such a solution. Low cost sources of N through biological N fixation hold promise for such areas. Nitrogen-fixing bacteria, free-living blue-green algae, and endosymbiotic blue-green algae associated with the water fern, azolla, are providing low cost sources of N in ricelands subjected to N stress (IRRI 1985).

The greatest gains in crop production are expected to come from the modification of plant genes to allow plants to tolerate stress, rather than from the modification of the plant's environment. These gains will occur through conventional plant breeding techniques and through those involving molecular genetics.

Through conventional plant-breeding techniques, plant breeders have over several decades developed cultivars that are tolerant to temperature extremes (Marshall 1982), drought and mineral stresses (IRRI 1985), and air pollutants (Reinert et al. 1982), and are resistant to insects (Painter 1951), Maxwell and Jennings 1980, Heinrichs 1986). Techniques have been developed to evaluate rice-breeding lines for a least 40 different abiotic and biotic stresses (IRRI 1985). However, among the various fiber and food crops there are numerous stresses for which tolerant cultivars are not available. In some cases it is because of a lack of tolerant germplasm for use as donors in crosses and the inability to transfer genes for tolerance from wild relatives to acceptable cultivars. Biotechnology is now providing a means to overcome these constraints.

Tissue culture techniques have been successfully applied in selecting plants that are tolerant to freezing, salinity, mineral stress, and water stress (Krizek 1984, Stavarek and Rains 1984). The embryo rescue technique with tissue culture has been used to transfer insect resistance from wild rice species to cultivated rice, *Oryza sativa* (L.) (Heinrichs 1986). In the future it should be possible to use recombinant DNA techniques for the transfer of genes for resistance to abiotic and biotic stresses from one plant species to another. This will make available, for breeding purposes, wild relatives of domestic crops that are rich sources as donors for stress resistance. Indeed, genetic engineering will play a vital role in the development of abiotic and biotic stress-resistant cultivars in the future.

The modification of genes which alters plants' ability to withstand various abiotic and biotic stresses will affect their suitability as hosts for insects. During the development of stress-resistant cultivars, the level of susceptibility to insects must be closely monitored, to select strains with equal or higher levels of insect resistance than currently grown commercial cultivars. The development of insect-resistant varieties should include the evaluation of breeding lines for tolerance to other stresses. In either case the failure to do so could result in the release of cultivars that are super susceptible to insect pests and other stresses.

In this chapter I have attempted to introduce the nature and importance of plant stresses. Discussions in the following chapters concentrate in more detail on the effect of the various abiotic and biotic stresses on plants and how changes in plant physiology, chemistry, and morphology affect their suitability as hosts for herbivorous insects and their natural enemies.

ACKNOWLEDGMENTS

I thank J. H. Benedict, A. M. Hammond, Jr., T. N. Hardy, T. O. Holtzer, P. R. Hughes, M. Kogan, D. M. Norris, T. J. Riley, and C. M. Smith for their review of the manuscript and B. M. Rheams and M. Thrasher for their assistance in preparation of the manuscript.

REFERENCES

Alstad, D. N., and G. F. Edmunds, Jr. 1982. Effects of air pollutants on insect populations. *Annu. Rev. Entomol.* 27: 369–384.

Baldwin, I. T., and J. C. Schultz. 1983. Rapid changes in tree leaf chemistry induced by damage: evidence for communication between plants. *Science* 221: 277–279.

Barr, T. N. 1981. The world food situation and global grain prospects. *Science* 214: 1087–1095.

Blum, A. 1985. Breeding crop varieties for stress environments. *CRC Critical Reviews in Plant Sciences* 2: 199–238.

Borlaug, N. E. 1981. Using plants to meet world food needs. *In* Woods, R. G. (ed.), *Future Dimensions of World Food and Population*. Westview, Boulder, CO, pp. 101–182.

Brown, L. R. 1970. *Seeds of Change*. Praeger, New York.

Brown, L. R. 1976. *Increasing World Food Output*. Arno, New York. (Reprint of 1965 ed. published by U.S. Dept. of Agriculture on Foreign Agricultural Economic Report No. 25, Washington.)

Brown, L. R. 1981. The worldwide loss of cropland. *In* Woods, R. G. (ed.), *Future Dimensions of World Food and Population*. Westview, Boulder, CO, pp. 57–96.

Campbell, B. C., B. G. Chan, L. L. Creary, D. L. Dreyere, L. B. Robin, and A. C. Waiss, Jr. 1984. Bioregulation of host plant resistance to insects. *In* Ory, R. L., and F. R. Rittig (eds.), *Bioregulators: Chemistry and Uses. ACS Symposium Series,* No. 257, pp. 193–203.

Cathey, H. M., and L. E. Campbell. 1982. Plant response to light quality and quantity. *In* Christiansen, M. N., and C. F. Lewis (eds.), *Breeding Plants for Less Favorable Environments*. Wiley, New York, pp. 213–257.

Christiansen, M. N. 1979. Organization and conduct of plant stress research to increase agricultural productivity. *In* Mussell, H., and R. C. Staples (eds.), *Stress Physiology in Crop Plants*. Wiley, New York, pp. 1–14.

Christiansen, M. N. 1982. World environmental limitations to food and fiber culture. *In* Christiansen, M. N., and C. F. Lewis (eds.), *Breeding Plants for Less Favorable Environments*. Wiley, New York, pp. 1–11.

Clark, R. B. 1982. Plant response to mineral element toxicity and deficiency. *In* Christiansen, M. N., and C. F. Lewis (eds.), *Breeding Plants for Less Favorable Environments*. Wiley, New York, pp. 71–142.

Dethier, V. G. 1947. *Chemical Insect Attractants and Repellents*. Blakiston, Philadelphia.

DiCosmo, F., and G. H. N. Towers. 1984. Stress and secondary plant metabolism in cultured plant cells. *In* Timmermann, B. N., C. Steelink, and F. A. Loewus (eds.), *Phytochemical Adaptations to Stress. Rec. Adv. Phytochem.*, Vol. 18. Plenum, New York, pp. 97–175.

Dudal, R. 1976. Inventory of the major soils of the world with special reference to mineral stress hazards. *In* Wright, M. J. (ed.), *Plant Adaptation to Mineral Stress in Problem Soils*. Cornell Univ. Agr. Exp. Sta., pp. 3–13.

Dyer, M. F., and V. G. Bokhari. 1976. Plant–animal interactions: studies of the effect of grasshopper grazing on bluegrama grass. *Ecology* 57: 762–772.

Fraenkel, G. S. 1953. The nutritional value of green plants for insects. *Trans. 9th Intl. Cong. Entomol.*, Amsterdam, 1951, 2: 90–100.

Franenkel, G. S. 1959. The raison d'être of secondary plant substances. *Science* 129: 1466–1470.

Gerzhenzon, J. 1984. Changes in the levels of plant secondary metabolites under water and nutrient stress. *In* Timmermann, B. N., C. Steelink, and F. A. Loewus (eds.), *Phytochemical Adaptations to Stress. Rec. Adv. Phytochem.*, Vol. 18. Plenum, New York, pp. 273–320.

Gillette, D. G. 1984. Concern about atmospheric pollution *In* Treshow, M. (ed.), *Air Pollution and Plant Life*. Wiley, Chichester, England, pp. 7–13.

Goodman, R. N. 1980. Defenses triggered by previous invaders: bacteria. *In* Horsfall, J. G., and E. B. Cowling (eds.), *Plant Disease*, Vol. 5. Academic Press, New York, pp. 305–317.

Grime, J. P. 1981. Plant strategies in shade. *In* Smith, H. (ed.), *Plants and the Daylight Spectrum*. Academic Press, New York, pp. 159–186.

Hamilton, R. I. 1980. Defenses triggered by previous invaders: viruses. *In* Horsfall, J. G., and E. B. Cowling (eds.), *Plant Disease*, Vol. 5. Academic Press, New York, pp. 279–303.

Heinrichs, E. A. 1986. Perspectives and directions for the continued development of insect-resistant rice varieties. *Agric., Ecosys. Environ.* 18: 9–36.

Hekstra, G. P. 1981. Toward a conservation strategy to retain world food and biosphere options. *In* Bach, W., J. Pankrath, and S. H. Schneider (eds.), *Food–Climate Interactions*. Reidel, Dordrecht, Holland, pp. 325–359.

Hopper, D. W. 1981. Recent trends in world food and population. *In* Woods, R. G. (ed.), *Future Dimensions of World Food and Population*. Westview, Boulder, CO, pp. 35–55.

Hughes, P. R., and J. A. Laurence. 1984. Relationship of biochemical effects of air pollutants on plants to environmental problems: insects and microbial interactions. *In* Koziol, M. J., and F. R. Whatley (eds.), *Gaseous Air Pollutants and Plant Metabolism*. Butterworths, London, pp. 361–377.

International Rice Research Institued (IRRI). 1985. *International Rice Research: 25 Years of Friendship*. IRRI, Los Baños, Laguna, Philippines.

Jaffe, M. J., and R. Biro. 1979. Thigmorphogenesis: the effect of mechanical perturbation on the growth of plants, with special reference to anatomical changes, the role of ethylene, and interaction with other environmental stresses. *In* Mussell, H., and R. C. Staples (eds.), *Stress Physiology in Crop Plants*. Wiley, New York, pp. 25–69.

Jaffe, M. J., and F. W. Telewski. 1984. Thigmorphogenesis: callose and ethylene in the hardening of mechanically stressed plants. *In* Timmermann, B. N., C. Steelink, and F. A. Loewus (eds.), *Phytochemical Adaptations to Stress. Rec. Adv. Phytochem.*, Vol. 18. Plenum, New York, pp. 79–95.

Kogan, M. 1986. Natural chemicals in plant resistance to insects. *Iowa State J. Res.* 60: 501–527.

Kozlowski, T. T. (ed.), 1968. *Water Deficits and Plant Growth,* Vol. 1. Academic Press, New York.

Kozlowski, T. T. (ed.). 1972. *Water Deficits and Plant Growth,* Vol. 2. Academic Press, New York.

Kozlowski, T. T. 1984. Extent, causes, and impacts of flooding. *In* Kozlowski, T. T. (ed.), *Flooding and Plant Growth.* Academic Press, New York, pp. 1–7.

Krizek, D. J. 1984. Introduction to the Symposium. *In* Proc. of the Symposium, Somatic cell genetics: prospects for development of stress tolerance. *HortScience* 19: 366–367.

Lee, J. E., and G. C. Taylor. 1986. Agricultural research: who pays and who benefits? *In* Crowley, J. J. (ed.), *Research for Tomorrow, 1986 Yearbook of Agriculture.* USDA, Washington, DC, pp. 14–21.

Levitt, J. 1972. *Responses of Plants to Environmental Stresses.* 1st ed. Academic Press, New York.

Levitt, J. 1980. *Responses of Plants to Environmental Stresses. Water, Radiation, Salt and Other Stresses,* Vol. 2. 2d ed. Academic Press, New York.

Lindh, G. 1981. Water resources and food supply. *In* Bach, W., J. Pankrath, and S. H. Schneider (eds.), *Food–Climate Interactions.* Reidel, Dordrecht, Holland, pp. 239–260.

Luckey, T. D. 1968. Insecticide hormoligosis. *J. Econ. Entomol.* 61: 7–12.

Marshall, H. G. 1982. Breeding for tolerance to heat and cold. *In* Christiansen, M. N., and C. F. Lewis (eds.), *Breeding Plants for Less Favorable Environments.* Wiley, New York, pp. 47–71.

Mattson, W. J. 1980. Herbivory in relation to plant nitrogen content. *Annu. Rev. Ecol. Syst.* 11: 119–161.

Maxwell, F. G., and P. R. Jennings (eds.). 1980. *Breeding Plants Resistant to Insects.* Wiley, New York.

McClure, M. S. 1983. Competition between herbivores and increased resource heterogeneity. *In* Denno, R. F., and M. S. McClure (eds.), *Variable Plants and Herbivores in Natural and Managed Systems.* Academic Press, New York, pp. 125–153.

McDaniel, R. G. 1982. The physiology of temperature effects on plants. *In* Christiansen, M. N., and C. F. Lewis (eds.). *Breeding Plants for Less Favorable Environments.* Wiley, New York, pp. 13–45.

Meigs, P. 1953. World distribution of arid and semiarid homoclimates. *Arid Zone Res.* 2: 203–210.

Mitchell, R. G., and H. G. Paul. 1974. Field fertilization of Douglas Fir and its effect on *Adelges cooleyi* populations. *Environ. Entomol.* 3: 501–504.

Mudd, J. B., and T. T. Kozlowski (eds.). 1975. *Responses of Plants to Air Pollution.* Academic Press, New York.

Norris, D. M. 1979. How insects induce disease. *In* Horsfall, J. G., and E. B. Cowling (eds.), *Plant Disease,* Vol. 4. Academic Press, New York, pp. 239–255.

Norris, R. F. 1982. Interactions between weeds and other pests in the agro-ecosystem. *In* Hatfield, J. L., and G. J. Thomason (eds.), *Biometeorology in Integrated Pest Management.* Academic Press, New York, pp. 343–405.

Painter, R. H. 1951. *Insect Resistance in Crop Plants.* Macmillan, New York.

Parsons, L. R. 1982. Plant responses to water stress. *In* Christiansen, M. N., and C. F. Lewis (eds.), *Breeding Plants for Less Favorable Environments.* Wiley, New York, pp. 175–192.

Pimentel, D. 1981. Food, energy and climate change. *In* Bach, W., J. Pankrath, and S. H. Schneider (eds.), *Food–Climate Interactions*. Reidel, Dordrecht, Holland, pp. 303–323.

Pimentel, D., E. C. Jerhune, R. Dyson-Hudson, S. Rochereav, R. Somie, E. Smith, D. Denman, D. Reipschneider, and M. Shepard. 1976. Land degradation: effects on food and energy resources. *Science* 94: 149–155.

Ponnamperuma, F. N. 1984a. Role of cultivar tolerance in increasing rice production on saline lands. *In* Staples, R. C., and G. H. Toenniessen (eds.), *Salinity Tolerance in Plants*. Wiley, New York, pp. 255–271.

Ponnamperuma, F. N. 1984b. Effects of flooding on soils. *In* Kozlowski, T. T. (ed.), *Flooding and Plant Growth*. Academic Press, New York, pp. 9–45.

Quizenberry, J. E. 1982. Breeding for drought resistance and plant water use efficiency. *In* Christiansen, M. N., and C. F. Lewis (eds.), *Breeding Plants for Less Favorable Environments*. Wiley, New York, pp. 193–212.

Reese, J. C. 1979. Interactions of allelochemicals with nutrients in herbivore food. *In* Rosenthal, G. A., and D. H. Janzen (eds.), *Herbivores: Their Interaction with Secondary Plant Metabolites*. Academic Press, New York, pp. 309–330.

Reese, J. C., and D. J. Schmidt. 1986. Physiological aspects of plant–insect interactions. *Iowa State J. Res*. 60: 545–567.

Reinert, R. A., H. E. Heggestad, and W. W. Heck. 1982. Plant response and genetic modification of plants for tolerance to air pollutants. *In* Christiansen, M. N., and C. F. Lewis (eds.), *Breeding Plants for Less Favorable Environments*. Wiley, New York, pp. 259–292.

Reynolds. R. R. 1940. Lightning as cause of timber mortality. *S. Forest Exp. Sta. Notes* 31: 1.

Rhoades, D. F. 1979. Evolution of plant chemical defense against herbivores. *In* Rosenthal, G. A., and D. H. Janzen (eds.), *Herbivores: Their Interaction with Secondary Plant Metabolites*. Academic Press, New York, pp. 3–54.

Rhoades, D. F. 1983. Herbivore population dynamics and plant chemistry. *In* Denno, R. F., and M. S. McClure (eds.), *Variable Plants and Herbivores in Natural and Managed Systems*. Academic Press, New York, pp. 155–220.

Ritchie, J. T. 1980. Plant stress research and crop production: the challenge ahead. *In* Turner, N. C., and P. J. Kramer (eds.), *Adaptation of Plants to Water and High Temperature Stress*. Wiley, New York, pp. 21–29.

Ryan, C. A. 1983. Insect-induced chemical signals regulating natural plant protection responses. *In* Denno, R. F., and M. S. McClure (eds.), *Variable Plants and Herbivores in Natural and Managed Systems*. Academic Press, New York, pp. 43–60.

Scheurer, S. 1976. The influence of phytohormones and growth regulating substances on insect development processes. *In* Jermy, T. (ed.), *The Host Plant in Relation to Insect Behavior and Reproduction*. Plenum, New York, pp. 255–259.

Schneider, S. H., and W. Bach. 1981. Interactions of food and climate: issues and policy considerations. *In* Bach, W., J. Pankrath, and S. H. Schneider (eds), *Food–Climate Interactions*. Reidel, Dordrecht, Holland, pp. 1–19.

Schoonhoven, L. M. 1972. Secondary plant substances. *In* Runeckles, V. C., and T. C. Tso. (eds.), *Structural and Functional Aspects of Phytochemistry. Rec. Adv. Phytochem.*, Vol. 5, pp. 197–224.

Scriber, J. M. 1984. Host-plant suitability. *In* Bell, W. J., and R. T. Carde (eds.), *Chemical Ecology of Insects*. Chapman and Hall, London, pp. 159–202.

Seigler, D. S., and P. W. Price. 1976. Secondary compounds in plants: primary functions. *Am. Nat*. 110: 101–105.

Simpson, G. M. 1981. *Water Stress on Plants*. Praeger, New York.

Slater, L. E. 1981. Dimensions of the world food and climate problem. *In* Bach, W., J. Pankrath, and S. Schneider (eds.), *Food–Climate Interactions*. Reidel, Dordrecht, Holland, pp. 21–46.

Smirnoff, W. A., and B. Bernier. 1973. Increased mortality of the Swaine jackpine sawfly, and foliar nitrogen concentration after urea fertilization. *Can. J. For. Res.* 3: 112–121.

Stavarek, S. J., and D. W. Rains. 1984. The development of tolerance to mineral stress. *In* Proc. of Symp., Somatic Cell Genetics: Prospects for Development of Stress Tolerance. *HortScience* 19: 377–382.

Suzuki, H. 1980. Defenses triggered by previous invaders: fungi. *In* Horsfall, J. G., and E. B. Cowling (eds.), *Plant Disease,* Vol. 5. Academic Press, New York, pp. 319–332.

Thalwitz, W. P. 1981. Strategies to deal with climate/food interactions in developed countries. *In* Bach, W., J. Pankrath, and S. H. Schneider (eds.), *Food–Climate Interactions*. Reidel, Dordrecht, Holland, pp. 465–475.

Timmermann, B. N., and C. Steelink. 1984. Introductory chapter. *In* Timmermann, B. N., C. Steelink, and F. A. Loewus (eds.), *Phytochemical Adaptations to Stress. Rec. Adv. Phytochem.,* Vol. 18. Plenum, New York, pp. 1–6.

Toenniessen, G. H. 1984. Review of the world food situation and the role of salt-tolerant plants. *In* Staples, R. C., and G. H. Toenniessen (eds.), *Salinity Tolerance in Plants*. Wiley, New York, pp. 399–413.

Treshow, M. 1970. *Environment and Plant Response*. McGraw-Hill, New York.

Treshow, M. 1984. Introduction. *In* Treshow, M. (ed.), *Air Pollution and Plant Life*. Wiley, Chichester, England, pp. 1–6.

van Emden, H. F., V. F. Eastop, R. D. Hughes, and M. J. Way. 1969. The ecology of *Myzus persicae*. *Annu. Rev. Entomol.* 14: 197–270.

Weinstein, L. H., and D. C. McCune. 1979. Air pollution stress. *In* Mussell, H., and R. C. Staples (eds.), *Stress Physiology in Crop Plants*. Wiley, New York, pp. 327–342.

White, T. C. R. 1984. The abundance of invertebrate herbivores in relation to the availability of nitrogen in stressed food plants. *Oecologia* 63: 90–105.

2

PLANT-MEDIATED EFFECTS OF SOIL MINERAL STRESSES ON INSECTS

D. Dale

Department of Entomology
Kerala Agricultural University
Vellayani
Kerala, India

1. INTRODUCTION

Plants are exposed to many biotic and abiotic stresses. Among these stresses soilborne factors—both chemical and physical—are of prime importance. Soil nutrients on either side of the optimum may act as a stress to plants. When fertilizer is added to the soil, plants have access to more nutrients and water. The water-holding capacity of the soil is increased by improving the humus. The resulting changes in the plant are mainly physiological, but its morphology and phenology may also be affected.

Plant growth is highly dependent on the chemical properties of the soil, and these have an important influence on insect herbivores. Numerous observations relate the patterns of insect incidence on plants to the nutritional status of soils on which the host plants grow. The consequent change in abundance of insect herbivores is variable, depending on the impact of altered host nutrition on insect fecundity and immigration. Host nutrition can also modify the plant's reaction to insects and its susceptibility to infestation and create environmental conditions more or less favorable to insects. A complex relationship very often exists between host plant quality and insect herbivore–predator/parasite interactions. Soil chemistry also plays a role in the visitation of plants by pollinators.

Even slight changes in the balance between the host plant and its insect herbivore can lead to marked fluctuations in all parameters of insect development and population dynamics. But insect response to plants in mineral stressed soils may often be modified by many extraneous factors such as moisture availability (Scriber 1977), allelochemics (Roehrig and Capinera 1983), carnivory (Al-Zubaidi and Capinera 1983), shade (Campbell 1984), and atmospheric humidity (Painter 1954).

Variations in amounts of soil nutrients can lead to changes in levels of genetic resistance of plants to insects. In a few cases nutrients, such as potash, themselves may impart temporary host plant resistance which is generally termed as "induced resistance." The relationship between fertilizer regimes and host plant resistance to insect pests has been reviewed by Singh (1970), Leath and Radcliffe (1974), Jones (1976), and Tingey and Singh (1980).

This chapter discusses the effect of soil chemical properties on the suitability of plants as hosts for insects and how insects respond to stressed plants, the variable effects in different ecosystems, and possible causes of variability in effects. The chapter is not an extensive review of the literature on the subject but is intended to cover, through examples, the role that soil minerals, primarily the macronutrients play in plant-mediated effects on insects.

2. EFFECT OF MINERAL STRESSES ON THE SUITABILITY OF PLANTS AS HOSTS FOR INSECTS

Among the various external factors provided by the soil are mechanical support, heat, air, water, and nutrients. Growth is dependent on a favorable combination of these factors plus light (Brady 1974). Seventeen elements have been shown to be essential for plant growth (Table 1). It is the nutrient elements supplied by the soil that usually act as stress agents and affect crop development. Of the seventeen essential elements, six are used in large quantities and are thus referred to as macronutrients (nitrogen, phosphorous, potassium, calcium, magnesium, and sulfur). Essential elements used by

Table 1. Essential Nutrient Elements and Their Sources

Essential Elements Used in Relatively Large Amounts		Essential Elements Used in Relatively Small Amounts	
Mostly from Air and Water	From Soil Solids	From Soil Solids	
Carbon	Nitrogen	Iron	Copper
Hydrogen	Phosphorus	Manganese	Zinc
Oxygen	Potassium	Boron	Chlorine
	Calcium	Molybdenum	Cobalt
	Magnesium		
	Sulfur		

From Brady 1974.

Note: Other minor elements, such as sodium, fluorine, iodine, silicon, strontium, and barium, do not seem to be universally essential, as are the 17 listed here, although the soluble compounds of some may increase crop growth.

Table 2. Functions in Higher Plants of Several Micronutrients

Micronutrient	Functions in Higher Plant Processes
Zinc	Formulation of growth hormones, promotion of protein synthesis, seed and grain maturation and production
Iron	Chlorophyll synthesis, oxidation–reduction in respiration, constituent of certain enzymes and proteins
Copper	Catalyst for respiration, enzyme constituent, chlorophyll synthesis, carbohydrate and protein metabolism
Boron	Protein synthesis, nitrogen and carbohydrate metabolism, root system development, fruit and seed formation, and water relations
Manganese	Nitrogen and inorganic acid metabolism, carbon dioxide assimilation (photosynthesis), carbohydrate breakdown, formation of carotene, riboflavin, and ascorbic acid
Molybdenum	Symbiotic nitrogen fixation and protein synthesis

From Brady 1974.

plants in relatively small amounts are referred to as micronutrients (iron, manganese, boron, molybdenum, copper, zinc, chlorine, and cobalt).

The role of micronutrients in plant growth is complicated and not well understood (Brady 1974). The range between a deficiency and toxicity due to micronutrient levels is in many cases narrow. There is increased concern regarding micronutrient levels in crop production as improved crop varieties and micronutrient fertilizer practices have increased the level of crop production and micronutrient removal. In addition, increased knowledge of plant nutrition has aided in the diagnosis of micronutrient deficiences that previously went unnoticed.

Information available suggests that the micronutrients are effective through certain enzyme systems (Table 2). Copper, iron, and molybdenum act as electron carriers in enzyme systems that bring about oxidation-reduction reactions in plants. Molybdenum is believed to be essential for nitrogen fixation, both symbiotic and nonsymbiotic, and must be present if nitrates are to be metabolized into amino acids and proteins. Zinc is included in the formation of some growth hormones, copper in respiration and iron utilization, boron in sugar translocation, and iron is essential for chlorophyll formation and protein synthesis in chloroplasts.

Among the various elements, deficiency of nitrogen is the most common of the various mineral stresses, and there is abundant literature on the effect of nitrogen levels on plant–insect interactions. The plant-mediated effects of the macronutrients, nitrogen, phosphorus, and potassium on insects are discussed in this section. The lack of literature on the plant-mediated effects of micronutrients on insects precludes a discussion on this subject.

Mineral stress in soil, both an excess and a deficiency of nutrients, may cause changes in the plant that are primarily physiological, but sometimes morphology and phenology of the plants are also affected. Macronutrients differ in their effects on plant growth. Of the mobile elements, nitrogen has the greatest impact.

Nitrogen is essential for all living organisms. The synthesis of cellular proteins, amino acids, nucleic acids, purine, and pyrimidine nucleotides is dependent upon this structurally, functionally, and environmentally important element. Of the macronutrients applied in commercial fertilizers, nitrogen has the quickest and most pronounced effect. Nitrogen is a regulator that governs the utilization of potassium, phosphorus, and other elements. A deficiency of nitrogen stunts plant growth, restricts root systems, and causes leaves to turn yellowish green. In severe cases leaves turn completely yellow and eventually fall off. The major physiological responses of the plant to low nitrogen availability are accelerated proteolysis, accumulation of soluble nitrogen and carbohydrates, and in some plants enhanced nectar production. An oversupply of nitrogen can also be harmful to certain plant species.

Nitrogen is particularly important to insects and mites, as there exists a significant difference between the nitrogen content of plants (averaging around 2%) and that of insects (reaching up to 7%). So for an insect to grow and reproduce, it has to successfully locate the source of high nitrogen in the plant, consume a sufficient amount, and efficiently utilize it. Nitrogen is the most widely studied soil and plant nutrient, and its effect on various aspects of herbivore abundance has been reviewed by McNeill and Southwood (1978) and Mattson (1980). The practical implications of plant nitrogen on insect invasion and population growth in cultivated crops were discussed by Scriber (1984). The review of Stark (1965) covers the forest flora and their insect pests.

Most soils are inherently deficient in nitrogen, partly because of the very low nitrogen in their parent materials and partly because this element is easily lost through leaching and evaporation. Even though soil nitrogen reserves are continually replenished by gas absorption and rain, nitrogen concentrations are often less than optimal for plant growth (Bohn et al. 1985). Nitrogen is the mineral nutrient that most often limits crop productivity.

Gaseous atmosphere contains 80% nitrogen. Free nitrogen in the atmosphere is converted to N-containing compounds by lightning and biological fixation. Biological fixation involves conversion of nitrogen to ammonia largely by free-living bacteria and blue-green algae. Ammonia is oxidized to nitrous acid by bacteria (*Nitrosomonas, Nitrobacter*) and then to nitric acid and nitrates. Nitrates are generally the most readily available source of nitrogen to plants, but ammoniacal nitrogen is also used.

Nitrogen is the most abundant soil-derived mineral element in plant tissues. Nitrogen content in many plants varies between 1 and 4% (Allen et al. 1974). High concentrations (3 to 7%) occur in young, actively growing tissues or in seeds, and as soon as plants become senescent, nitrogen levels drop sharply. Minimal concentrations (0.5 to 1.5%) occur when the tissues abscise (Mattson 1980). The nitrogen content of phloem and xylem sap exhibits seasonal and plant-to-plant variations. Phloem sap is at least 10 times, sometimes even up to 100 times, richer in nitrogen than xylem sap.

Nitrogen fertilization seems to have the greatest effect on the levels of soluble nitrogen compounds. Amino acid and amide levels invariably increase (Hoff et al. 1974), though an occasional rise of inorganic nitrogen levels has also been reported (McDole and McMaster 1978). The most remarkable change related to the increase in soluble nitrogen content is a decline in the C/N ratio. Rice plants grown in a high nitrogen solution had 1.4 times as much nitrogen as the plants in a low nitrogen solution, and C/N ratios were 13 in the high nitrogen but 20 in the low nitrogen plants (Ishii and Hirano 1959). The effect of a change in the ratio of proteins to carbohydrates in favor of proteins in forest trees is often detrimental to insect pests. It has been suggested that nutritional status decreases as a result of this phenomenon. However, several authors (Eidmann 1963, Merker 1969) have reservations about this hypothesis. Lunderstädt and Hoppe (1975) found no evidence that the ratio between carbohydrates and proteins was a determining factor in decreasing the nutritional value of spruce needles for larvae of the sawfly, *Gilpinia hercyniae* Htg.

Certain nitrogenous fertilizers may lead to a condition of imbalance in the composition of nutrients in the host plant in such a way as to adversely affect the life activities of insect herbivores. Despite Fraenkel's (1953) original contention that the nutritional requirements of insects are "essentially similar," numerous investigators (Painter 1958; Auclair 1963, 1965, 1969; House 1965, 1966, 1969; Mehrotra et al. 1972) have emphasized the importance of a nutritionally balanced diet for insects.

The effect of nitrogen fertilization on insect incidence varies depending on the insect species, host plant species, soil fertility in regard to nitrogen and other elements, and possibly other environmental factors. In 23 studies involving nine crops in India, nitrogen fertilization increased insect incidence in 17 trials, decreased incidence in one, and had no effect in five trials (Singh and Agarwal 1983). Carrow and Betts (1973) found that population growth of the balsam woolly aphid, *Adelges piceae* (Ratzeburg), was higher on urea-fertilized, but lower on ammonium nitrate-fertilized trees, than on unfertilized trees. They attributed these different responses of the aphid to changes in amino acid composition of the host trees that varied between the two forms of nitrogenous fertilizer. Auerbach and Strong (1981) traced the poor performance of the herbivore *Caligo memnon* to the altered nutrient composition of *Heliconia imbricata* (Kuntze) which resulted from fertilization. In general, the form of nitrogenous fertilizer is less responsible for plant susceptibility to insects than to plant pathogens.

The forms in which nitrogen is transported and stored in plants also differs. Trees tend to transport it as organic nitrogen (specific proteins and/or amino acids), whereas herbs and shrubs transport it as inorganic nitrogen (Terman et al. 1976, Mattson 1980, Pate 1980). Plants of the Chenopodiaceae,

Compositae, Cruciferae, and Solanaceae families tend to store nitrogen as NO_3, sometimes in amounts up to 5% of tissue dry weight (Pate 1980). Levels of NO_3 even lower than this can be harmful to insect herbivores (Manglitz et al. 1976). However, the implications of these phenomena—transport and storage of nitrogen—in explaining fertilizer-induced differences of insect responses on plants are as yet not clear (Scriber 1984).

Although nitrogenous fertilizers can modify the mechanisms of host plant resistance, it is usually not certain whether these effects are due to changes in the contents of allelochemics, nutrients, or water. The principal role that nitrogen fertilization plays in pest management is to help plants compensate for lost or damaged parts, rapidly pass critical developmental stages, and escape attack (Jones 1976).

Increased rates of nitrogen fertilizers make the plant more succulent by increasing tissue softness and by increasing the water content. Plant tissues become softer as the carbohydrates are diverted for protein synthesis rather than the construction of cell walls (Tisdale and Nelson 1975). Plants grown with excessive amounts of nitrogen usually have dark green leaves and a lush foliage, a poorly developed root system, and therefore a high shoot-to-root ratio. Excess nitrogen can cause fruits (e.g., tomato) to split as they ripen and delay flowering and seed formation (e.g., cotton).

The change in the phenology of forest trees as affected by the application of nitrogenous fertilizers alters their suitability as a host for certain insects. Thalenhorst (1972) showed that buds opened earlier on spruce treated with nitrogen fertilizer than on untreated trees. The females of *Pristiphora abietina* (Hartig) lay their eggs in spruce buds, where the scales have already fallen off, but the needles are still compactly bundled together. Because fertilization affects the timing of this stage, the females are forced to disperse to trees with late development. Such a change in phenology can lead to the mortality of less mobile insects or larval stages. The resistance of certain spruces to the development of galls also depends in part on faulty timing between budburst and the life cycle of the gall-forming aphids (Bischoff et al. 1969).

Second to nitrogen, phosphorus is the most limiting element in soils. Phosphorus is an essential component of all living matter. A lack of phosphorus is especially serious as it prevents other nutrients from being acquired by plants (Brady 1974). It is mostly concentrated in the younger parts of the plant, flowers, and the seeds. It is necessary for photosynthesis and is involved in many energy-transfer reactions. Phosphorus is a part of the nucleus of the cell and is also present in the cytoplasm, where it is concerned with the organization of cells and the transfer of hereditary characteristics. Phosphorus does not move appreciably, and hence it accumulates primarily in the upper crest of the soil. The relative amount of phosphorus available to plants to the total content of the nutrient in the soil varies with different types of soil. Phosphorus occurs naturally in soils as phosphates of calcium, iron, and aluminum.

Plants deficient in phosphorus are stunted and are often dark green in color. Plant maturity is often delayed. Physiologically, phosphorus deficiency leads to an inhibition of protein metabolism and auxin production. If excess phosphorus is given to a plant, root growth is stimulated with respect to shoot growth.

In comparison to nitrogen, there is a dearth of literature on the plant-mediated effects of phosphorus on insect pests. However, the few published reports indicate that the effects are minor. Singh and Agarwal (1983) in reporting on the results of seven studies in India, indicated that in six, phosphorus did not affect insect populations in sugarcane, maize, mustard, and sunflower, but phosphorus did have a negative effect on the shootfly *Atherigona soccata* Rond in sorghum.

Plants need large amounts of potassium. Potassium exerts a general balancing effect on both nitrogen and phosphorus, and thus is an important component of a mixed fertilizer (Brady 1974). Potassium is essential for starch formation, for the translocation of sugars, and for the development of chlorophyll and is involved in protein synthesis, cell division, and growth. It is seldom deficient in soils which have relatively high pH values; acidic soils are often deficient in potassium. Deficiency symptoms of potassium in plants are first seen on the older, lower leaves. In dicots these leaves become chlorotic, and dark-colored, necrotic lesions develop. In many monocots the cells at the tips and margins of the leaves die first, and then necrosis spreads toward the younger lower parts at the base of the leaf. A low potassium supply frequently favors the development of insects and mites, whereas optimum or high potassium has a depressant or neutral effect. Potassium deficiency has certain effects that predispose the plant to herbivore attack (Baskaran et al. 1985). For example:

1. Soluble carbohydrates and reducing sugars accumulate.
2. Starch and glycogen synthesis is impaired.
3. Amino acids accumulate.
4. Protein synthesis is blocked.
5. The utilization of respiratory substrates is retarded.
6. Oxidative phosphorylation and photophosphorylation rates are decreased.

Increasing levels of potassium fertilizer generally appear to have a negative influence on insect populations. A possible reason for the reduction of pests by increased potassium may be due to a higher proteogenesis in plants, a physiological phenomenon correlated with elimination of amino acids, and reducing sugars in the sap which are otherwise favorable for the reproduction of sucking pests (Chaboussou 1972). Further, increase in the sclerenchymatous layer and silica content may also act as a mechanical barrier in plants that receive high potassium rates (Vaithilingam 1975).

Subramanian and Balasubramanian (1976) reported that high doses (200 and 250 kg/ha) of muriate of potash reduced the incidence of all the pests of rice they have studied; green leafhopper *Nephotettix* sp., brown plant-hopper *Nilaparvata lugens* Stål), leaffolder *Cnaphalocrocis medinalis* (Guenee), whorl maggot *Hydrellia philippina* Ferino, and thrips *Stenchaetothrips biformis* Bagnall. Similarly, potash reduced the incidence of the yellow stem borer of rice, *Scirpophaga incertulas* (Walker) (John and Thomas 1980). In the case of *N. lugens*, feeding activity was less on those rice plants that received the high levels of potassium than on the plants that received no potassium (Vaithilingam et al. 1976). But Israel and Prakasa Rao (1967) seldom could achieve a significant correlation between the amount of potassium (0, 5, 10, 20, and 25 ppm) in the nutrient supplied to the rice plants and the degree of infestation by the gall midge *Orseolia oryzae* (Wood-Mason). Singh and Agarwal (1983) reported the results of 17 trials involving 15 insect species on five crops in India. In only one trial did potassium increase the incidence of an insect, whereas in 12 other trials the nutrient decreased the population and in four, potassium had no effect at all.

Joint action of various plant nutrients can cause varying responses in plants and their insect herbivores. Regupathy and Subramanian (1972) reported increased plant growth and altered mineral metabolism in rice plants consequent to increased rates of NPK fertilizers. These changes in the plants favored the incidence and multiplication of the gall midge. The high doses of N, P, and K fertilizers increased N, K, Mn, and water contents of leaves and decreased P and Fe. Moreover the carbohydrate level and carbohydrate/N ratio were lowered. A rare study was carried out by Michael Raj and Morachan (1973) to assess the incidence of yellow stem borer and leaffolder when rice plants were treated with two types of NPK fertilizer—straight and mixed—and three rates. Different types of fertilizers exerted little influence on the stem borer population, whereas a significant increase in the incidence was noticed due to increased rate of fertilization. But in the case of the leaffolder, both fertilizer types and rates significantly influenced incidence.

When different plant nutrients are simultaneously applied, they can modify the elemental composition of plants, depending on the type of interaction among the nutrients. For example, as nitrogen and potassium are independent in their action, a significant increase in nitrogen and potassium in plant tissues was observed by Eaton et al. (1973) after the joint application of these two elements. With regard to herbivory high nitrogen and high N/K ratios often stimulate insects and mites (Perrenoud 1976). So it follows that increasing nitrogen should be balanced with potassium. In addition to effects on insect pests, plants receiving balanced fertilizer treatment are better able to withstand herbivory and recover from injury faster. It is not uncommon to observe an inverse relationship between nitrogen and phosphorus where the phosphorus content in plants decreases when nitrogenous and phosphatic fertilizers are jointly applied to plants (Smith 1964). The impact on insect herbivory of fertilizer-induced phytochemical changes is seldom determined.

Phenological changes due to mineral fertilization can alter the degree of damage by insects in crop plants. Injury to the barley seedlings from the frit fly, *Oscinosoma pusilla* Meig., is most damaging prior to the tillering stage when the main stem is attacked. But injury to the lateral stems lowers yields much less. Barley grew much more rapidly after mineral fertilizers were added to the soil and the frit fly generally attacked the lateral stems, whereas in the control, it mostly attacked the main stem (Gurevich et al. 1971). Some 22 to 54% of the plants of the nonfertilized plots suffered injury to the main stem. But after soil applications of NPK, NP, and P, the number of plants with an injured main stem did not exceed 9%.

3. INSECT RESPONSES TO MINERAL-STRESSED HOST PLANTS

The magnitude and type of effects of mineral stress upon insect populations vary significantly depending on many factors. However, some general and common effects are discussed here.

3.1 Food Preference

Various authors report that mineral stresses in plants alter the food preferences of insect herbivores. The squash bug, *Anasa tristis* (De Geer), preferred to feed on varieties of *Cucurbita pepo* (L.) grown in media deficient in phosphorus, potassium, and sulfur to those receiving complete nutrition (Benepal and Hall 1967). The feeding preference of the bug seemed to increase due to the high content of free amino acids in the mineral-deficient plants (except those lacking nitrogen).

Mineral deficiencies may also confer a nonpreferred status on a normally accepted food plant by insects. The flea beetle, *Agasicles hygrophila* Selman and Vogt, significantly preferred alligatorweed that received complete nutrition to plants deficient in phosphorus, calcium, or magnesium.

Nitrogen stress in the host plant at the sensitive period of insect reproduction can lead to a drastic reduction in populations. A number of strategies have evolved to buffer the insects against this possible ecological problem. Varying feeding sites on the same host plant and movement to other nutritionally better plants are the most important among them (McNeill and Southwood 1978). On runner beans, for example, *Aphis fabae* Scopoli colonies move along with the growth of the host plant and eventually congregate on the flowering shoots and pods. The grass bug, *Leptopterna dolabrata* (L.), changes its feeding site from leaves of *Holcus mollis* L. to young seeds of *H. lanatus* L. This change of host plants occurs at a time when insects are undergoing reproductive maturity and when demand for nitrogenous chemicals is at its peak. Insects that happen to continue feeding on leaves are weak and have a lower reproductive capacity. Prestidge (1982) reported that the occurrence of certain insect species on a given plant is related to a particular nitrogen level of the host plant. Some insects switch host plant species when nitrogen concentrations change; in this way interspecific competition is avoided.

3.2 Consumption, Digestion, and Utilization of Food Plants

Plants generally contain lower amounts of essential nutrients than insects, particularly proteins and their constituent amino acids, (McNeill and Southwood 1978). Russell (1947) computed the mean nitrogen content of nearly 400 species of plants to be 2.14%, which is considerably below the level needed for normal growth and development of insects. Many plants have evolved mutualistic relations with nitrogen-fixing bacteria or blue-green algae to enhance their nutrient status under stress conditions. Approximately 10,000 species of legumes and at least 160 species of nonleguminous plants harbor nitrogen-fixing symbionts (Delwiche 1978). The nitrogen contents of these plants reaches 5%.

Insects often consume more food to compensate for the low quality of host plants on which they live (Mattson 1980, Slansky and Scriber 1985). Feeding rate of the rice brown planthopper *N. lugens* on rice increased about threefold on plants receiving nitrogen fertilizer in the form of $(NH_4)_2SO_4$ in comparison to those on plants that received no nitrogen fertilizer (Hiroo et al. 1977). Certain caterpillars have been shown to increase their total food consumption and rate of feeding on plants deficient in nitrogen to maintain growth parameters equal to those on plants with high nitrogen concentrations (Auerbach and Strong 1981). A point of ecological significance in this connection is that as the larvae spend more time feeding, they are more exposed to natural enemies (Slansky and Scriber 1985).

In addition to an increased consumption rate, insects resort to various other mechanisms to cope with an inadequate nitrogen supply in their food. These mechanisms are (1) specialized alimentary canals and digestive systems, (2) symbiosis with microorganisms, (3) occasional cannibalism and predation, (4) changing plant parts and plant species for feeding, (5) regulation of host plant metabolism, and (6) evolution of larger body size. These points have been discussed in the review by Mattson (1980). Hill (1976) has shown that adult leafhoppers of *Dicranotropis hamata* (Boheman), *Adarrus ocellaris* (Fallen), *Diplocolenus abdominalis* (Fab.), *Recilia coronifera* (Marshall), and *Zyginidia scutellaris* (Herrich-Schaeffer) preferentially feed upon the developing flowers, seeds, and flower stems that have a high nitrogen content.

The digestive capacity of an insect seems to increase with the nitrogen level of its food (McNeill and Southwood 1978, Scriber 1984). Insects feeding on nitrogen-rich foods such as seeds, phloem sap, and pollen have relatively high AD (approximate digestibility) values compared to those feeding on nutrient-poor foliage. *Cephaloleia consanguinea* Baly larvae on their normal and fertilized hosts differed significantly in their nitrogen consumption rates and in nitrogen utilization efficiencies (Auerbach and Strong 1981). The larvae on fertilized hosts had a slightly higher AD but lower ECD (efficiency of conversion of digested food), resulting ultimately in similar growth efficiencies on the two hosts.

The efficiency with which a consumer can convert ingested food into its own biomass (ECI) is undoubtedly related to the nitrogen content of the foodstuff. The usual positive relationship between nitrogen and water content makes it difficult to separate the effect of each variable. Moreover, their effects are confounded with ecological factors (Mathavan and Pandian 1975) and insect body size (Iversen 1979). Sometimes even certain allelochemics such as DIMBOA (2,4-dihydroxy-7-methoxy [2*H*]-benzoxazin-3 [4*H*]one) may cause a decline in the efficiency of conversion of ingested food (Scriber 1984).

For leaf-chewing insects, a common response to low nitrogen availability is a high ingestion rate, which itself is often correlated with low efficiency of food use (McNeill and Southwood 1978, Auerbach and Strong 1981). But Al-Zubaidi and Capinera (1984), in their study with beet armyworms, reported that nitrogen level had no effect on NCR (nitrogen consumption rate), but it was inversely related to NAR (nitrogen accumulation rate) and NUE (nitrogen utilization efficiency) indexes. This suggests that the insects extracted proportionately more nitrogen from diets poorer in the nutrient to satisfy their biological demands.

Growth rates and nitrogen accumulation rates are likely to vary in different herbivores even with the same host plant. For example, the hispines had lower relative consumption and growth rates than the lepidopterans on all hosts studied (Auerbach and Strong 1981). Growth rates and nitrogen accumulation rates for the hispine species were not affected by host plant fertilization.

Food type affects food utilization efficiencies (Waldbauer 1968, Slansky and Feeny 1977). Many studies (Feeny 1970, Morrow and Fox 1980, Al-Zubaidi and Capinera 1984) have been conducted to determine the effects of both host plant and nitrogen content of the diet on nitrogen utilization efficiency of insects. They have clearly shown an inverse relation between nitrogen utilization efficiency and nitrogen content of the diet or host plant.

3.3 Insect Growth and Development

Growth is change in size of an individual; development is change in form. Both of these life parameters of insects are variously affected by the quality and quantity of food they consume.

The size of insects is generally influenced by crowding and genetic makeup (Richards and Myers 1980). But studies of Myers and Post (1981) have clearly indicated that the quality of the host plant can also affect insect size, although this is not universally true. Zitzman (1984) found that under controlled conditions biomass increments of the Colorado potato beetle were negatively correlated with the level of nitrogen applied, whereas under less controlled conditions biomass increments were positively correlated with nitrogen. Almost similar results obtained by Jansson and Smilowitz (1985) were explained through reasons such as emigration of early instar larvae and the presence of allelochemics which are factors specific to the particular host plant. Larvae of the cinnabar moth reared on fertilized plants produced significantly larger pupae than those reared on unfertilized plants. Body size of insects has important effects on survivorship and reproductive parameters, including metabolic rate (Phillipson 1981), mating success (Thornhill and Alcock 1983), and body temperature (Willmer and Unwin 1981). Moreover, there is a positive relationship between the size of female insects and their fecundity (Myers and Post 1981, Brewer et al. 1985). Such variation in fecundity may affect not only individual reproductive success but even the overall population of the herbivore species both in time and space.

Insects exhibit differences in the duration of larval instars when feeding on varying quality of food. These differences are more pronounced at low temperatures. Usually larval development period is inversely related to nitrogen level found in the food (Al-Zubaidi and Capinera 1984). But Prestidge (1982) showed that high levels of nitrogen in the food can lengthen instar duration in the case of a few leafhoppers feeding on *Holcus lanatus*. However, Barker and Tauber (1951) observed no significant change in the developmental period of the green peach aphid *Myzus persicae* (Sulzer), when it was reared on *Nasturtium* plants deficient in nitrogen, phosphorus, potassium, calcium, and magnesium.

Growth efficiency of many chewing insects is positively related to plant nitrogen level (Fox and Macauley 1977, Slansky and Feeny 1977). An increase in growth efficiency should normally translate into an increase in insect population. But many factors of varying causes intervene before a population explosion is reached. For example, cabbage butterfly larvae decrease their consumption rate in response to increased nitrogen content in the host plant (Slansky and Feeny 1977). While growth efficiency increases, growth rate does not.

In ecosystems such as forest and salt marsh, where plants are just marginally sufficient for the herbivores, insects do not exhibit a feedback regulation of food consumption (Fox and Macauley 1977). Species of *Eucalyptus* containing less than 0.9% nitrogen (dry weight basis) do not normally support the growth of beetle larvae, but those with slightly greater nitrogen contents result in increased larval growth efficiency, nitrogen accumulation, and growth rate. A similar pattern of responses to nitrogen-enriched salt marsh grasses would account for the greater size and survivorship of the mirid, *Trigonotylus* sp., and the cicadellid, *Amphicephalus simplex* (Van Duzee), in a study conducted by Vince et al. (1981).

The physiological basis for the response of growth parameters to food quality rests on the difference in the elemental composition between plants and insects. In most cases nitrogen remains a limiting factor relative to the carbon and energy available to the herbivores. Presumably an increase in nitrogen content would allow for a more efficient use of energy intake and an increased biomass production. Increased protein levels enhance the nutritional value of the host plant to insect herbivores. Because rice stem borer *Chilo suppressalis* (Walker) larvae require a very high level of dietary protein for their optimal growth (Ishii and Hirano 1957), the protein-rich plants are nutritionally more favorable for their growth and development.

3.4 Fecundity and Oviposition

The capacity of an insect species to produce a given number of offspring depends primarily on the number of ovarioles, their structure, and life span of the adult females (Huffaker and Rabb 1984). Even though these factors are genetically governed, they can be modified by the quality and quantity of food obtained during critical periods of ovarian development. A well-known observation is that insects which feed on more protein either as late-instar larvae or young adults, or both, produce more eggs than those that take in little.

Most of the sap-feeding insects respond to host-plant quality, rather than quantity. Both Hinckley (1963) and Fennah (1969) found a positive correlation between leafhopper fecundity and host plant soluble nitrogen levels. Metcalfe (1970) correlated the increased fecundity of *Saccharosdyne saccharivora* Westw. to an increase in plant nitrogen levels following nitrogen fertilization. Reports of Kalode (1974), Pathak (1975), and Dyck et al. (1979) on increases in rice brown planthopper, *N. lugens*, fecundity with addition of nitrogenous fertilizers reflect the fundamental problem of many studies in that they have no supporting plant nutrient data.

All leaves appear to provide similar environments to less mobile insects such as aphids, planthoppers, and leafhoppers; but they provide different quantities of resources. Photosynthate production of a leaf is proportional to leaf area. The assertion that the photosynthate from a leaf determines the performance of these insects is supported by experiments in which *Pemphigus* aphids on fertilized trees have been shown to have higher fecundities (Whitham 1978).

Hill (1976) determined nitrogen budgets for a number of leafhoppers in an attempt to determine periods of high nitrogen utilization in the insect's life cycle. He concluded that host plant nitrogen levels were normally adequate to ensure nymphal and adult survival but that the period of egg maturation created a very high nitrogen demand, and this was the most likely period of nitrogen stress. The pattern of exploitation of food resources is dependent on the levels of available nitrogen in the food (Prestidge 1982). Reproduction is also very dependent on the correct level of nitrogen in the food during maturation and nitrogen utilization efficiency is also tied to egg production. However, egg production need not be maximum at the highest level of available nitrogen.

There are reports that minerals in the host plant do not influence the fecundity of certain herbivore species. Taylor et al. (1952) found that the reproductive capacity of the pea aphid, *Acyrthosiphon pisum* (Harris), and the potato aphid, *Macrosiphum euphorbiae* (Thomas), was not significantly affected by feeding on pea or potato plants grown with high and low levels of nitrogen, phosphorus, and potassium. Similarly, no statistically significant differences could be obtained among the individual dry weights of eggs deposited when female leafhoppers fed on *H. lanatus* at different nitrogen levels (Prestidge 1982).

The host quality of many larval defoliators influences egg production of their imagines. Allen and Selman (1955) found that the fecundity of *Phaedon cochleariae* (F.) was significantly reduced when the beetles fed on watercress leaves deficient in nitrogen, phosphorus, potassium, or iron. The rate of egg production in this case was related to protein consumption by the beetles. Similarly, alligatorweed plants receiving complete mineral nutrition evoked a stronger feeding response in flea beetles, which resulted in enhanced fecundity in the females. Fecundity was also greater for moths of *Spodoptera frugiperda* (J. E. Smith) that originated from larvae that fed on fertilized bermuda grass than for moths that emerged from larvae that fed on unfertilized grass (Lynch 1984). Brewer et al. (1985) have reported that the mean number and weight of eggs laid by the western spruce budworm, *Choristoneura occidentalis* Freeman were highest when the larvae fed on foliage from the mid-range nitrogen level (200 ppm ammonium nitrate in aqueous solution). But egg viability, as measured by eclosion, was not affected by nitrogen treatment. Thus it appears that the major influence of foliar nitrogen on egg production is that it affects the number of eggs laid rather than the quality of individual eggs.

The positive relationship between plant water content and ovipositional preference by insects has been implicated in many studies (Benepal and Hall 1967). As the soil nutrient status, especially that of nitrogen, is directly linked to the water content of plants, it seems logical to presume that soil nutrients can influence the ovipositional behavior of gravid insects. There are reports (Hirano and Kiritani 1975, Wolfson 1980) documenting that plants receiving higher nitrogen levels are more succulent and thus attract more female moths to oviposit on them. Fujimura (1961) reported that rice plants supplied with additional nitrogen received six to seven times as many egg masses of the stem borer, *Chilo suppressalis* (Walker), as those conventionally fertilized. This is further supported by the fact that the annual consumption of ammonium sulfate as fertilizer and area infested by rice stem borer in Japan were positively related.

3.5 Insect Survival

The important role of nitrogen in protein synthesis and the significant variation in concentrations between plant and animal tissues make it a principal determinant of larval survival and development in insects. The suitability of a host plant as a food source for insects is often positively related to its nitrogen concentration (Mattson 1980). More larvae die as the degree of deficiency increases. However, the negative effects of excess nitrogen are less well known, although such effects have also been reported for many insects (Smirnoff and Bernier 1973, Morrow and Fox 1980, Archer et al. 1982, Prestidge 1982, Brewer et al. 1985) and mites (Jackson and Hunter 1983a).

Nitrogen required for protein synthesis and for growth is often a limiting factor for early instar insect larvae (White 1984). But those feeding on resources low in nitrogen can compensate for poor quality by increased feeding, as reported by Wagner and Blake (1983). However, in nature this strategy may also lead to increased mortality, as larvae are exposed to predation, parasitism, diseases, and adverse weather conditions for a longer time period (Fellin and Dewey 1982).

Smith and Northcott (1951) have shown that the grasshopper, *Melanoplus mexicanus* Sauss, failed to develop normally on wheat when low soil nitrogen levels led to low levels of leaf protein. Myers and Post (1981) tested the relationship between the survival of field populations of the cinnabar moth larvae and plant protein. Average larval survival and population fluctuations were significantly correlated to percent protein in plants.

Deficiency of other minerals in the host plant may also negatively affect insect survival. For example, copper and zinc deficiency greatly increased the mortality rate of cotton leafworm, *Alabama argillacea* Hubner (Creighton 1938). Taylor et al. (1952) found that larvae of the European corn borer, *Ostrinia nubilalis* (Hubner), grew faster and had higher survival on control corn plants than on plants deficient in nitrogen, phosphorus, and potassium.

3.6 Population Growth

Several reports (Taylor 1955, 1962; Jansson and Smilowitz 1986) have suggested that variation in amino-N concentrations in plants might influence the population dynamics of sap-feeding insects. For example, green peach aphids were more abundant and increased in numbers more rapidly on the lower leaves of potato because of elevated concentrations of free amino acids in the phloem that were produced during senescence (Taylor 1962).

Van Emden et al. (1969) reviewed examples where deficiencies of nitrogen, phosphorus, and potassium in the soil increased the total and soluble nitrogen content of plant tissues and had a positive effect on the population growth of aphids. Deficiencies of other ions (Webster 1959, Labanauskas and Handy 1970, Court et al. 1972, Lerer and Bar-Akiva 1976, Rendig et al.

1976), including trace elements (Steward et al. 1959, Cowling and Bristow 1979, Ghildiyal et al. 1977) also have the same effect of increasing the titer of soluble nitrogen in plant tissues. Insects other than aphids (Merino and Vazquez 1966, Metcalfe 1970, Sharma 1970) and mites (Rodriguez 1951, 1960; Markkula and Tiittanen 1969; Tulisalo 1971) also have been reported to respond favorably to the increased soluble nitrogen content in plants.

Many amides and amino acids have been implicated as responsible for increasing insect populations. For example, van Emden (1973) found that the levels of amide (glutamine + asparagine) in the leaves of host plants were positively correlated to aphid increases. Levels of other amino acids that have been reported as important in population increases are threonine and glutamic acid for *Brevicoryne* and methionine and leucine for *Myzus*. Sogawa (1971) and Cheng and Pathak (1972) linked fecundity levels and consequent population increases of rice leafhoppers to the contents of asparagine, an amino acid constituent of the amide fraction in rice plants.

Our present knowledge of amino acid biosynthesis suggests that there are three families of amino acids, each of which arises from a single common presursor. Group I comprises the amide forms, glutamine + asparagine, and their close allies, that show rapid fluctuations in their concentrations within the plant system. This group has the greatest effect on population growth. Group II amino acids, on the other hand, include some toxic chemicals that have no significant impact on insect population growth. Group III amino acids, even though they are essential components of the insect diet, are relatively constant in their concentration in the plant and thus are of lesser significance in their effects on insect populations.

In agroecosystems, application of nitrogenous fertilizers, within an optimum range, may lead to a corresponding increase in insect populations as the amount of fertilizer is increased. But nitrogen above or below this range may often be detrimental to phytophagous pests (Maxwell and Harwood 1960, Jansson and Smilowitz 1986). Where the application of nitrogen alleviates a deficiency of the nutrient in the soil, it might be expected to decrease the level of soluble nitrogen in the plant tissues and thus reduce the chance of success of herbivorous feeding. Trees growing on infertile soils usually harbor higher populations of insect pests (Bevan 1969) and the correction of deficiencies in these soils often decreases pest populations (Stark 1965, Bischoff 1967) apparently as a result of the young larvae failing to survive (Nef 1967, Carrow and Graham 1968). On the other hand, where the application of nitrogen to the soil is sufficient to increase the general level of nitrogen in the plant, an increase in insect feeding and population growth might be expected (van de Vrie and Delver 1979, Vince et al. 1981, Archer et al. 1982). *N. lugens* weight, feeding rate, and population growth increased with increasing levels of nitrogen fertilizer in the form of $(NH_4)_2SO_4$ on both susceptible and resistant varieties (Fig. 1) (Heinrichs and Medrano 1985).

In addition to amino-N, nitrate-N in plants has also been implicated in regulating herbivore populations. Greater aphid densities were associated

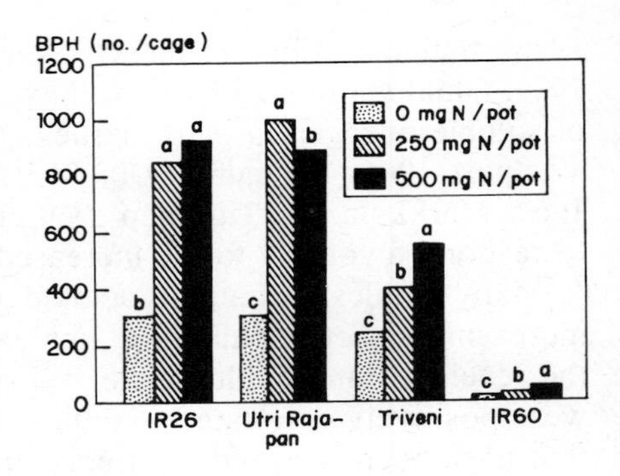

Figure 1. Brown planthopper (BPH) *N. lugens* biotype 2 population growth on selected rice varieties grown under different N fertilizer $[(NH_4)_2SO_4]$ rates. IR26 is susceptible to *N. lugens*, Utri Rajapan has tolerance but no antibioisis, Triveni has moderate levels of tolerance and antibiosis, and IR60 has high levels of antibiosis. Bars within a variety with the same letter are not significantly different at the 5% level by Duncan's multiple range test. (From Heinrichs and Medrano 1985)

with crops having low nitrate concentrations in plant sap in the early stages of plant growth (Henderson and Perry 1978). In sweet clover, Manglitz et al. (1976) reported that nitrate-N reduced feeding by the weevil *Sitona cylindricollis* Fahraeus, and consequently checked the population. But in potato, even though the lower leaves contained higher nitrate-N concentration, green peach aphid could develop more rapidly and increase its population (Jansson and Smilowitz 1986). It appears that some other factors might also be responsible for the populations of these insects.

Population increases of insect herbivores consequent to increased nitrogen applications are explained in other ways as well. Vereijken (1979) attributed the increased rate of *Sitobion avenae* F. multiplication on wheat to reductions in the proportion of alates produced and in their migration. Another reason for the high aphid populations may have been that nitrogen prolonged leaf duration allowing more time for multiplication (Prew et al. 1983).

Massive application of potassium to the soil causes reduction in certain insect pests (Chaboussou 1976). Use of K_2SO_4 and KNO_3 as potassic fertilizers to citrus led to a drop in population of coccids *Lepidosaphes beckii* (Newman) and *Saissetia oleae* Olivier. The decline in population of these two sucking insects was correlated to the changes in the physiology of the host plant, especially to an increase in the K/Ca + Mg ratio in the trees.

3.7 Wing Morphism

Many environmental cues trigger a developmental switch that determines adult wing form among planthoppers and aphids. Crowding and host plant quality are among the most important. Subtle changes in the amino-N balance and in the concentration of amino acids, amides, and salts—all of which are liable to change according to the nutrient status of the soils on which the plants grow—are known to affect the wing form of sap-feeding insects (Kisimoto 1965, Sutherland 1967, Dadd 1968, Harrewijn 1976, Carter and Cole 1977, Scheurer 1976). For example, McNeill and Southwood (1978) have suggested that very low methionine levels in *Holcus* in June may be responsible for alate production in *Holcaphis holci* Hille Ris Lambers.

Shull (1930) concluded that it was probably the chemical composition of the host plants that initiated the production of alate aphids. But as he did not analyze the test plants, his results must be considered with caution. Later Evans, in 1938, clearly demonstrated through chemical analysis of plants that a highly significant negative relationship existed between the number of alate aphids and the protein nitrogen content of the plants.

Denno et al. (1985) investigated the wing form response of *Prokelisia marginata* (Van Duzee), a planthopper that feeds on the phloem sap of the intertidal marshgrass, to crowding and plant nutrition. They found that the incidence of macroptery was density dependent in females but not in males. There was no consistent main effect of host plant nutrition on wing morphism. However, there was a significant crowding × host nutrition interaction, which suggests that the density-dependent migration response observed at higher hopper densities is dampened if planthoppers feed on nutritionally superior hosts.

Although fertilization leads to superior host quality that has a brachypterizing effect on planthoppers, there is reason to conclude that the enhanced nutritional quality of stressed or senescing host plants (Rahman et al. 1971; White 1974, 1976; Harborne 1977; Mali and Mentha 1977) may trigger macroptery in insects. This was reported for *N. lugens* females which were induced into macroptery on wilted rice plants.

3.8 Cannibalism

Some herbivorous insects, in addition to their host plant, have a second source of nitrogen which is often overlooked, facultative cannibalism. This behavior of intraspecific predation enables them to obtain additional high quality nitrogen (Mattson 1980). Cannibalism enhances the survival and reproductive capacity of these phytophagous insects. Al-Zubaidi and Capinera (1983) in evaluating the effects of dietary nitrogen levels on the cannibalistic behavior of the beet armyworm, *Spodoptera exigua* (Hübner), found that there was a significant inverse relationship between foliar nitrogen content and percent cannibalism. Fecundity was significantly increased by cannibalistic behavior of female larvae. Thus larval compensation for poor quality food by increased cannibalism is to be considered as an important factor in herbivore population dynamics.

Lycaenids are a unique group characterized by their preference for nitrogen-rich food sources such as terminal foliage, buds, shoots, flowers, fruits, and seeds. Their propensity for cannibalism may relate to their normal diet of nitrogen-rich plant tissues; they may lack the digestive systems necessary for switching to more nutrient-poor plant tissues when their primary food source becomes scarce (Mathews 1976).

3.9 Beneficial Insects

The performance of pollinating insects is significantly related to floral characters such as color, turgor, quantity of nectar, and the chemical constituents in the nectar. All of these parameters are modified by the physiological changes of plants as a result of mineral stress in the soil and/or fertilizer application. A delicate balance between pollinators and flower quality is critical to the coevolutionary ecology and practical management of the insect species involved (Heinrich 1979, Gojmerac 1980).

Nectar is a plant secretion derived from phloem sap, and sugars are the major contituents. Amino acids, proteins, lipids, antioxidants, alkaloids, vitamins, organic acids, and minerals may be present in the nectar, and all probably have some role in pollination (Kevan and Baker 1983). Although sugar may be most critical for honey bees, other pollinators such as many species of flies are attracted to amino acids and odors (Baker et al. 1978). Baker et al. (1978) have surveyed floral nectars for the amounts of amino acids and found that those with the most amino acids are taken by insects that do not ingest pollen. All 20 protein amino acids can be found in nectars, but all are not equally available (Kevan and Baker 1983). Apart from the nutritive function, amino acids probably act as feeding stimulants and taste modifers. Potassium at 1500 ppm deters bees from taking onion nectar (Waller et al. 1972).

Nitrogen fertilization generally has a deleterious effect on nectar production and the ''pollination index'' (Maurizio 1975, Shasha'a et al. 1976). The reduced nectar secretion at high nitrogen levels may be related to increased vegetative growth, whereas low levels of nitrogen would favor limited growth, accumulation of excess sugars, and copious nectar secretion (Shuel 1957).

Silk, produced by the silkworm, *Bombyx mori* L., consists of 75% fibroin, a tough elastic protein, and 25% gelatinous protein, siricin. An enhancement in the quality of mulberry leaves (in terms of protein or amino nitrogen) through fertilization increases silk production (Legay 1958).

3.10 Biological Control of Weeds

Biocontrol operations against weeds often fail because of a decline in the population of the insect agents on nutritionally poor host plants. Future attempts against weeds might benefit if fertilizers are judiciously used to increase the susceptibility of weeds to their insect predators.

Control of the prickly pear in Australia using *Cactoblastis cactorum* (Berg.) is one of the classical examples of biocontrol of weeds. Plants growing in poor soils were abnormally yellow and contained only about half as much nitrogen as those in fertile soils. Under these conditions the biocontrol operations were almost a failure. But fertilization of plants in some regions of Australia enhanced the feeding activity and survival of insects which led to the suppression of the noxious weed (Dodd 1940).

The alticine flea beetle, *Agasicles hygrophila* Selma and Vogt, was successfully introduced into southeastern United States from Argentina as a biocontrol agent of alligator weed, *Alternanthera philoxeroides* (Mart.) Griseb. Although most plants in aquatic plant communities are attacked by the flea beetle, some appeared not preferred and escaped severe feeding damage. In a study conducted by Maddox and Rhyne (1975) feeding preference, rate, and fecundity were all significantly affected by the mineral deficiencies of the host plant.

Because of its potential as a biotic suppressant on tansy ragwort *Senecio jacobaea* L., a pasture weed, the cinnabar moth *Tyria jacobaeae* L., a native to western Europe and Britain, has been introduced to North America, New Zealand, and Australia. Myers and Post (1981) found a positive correlation between the percentage of nitrogen in the plants and the coefficient of variation of moth population density. Moths tended to be larger and produced more eggs in areas with food plants of higher nutritional status. Food plants with high nitrogen levels increased larval survival and moth fecundity which allowed the insects periodically to overexploit their food supply. As a result population fluctuations occurred.

The floating fern, *Salvinia molesta* Mitchell, originated in southeastern Brazil and has been carried by people to many tropical and subtropical countries where it has become a serious weed. Field studies indicate that raising the concentration of nitrogen in plant issues from near 1 to 1.75% by fertilization resulted in more rapid increase in populations of the biocontrol agent, *Cyrtobagous* sp. (Room and Thomas 1985). Peak abundance of the insects occurred in sequence from the +N, to the +NP, to control treatments, with there being approximately twice as many insects present in the fertilized treatments as in the control. Total number of damaged buds was between two and three times as great in the fertilized areas as in the control.

3.11 Natural Enemies of Insect Pests

Little is known about host plant–herbivore–predator/parasite interactions as they relate to the nutritional status of soil. When soil is rich in nitrogen, the growing season for crops such as cotton is prolonged and the plants remain succulent longer and fruit later in the season. Delay in maturity of a crop for a period of time, however short it may be, can lead to a rapid buildup of herbivorous insects and their predators. Results of Adkisson (1958) indicated that predator populations were significantly higher in fertilized fields of cotton. Increases or decreases in pest populations in a fertilized field in late season are hence the ultimate result of the interaction between a pest and its natural enemies.

Predators may indirectly influence plant quality and affect the population dynamics of insect herbiovres. Vince et al. (1981) concluded that potential increases in the population of the delphacid, *P. marginata*, in response to fertilization of its host plant can be checked by the increased abundance of spiders. Because at high densities, planthoppers can reduce the nutritional quality of the host plant, and predatory spiders can suppress planthopper abundance, it is plausible to conclude that spiders may indirectly influence host plant quality.

Phytophagous insects often reduce their feeding time when they are on nutrient-rich host plants. This has great ecological significance in that herbivores in such ecosystems are less exposed to their natural enemies and disease-causing organisms. Specific examples have already been given in other sections of this chapter.

Smirnoff and Bernier (1973) reported that infection of Swaine jack-pine sawfly larvae with the virus, *Borrelinavirus swainei*, developed much more rapidly in larvae placed on fertilized trees. As virus replication is a function of the amount and composition of nitrogenous constituents assimilated by the larvae, extra protein and amino acids in the fertilized trees would have caused the added infectivity of the viral particles. In fact, Durzan and Steward (1967) had previously shown that the amino acid composition of needles of jack-pine seedlings varied with the nitrogen composition of the media on which they grew.

3.12. Structure of Herbivorous Insect Communities

Plant nutritional quality is of significant importance in governing the population of insect herbivores, with life history characteristics and natural enemies interacting to structure the community. Vince et al. (1981) studied factors regulating the numbers and species of herbivorous insects in a salt marsh ecosystem. Nutrient additions did not change the composition of the high marsh herbivore community but led to increased diversity in the low marsh. The low marsh insect community responded with greater numbers of mirids, cicadellids, and grasshoppers, although normally dominated by the delphacid, *P. marginata*. The rate of increase, however, differed among herbivore species and was positively related the number of generations in a year.

Moore and Clements (1984) reported that nitrogen fertilization alters the population ratios of dipteran stem borer species on perennial ryegrass. This may have been caused by the differential ovipositional response or to differential mortality of the early instar larvae within a narrow range of plant nitrogen. The composition of stem borer species may also be related to differential larval survival at various levels of plant nitrogen content, as Prestidge (1980) showed for a leafhopper species.

4. PATTERNS OF STRESS EFFECTS IN SELECTED ECOSYSTEMS

Insect herbivore populations respond to changes in host plant quality, and these changes may be moderated by factors specific to a particular ecosystem. Among the mechanisms that bring about a change in plant quality, soil nutrient stress is probably the most important. A few ecosystems are chosen as examples to show how stress effects vary in different ecosystems.

4.1 Forest Ecosystems

There are many types of forests that differ in respect to primary production and nutrient cycling. In these ecosystems insects act as regulators of primary production and thus perform a vital function in ecosystem dynamics (Mattson and Addy 1975).

The principal nitrogen reserve in most forest ecosystems is the organic matter associated with litter decomposition in the surface layers of the soil. Because the acidity and high tannin content of these soil layers may severely restrict nitrification, the main form of nitrogen provided to the plants may be ammonium, not nitrate (Jordan et al. 1979). But there are, however, other forest ecosystems in which nitrate may be regularly available in reasonably high amounts, or in which nitrate arises seasonally or after events such as fire, clearing, or felling of trees. Under these ephemeral conditions of nutrient availability, trees and shrubs may have to deploy their aerial parts for storing or reducing nitrate nitrogen. The ability to do this has been recently demonstrated in trees of certain West African ecosystems (Stewart and Orebamjo 1983).

Basically there are two situations in forests where insect outbreaks very often occur. In the first case insects appear to act as scavengers and attack weakened trees. These conditions are common on poor sites, or where the trees are crowded and/or declining in vigor from competition or maturity. The second case is density dependent in which favorable conditions, in terms of quality and quantity of food for insect herbivores, exist. In some forests, trees of the same species and age often occur either naturally or by planting, a condition akin to monoculture in the agroecosystems. At some point in development these trees uniformly become susceptible to an insect pest.

The sensitivity of forest insects to variations in host plant quality has not been amply studied. Nevertheless, there is circumstantial evidence to infer that insects can sense and respond to even subtle changes in host tree quality brought about by a complex of factors, including nutrient stress, aging, and moisture scarcity. This has led insect ecologists to conclude that the populations of forest insects increase inversely to host plant vigor (Mattson and Addy 1975). Many reports indicate that massive insect outbreaks typically start in middle-aged to old forests, which are in the waning years of their productivity. However, other scattered outbreaks seem to occur in forests of all ages, but those cases typically are associated with the least vigorous, slowest growing trees.

Soil nutrient levels and incidence of forest insects are very often inversely related (Table 3). Zwolfer (1957) observed that outbreaks of defoliating insects in Europe usually appear in forests of poor soils in which impoverished

Table 3. Effects of Host Plant Nutrients on Insects and Mites in Forest and Grassland Ecosystems

Plant	Insect/Mite	Insect/Mite Response	Reference
Conifer	*Pityokteines curvidens* Germ., *Ips typographus* L.	Application of calcium ammonium nitrate (1000 kg/ha) increased osmotic pressure of the trees, thus repelling bark beetles	Merker 1969
	Fiorinia externa Ferris	Less mortality of nymphs, faster development, and more progeny produced on host plants rich in N	McClure 1980
Fir	*Bupalus piniarius*, *Rhyacionia buoliana* Dennis Schiffermuller, *Coleophora laricella* Hb., *Diprion pini* (L.), *Pristiphora abietina* Christ, *Brachyderes incanus* L.	N, P, and Ca decreased insect populations; K and certain trace elements increased populations	Thalenhorst 1972
	P. abietina	N treatment did not change population	Gussone and Zottl 1975
	Choristoneura occidentalis Freeman	Female budworms from trees high in foliar N weighed less	Redak and Cates 1984
	C. occidentalis	Larval mortality was higher, and development time longer, at both upper and lower extremes of N treatment	Brewer et al. 1985
Grass	Stem-boring dipterans in grassland	N fertilization had no effect on populations	Henderson and Clements 1977
	Mirids, cicadellids, delphacids and grasshoppers in salt marsh	Greatest increase in herbivore populations occurred where nitrogen content of plants increased	Vince et al. 1981

	Prokelisia marginata (Van Duzee) on marsh grass	Density-dependent migration response observed at high densities was dampened when planthoppers fed on nutritionally superior hosts	Denno et al. 1985
Hemlock	*F. externa*	Less mortality and higher percentage of fecund females on fertilized trees	McClure 1977
	F. externa	Development was faster and fecundity higher on trees with more N	McClure 1980
Lupine	*D. pini*	N, P, and Ca reduced the number of pupae	Schwenke 1960
Mangrove plants	Many herbivores	Insects were more abundant in high nutrient area; mangrove skipper, *Phocides pigmalion,* attained a higher growth efficiency	Onuf et al. 1977
Oak	*Lasiocampa quercus* (L.)	N and Ca increased larval mortality	Merker 1961
	Eulecanium corni Bche., *E. rufulum* Ckll.	N increased scale populations while the effect was partially counteracted by K	Brüning and Uebel 1968, 1971
Pine	*Lymantria dispar* (L.)	N, P, and Ca increased larval mortality	Büttner 1956
	P. abietina	Lime and lime + N reduced the number of cocoons	Merker 1958
	B. pinarius	N and Ca increased larval mortality	Oldiges 1959
	R. buoliana	Population increased with tree fertilization	Merker and Büttner 1959
	Lymantria monacha (L.)	Ca increased larval mortality	Merker 1960
	R. buoliana	Population increased with N fertilizer	Eidmann and Ingestad 1963
	R. buoliana	Population decreased with fertilization	Schindler and Baule 1964
	R. buoliana	Population increased with fertilization	Burzynski 1966
	L. dispar, L. monacha	N, P, and Ca decreased populations; K and certain minor elements increased populations	Baule and Fricker 1967
	R. buoliana	Population decreased after K treatment	Nef 1967
	R. buoliana	Population reduced after application of fertilizer	Schindler 1967

Table 3. *(continued)*

Plant	Insect/Mite	Insect/Mite Response	Reference
	R. buoliana	N had no effect on population density; P led to a decrease; (K + P) caused a more evident decrease	Pritchett and Smith 1972
	Neodiprion swainei Midd.	Larval mortality increased with N fertilization	Smirnoff and Bernier 1973
	B. incanus	Compost caused 50% reduction in population	Wellenstein 1973
	Neodiprion sertifer (Geoffroy)	Larval performance was better on nonstressed trees	Larsson and Tenow 1984
Spruce	*P. abietina*	N treatment did not alter population	Schwerdtfeger 1970
	Oligonychus ununguis (Jacobi)	NPK and PK promoted population development	Thalenhorst 1972
	Eucosma tedella	NK, NP, and NPK decreased population; PK promoted	Thalenhorst 1972
	Pristiphora ambigua	All fertilizer combinations stimulated pest development	Thalenhorst 1972
	Sacchiphantes abietis	Fertilization had no effect	Thalenhorst 1972
	C. occidentalis	In outbreak areas soils had no volcanic ash, low available moisture, and low extractable Al, but the plant tissue contained high Na and P, and low Ca	Kemp and Moody 1984

microfauna are found. Shepherd (1959) concluded that abnormal increases in populations of the spruce budworm, *Choristoneura fumiferana* (Clem.), are restricted to drier, poorer sites. Furthermore outbreaks are often preceded by predisposing extrinsic stresses to plant systems, such as pollution, drought, or excessive moisture. There is much evidence to suggest that when plants are subjected to environmental stress, there is an increase in the soluble nitrogen levels and a decrease in the concentrations of secondary compounds of their leaf tissues. This may increase their palatability to insects (White 1969, 1974, 1976). The total quantity of available food is therefore a function of both plant quality and plant biomass (White 1978).

The nutritional requirements of insects are probably most critical during the early and late feeding stages. White (1974) speculated that nitrogen concentrations in pine trees are a determining factor in the survival of early instar larvae which is a key stage in the population dynamics of many important defoliating insects. Sugar concentrations apparently are not as critical in survival of these larvae as sugar is usually sufficiently abundant (Feeny 1970). For late-instar larvae of the eastern spruce budworm, however, increased sugar levels may produce increased feeding, growth, survival, and fecundity (Harvey 1974).

Fertilizer applications in forests generally seem to cause changes in food quality that reduce insect feeding and survival (Stark 1965). This may be due to nutrient imbalances in the chemical composition of the host plants caused by fertilizers (House 1965). Merker (1962a, b, c, 1963) believes that fertilizer elements are incorporated directly into various insect tissues, often with deleterious effects on the herbivores. Using labeled compounds, he could show that calcium, phosphorus, potassium, and nitrogen were directly incorporated in the Malpighian tubules, intestine, integument, and other organs of beetles. However, his studies did not explain why only defoliators are adversely affected by fertilizer applications and not sucking insects. Schwenke (1960, 1961, 1962) reported that fertilizers lowered the sugar content of foliage, and when it was reduced, sugar became a limiting nutrient for defoliators. He claimed that sap-feeding insects such as aphids, scales, and mites are more severely affected by turgor pressure of host plants and, because pressure increased with fertilizer application, the increase of these pests may be facilitated by fertilization. There are also references in the literature reporting fertilization to increase the weight and the fecundity of some forest insects (Bakke 1969).

4.2 Agroecosystems

Terrestrial agriculture represents the most manipulated of all the nonurban ecosystems, in which the pathways of energy and matter are directed almost solely toward humans and where humans maintain a high level of input to keep the system stable in order to maximize the yield of preferred crops (Simmons 1981). Agricultural systems are divided into "shifting" and "sedentary'' types. Shifting cultivation is mostly restricted to tropical forests, savannas, and grasslands. The crops are planted in a mosaic of varying heights, and times of planting vary so that the green cover of the plot remains as complete as possible throughout the year in order to reduce leaching by heavy rainfall. Burning, which follows the clearing of natural vegetation, mineralizes organic matter and allows its uptake by the crops. Plots are usually discarded when nutrient drain reduces soil fertility or when competition to the crops from pests and weeds reaches an uneconomic level (Watters 1960). Sedentary agriculture represents the permanent manipulation of an ecosystem: the natural biota are removed and replaced with domesticated flora. The nutrients in this system must be replenished at intervals by the addition of organic manures and chemical fertilizers.

Agroecosystems are ideal habitats for certain species of insect herbivores. Crop cultivars that have been bred for high yields on soils treated with manures and fertilizers are grown in these ecosystems throughout the year. But when these cultivars are grown under suboptimal conditions, they may become highly susceptible to insect infestations.

Fertilization of crop plants may change patterns of host selection, growth, and reproduction of insect herbivores and can also alter plant microclimate. These changes are likely to modify the role of a crop as a host for the insect pest and its natural enemies (Altieri and Whitcomb 1979). The extent of the fertilizer-induced effects on the physiology and morphology of the plants and subsequent effects on herbivorous insects depends on the ''feeding guild'' to which the insect herbivore belongs (Slansky and Scriber 1985).

The amounts and forms of nutrients in soil and plants of agroecosystems may also be changed by the application of pesticides (Pimentel and Edwards 1982). Where pesticides drastically affect the soil microflora, organic matter may not decompose and be incorporated into the soil, resulting in degradation of soil structure and poor water percolation. Herbicides reduce vegetative cover and soil organic matter, often resulting in soil erosion. These pest-related factors may modify and compound the effects of nutrient-stress on insect herbivores mediated through their host plants.

Although it is generally presumed that application of excess nitrogenous fertilizers and increased plant-N enhances insect pest populations and consequent crop damage, there is ample evidence of cases indicating the reverse. This is also the case for other plant nutrients as Table 4 indicates.

4.3 Salt Marshes and Mangrove Swamps

The boundaries between the sea and land are not always clear-cut, and throughout the world transitional zones exist. In the temperate regions these zones are represented by salt marshes and in the tropical regions, mangrove swamps flourish on many coasts. Both areas are highly productive and present a habitat that has a unique flora and a populous fauna. Plants here

Table 4. Effects of Host Plant Nutrients on Insects and Mites in Managed Agroecosystems

Plant	Insect/Mite	Insect/Mite Response	Reference
Alfalfa	*Therioaphis maculata* (Buckton)	Fertilizers had no effect on population	McMurtry 1962
	T. maculata	Host plant resistance was significantly reduced by low Ca or K, or excess Mg or N and increased in plants receiving low levels of P; sulfur did not affect plant resistance	Kindler and Staples 1970
Apple	Aphids	Populations increased when K was decreased	Jancke 1933
	Panonychus ulmi	Populations were higher in the plots having high leaf N	Hamstead and Gould 1957
	P. ulmi (Koch), *Tetranychus urticae* Koch	Increasing N levels increased populations	Rodriguez 1958
	P. ulmi	N content of the leaves and mite development were positively correlated	Breukel and Post 1959
	T. urticae	Increase in the total N content in the leaves resulted in increased egg production and greater longevity	Storms 1969
	P. ulmi	Phytophagous and predatory mites increased with N	van de Vrie 1972
	T. urticae	N deficiency increased preimaginal development time, preoviposition period, and decreased female weight, fecundity, and oviposition rate	Wermelinger et al. 1985
Barley	*Rhopalosiphum maidis* (Fitch)	All biotypes were heaviest on plants grown with a full complement of nutrients	Singh and Painter 1964
	Oscinosoma pusilla Meig., *Brachycolus noxius* Mord.	Mineral fertilizers, especially P and NPK, reduced plant injury	Gurevich et al. 1971
Broadbalk	Midges	N, P, K, and dung had no long-term effect on population	Johnson et al. 1969

Table 4. *(continued)*

Plant	Insect/Mite	Insect/Mite Response	Reference
Brussels sprouts	*Brevicoryne brassicae* (L.), *Myzus persicae* (Sulzer)	Aphids increased with N and decreased with K; high N/K ratio favored aphids	van Emden 1966
Cabbage	*B. brassicae*	A negative correlation between N in host plant and wing formation in the aphids	Evans 1938
	Pieris brassicae (L.)	Deficiencies of N and Fe slowed larval growth: P and K were found to be essential	Allen and Selman 1957
	Lipaphis erysimi (Kalt.)	N had no effect on reproduction	Kundu and Pant 1967
Cacao	Mealybugs	Infestation was positively related to N and negatively to K	Fennah 1959
Chrysanthemum	*Liriomyza pusilla* (Meig.)	Damage was greater under high levels of N in the leaf tissue	Woltz and Kelsheimer 1959
Citrus	*Chrysomphalus ficus*	Population was higher in overmanured fields	Schweig and Grunberg 1936
	Aonidiella aurantii (Maskell), *Lepidosaphes beckii* (Newman)	Excess N, P, and K increased resistance of citrus seedling to red scale and decreased resistance to purple scale	Salama et al. 1972
	Parlatoria zizyphus (Lucas), *Icerya purchasi* Maskell	High rates of N or P caused an increase in populations; Ca had no effect	Salama et al. 1985
Corn	*Blissus leucopterus* (Say)	Corn planted with soybeans was less severely attacked by the pest due to a rise of N in the soil	Haseman 1946
	Ostrinia nubilalis (Hübner)	Larval survival rates increased on plants grown in fertilized fields	Patch 1947
	O. nubilalis	High and low levels of N had little effect; high K reduced populations	Taylor et al. 1952
	Sitophilus oryzae (L.)	Increases of N, P, and K had no effect	Eden 1953
	O. nubilalis	Larval weight, survival, leaf injury, and number of larval galleries increased as the amount of N increased	Cannon and Ortega 1966
	O. nubilalis	N increased survival	Scott et al. 1965
	Chilo zonellus Swin.	Heavier infestations with high levels of N	Singh et al. 1968
	Chilo partellus (Swin.)	Increase in N caused higher borer infestation	Singh and Singh 1969

	Diatraea saccharalis (F.)	Fertilized plants suffered greater insect damage	Parisi et al. 1973
	C. partellus	Response of borer incidence to N fertilization varied in the different varieties	Sarup et al. 1978
Cotton	*Empoasca devastans* Dist.	Fertilizers had little effect	Parnell 1927
	E. devastans	High N stimulated reproduction	Sloan 1938
	Aphis gossypii Glov.	N increased population when plants were treated with calcium arsenate, but no appreciable difference was observed when insecticide was not used	McGarr 1942
	E. devastans	Insects thrived better on well-manured plots	Balasubramanian and Iyengar 1950
	Pectinophora gossypiella (Saunders)	Insect attack was more severe in fertilized fields	El-Gabaly 1952
	Heliothis zea (Boddie)	N, P, and K application significantly increased the population	Adkisson 1958
	E. devastans, Bemisia tabaci (Genn.)	N increased populations of both pests	Joyce 1958
	E. devastans	N and N + P increased population, whereas P alone and cow manure had no effect	Jayaraj and Venugopal 1964
	H. zea	Infestation increased with increase in N levels, especially when fertilizer was applied at time of planting	Kumar et al. 1982
	H. zea	N increased larval weight, assimilation efficiency of food, and insect development	Zeng et al. 1982
Cucumber	*P. ulmi*	High rates of mineral fertilizers increased populations	Bondarenko 1950
	T. urticae	Mites increased with N and K, and sometimes with P	Le Roux 1954
	Acyrthosiphon pisum (Harris), *M. persicae, T. urticae*	Population increased by K	Markkula et al. 1969
	B. tabaci	Unfertilized plots and those that received high N caused highest pest incidence; P + K reduced the number of nymphs	El-Beheidi and Gouhar 1971

Table 4. *(continued)*

Plant	Insect/Mite	Insect/Mite Response	Reference
Eggplant	*Leucinodes orbonalis* Guen.	Infestation was highest in high N plots; P and K promoted early maturity and hardening of the plant tissues which enabled the plants to escape borer injury	Mehto and Lall 1981
Grapefruit	*Lepidosaphes beckii* (Newman)	Mg deficiency reduced scale population	Thompson 1942
Linseed	*Dasineura lini* Barnes	No differences in populations in plots receiving different N rates	Jakhmola et al. 1973
Oats	*Oscinella frit* (L.)	N and P had no effect on tiller attack; N reduced attack on grain	Cunliffe 1928
	Schizaphis graminum (Rondani)	N decreased the number of aphids per leaf	Arant and Jones 1951
	S. graminum	Deficiencies of N, P, K, Ca, and Mg reduced fecundity	Blickenstaff et al. 1954
	Rhopalosiphum fitchii (Sanderson)	N increased aphid multiplication	Coon 1959
Oil palm	*T. urticae*	Population was highest on plants without boron	Rajaratnam and Hock 1975
Peach	*T. urticae*	Fertilizer treatments increased populations	Garman and Kennedy 1949
	Synanthedon exitiosa (Say)	N increased infestation probably through favorable conditions for oviposition and development	Smith and Harris 1952
Peanut	*Aphis craccivora* Koch	Highest fecundity was observed when plants were grown on full set of amino acids	Waghray and Singh 1965
Pear	*Psylla pyricola* Foerster	Egg and nymph density increased on trees receiving higher N levels	Pfeiffer and Burts 1983
Pearl milllet	*Spodoptera frugiperda* (J. E. Smith)	Host plant resistance was affected by fertilizer combinations	Leuck 1972
Peas and beans	*Aphis rumicis* L.	K and Mg favored aphid multiplication	Davidson 1925
	T. urticae	Fertilizer treatments induced high populations	Garman and Kennedy 1949
	Acyrthosiphon pisum (Harris)	Deficiencies of N, P, K, Ca, and Mg reduced fecundity	Barker and Tauber 1951

	T. urticae	Increase in population was related to high N and low K	Fritzsche et al. 1957
	A. pisum	High and low levels of N had no effect; high K reduced populations	Taylor et al. 1952
	Bryobia praetiosa Koch	High N, medium P, and medium K increased populations	Morris 1961
	T. urticae	Mites increased with N, and decreased with P, and total carbohydrates	Henneberry 1962
	T. urticae	Mites increased with N	Henneberry and Shriver 1964
	Tetranychus atlanticus McGregor	N increased mites; very high P resulted in higher population; K had no effect	Cannon and Connell 1965
	Aphis fabae Scopoli	Levels of N decreased number of aphids per leaf	Banks and Macaulay 1970
	Sitona lineatus (L.)	Peat, N, P, and K had no effect	Tulisalo and Markkula 1970
	Epilachna varivestis Mulsant	Correlation between percentage of leaf protein and insect numbers on non-nodulating but not on nodulating soybeans	Todd et al. 1972
	B. tabaci	N and P increased whitefly populations	Yein and Singh 1982
	Empoasca terminalis Dist., *E. motti* Pruthi, *Amrasca biguttula* (Ishida)	Population tended to increase due to fertilizer application	Gatoria and Singh 1984
	T. urticae	N deficiency affected growth, fecundity, and oviposition rate	Wermelinger et al. 1985
Pecan	*Eotetranychus hicoriae* (McGregor)	Largest population developed on seedlings with N at 200 ppm (recommended level)	Jackson and Hunter 1983a, b
Petunia	*Trialeurodes vaporariorum* (Westwood)	Insects thrived best on plants supplied with full nutrients; absence of Fe or K reduced larval development	Haseman 1946
Potato	*M. persicae*	High K reduced population	Quanjer 1929
		K increased population	Ross et al. 1948
		N increased population growth rate	Jansson and Smilowitz 1986

Table 4. *(continued)*

Plant	Insect/Mite	Insect/Mite Response	Reference
	M. persicae, *Aphis rhamni* Boyer de Fonscolombe, *Macrosiphum euphorbiae* (Thomas), *Aulacorthum solani* (Kalt.)	N and P increased, high K reduced, populations	Broadbent et al. 1952
	M. euphorbiae	High K reduced population; high and low N levels had little effect	Taylor et al. 1952
	Peach aphid	Population decreased with K	Volk et al. 1952
	M. euphorbiae	N, P, and K had little effect	Volk and Bode 1954
	M. persicae	Aphid infestations increased on plants that received high N/K ratio	Vijverberg 1965
	Leptinotarsa decemlineata (Say)	Egg masses were generally more abundant in the lowest and highest N treatments; developmental rates were negatively correlated with foliar concentrations of N	Jansson and Smilowitz 1985
Radish	*T. urticae*	N, P, and K increased mite infestation	Mellors and Propts 1983
Rice	*Chilo suppressalis* (Walker)	High N level favored the pest	Ishii and Hirano 1958
	Nephotettix bipunctatus F., *Nilaparvata lugens* (Stål)	N increased population	Ananthanarayanan and Abraham 1956
	N. lugens	N increased population	Abraham 1957
	C. suppressalis	K and P had no effect	Hirano and Ishii 1959, 1961
	C. suppressalis	Plants with more N received more egg masses	Fujimura 1961
	Scirpophaga incertulas (Walker)	N increased population	Ghosh 1962
	Orseolia oryzae (Wood-Mason)	Pest incidence was proportional to K levels	Israel and Prakasa Rao 1967
	Lissorhoptrus oryzophilus Kuschel	N increased population	Bowling 1963
	S. incertulas	N increased population; P had no effect	Varadharajan and Nagaraja Rao 1965

O. oryzae, Cnaphalocrocis medinalis (Guenee)	N, P, and K favored both species	Regupathy and Subramanian 1972
S. incertulas, C. medinalis	N, P, and K increased pest incidence	Michael Raj and Morachan 1973
O. oryzae	N increased population	Narayanan et al. 1973
Sitotroga cerealella (Olivier)	Susceptibility of grains in storage was not related to the N applied in the field	Saradamma et al. 1973
C. medinalis	Infestation increased with increase in N rates	Chandragiri et al. 1974
N. lugens	Population increased with N fertilization	Kalode 1974
C. medinalis	Pest incidence positively correlated to the levels of N	Subbaiah and Morachan 1974
Nephotettix nigropictus (Stål)	N had no effect	Krishnaiah and Bari 1975
N. lugens	Population enhanced with more N	Pathak 1975
Nephotettix virescens Distant, *C. medinalis, Hydrellia sasakii* Yuasa et Isitani, *Stenchaetothrips biformis* (Bagnall)	High K application (200 and 250 kg/ha) reduced the incidence of all the pests studied	Subramanian and Balasubramanian 1976
N. lugens	Feeding was less on plants receiving high K	Vaithilingam et al. 1976
N. lugens	Reproduction increased with the addition of N	Dyck et al. 1979
N. lugens	Population increased on plants which received high fertilizer rates	Mochida et al. 1979
N. lugens	N increased population	Pillai et al. 1979
S. incertulas	K reduced incidence	John and Thomas 1980
C. medinalis, N. lugens	N increased population	Subramanian et al. 1980
C. medinalis	High N had no effect	Kushwaha and Sharma 1981
S. incertulas	N levels and stem borer incidence were positively correlated	Saroja and Raju 1981
C. medinalis	N increased pest damage	Saroja et al. 1981
H. philippina	Potash had no influence on pest incidence	Sathiyanandam and Subramanian 1981
C. medinalis	Incidence positively correlated with N levels	Upadhyay et al. 1981

Table 4. *(continued)*

Plant	Insect/Mite	Insect/Mite Response	Reference
	C. medinalis	Increased N levels caused a corresponding increase in populations	Balasubramanian et al. 1983
	Diopsis macrophthalma Dalman, *Sesamia calamistis* Hmps.	Pest damage was most severe at 120 kg N/ha	Ho and Kibuka 1983
	C. medinalis	Incidence increased with N applications	Sathasivan et al. 1983
	N. lugens	N increased population; P and K did not	Uthamasamy et al. 1983
	N. lugens	Insect weight, feeding rate, and population increased with N applications	Heinrichs and Medrano 1985
	C. medinalis	N increased population	Rekhi et al. 1985
	S. incertulas, O. oryzae	Increased N levels increased *S. incertulas* but not *O. oryzae* population	Swaminathan et al. 1985
Rye	*S. graminum*	Deficiencies of N, P, K, Ca, and Mg reduced fecundity	Blickenstaff et al. 1954
Ryegrass	*S. graminum*	Mineral deficiency reduced fecundity	Blickenstaff et al. 1954
	Listronotus bonariensis Kuschel	High N resulted in most insect damage	Hunt and Gaynor 1982
	Oscinella frit (L.), *O. vastator, Geomyza tripunctata* Fallén	More larvae in plots treated with ammonium sulphate than in those treated with calcium nitrate	Moore and Clements 1984
Safflower	*Heliothis peltigera* Schiff, *H. armigera* Hübner	Adequate application of fertilizers (40 kg N and 30 kg P/ha) reduced pest incidence	Das and Sanghi 1983
Sorghum	*B. leucopterus*	Host plant resistance reduced by N, and increased by P	Dahms and Fenton 1940
	B. leucopterus	Bugs laid more eggs on plants growing on high N, or low P	Dahms 1947
	Rhopalosiphum maidis (Fitch)	High N increased aphid multiplication	Branson and Simpson 1966
	S. graminum	Tolerance, more than antibiosis, was changed by nutrients	Schweissing and Wilde 1979

Spinach	*Heliothrips haemorrhoidalis* (Bouche)	Thrips injury was almost absent at the highest levels of N and Ca	Wittwer and Haseman 1945
	H. haemorrhoidalis	Population highest on plants receiving low N; Ca level had no effect	Haseman 1946
Strawberry	*T. urticae*	N level in host leaves affected the number of mites produced but not duration of development	Hamai and Huffaker 1978
Sugar beet	*Eutettix tenellus* (Bak.)	High N stimulated reproduction	Mumford and Hey 1930
	Aphis fabae Scopoli	K reduced incidence	Stapel 1958
	M. persicae	No difference in populations on plants with and without N	Heathcote 1974
	Spodoptera exigua (Hübner)	Inverse correlation between foliar nitrogen and percent cannibalism among insects	Al-Zubaidi and Capinera 1983
	S. exigua	Larval development time and mortality decreased and egg production increased by N application	Al-Zubaidi and Capinera 1984
Sugarcane	*Bemisia tabaci* (Genn.)	Severe attack on leaves with low N	Mathur 1941
	Perkinsiella saccharicida Kirkaldy	Insect fecundity and host plant soluble N levels positively correlated	Fennah 1969
	Diatraea saccharicola (F.)	K reduced insect damage	Hatmosoewarno 1970
	P. saccharicida	Fecundity increased with increase in N levels	Metcalfe 1970
Sunflower	Leaf caterpillar and capitulum borer	Pest incidence was the least on plants fertilized with K	Rangarajan et al. 1974
Tea	Shothole borer	Population increased by N; P and K had little effect	Gadd 1943
Tobacco	*M. persicae*	Population was decreased by N and increased by K	Michael 1963
	M. persicae	P generally favored aphids	Michael and Chouteau 1963
	M. persicae	Growth and numbers increased in N and K treatments	Wooldridge and Harrison 1968
	Manduca sexta (L.)	Higher N rates favored pest multiplication	Reagan et al. 1978
	M. sexta, Epitrix hirtipennis (Melsheimer)	Populations were highest on plants receiving low P, high N, and K	Semtner et al. 1980
Tomato	*T. vaporariorum*	Plants supplied with all nutrients were least attractive; lack of P or Mg increased attractiveness	Haseman 1946

Table 4. *(continued)*

Plant	Insect/Mite	Insect/Mite Response	Reference
	T. urticae	Mite populations and concentrations of soil soluble salts were positively correlated	Rodriguez and Neiswander 1949
	T. urticae	Populations increased as the concentration of the major nutrients increased	Rodriguez 1951
	T. urticae	Population increased with N	Rodriguez and Rodriguez 1952
	T. vaporariorum	N had no influence, but P increased fecundity	Hussey and Gurney 1960
Velvet grass	*Dicranotropis hamata* Boheman, *Elymana sulphurella* Zetterstedt, *Eucelis incisus* Kirschbaum, *Zyginidia scutellaris* Herrich-Schaeffer	Food consumption and instar durations altered with nutrient changes; egg production and adult longevity not affected	Prestidge 1982
Watercress	*M. persicae*	Plant nutrients had no effect on aphids	Barker and Tauber 1951
	Phaedon cochleariae (F.)	Low egg production of the beetles fed on leaves deficient in N, P, K, or Fe	Allen and Selman 1955
Weeds			
Prickly pear	*Cactoblastis cactorum* (Berg)	Plants on poor soils were chlorotic, and insect survival was low	Dodd 1940
Alligatorweed	*Agasicles hygrophila* Selman and Vogt	Plants receiving complete nutrition were preferred for feeding and oviposition over deficient plants	Maddox and Rhyne 1975

Tansy ragwort	*Tyria jacobaeae* (L.)	Higher N levels increased larval survival, moth fecundity, and consequent moth populations	Myers and Post 1981
Redroot pigweed and lamb's-quarters	*Spodoptera exigua* (Hübner)	Larval development time and mortality decreased, and egg production increased, by N application	Al-Zubaidi and Capinera 1984
Bermuda grass	*S. frugiperda*	Fecundity was positively correlated to level of fertilization	Lynch 1984
Salvinia molesta	*Cyrtobagous* sp.	Population increased with N application	Room and Thomas 1985
Wheat	*S. graminum*	N was essential for normal development; lack of Ca, P, K, or Mg did not cause serious ill effects	Haseman 1946
	Melanoplus mexicanus Sauss.	Grasshoppers failed to develop normally on plants grown in soil low in N	Smith and Northcott 1951
	S. graminum	Aphids per gram of plant varied inversely with the amount of N applied	Daniels 1957
	Mayetiola destructor (Say)	Population increased with N as a result of tillering	Okigbo and Gyrisco 1962
	Leptohylemyia coarctata	K-fertilized plants could withstand insect attack	Johnson et al. 1969
	S. oryzae	Infestation in storage maximum for grains from field plots with no N fertilizer	Chakrabarthy and Mathew 1972
	Metopolophium dirhodum, Sitobion avenae, Rhopalosiphum padi (L.)	Greater aphid densities were associated with low nitrate concentrations in plant sap	Henderson and Perry 1978
	R. padi, S. avenae, M. dirhodum	Aphid populations highest on plants receiving high N	Prew et al. 1983
Willow	*Tuberolachnus salignus*	Populations increased on plants receiving high N/K nutrition	Vijverberg 1965

are especially adapted to tolerate the presence of salt and periodic inundation by sea water.

On the borders of the sea, salt marshes display some of the characteristics of terrestrial ecosystems. Vascular plants dominate, but little of this vegetation is directly consumed by herbivores (Teal 1962). Along the North American Atlantic coast, extensive monocultures of an intertidal grass, *Spartina alterniflora* Loisel., are common and they appear to be persistent and relatively unchanging in time.

Vince et al. (1981) tested the hypothesis that nutritional quality of salt marsh grasses limits herbivore abundance. The technique used was chronic nutrient enrichment of salt marsh plots. Fertilization resulted in an increased standing crop of low and high marsh insect herbivores. However, the greatest insect populations occurred where grass nitrogen content, as well as the standing crop, had increased. The herbivore increases were largely due to *in situ* changes in survivorship and fecundity in response to higher plant nitrogen content.

The distribution and abundance of mangrove swamps on tropical shores are dependent on the interaction of frequency of tidal flooding, soil salinity, and waterlogging of soil (Ananthakrishnan 1982). Mangrove swamps are very extensive along the river belts entering the Bay of Bengal, the Straits of Malacca in southern Borneo, New Guinea, Thailand, and adjoining territories of the East. The mangrove roots trap sediment that accumulates to form muddy banks; these present a new substrate for more trees to colonize.

Mineral cycles in mangrove swamps are of the "open type" as the primary source of nutrients, the sediment, is constantly washed away by tidal flushing and seasonal flooding. Soil quality in the mangrove areas at river mouths is dependent on the nutrient status of the alluvium brought down by rivers. The alluvium undergoes a series of biological transformations, involving the activity of bacteria, blue-green algae, green algae, and diatoms. These organisms convert ammonia into nitrate and release phosphates from their bound forms.

The swamp along peninsular Florida's (USA) coastal region is an example of a mangrove ecosystem. Odum and Heald (1972) demonstrated that detritus was the basis of the estuarine food webs on which coastal commercial and sport fish stocks depend. The major effect of more people and more intensive agriculture is higher loads of plant nutrients in surface waters. Onuf et al. (1977) studied the effect of nutrient enrichment of red mangroves, *Rhizophora mangle* L., by comparing two islands in the Indian River at Ft. Pierce. One island (high nutrient) had and another (low nutrient) did not have a breeding colony of pelicans and egrets. Both leaves and fruits at the high nutrient island were richer in nitrogen. There was also proportionately greater stimulation of herbivory by insects (five lepidopterans and a scolytid) in response to nutrient enrichment. But the difference in herbivory disappeared when the birds seasonally migrated away from their nesting areas at the high nutrient site.

4.4 Grassland Ecosystems

Grasses are the dominant plants over much of the world. They are mostly found where rainfall is too low to support tree growth, yet above the level at which semidesert conditions prevail. Herbivorous insects are important in the economy of the grasslands because grasses stimulate soil fertility and help to prevent soil erosion.

Among the insects in grassland ecosystems, it is only the British grassland leafhoppers that have been studied in detail. Studies of Auchenorrhyncha fauna (Morris 1971) in calcareous grasslands clearly show the influence of vegetation structure on abundance and on species diversity. Timings of grazing and cutting of grass are of prime importance in the population dynamics of grassland leafhoppers. Further work of Andrzejewska (1965) demonstrated that the large doses of mineral fertilizers (680 kg NPK per ha) produced an approximately fourfold increase in biomass of grass, with a corresponding 1.5-fold increase in the biomass of phytophagous insects compared to those in the control plots. The nutritive level of plants also increased, their nitrogen content being 4.2% compared with 2.9% in the controls.

A comparison of species diversity of leafhoppers on acidic grasslands and on calcareous soils has shown that insect fauna are more diverse in acidic grasslands (Morris 1971). The indexes of species diversity in the Silwood, England, area (acidic grassland) ranged between 3.2 and 3.9, whereas the maximum at Clare, Ireland (calcareous grassland), was only 2.6.

Seasonal and interseasonal variations of the nutritional status of a host plant have a great impact on the population ecology of grassland leafhoppers. McNeill (1973) working on a mesophyll-feeding mirid, *Leptoterna dolabrata* L., recorded large seasonal fluctuations in the total nitrogen levels in the host plant, *H. mollis*, and found this to be an important factor in the population dynamics of the insect.

Nymphs of grassland Auchenorrhyncha develop mostly upon leaves containing low soluble nitrogen levels (Hill 1976). However, adults of many species appear in the field at times when high nitrogen food sources are becoming available. Waloff (1980) found that in hot summer and in periods of prolonged drought, *H. mollis* failed to flower, and many females of *Errastunus ocellaris* (Fallen) and *Elymana sulphurella* (Zetterstedt) died without reaching maturity. In the second generation of *E. ocellaris*, maturation and oviposition are spread over the period of autumn flush of grass growth. It is probable that in late autumn soluble nitrogen levels are again high in the senescing leaves.

5. SOURCES OF VARIATION IN EFFECTS

Tables 3 and 4 clearly demonstrate that a great deal of variation exists among the effects on insects caused by nutrient changes in soil that are transmitted through insects' host plants. There are many causes for the variation. Among these are soil, plant, and insect factors and biological nitrogen fixation.

5.1 Soil Factors

5.1.1 Sources, Forms, and Availability of Nitrogen

The nitrogen content in the host plant in agroecosystems may depend on how, when, how much, and in what form the nitrogenous fertilizers are applied (Macleod and Suzuki 1967, Carrow and Betts 1973) and the interactions with other environmental factors to which the plants are exposed (Steward et al. 1959). In natural ecosystems most of the higher plants depend on the microbial breakdown of soil organic matter for their nitrogen supply. In a few soils this results in an appreciable accumulation of amino acids, and some of them may be readily taken up and utilized by plants. But there are potentially only two major sources of nitrogen: ammonium and nitrate ions. These result from ammonification and nitrification processes, both of which are mediated by heterotrophic microorganisms. So any environmental factor that affects the microorganisms will directly determine the conversion processes and supply of nutrients for plant growth. In many soils there is a marked seasonality in the rate of supply of ammonium and nitrate ions. For example, in temperate zones there is usually a peak of availability in late winter or early spring, whereas in tropical systems it is often associated with the start of the rains following the dry season.

Ammonification and nitrification are variously affected by environmental factors. Low temperature, high acidity, and anaerobic conditions depress nitrification to a greater extent than ammonification. Conversely, in tropical habitats nitrate is often the dominant form of nitrogen, and in some soils ammonium is almost undetectable (Greenland 1958).

Generally, fewer plant species prefer ammonium to nitrate. The choice in many cases is dependent on the acidity of the medium and age of the plant. A further factor is seasonal variation in utilization and uptake of nitrogen; Grasmanis and Nicholas (1971) found that in the case of apple trees the annual uptake of ammonium and nitrate was almost equal, but during winter ammonium was taken up in much greater amounts than nitrate. This was ascribed to the effect of low temperature on nitrate reduction and on the absorption processes.

Plants growing on nitrate or ammonium show differences in their chemical constitution that may be related to the form of nitrogen present in the soil. Vladimov (1945) has suggested that nitrate nutrition should result in the accumulation of oxidized compounds such as organic acids, whereas ammonium nutrition should favor accumulation of reduced compounds. Plants grown on ammonium usually accumulate amino acids and amides and contain much lower levels of organic acids than nitrate grown plants (Kirkby 1968). This observation has important consequences for insect herbivores.

Indeed, Moore and Clements (1984) have shown that the form of nitrogen source may change plant-mediated effects on insects such as herbivore phenology and composition in an ecosystem. Plots of perennial ryegrass receiving nitrogen as ammonium sulphate contained more stem-boring dipterous larvae than those that received a calcium nitrate treatment. Many other grassland pest populations are as well affected by the form and levels of nitrogen (Prestidge 1980). Nitrogen supplied from $(NH_4)_2SO_4$ solutions caused leaf epinasty and an increase in the rate of ethylene evolution of tomato plants (Corey et al. 1987). Epinasty can alter the leaf as a habitat for foliage feeding insects. Ethylene is a hormone that influences plant responses to various subsequent environmental stresses (see Chapter 10).

The seasonality in the supply of nitrate and ammonium in many soils may to some extent influence the growth strategies of plants, particularly those of annuals. The absolute nitrogen-supplying power of most arctic soils may be an important determinant for the scarcity of annuals in arctic regions. But perhaps the most intriguing aspect of the nitrogen supply is the ecological consequence of different nitrogen sources. The differential availability of nitrogenous ions in soils raises the prospect that plants may be principally adapted to utilize only one nitrogen source, and this may result in the form of available nitrogen being an important factor limiting species distribution.

Nitrogen availability is a key determinant in the successional events of plant communities. The most intriguing aspect of succession does not involve the absolute rate of supply of the element but instead its form. If plant species do vary in their ability to utilize nitrate and ammonium, then the form in which nitrogen becomes available may determine which species should grow in a particular soil. Controversially, a species may determine which species can coexist with it by affecting not only the absolute amount of nitrogen available but also its form.

5.1.2 Soil pH

Because plants respond so markedly to their chemical environment, the importance of the soil pH has long been recognized (Brady 1974). Three conditions are possible: acidity, neutrality, and alkalinity. Soil acidity is common in all regions where precipitation is sufficiently high to leach appreciable amounts of exchangeable bases from the surface layers of soils. It is of widespread and common occurrence. Alkaline soils are characteristic of most arid and semiarid regions where the presence of salts, especially calcium, magnesium, and sodium carbonates gives a preponderance of hydroxyl ions over hydrogen ions in the soil solution.

Plants grow on soil over a pH range of 3 to 9, and the extremes cause a stress to plants. When the soil is highly acidic, the activity of nitrifying bacteria is inhibited, nodulation is reduced, and minerals such as phosphorus and molybdenum become largely unavailable (Buckman and Brady 1969). The less abundant calcium in acid soils may limit plant growth. The relatively high concentrations of available aluminum in many acidic soils can also inhibit plant growth by the detrimental effects on phosphate availability, by inhibiting absorption of iron, and by direct toxic effects on plant metabolism (Salisbury and Ross 1978). The pH also alters the solubility of certain elements in the soil and their absorption by plants. Iron, zinc, copper, and manganese are less soluble in alkaline than acidic soils.

Few studies have been conducted on the plant-mediated effects of soil pH on insect herbivores. A comparison between acidic grassland in Silkwood Park and alkaline chalk grasslands (Waloff and Solomon 1973) has shown that the leafhopper populations in the acidic grassland are richer in number of species. Foliar damage to whorl stage sorghum by the fall armyworm *S. frugiperda* was greater in plants growing in acidic soils (pH below 5.4) than in plants grown in soils with more optimal pH levels (pH above 6.0) (Gardner and Duncan 1982). Adverse soil pH retarded plant development and growth rate which caused the plants grown on more acidic soils to remain in the whorl stages longer than plants grown on less acidic soils, thus extending the period in which the fall armyworm attacks the foliage.

5.1.3 Salinity

The importance of salinity in crop production is discussed in Chapter 1. Salinity is considered to be a problem where the concentration of sodium chloride, sodium carbonate, sodium sulphate, or salts of magnesium are present in excess. It is, however, debatable as to the exact concentration where the salinity problem first appears—where glycophytic conditions became halophytic, and vice versa. About 0.5% sodium chloride in the soil solution is the suggested level (Chapman 1975).

Changes in plants induced by salinity vary considerably depending on the plant species, stage of plant development, and the influence of external factors such as edaphic conditions, species of salt involved, and climatic conditions such as heat and humidity (Poljakoff-Mayber and Gale 1975). There are two hypotheses to explain the effect of salinity on plants. The first is that salinity decreases the water available to the plant by decreasing the osmotic potential at the root surfaces, and second, that an excess of certain ions such as sodium or chloride directly exerts toxic effects (Terry and Waldron 1984). Levitt's (1980) list of metabolic changes associated with salt stress includes respiratory rate, photosynthetic rate, protein synthesis, nucleotide synthesis, and enzyme activity. Those metabolic changes influence many aspects of plant growth and induce changes in plant chemistry and anatomy.

Salinity induces a number of plant responses that can alter the suitability of a plant as a host for insects. Salt stress inhibits the growth of the various plant parts throughout the plant cycle and causes stunting of glycophytes. Saline soil inhibits root growth and production of new roots in citrus seedlings (Blaker and MacDonald 1986) and growth of sugar beet leaves (Terry and Waldron 1984). Sensitivity of bean leaves to salinity stress varies with leaf age with the younger leaves being most sensitive (Poljakoff-Mayber 1975). Saline conditions also decrease the diameter of bean stems due to a reduction in vascular tissue.

Salinity has distinct effects on plant succulence and anatomy. According to Rains (1972), plants exposed to chloride salinity have a greater degree of succulence than plants exposed to sulphate salinity. Succulence results from the development of larger cells in the spongy mesophyll and the presence of a multilayer palisade tissue. In general, salinity induces changes in the number and size of stomata, thickening of the cuticle, earlier occurrence of lignification, extensive development of tyloses, and changes in the number and diameter of xylem vessels (Poljakoff-Mayber 1975). Under saline conditions greater growth of intervein tissue of tomato leaves as compared to the vein system caused leaves to become waxy and rippled.

Salinity has varied effects on plant biochemistry, and thus the quality of the plant as food for insects. Nitrogen metabolism is affected by salinity. Synthesis of new proteins decreases while hydrolysis of storage proteins increases, resulting in an accumulation of amino acids and other metabolic products such as phenylalanines, flavones, alkaloids, nicotine, aspartic acid, proline, and lysine (Levitt 1972). Salinity induces distinct protein changes in the root and shoot tissues of barley seedlings. Salinity inhibits the synthesis of a majority of shoot proteins and induces synthesis of new proteins of low molecular weight in both shoots and roots. High levels of organic acids, including oxalic acid, have been reported in a number of plant species associated with dry climates and high soil salinity (Rains 1972).

Studies that link the plant mediated salinity effects with the suitability of salinity stressed plants as hosts for insects are lacking. However, limited studies on the susceptibility of saline stressed plants to pathogens have been conducted. MacDonald (1984), for example, reported that salinity decreased the active defense response of chrysanthemums to Phytophthora root rot. Similar studies with insects are needed.

Despite the lack of information on plant-mediated effects of salinity on insects extrapolations are suggested. For example, plant growth inhibition by salinity results in a decrease in the plant biomass, and thus a decrease in the amount of food available to herbivorous insects. Decreased stem diameter could result in the stem being too small to accommodate a stem-boring insect species. Succulence generally increases the suitability of plant tissue for chewing insects. Changes in leaf anatomy such as the formation of waxy leaves can alter the suitability of the leaf as a habitat for small insects. Changes in the size of cells and thickness of leaves would be expected to affect the activity of leaf-mining insects. Biochemical changes caused by salinity alter the plant as a host for insects by affecting levels of allelochemicals and nutrients such as amino acids. Oxalic acid in rice, for example, acts as a sucking inhibitor of the brown planthopper *N. lugens* (Yoshihara et al. 1980).

5.2 Plant Factors

5.2.1 Intraplant Variation

Individual plants can become a highly heterogeneous habitat for insect herbivores as a result of environmental factors (Whitham 1981). The demand of a plant for a given nutrient changes with time as it is influenced by various environmental factors. Moisture stress, for example, has a tremendous impact on both carbon and nitrogen metabolism (Bidwell and Durzan 1975). The effects vary by plant species, type of plant tissue, and duration of stress. Typically, as stress increases, total plant nitrogen gradually increases (Bates 1971, Viets 1972). In many cases nitrogen content may increase in one plant part while decreasing in another (Dina and Klikoff 1973). If stress continues to be severe, the levels of amino acids and soluble proteins increase due to lysis of structural proteins and the products are translocated to the young fast-growing tissues (Baskin and Baskin 1974, White 1978).

Plant injury, both by physical and biotic causes, alters plant composition. Defoliation and pathogenic infections consistently increased total protein nitrogen (McNaughton 1979) and the enzymes related to phytoalexin reactions (Green and Ryan 1972, Ryan and Green 1974, Mayer and Harel 1979). Simultaneously, there are often increases of the aromatic amino acids (Kosuge 1969), nonprotein amino acids (Rice 1974), total amino acids (Parker and Patton 1975, Roberts et al. 1976), and phenols and terpenes (Rice 1974, DeVerall 1977, Harborne 1977).

Herbivores themselves can change the physiology of the plant part on which they feed, thus further increasing intraplant variation (see Chapters 9 and 10). Also insects and pathogens may induce changes in the defensive chemistry of the plant. Superior competitors can push the inferior ones to less suitable regions of a plant where nutrients are either less available or of poor quality (McClure 1980).

There are reports that competition among plants in an ecosystem influences the ontogeny of a plant. Mitchell and Chandler (1939) reported decreasing foliar nitrogen concentrations with increasing density of trees. Similarly, the nitrogen content of *Carex* meadow plants was negatively correlated with plant population in unit area (Auclair 1977). Thinning rank forest stands usually increases foliar nitrogen in the remaining trees (Piene 1978), except in a situation where nitrogen is not a limiting resource.

5.2.2 Water Content

The suitability of a host plant as food for insects is governed by a complex interaction of factors. Moisture stress can significantly alter food quality through its effect on nitrogen uptake and metabolism (Mattson and Addy 1975). A shortage in water has been considered to cause fundamental changes in plant nitrogen that lead to conditions suitable for insect outbreaks (White 1974, 1976, 1978).

Changes in plant water affect plants at a basic developmental level, often leading to complex changes in the plant's physiology. Leigh et al. (1970) provided an example. They found that cotton susceptibility to lygus bugs is determined more by the interaction between irrigation and fertilizer application than by their individual direct effects. Similarly, Wolfson (1980) showed that reducing soil nitrogen available to cabbage reduced foliar water content. Oviposition preference of the imported cabbage butterfly *Pieris rapae* (L.) was more strongly correlated with high foliage water content than with nitrogen availability or the allelochemic content of the plants.

5.2.3 Chemical Defenses of Plants

One of the causes of discrepancy in results with respect to nutrient-stress studies is the effects of nitrogen on concentrations of defense chemicals that reduce the fitness of insect herbivores. Almost all plants contain some chemical substances to defend against herbivory. These may be either "quantitative" defenses (digestibility-reducing substances) prevalent in "apparent" plants, such as long-lived trees, or "qualitative" chemical defenses (toxins), such as alkaloids or cyanogenic glycosides, which are present in "unapparent" early successional herbaceous plants (Feeny 1976, Futuyma 1976, Rhoades and Cates 1976). Many lepidopterous insects that feed on plants with qualitative defenses are either monophagous or oligophagous, whereas others exploiting plants with quantitative defenses tend to be more polyphagous (Futuyma 1976, Slansky 1976).

The quantitative defenses are not basically nitrogen based but are carbon-based phenolics, tannins, and various terpenoid compounds (Harborne 1977). These digestibility-reducing allelochemics are usually associated with nitrogen-poor communities such as tropical sands, bogs, and climax forests; plant tissues such as mature leaves, stems, and bark; and slow-growing woody perennials (Mattson 1980). Black spruce, which is a dominant bog tree in North America and is sufficiently hardy to withstand hostile nitrogen-poor soil conditions, contains abundant terpenes (von Rudloff 1975) and phenolic compounds (Hanover 1975), as do many species of *Eucalyptus* that are apparently adapted to soils poor in nitrogen and phosphorus (Macauley and Fox 1980, Morrow and Fox 1980).

The qualitative defenses, on the other hand, are usually associated with nitrogen-rich plant tissues (newly developing tissues), reproductive organs like seeds, and succulent herbaceous plants. The nitrogen-based toxins are usually found in highest concentration in early season foliage or in ephemeral plants (Cooper-Driver et al. 1977, Mattson 1980). The legumes, for example, most of which harbor nitrogen-fixing symbionts and therefore have abundant nitrogen, have a vast set of nitrogen-based allelochemics: nonprotein amino acids, protease inhibitors, phytohemagglutinins, cyanogenic glycosides, alkaloids, antivitamin factors, metal-binding factors, and so on (Harborne 1977; Liener 1973, 1977; McKey 1979).

Among the nitrogen-based toxicants, alkaloids as a class are considered to be most affected by changes in the nutrient status of soils. Alkaloids increase in amount at the time when the protein is decreasing through catabolism. The loss of protein is accompanied by a corresponding increase in soluble nitrogen. It is possible to produce two- to 10-fold increases of alkaloids when the plants are fertilized with high levels of nitrogen. Nowacki et al. (1976) demonstrated that generally plants that produce alkaloids originating from amino acids react to increased nitrogen fertilization with added alkaloid synthesis.

A plant's defense profile is mostly governed by the chemicals that are either readily available or in surplus of high-priority metabolic needs. Early in the season and in early successional habitats, plants may translocate compounds that can later be used for growth. In young leaf tissue carbon is a limiting element as the photosynthetic process has not reached its peak and been stabilized. So the use of nitrogen-based secondary substances for defense may be a plant strategy for maximizing allocation demands and maintaining defensive chemistry. But later in the season, supply of carbon exceeds demand, and a shift to the production of carbon-based allelochemics takes place. This "availability hypothesis" (Mattson 1980) suggests that nitrogen-based defenses occur when nitrogen supply is greater than the amount needed for growth. But when nitrogen becomes scarce, plants shift to carbon-based chemical principles.

Metabolic pathways that lead to plant growth and allelochemic production are often competitive (Mattson 1980). For example, increasing plant nitrogen through fertilization can lower phenolic content and lignification (Kiraly 1976, Trolldenier and Zehler 1976), whereas reducing it through growing plants in soils deficient in nitrogen and phosphorus has increased levels of phenolics (Hahlbrock and Grisebach 1979, Stafford 1974). Because the aromatic amino acids like tyrosine and phenylalanine are the basic precursors for many phenolics and proteins, protein synthesis is the rate-limiting step in phenolic synthesis due to its preemption of the amino acids (Phillips and Henshaw 1977).

The response of insect herbivores to allelochemics, however, is not limited to toxicity and disruption of the digestive process alone. Indeed, some phytophages depend on the secondary plant substance to select their host plant, and they may use it as a phagostimulant, or even sequester it to use against their own natural enemies (Schoonhoven 1973). Plant allelochemicals also affect the availability of nitrogen for phytophagous insects by forming nonutilizable chemical complexes with proteins (Cates and Rhoades 1977) and through effects on the nutritional physiology of insects (Reese 1978).

Vegetation on low nutrient soils contains relatively high concentrations of chemicals deterrent to herbivores. The cost of herbivory to a plant should be reflected in the magnitude of the plant's investment in chemical defenses. Janzen (1974) reasoned that in two African rain forests the cost of replacing materials eaten by herbivores would be greater in areas of nutrient-deficient soils than for plants growing on sites richer in nutrients. Evaluating this prediction using insect nutrients would be worthwhile.

5.3 Insect Factors

5.3.1 Intraspecific Variation in Herbivores

Superficially, a species of insect herbivore may seem to behave uniformly in its responses toward a host plant. But there is a growing body of evidence to indicate that herbivore populations routinely harbor genetic variation for host-utilization traits (Mitter and Futuyma 1983).

The term "biotype" is used in agricultural entomology as a loose synonym for genotype. Biotypes are populations of an insect species that differ in their response to a species or cultivar of plant, or in insecticide resistance, or in features such as coloration. Saxena and Pathak (1979) have clearly shown that the responses of three biotypes of the rice brown planthopper, *Nilaparvata lugens* (Stål), differ significantly in their orientation and feeding responses, metabolic utilization of food, growth rate, survival, and oviposition on rice varieties. Biotypes may be sympatric or allopatric, and as Claridge and den Hollander (1980) have pointed out, each recognized biotype may be a genetically heterogeneous entity.

Biotypes, if present in the biome, can cause a gradual change in plant-insect relationships over time. Pathak and Heinrichs (1982), working with the brown planthopper, *N. lugens*, increased the ability of a laboratory colony to survive on two resistant rice cultivars by mass selection of a portion of the colony on each cultivar for seven generations.

There is considerable evidence (Briggs 1965, Cartier and Painter 1956) that various biotypes of aphids respond differently to the same host plant. The four biotypes of the corn leaf aphid, *Rhopalosiphum maidis* (Fitch), had significantly different body weights when they were reared on Reno barley in nutrient solutions (Singh and Painter 1964). It is possible that the biotypes have different nutritional requirements and thus respond differently to plant-mediated mineral stresses.

It is interesting to note that even male and female insects of the same species respond to plant-mediated effects of mineral stress differently. Females of the spruce budworm, *Choristoneura occidentalis* Freeman, weighed less on Douglas fir trees high in foliar concentrations of beta-pinene and total nitrogen than those from trees lacking these chemicals, whereas males responded similarly to trees high in foliar concentrations of alpha-pinene, terpinolene, citronellyl acetate, and terpenoid esters. The authors (Redak and Cates 1984) attributed this observation to a possible difference in foliage requirements of the sexes.

5.3.2 Variable Herbivores

Many insect-borne factors determine the way in which various insect groups respond to plant-mediated soil mineral stress. Mode of feeding is an important characteristic that determines the performance of an insect herbivore. Most chewing insects are nonselective; they normally ingest macerated whole leaf tissue. Sucking insects are more selective, and growth efficiency has been shown to correlate positively with nitrogen level of the food (Fox and Macauley 1977, Slansky and Feeny 1977). An increase in individual growth efficiency may translate to a larger herbivore population, but this only seldom happens. For example, larvae of the cabbage butterfly decrease their consumption rate as the nutrient content in the plant increases (Slansky and Feeny 1977), while growth efficiency increases, growth rate, and presumably fecundity, do not.

Sap-sucking insects have a broad spectrum in the choice of plant tissue for feeding. Some leafhoppers (Cicadellidae) feed on the contents of mesophyll cells, using their short stylets with barbed apices. In contrast, other Homoptera feed on the deeper vascular tissues, the phloem and the xylem. Spittlebugs mostly feed on xylem sap. Sucking insects have high consumption rates, and rather elaborate and efficient alimentary canals for concentrating nitrogenous nutrients from a diluent of water and soluble carbohydrates. Many sucking insects also harbor obligatory endosymbionts that fix atmospheric nitrogen as well as convert part of it into vitamins, amino acids, and proteins (Blackman 1974, Miller and Kosztarab 1979).

The distribution, among phytophagous insects, of various types of detoxification systems increases their variability still further. The mixed-function oxidases are enzymes that detoxify an array of toxicants, including many plant chemicals. Highly polyphagous herbivores are always exposed to a large number of such substances, and indeed their titers of the mixed-function oxidase tend to be high (Krieger et al. 1971). Many Lepidoptera have alkaline midguts, which apparently are an adaptation to plant tannins since tannin-protein complexes are easily dissociated at an alkaline pH (Berenbaum 1980).

It is hypothesized that monophagous and oligophagous insects use their host plants as food more efficiently than polyphagous species do (Dethier 1954, Gordon 1959, Waldbauer 1968). Presumably, this would be the result of the close adaptation of mono- and oligophagous species to the chemistry of their host plants and the extra cost to be expended by the polyphagous insects in detoxification systems (Kreiger et al. 1971, Brattsten et al. 1977, Blau et al. 1978). However, tests of this hypothesis have often produced conflicting results (Rhoades and Cates 1976; Schroeder 1976, 1977; Scriber and Feeny 1979). This anomaly may be partly attributable to the heterogeneity of defense compounds postulated to occur among plants of different growth forms. Scriber and Feeny (1979) compared food utilization of polyphagous and oligophagous Lepidoptera on different plants and concluded that plant growth form is a more important factor governing the efficiency of utilization than the range of an insect's food spectrum.

A study of the structure of insect communities in a salt marsh and their rate of increase by fertilization has clearly demonstrated that herbivore characteristics such as overwintering and number of generations per year have a significant effect on the ability of populations to adjust to new resources (Vince et al. 1981). Incorporation of an inactive phase in the insect's life pattern enables it to synchronize feeding with optimum resource availability. The univoltine life history of the cicadellid *A. simplex* restricted the rate at which the population responded to marsh fertilization. But bi- and trivoltine herbivores, especially those with overwintering nymphs, have the potential for higher and faster rates of population increase in response to nutrient enrichment. Plant nutrient levels affect the populations of chewing insects as well as the level of their reproductive capacity. In univoltine species such as the winter moth, this means that a host plant's nutrient status in one year may influence the population of the insect the next year (Wint 1983).

In a study with two aphid species van Emden (1966) has shown how feeding behavior may determine the response to mineral stresses in plants. The response of *M. persicae* as measured by fecundity and reproductive rate was correlated with changes in levels of nitrogen and potassium, while *Brevicoryne brassicae* (L.) showed a markedly lower response. The feeding behavior of the two insects varies significantly. Differences in the normal feeding sites of the two aphids may be related to soluble nitrogen levels as well as phloem pressure of the plant. *B. brassicae* occupies a leaf stratum where conditions are fairly constant over long periods, whereas *M. persicae* occupies a more temporary site where both phloem pressure and soluble nitrogen levels are constantly changing. *M. persicae* is more restless and migratory and exploits a succession of temporary rich sources of soluble nitrogen and the mean fecundity, and rate of reproduction of *M. persicae* may be due to such behavior. There may also be an indirect effect on field survivorship of *M. persicae* as shorter bouts of feeding on high nitrogen parts of different plants could mean less exposure to natural enemies. This points to a complex relationship between nutrient enrichment and predatory–prey interactions.

Although most phytophagous species seem not to compete for food, interspecific competition may be significant under limited resource conditions such as outbreaks. There is ample evidence that competition is especially common among Hemiptera (Strong et al. 1984). Unlike endemic communities in which fitness of herbivores is often governed by an intimate coevolutionary relationship with the host plant, the success of introduced insects is more often a function of their compatibility with the phenology of the new host plant and their ability to adjust to the density-dependent changes in the quality and quantity of food. The extent to which competition as a factor may contribute in resource-stressed ecosystems is yet to be assessed.

5.3.3 Symbiosis with Microorganisms

The extensive utilization of microorganisms by insect herbivores modifies the level of constraint that mineral stressed plants exert on insects and the insect–plant relationship. Through symbiosis even marginal resources, which are not exploitable under normal conditions, become usable for insects, and no visible sign of nutrient deficiency is apparent. Symbiosis is very common among insects attempting to eke out an existence on the foilage of nitrogen-deficient species (Mattson 1980). Gut symbionts of many phytophagous insects provide them with amino acids that are normally not available or are insufficient in their food. A common symbiont in many aphids is *Azotobacter* which fixes atmospheric nitrogen in conditions of nitrogen scarcity (Buchner 1965). Mittler (1970) has shown that in *M. persicae* the gut microorganisms provide all the essential amino acids except four which are to be supplied in the diet. Van Emden (1966) speculated that the comparatively lower response of *B. brassicae* than that of *M. persicae* to nutrient treatments was related to a greater ability of *B. brassicae* symbionts to supplement the nitrogen economy.

5.4 Biological Nitrogen Fixation

The extent to which nitrogen-deficient soils affect plant chemistry, and thus their suitability as hosts for insects, depends partially on the ability of the plants to fix nitrogen. Stress-adapted plant species with nitrogen-fixing symbionts that survive on very low levels of soil nitrogen are probably similar in nitrogen content (2 to 5% dry weight) to nitrophilous species that typically occupy rich ecosystems. Approximately 10,000 species of legumes and 160 species of nonleguminous plants harbor nitrogen-fixing symbionts (Delwiche 1978, Stewart 1977, Torrey 1978).

Nitrogen fixation in legumes occurs by the bacterium *Bradyrhizobium*. The legume provides the bacteria with carbohydrates, which they oxidize. Some electrons and ATP obtained during this oxidative process are used to reduce nitrogen to ammonium. The ion is then converted into other nitrogenous compounds, most of which are absorbed by the nodular cells and translocated to other organs.

Some tropical species of grain legumes such as *Vigna, Phaseolus*, and *Cajanus* specialize in forming ureides as major products of nitrogen fixation and may store these compounds in stems and petioles during vegetative growth (Pate et al. 1980). Other grain legumes (e.g., *Lupinus, Pisum, Vicia*), largely of temperate origin, tend to form asparagine and glutamine as major products of nitrogen fixation and store them in their stems and leaves (Pate and Atkins 1983). These amino compounds are mostly transported through the phloem. The differential chemistry and transport routes of the end products of nitrogen fixation in these two groups of legumes may have great ecological significance in terms of insect herbivory, but no study has so far been undertaken in this regard.

Many factors influence the rate of nitrogen fixation. In general, those factors that speed up photosynthesis, such as adequate moisture and temperatures, bright sunshine, and high levels of carbon dioxide, stimulate nitrogen fixation. The stage of plant growth also influences nitrogen fixation.

Soybeans and peanuts fix more nitrogen after flowering when the demand for nitrogen in the developing seeds and fruits is at its peak. Fertilization of these crops with NO_3, NH_4^+, or urea at the later stages of plant growth does not increase yield because nitrogen fixation is decreased in proportion to the added amount of fertilizer nitrogen absorbed. This decrease probably results from inhibition of nitrogenase synthesis and more rapid senescence of the nodules by ammonium ions or organic nitrogen compounds formed from the fertilizers (Salisbury and Ross 1978).

The ability of a plant to fix nitrogen allows it to be a suitable host to insects, even when growing in nitrogen-deficient soils. Even when nitrogen fertilizer is added, nitrogen-fixing plants may be more nutritious than those that do not fix nitrogen. Todd et al. (1972) studied the trend of Mexican bean beetle (*Epilachna varivestis* Mulsant) infestations in relation to nitrogen fertilization on nodulating and non-nodulating soybeans. Nodulating soybeans had significantly higher numbers of beetles than non-nodulating soybeans at all nitrogen fertilization levels except the highest. Chemical analysis revealed that nodulating soybeans had 23% leaf protein at nitrogen fertilizer rates of 0 to 236 kg/ha, whereas protein content in non-nodulating soybeans increased from 13 to 23%, respectively.

Insects can serve as biotic stresses that reduce the nitrogen-fixing ability of plants, which in turn reduces the quality of such plants as hosts for insects. Three types of insect-induced stress have been reported to adversely affect nitrogen production in soybean plants. Nodule destruction by larvae of the bean leaf beetle *Cerotoma trifurcata* Forster reduced N_2-fixation twofold (Layton 1983). Stem girdling by the three-cornered alfalfa hopper *Spissistilus festinus* Say reduces N_2-fixation by restricting translocation of photosynthate to the root (Hicks et al. 1984). Defoliation is another type of indirect damage that can have a significant impact on N_2-fixation rate. Layton and Boethel (1987) determined the relationship between percentage defoliation by the soybean looper *Pseudoplusia includens* (Walker) and N_2-fixation in soybeans. Nitrogen-fixing ability measured by C_2H_2—C_2H_4 assay decreased linearly in response to increased defoliation levels. The decrease was caused by a reduced number and weight (Fig. 2), and efficiency (Fig. 3) of nodules. The deleterious effects of defoliation on N_2-fixation was attributed to a reduced photosynthate supply that (1) depressed root and nodule development and (2) reduced the energy supplied to the N_2-fixing bacteria which is needed to drive the N_2-fixing process. It is thus evident that in nitrogen-deficient soils, insect-induced defoliation of N_2-fixing plants can serve as a biotic stress that in turn causes mineral stress to be more acute than would be the case in the absence of the defoliators.

6. CONCLUSIONS

Soil–plant–insect interactions are extremely complex and are always affected by various intrinsic and extrinsic factors. Soil is not an inert medium

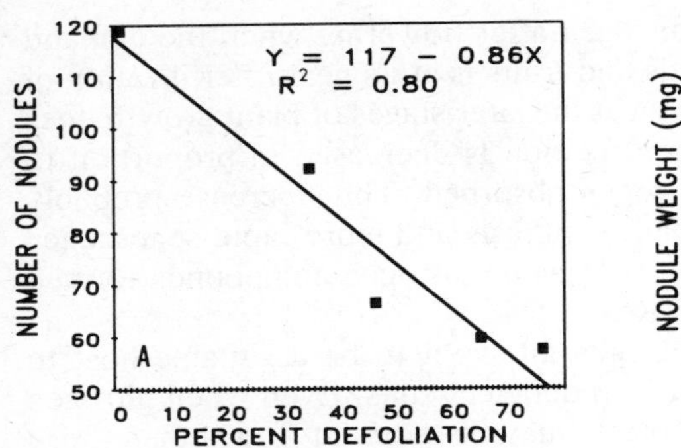

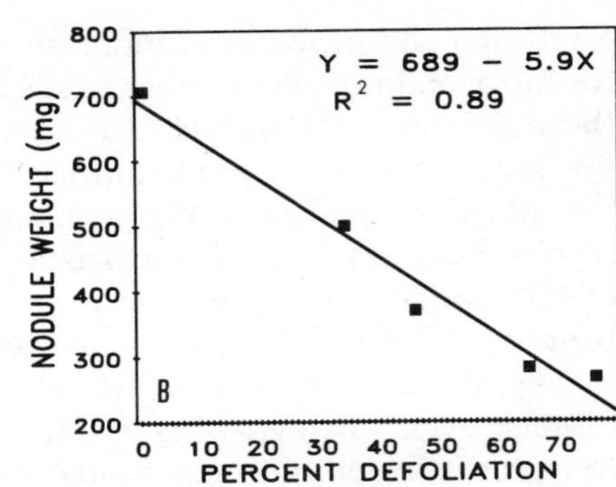

Figure 2. Effect of increasing defoliation on nodule number (*A*) and weight (*B*) per experimental unit; squares represent means. Regression equations were calculated from individual data points. (From Layton and Boethel 1987)

on which plants grow but harbors microflora and many organisms. The extent of nutrient stress, availability of plant nutrients from the soil, and soil physical structure are all regulated by the activity of the flora and fauna of the soil. So it is wise to consider the relationships among soil, plant, and insects as an active interplay of three biological systems that are themselves prone to be acted upon by many ecological factors. Naturally, there is always difficulty in weaning out the direct linear relation among nutrient stress in the soil, changes in the plant consequent to stress, and the array of alterations that take place in the biology of insect herbivores.

Criteria of stress measurement by various workers are not uniform and are often incorrect. It is the quantity of actual available nutrients that is to be considered for computing stress. But in many studies, graded doses of fertilizers are simply correlated to their plant-mediated effects on insects.

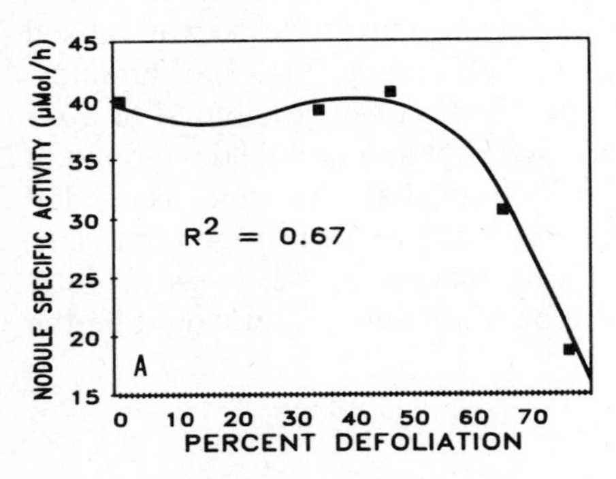

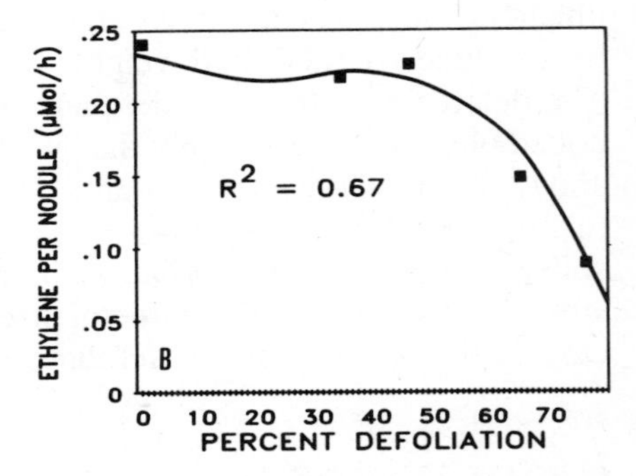

Figure 3. Relationship of (*A*) nodule specific activity (μmol ethylene per g dry nodules) to percent defoliation and (*B*) ethylene per nodule to percent defoliation; squares represent means. Regression equations were calculated from individual data points. (From Layton and Boethel 1987)

Moreover it is advisable to work with a single plant nutrient at a time rather than studying the effects of a spectrum of macro- and microelements, in which case interpretation of the complex effects on insects becomes difficult and ambiguous.

The conditions under which experiments have been conducted, the results of which are cited in this chapter, were extremely diverse. Very often there is an attempt to correlate directly the nutrient status of the soil and insect populations without consideration of the effects that the nutrient status has on the physiology of the host plant. In this regard insect responses to the allelochemics, nutrients, and water in the plant are also to be monitored (Scriber 1984). We must also consider the physiological effects of pesticides on the susceptibility of plants to pests (Oka and Pimentel 1976, Tingey and Singh 1980) (see Chapter 6). The role of pesticides in altering the mineral stress status of soils through interacting with soil microflora cannot be overlooked (Alexander 1981).

In order to eliminate the latent effect of nutrients in soil, controlled experiments in pots, sand, or water culture are needed. But such experiments have inherent drawbacks in terms of limited volume and difficulties in adjusting the pH of the medium. Moreover the results from these studies are most often difficult to extrapolate into field conditions. Rarely are field trials exclusively undertaken to assess the fertilizer–pest relationships.

The problems of plant analyses are often considerable to entomologists who find relatively little data on the nutrient status of plants at varying stages of growth (McNeill and Southwood 1978). For analysis, researchers must choose those components of plants that have the maximum impact on the biological parameters of insects. Analysis of individual amino acids in plants may be more meaningful than assaying the total nitrogen. Changes in the nutrient content of assimilating organs and tissues of the plant must also be monitored after the initial attack and settlement of insects. Such insect-induced alterations in phytochemistry are to be taken into consideration in the final computation of cause–effect relationships.

Fertilization of crop plants and forest trees induces physiological, morphological, and phenological changes in plants. Nitrogen application usually increases insect populations. But the spectacular decrease in the number of forest insects after fertilizing the trees with nitrogen is contrary to the general experience with field crops. The reasons for this difference are uncertain. The effects of phosphorus, potassium, and minor and trace elements are even less clear than that of nitrogen. Similarly, the impact of organic manures on the population dynamics of insect pests is also obscure.

The long chain of interactions initiated by nutrient stress often does not stop at the insect herbivore level. It passes further to the parasites, predators, and hyperparasites. Many results suggest that the abundance of these natural enemies of insect herbivores is regulated by nutrient stress in soil, mediated through plant and insect herbivores (White 1984). Nutrient status of the soil also affects visitation of pollinators to plants (Scriber 1984).

The foregoing discussion clearly demonstrates that our present knowledge on the basic nature of soil minerals–plant–insect interactions is weak. More concerted research involving the integrated efforts of soil scientists, biochemists, plant physiologists, and entomologists is needed to elucidate more fully the various interactions in order that various fertilizer management practices can be exploited in the control of insect pests.

ACKNOWLEDGMENTS

I am grateful to my wife, Dr. S. T. Mercy, and to my former student and current colleague, Mr. Thomas Biju Mathew, for their help in the compilation of the references.

REFERENCES

Abraham, E. V. 1957. A note on the influence of manuring on the incidence of the fulgorid, *Nilaparvata lugens* S. on paddy. *Madras Agric. J.* 44: 529–532.

Adkisson, P. L. 1958. The influence of fertilizer applications on populations of *Heliothis zea* (Boddie) and certain insect predators. *J. Econ. Entomol.* 51: 757–759.

Alexander, M. 1981. Biodegradation of chemicals of environmental concern. *Science* 211: 132–138.

Allen, M. D., and I. W. Selman. 1955. Egg-production in the mustard beetle, *Phaedon cochleariae* (F.) in relation to diets of mineral-deficient leaves. *Bull. Entomol. Res.* 46: 393–397.

Allen, M. D., and I. W. Selman. 1957. The response of larvae of the large white butterfly (*Pieris brassicae* [L.]) to diets of mineral-deficient leaves. *Bull. Entomol. Res.* 48: 229–242.

Allen, S. E., H. M. Grimshaw, J. A. Parkinson, and C. Quarmby. 1974. *Chemical Analysis of Ecological Materials*. Blackwell Scientific, Oxford.

Altieri, M. A., and W. H. Whitcomb. 1979. The potential use of weeds in the manipulation of beneficial insects. *HortScience* 14: 12–18.

Al-Zubaidi, F. S., and J. L. Capinera. 1983. Application of different nitrogen levels to the host plant and cannibalistic behavior of beet armyworm, *Spodoptera exigua* (Hübner) (Lepidoptera: Noctuidae). *Environ. Entomol.* 12: 1687–1689.

Al-Zubaidi, F. S., and J. L. Capinera. 1984. Utilization of food and nitrogen by the beet armyworm, *Spodoptera exigua* (Hübner) (Lepidoptera: Noctuidae), in relation to food type and dietary nitrogen levels. *Environ. Entomol.* 13: 1604–1608.

Ananthakrishnan, T. N. 1982. *Bioresources Ecology*. Oxford and IBH, New Delhi.

Ananthanarayanan, K. P., and E. V. Abraham. 1956. Bionomics and control of rice jassid, *Nephotettix bipunctatus* F. and rice fulgorid *Nilaparvata lugens* S. in Madras. *In Proc. 6th Sci. Work Conf. Agric. College and Res. Inst.*, Coimbatore, India, pp. 33–44.

Andrzejewska, L. 1965. Stratification and its dynamics in meadow communities of Auchenorrhyncha (Homoptera). *Ekologia Polska* A 13: 685–715.

Arant, F. S., and C. M. Jones. 1951. Influence of lime and nitrogenous fertilizers on populations of greenbug infesting oats. *J. Econ. Entomol.* 44: 121–122.

Archer, T. L., A. B. Onken, R. L. Matheson, and E. D. Bynum, Jr. 1982. Nitrogen fertilizer influence on greenbug (Homoptera: Aphididae) dynamics and damage to sorghum. *J. Econ. Entomol.* 75: 695–698.

Auclair, J. L. 1963. Aphid-feeding and nutrition. *Annu. Rev. Entomol.* 8: 439–490.

Auclair, J. L. 1965. Feeding and nutrition of the pea aphid, *Acyrthosiphon pisum* (Homoptera: Aphididae), on chemically defined diets of various pH and nutrient levels. *Ann. Entomol. Soc. Am.* 58: 855–875.

Auclair, J. L. 1969. Nutrition of plant-sucking insects on chemically defined diets. *Entomol. Exp. Appl.* 12: 623–641.

Auclair, N. D. 1977. Factors affecting tissue nutrient concentration in a grey meadow. *Oecologia* 28: 233–246.

Auerbach, M. J., and D. R. Strong. 1981. Nutritional ecology of *Heliconia* herbivores: experiments with plant fertilization and alternative hosts. *Ecol. Monogr.* 51: 63–83.

Baker, H. G., P. A. Opler, and I. Baker. 1978. A comparison of the amino-acid complements of floral and extra floral nectars. *Bot. Gaz.* 139: 322–332.

Bakke, A. 1969. The effect of forest fertilization on the larval weight and larval density of *Saspeyresia strobilella* (L.) (Lepidoptera: Tortricidae) in cones of Norway spruce. *Z. Angew. Entomol.* 63: 451–453.

Balasubramanian, R., and N. K. Iyengar. 1950. The problem of jassids on American cotton in Madras with special reference to black soils of ceded districts. *Indian Cotton Gr. Rev.* 4: 199–211.

Balasubramanian, P., S. P. Palaniappan, and M. Gopalan. 1983. Effect of carbofuran and nitrogen on leaf folder incidence. *Intl. Rice Res. Newsl.* 8(5): 13–14.

Banks, C. J., and E. D. M. Macaulay. 1970. Effects of varying the host plant and environmental conditions on the feeding and reproduction of *Aphis fabae*. *Entomol. Exp. Appl.* 13: 85–96.

Barker, J. S., and O. E. Tauber. 1951. Fecundity of and plant injury by the pea aphids as influenced by nutritional changes in the garden pea. *J. Econ. Entomol.* 44: 1010–1012.

Baskaran, P., P. Narayanasamy, and A. Pari. 1985. The role of potassium in incidence of insect pests among crop plants, with particular reference to rice. *In Role of Potassium in Crop Resistance to Insect Pests, Res. Ser. No. 3, Potash Res. Inst. of India*, Guragaon, Haryana, India, pp. 63–68.

Baskin, C. C., and J. M. Baskin. 1974. Responses of *Cestragalus tennesseensis* to drought changes in free amino acids and amides during water stress and possible ecological significance. *Oecologia* 17: 11–16.

Bates, T. E. 1971. Factors affecting critical nutrients concentrations in plants and their evaluation: a review. *Soil Sci.* 112: 116–130.

Baule, H., and C. Fricker. 1967. Die Dungung von Waldbaumen. *Bayerischer Landwirtschafts-verlag*, München, Basel, Wien.

Benepal, P. S., and C. V. Hall. 1967. The influence of mineral nutrition of varieties of *Cucurbita pepo* L. on the feeding response of squash bug *Anasa tristis* DeGeer. *Proc. Am. Soc. Hortic. Sci.* 90: 304–312.

Berenbaum, M. 1980. Adaptive significance of midgut pH in larval Lepidoptera. *Am. Nat.* 115: 138–146.

Bevan, D. 1969. Philosophy of forest insect control in Britain. *J. Sci. Fd. Agric.* 20: 505–506.

Bidwell, R. G. S., and D. J. Durzan. 1975. Some recent aspects of nitrogen metabolism. *In* Davies, P. J. (ed.), *Historical and Current Aspects of Plant Physiology: A Symposium Honoring R. C. Stewart*. New York State College of Agriculture and Life Sciences, Ithaca, pp. 152–225.

Bischoff, M. 1967. Investigation on feeding damage by *Brachydores incanus* L. on fertilized test plots. *Forst-u.-Holzw.* 2: 4–8. (In German).

Bischoff, M., S. B. Ewert, and W. Thalenhorst. 1969. Untersuchungen über die Abhängigkeit der Befallsstärke der Gallenlaus *Sacchiphantes abietis* (L.) vom Austreibetyp der Fichte. *Z. Angew. Entomol.* 64: 65–85.

Blackman, R. 1974. *Aphids*. Ginn, London.

Blaker, N. S., and J. D. MacDonald. 1986. The role of salinity in the development of Phytophthora root rot of citrus. *Phytopathology* 76: 970–975.

Blau, P. A., P. Feeny, L. Contardo, and D. S. Robson. 1978. Allyl glucosinolate and herbivorous caterpillars: a contrast in toxicity and tolerance. *Science* 200: 1296–1298.

Blickenstaff, C. C., D. D. Morey, and G. N. Burton. 1954. Effect of rates of nitrogen application on greenbug damage to oats, rye, and ryegrass. *Agron. J.* 46: 338.

Bohn, H. L., B. L. McNeal, and G. A. O'Connor. 1985. *Soil Chemistry*. 2d ed. Wiley, New York.

Bondarenko, N. V. 1950. The influence of shortened day on the annual cycle of development of the common spider mite. *Dokl. Akad. Nauk SSSR* 70: 1077–1080.

Bowling, C. C. 1963. Effect of nitrogen levels on rice water weevil populations. *J. Econ. Entomol.* 56: 826–827.

Brady, N. C. 1974. *The Nature and Properties of Soils*. 8th ed. Macmillan, New York.

Branson, T. F., and R. G. Simpson. 1966. Effects of a nitrogen-deficient host and crowding on the corn leaf aphid. *J. Econ. Entomol.* 59: 290–293.

Brattsten, L. B., C. F. Wilkinson, and T. Eisner. 1977. Herbivore-plant interactions: mixed-function oxidases and secondary plant substances. *Science* 196: 1349–1352.

Breukel, L. M., and A. Post. 1959. The influence of the manurial treatment of orchards on the population density of *Metatetranychus ulmi* (Koch) (Acarina: Tetranychidae). *Entomol. Exp. Appl.* 2: 38–47.

Brewer, J. W., J. L. Capinera, R. E. Deshon, Jr., and M. L. Walmsley. 1985. Influence of foliar nitrogen levels on survival, development, and reproduction of western spruce budworm, *Choristoneura occidentalis* (Lepidoptera: Tortricidae). *Can. Entomol.* 117: 23–32.

Briggs, J. B. 1965. The distribution, abundance, and genetic relationships of four strains of the rubus aphid (*Amphorophora rubi* [Kalt.]) in relation to raspberry breeding. *J. Hortic. Sci.* 40: 109–117.

Broadbent, L., P. H. Gregory, and T. W. Tinsley. 1952. The influence of planting date and manuring on the incidence of virus diseases in potato crops. *Ann. Appl. Biol.* 39: 509–524.

Brüning, D., and E. Uebel. 1968. Düngung und Populationsdichte von Napfschildläusen. *Allg. Forstz.* 23: 536–537.

Brüning, D., and E. Uebel. 1971. Manuring and the population density of scale insects. *Potash Rev.* 22/20.

Buchner, P. 1965. *Endosymbiosis of Animals with Plant Microorganisms*. Wiley, New York.

Buckman, H. O., and N. C. Brady. 1969. *The Nature and Properties of Soils*. Macmillan, New York.

Burzynski, J. 1966. Observations on the occurrence of insect pests in fertilized young plantations on dunes. *Sylwan* 110(8): 43–51. (In Polish with English summary).

Büttner, H. 1956. Die Beeinträchtigung von Raupen einiger Forstschädlinge durch mineralische Düngung der Futterpflanzen. *Naturwissenschaften* 43: 454–455.

Campbell, C. A. M. 1984. The influence of overhead shade and fertilizers on the Homoptera of mature upper Amazon cocoa trees in Ghana. *Bull. Entomol. Res.* 74: 163–174.

Cannon, W. N., Jr., and W. A. Connell. 1965. Populations of *Tetranychus atlanticus* (Acarina: Tetranychidae) on soybean supplied with various levels of nitrogen, phosphorus, and potassium. *Entomol. Exp. Appl.* 8: 153–161.

Cannon, W. N., Jr., and A. Ortega. 1966. Studies of *Ostrinia nubilalis* larvae (Lepidoptera:

Pyraustidae) on corn plants supplied with various amounts of nitrogen and phosphorus. I. Survival. *Ann. Entomol. Soc. Am.* 59: 631–638.

Carrow, J. R., and R. E. Betts. 1973. Effect of different foliar-applied nitrogen fertilizers on balsam woolly aphid. *Can. J. For. Res.* 3: 122–139.

Carrow, J. R., and K. Graham. 1968. Nitrogen fertilization of the host tree and population growth of the balsam woolly aphid *Adelges piceae* (Homoptera: Adelgidae). *Can. Entomol.* 100: 478–485.

Carter, C. I., and J. Cole. 1977. Flight regulation in the green spruce aphid (*Elatobium abietinum*). *Ann. Appl. Biol.* 86: 137–157.

Cartier, J. J., and R. H. Painter. 1956. Differential reactions of two biotypes of the corn leaf aphid to resistant and susceptible varieties, hybrids, and selections of sorghums. *J. Econ. Entomol.* 49: 489–508.

Cates, R. G., and D. F. Rhoades. 1977. Patterns in the production of antiherbivore defenses in plant communities. *Biochem. Syst. Ecol.* 5: 185–193.

Chaboussou, F. 1972. The role of potassium and of cation equilibrium in the resistance of the plant towards diseases. *Potash Rev.* 23/39.

Chaboussou, F. 1976. Cultural factors and the resistance of citrus plants to scale insects and mites. *In Fertilizer Use and Plant Health, Proc. 12th Colloq.*, Intl. Potash Inst., CH-3048 Worblaufen-Bern, Switzerland, pp. 259–280.

Chakrabarthy, D. D., and G. Mathew. 1972. Effect of host plant nutrition on the susceptibility of the seeds of several exotic wheat varieties to *Sitophilus oryzae*. *Bull. Grain. Technol.* 10: 116–119.

Chandragiri, K. K., R. Velusamy, I. P. Janaky, and M. S. Ramakrishnan. 1974. Effect of different levels of nitrogen on rice leaf roller (*Cnaphalocrocis medinalis* Guenee) incidence. *Madras Agr. J.* 61: 717.

Chapman, V. J. 1975. The salinity problem in general, its importance, and distribution with special reference to natural halophytes. *In* Poljakoff-Mayber, A., and J. Gale (eds.). *Plants in Saline Environments*. Springer-Verlag, New York, pp. 7–24.

Cheng, C. H., and M. D. Pathak. 1972. Resistance to *Nephotettix virescens* in rice varieties. *J. Econ. Entomol.* 65: 1148–1153.

Claridge, M. F., and J. den Hollander. 1980. The biotypes of the rice brown planthopper *Nilaparvata lugens*. *Entomol. Exp. Appl.* 27: 23–30.

Coon, B. F. 1959. Aphid populations on oats grown in various nutrient solutions. *J. Econ. Entomol.* 52: 624–626.

Cooper-Driver, G. A. 1977. Seasonal variation in secondary plant components in relation to the palatability of *Pteridium equilinum*. *Biochem. Syst. Ecol.* 5: 177–183.

Corey, K. A., A. V. Barker, and L. E. Craker. 1987. Ethylene evolution by tomato plants under stress of ammonium toxicity. *HortScience* 22: 471–473.

Court, R. D., W. T. Williams, and M. P. Hegarty. 1972. The effect of mineral nutrient deficiency on the content of free amino acids in *Setaria sphacelata*. *Aust. J. Biol. Sci.* 25: 77–87.

Cowling, D. W., and A. W. Bristow. 1979. Effects of SO_2 on sulphur and nitrogen fractions and on free amino acids in perennial ryegrass. *J. Sci. Food Agric.* 30: 354–360.

Creighton, J. T. 1938. Factors influencing insect abundance. *J. Econ. Entomol.* 31: 735–739.

Cunliffe, N. 1928. Studies on *Oscinella frit* L.: observations on infestations, yield, susceptibility, recovery power, the influence of variety on the rate of growth of the primary shoot of the oat and the reaction to manurial treatment. *Ann. Appl. Biol.* 15: 473–487.

Dadd, R. H. 1968. Dietary amino acids and wing determination in the aphid, *Myzus persicae*. *Ann. Entomol. Soc. Am.* 61: 1201–1210.

Dahms, R. G. 1947. Oviposition and longevity of chinch bugs on seedlings growing in nutrient solutions. *J. Econ. Entomol.* 40: 841–845.

Dahms, R. G., and F. A. Fenton. 1940. The effect of fertilizers on chinch bug resistance in sorghum. *J. Econ. Entomol.* 33: 688–692.

Daniels, N. E. 1957. Greenbug populations and their damage to winter wheat as affected by fertilizer applications. *J. Econ. Entomol.* 50: 793–794.

Das, S. K., and N. K. Sanghi. 1983. Effect of fertilizers on pest damage in sunflower. *Indian J. Agric. Sci.* 53: 616–617.

Davidson, J. 1925. Biological studies of *Aphis rumicis* Linn: factors affecting the infestation of *Vicia faba* with *Aphis rumicis*. *Ann. Appl. Biol.* 12: 472–507.

Delwiche, C. C. 1978. Legumes—past, present, and future. *BioScience* 28: 565–570.

Denno, R. F., L. W. Douglass, and D. Jacobs. 1985. Crowding and host plant nutrition: environmental determinants of wing-form in *Prokelisia marginata*. *Ecology* 66: 1588–1596.

Dethier, V. G. 1954. Evolution of feeding preferences in phytophagous insects. *Evolution* 8: 33–54.

DeVerall, B. J. 1977. *Defense Mechanisms of Plants*. Cambridge Univ. P., Cambridge.

Dina, S. J., and L. G. Klikoff. 1973. Effect of plant moisture stress on carbohydrate and nitrogen content of big sagebrush. *J. Range Management* 26: 207–209.

Dodd, A. J. 1940. *The Biological Campaign Against Prickly Pear*. Commonwealth Prickly Pear Board, Brisbane, Australia.

Durzan, D. J., and F. C. Steward. 1967. The nitrogen metabolism of *Picea glauca* (Moench) Voss and *Pinus banksiana* Lamb. as influenced by mineral nutrition. *Can. J. Bot.* 45: 695–710.

Dyck, V. A., B. C. Misra, S. Alam, C. N. Chen, C. Y. Hsieh, and R. S. Rejesus. 1979. Ecology of the brown planthopper in the tropics. *In Brown Planthopper: Threat to Rice Production in Asia*. Intl. Rice Res. Inst., Manila, Philippines, pp. 61–98.

Eaton, J. S., G. E. Likens, and F. H. Bormann. 1973. Throughfall and stemflow chemistry in a northern hardwood forest. *J. Ecol.* 61: 495–508.

Eden, W. G. 1953. Effects of fertilizer on rice weevil damage to corn at harvest. *J. Econ. Entomol.* 46: 507–510.

Eidmann, H. H. 1963. Über die Beziehungen zwischen Boden und Forstschädlingen. *Anz. Schädlingskde.* 36: 185–188.

Eidmann, H. H., and T. Ingestad. 1963. Ernahrungszustand, Zuwachs und Insektenbefall in einer Kiefernkultur. *Stud. Forst. Suecica Nr* 12: 1–22.

El-Beheidi, M., and K. A. Gouhar. 1971. The effect of nitrogen, phosphorus, potassium, and some selective insecticides on whitefly, *Bemisia tabaci* infestation and squash yield. *Beitr. Trop. Subtrop. Lantwirtsch Tropanveterinarmed* 9: 159–166.

El-Gabaly, M. M. 1952. Effect of fertilizer treatments on the yield of "Mounufi" cotton in the northern part of the delta. *Trans. Intl. Soc. Sci. Joint Meeting, Dublin*, 11: 261–262.

Evans, A. C. 1938. Physiological relationships between insects and their host plants. I. Effect of the chemical composition of the plant on reproduction and production of winged forms in *Brevicoryne brassicae* L. (Aphididae). *Ann. Appl. Biol.* 25: 558–572.

Feeny, P. P. 1970. Seasonal changes in oak leaf tannins and nutrients as a cause of spring feeding by winter moth caterpillars. *Ecology* 51: 565–581.

Feeny, P. P. 1976. Plant apparency and chemical defense. *Rec. Adv. Phytochem.* 10: 1–40.

Fellin, D. G., and J. E. Dewey. 1982. Western spruce budworm. *U.S. Dept. Agric. For. Serv. Forest Insect and Disease Leaflet*, 53 p.

Fennah, R. G. 1959. Nutritional factors associated with the development of mealy bugs on Cacao. *In A Report on Cacao Research*. Imperial College of Agriculture, St. Augustine, Trinidad, pp. 18–28.

Fennah, R. G. 1969. Damage to sugar cane by Fulgoroidea and related insects in relation to the metabolic state of the host plant. *In* Williams, J. R. (ed.), *Pests of Sugar Cane*. Elsevier, Amsterdam, pp. 367–389.

Fox, L. R., and B. J. Macauley. 1977. Insect grazing on *Eucalyptus* in response to variation in leaf tannins and nitrogen. *Oecologia* 29: 145–162.

Fraenkel, G. 1953. The nutritional value of green plants for insects. *Trans. IX Intl. Cong. Entomol. Amsterdam* (1951) 2: 90–100.

Fritzsche, R., H. Wolffgang, and H. Opel. 1957. Investigations on the dependence of spider mite increase on the state of nutrition of the food plants. *Z. PflErnahr. Düng*. 78: 13–27.

Fujimura, T. 1961. Relation between the quantity of nitrogen application to rice plant and the oviposition of rice stem borer, *Chilo suppressalis* Walker. *Agric. Res. Chugo.-ku* 23: 47–49.

Futuyma, D. J. 1976. Foodplant specialization and environmental predictability in Lepidoptera. *Am. Nat.* 110: 285–292.

Gadd, C. H. 1943. Does manuring reduce the damage caused by shot-hole borer? *Tea Quart.* 16: 30–39.

Gardner, W. A., and R. R. Duncan. 1982. Influence of soil pH on fall armyworm (Lepidoptera: Noctuidae) damage to whorl-stage sorghum. *Environ. Entomol.* 11: 908–912.

Garman, P., and B. H. Kennedy. 1949. Effect of soil fertilization on the rate of reproduction of the two-spotted spider mite. *J. Econ. Entomol.* 42: 157–158.

Gatoria, G. S., and H. Singh. 1984. Effect of insecticidal and fertilizer applications on the jassid complex in green gram, *Vigna radiata* (L.) Wilczek. *J. Entomol. Res.* 8: 154–158.

Ghildiyal, M. C., M. Pandey, and G. S. Sirohi. 1977. Proline accumulation under zinc deficiency in mustard. *Curr. Sci.* 46: 792–793.

Ghosh, B. H. 1962. A note on the incidence of stem borer, *Schoenobius incertulas* Wlk., on boro paddy under nitrogen fertilizers. *Curr. Sci.* 31: 472–473.

Gojmerac, W. 1980. *Bees, Beekeeping, Honey and Pollination*. AVI, Westport, CT.

Gordon, H. T. 1959. Minimal nutritional requirements of the German roach, *Blatella germanica* L. *Ann. N. Y. Acad. Sci.* 77: 290–351.

Grasmanis, V. O., and D. J. D. Nicholas. 1971. Annual uptake and distribution of N^{15}-labeled ammonia and nitrate in young Jonathan/MM104 apple trees grown in solution cultures. *Plant and Soil* 35: 95–112.

Green, T. R., and C. A. Ryan. 1972. Wound-induced proteinase inhibitor in plant leaves: a possible defense mechanism against insects. *Science* 175: 776–777.

Greenland, D. J. 1958. Nitrogen fluctuations in tropical soils. *J. Agric. Sci.* 50: 82–92.

Gurevich, S. M., L. A. Gulidova, and N. D. Slavko. 1971. Effect of mineral fertilizers on plant resistance to disease and pests. *Sov. Soil Sci.* (Engl. transl. *Pochvovedenie*) 3: 141–147.

Gussone, H. A., and H. W. Zöttl. 1975. Die Wirkung jahreszeitlich verschiedener Düngung auf junge Fichten. *Forstw. Cbl.* 94: 334–343.

Hahlbrock, K., and H. Grisebach. 1979. Enzymatic controls in the biosynthesis of lignin and flavonoids. *Annu. Rev. Plant. Physiol.* 30: 105–130.

Hamai, J., and C. B. Huffaker. 1978. Potential of predation by *Metasiulus occidentalis* in compensating for increased, nutritionally induced, power of increase of *Tetranychus urticae*. *Entomophaga* 23: 225–237.

Hamstead, E. O., and E. Gould. 1957. Relation of mite populations to seasonal leaf nitrogen levels in apple orchards. *J. Econ. Entomol.* 50: 109–110.

Hanover, J. W. 1975. Physiology of tree resistance to insects. *Annu. Rev. Entomol.* 20: 75–95.

Harborne, J. B. 1977. *Introduction to Ecological Biochemistry*. Academic Press, New York.

Harrewijn, P. 1976. Host-plant factors regulating wing production in *Myzus persicae*. *Symposia Biologica Hungarica* 16: 79–83.

Harvey, G. T. 1974. Nutritional studies of eastern spruce budworm (Lepidoptera: Tortricidae). I. Soluble sugars. *Can. Entomol.* 106: 353–365.

Haseman, L. 1946. Influence of soil minerals on insects. *J. Econ. Entomol.* 39: 8–11.

Hatmosoewarno, S. 1970. Fertilizing in relation to pests and diseases attack. *Sugarcane Pathol. Newsl.* 5: 21–24.

Heathcote, G. D. 1974. The effect of plant spacing, nitrogen fertilizer and irrigation on the appearance of symptoms and spread of virus yellows in sugarbeet crops. *J. Agric. Sci.* 82: 53–60.

Heinrich, B. 1979. *Bumblebee Economics.* Harvard Univ. P., Cambridge, MA.

Heinrichs, E. A., and F. G. Medrano. 1985. Influence of nitrogen fertilizers on the population development of brown planthopper. *Intl. Rice Res. Newsl.* 10(6): 20–21.

Henderson, I. F., and R. O. Clements. 1977. Stem-boring Diptera in grassland in relation to management practice. *Ann. Appl. Biol.* 87: 524–527.

Henderson, I. F., and J. N. Perry. 1978. Some factors affecting the build-up of cereal aphid infestations in winter wheat. *Ann. Appl. Biol.* 89: 177–183.

Hennebery, T. J. 1962. The effect of host-plant nitrogen supply and age of leaf tissue on the fecundity of the two-spotted spider mite. *J. Econ. Entomol.* 55: 799–800.

Hennebery, T. J., and D. Shriver. 1964. Two-spotted spider mite feeding on bean leaf tissue supplied with various levels of nitrogen. *J. Econ. Entomol.* 57: 377–379.

Hicks, P. M., P. L. Mitchell, E. P. Dunigan, L. D. Newsom, and P. K. Bollich. 1984. Effect of threecornered alfalfa hopper (Homoptera: Membracidae) feeding on translocation and nitrogen fixation in soybeans. *J. Econ. Entomol.* 77: 1275–1277.

Hill, M. G. 1976. The population and feeding ecology of 5 species of leafhoppers (Homoptera) on *Holcus mollis.* Ph.D. dissertation. University of London.

Hinckley, A. D. 1963. Ecology and control of rice planthoppers in Fiji. *Bull. Entomol. Res.* 54: 467–481.

Hirano, C., and S. Ishii. 1959. Effect of fertilizers on the growth of larvae of the rice stem borer, *Chilo supressalis.* III. Relation between application of phosphorus fertilizer and the growth of larvae. *Jap. J. Appl. Entomol. Zool.* 3: 86–90.

Hirano, C., and S. Ishii. 1961. Effect of fertilizers on the growth of larvae of the rice stem borer, *Chilo supressalis* Walker. IV. Growth responses of larvae to the rice plant supplied with potassium fertilizer at different levels. *Jap. J. Appl. Entomol. Zool.* 5: 180–184.

Hirano, C., and K. Kiritani. 1975. Paddy ecosystem affected by nitrogenous fertilizer and insecticides. *In Science for Better Environment. Proc. Intl. Cong. on the Human Environ. (HESC) (Kyoto)*, pp. 197–206.

Hiroo, K., K. Mujo, and S. Ishii. 1977. Feeding activity of the brown planthopper, *Nilaparvata lugens* Stål, on rice plant manured with different levels of nitrogen. *Jap. J. Appl. Entomol. Zool.* 21: 110–112.

Ho, D. T., and J. G. Kibuka. 1983. Effect of nitrogen and plant density on rice stem borer infestation in western Kenya. *Intl. Rice Res. Newsl.* 8(5): 17–18.

Hoff, J. E., G. E. Wilcox, and C. M. Jones. 1974. The effect of nitrate and ammonium nitrogen on the free amino acid composition of tomato plants and tomato fruit. *J. Am. Soc. Hortic. Sci.* 99: 27–30.

House, H. L. 1965. Effects of low levels of the nutrient content of a food and of nutrient imbalance on the feeding and the nutrition of a phytophagous larva, *Celerio euphorbiae* (L.) (Lepidoptera: Sphingidae). *Can. Entomol.* 97: 62–68.

House, H. L. 1966. The role of nutritional factors in biological control. *Can. Entomol.* 98: 1121–1134.

House, H. L. 1969. Effects of different proportions of nutrients on insects. *Entomol. Exp. Appl.* 12: 651–669.

Huffaker, C. B., and R. L. Rabb. 1984. *Ecological Entomology*. Wiley, New York.

Hunt, W. F., and D. L. Gaynor, 1982. Argentine stem weevil effects on Nui and Ranui ryegrasses grown with two levels of nitrogen/water nutrition. *N. Z. J. Agric. Res.* 25: 593–599.

Hussey, N. W., and B. Gurney. 1960. Some host plant factors affecting fecundity of whiteflies. *In Rep. Glasshouse Crops Res. Inst. 1959*, pp. 99–103.

Ishii, S., and C. Hirano. 1957. Effect of various concentrations of protein and carbohydrates in a diet on the growth of the rice stem borer larva. *Jap. J. Appl. Entomol. Zool.* 1: 75–79.

Ishii, S., and C. Hirano. 1958. Effect of fertilizers on the growth of the larva of the rice stem borer *Chilo suppressalis* Walker. I. Growth response of the larva to rice plants cultured in different nitrogen-level soils. *Jap. J. Appl. Entomol. Zool.* 2: 198–202.

Ishii, S., and C. Hirano. 1959. Effect of fertilizers on the growth of the larva of the rice stem borer *Chilo suppressalis* Walker. II. Growth of the larva on rice plants cultured in nutrient solutions of different nitrogen levels. *Jap. J. Appl. Entomol. Zool.* 3: 16–22.

Israel, P., and P. S. Prakasa Rao. 1967. Influence of potash on gall-fly incidence in rice. *Oryza* 4: 85–86.

Iversen, T. M. 1979. Laboratory energetics of larvae of *Sericostom personatum* (Trichoptera). *Holarct. Ecol.* 2: 1–5.

Jackson, P. R., and P. E. Hunter. 1983a. Effects of nitrogen-fertilizer level on development and populations of pecan leaf scorch mite (Acarina: Tetranychidae). *J. Econ. Entomol.* 76: 432–435.

Jackson, P. R., and P. E. Hunter. 1983b. Pecan leaf scorch mite population dynamics, management, and effects on pecan seedlings. *Misc. Pub. Entomol. Soc. Am.* 13: 35–45.

Jakhmola, S. S., U. K. Kaushik, and P. K. Kaushal. 1973. Note on the effect of date of sowing and nitrogen levels on the infestation of linseed-bud fly, *Dasyneura lini* Barnes (Diptera: Cecidomyiidae). *Indian J. Agric. Sci.* 43: 621–623.

Jancke, O. 1933, Über den Einfluss der Kalidüngung auf die Anfälligkeit der Apfelbäume gegen Blutlaus, Blattlaus und Mehltau. *Arb. Biol. Reichanst. (Berlin)* 20: 291–302.

Jansson, R. K., and Z. Smilowitz. 1985. Influence of nitrogen on population parameters of potato insects: abundance, development, and damage of the Colorado potato beetle, *Leptinotarsa decemlineata* (Coleoptera: Chrysomelidae). *Environ. Entomol.* 14: 500–506.

Jansson, R. K., and Z. Smilowitz. 1986. Influence of nitrogen on population parameters of potato insects: abundance, population growth, and within-plant distribution of the green peach aphid, *Myzus persicae* (Homoptera: Aphididae). *Environ. Entomol.* 15: 49–55.

Janzen, D. H. 1974. Tropical blackwater rivers, animals, and mast fruiting by the Dipterocarpaceae. *Biotropica* 66: 69–103.

Jayaraj, S., and M. S. Venugopal. 1964. Observations on the effect of manures and irrigation on the incidence of the cotton leaf hopper, *Empoasca devastans* Dist. and the cotton aphid, *Aphis gossypii* G., at different periods of crop growth. *Madras Agric. J.* 57: 189–196.

John, P. S., and M. J. Thomas. 1980. Effect of potash nutrition on infestation by the yellow borer, *Tryporyza incertulas*. *Agric. Res. J. Kerala* 18: 107–108.

Johnson, C. G., J. R. Lofty, and D. J. Cross. 1969. Insect pests on Broadbalk. *Rep. Rothamsted Exp. Sta. for 1968, Pt.* 2: 141–156.

Jones, F. G. W. 1976. Pests, resistance and fertilizers. *In Fertilizer Use and Plant Health*. Intl. Potash Inst., CH-3048 Worblaufen-Bern, Switzerland, pp. 233–258.

Jordan, C. F., R. L. Todd, and G. Escalante. 1979. Nitrogen conservation in a tropical rain forest. *Oecologia* 39: 123–128.

Joyce, R. J. V. 1958. Effect on the cotton plant in the Sudan Gezira of certain leaf feeding pests. *Nature* 182: 1463–1464.

Kalode, M. B. 1974. Recent changes in relative pest status of rice insects as influence by cultural, ecological, and genetic factors. Paper presented at the Intl. Rice Res. Conference. *Intl. Rice Res. Inst.*, Manila, Philippines, April 1974.

Kemp, W. P., and U. L. Moody. 1984. Relationships between regional soils and foliage characteristics and western spruce budworm (Lepidoptera: Tortricidae) outbreak frequency. *Environ. Entomol.* 13: 1291–1297.

Kevan, P. G., and H. G. Baker. 1983. Insects as flower visitors and pollinators. *Annu. Rev. Entomol.* 28: 407–453.

Kindler, S. D., and R. Staples. 1970. Nutrients and the reaction of two alfalfa clones to the spotted alfalfa aphid. *J. Econ. Entomol.* 63: 938–940.

Kiraly, Z. 1976. Plant disease resistance as influenced by biochemical effects of nutrients in fertilizers. *In Fertilizer Use and Plant Health*, Intl. Potash Inst., CH-3048 Worblaufen-Bern, Switzerland, pp. 33–46.

Kirkby, E. A. 1968. Influence of ammonium and nitrate nutrition on the cation-anion balance and nitrogen and carbohydrate metabolism of white mustard plants grown in dilute nutrient solutions. *Soil Sci.* 105: 133–141.

Kisimoto, R. 1965. Studies on the polymorphism and its role playing in the population growth of the brown planthopper, *Nilaparvata lugens* Stål. *Bull. Shikoku Agric. Exp. Sta.* 13: 1–106.

Kosuge, T. 1969. The role of phenolics in host response to infection. *Annu. Rev. Phytopathol.* 7: 195–222.

Krieger, R. I., P. P. Feeny, and C. F. Wilkinson. 1971. Detoxification enzymes in the guts of caterpillars: an evolutionary answer to plant defenses? *Science* 172: 579–581.

Krishnaiah, K., and M. A. Bari. 1975. Note on the influence of nitrogen levels on the susceptibility of paddy variety Taichung (native) 1 to green leaf hopper, *Nephotettix nigropictus* Stål. *Indian J. Entomol.* 37: 311–313.

Kumar, V., V. B. Ogunlela, and S. Mustara. 1982. Response of late-sown cotton to levels and times of nitrogen application in Nigeria. *Indian J. Agric. Sci.* 52: 578–583.

Kundu, G. G., and N. C. Pant. 1967. Studies on *Lipaphis erysimi* Kalt. with special reference to insect–plant relationship. II. Effect of various levels of N, P, and K on fecundity. *Indian J. Entomol.* 29: 285–289.

Kushwaha, K. S., and S. K. Sharma. 1981. Relationship of date of transplanting, spacings, and levels of nitrogen on the incidence of rice leaf folder in Haryana. *Indian J. Entomol.* 43: 338–339.

Labanauskas, C. K., and M. F. Handy. 1970. The effect of iron and manganese deficiencies on accumulation of non-protein amino acids in macadamia leaves. *J. Am. Soc. Hortic. Sci.* 95: 218–223.

Larsson, S., and O. Tenow. 1984. Effects of stressed trees on the population biology of *Neodiprion sertifer* (Hym., Diprionidae). *17th Intl. Cong. Entomol., Hamburg, FRG*, August 20–21, 1984.

Layton, M. B. 1983. The effects of feeding by bean leaf beetle larvae, *Cerotoma trifurcata* (Forster), on nodulation and nitrogen fixation of soybeans. M. S. thesis. Louisiana State University, Baton Rouge.

Layton, M. B., and D. J. Boethel. 1987. Reduction in N_2 fixation by soybean in response to insect-induced defoliation. *J. Econ. Entomol.* 80: 1319-1324.

Leath, K. T., and R. K. Radcliffe. 1974. The effect of fertilization on disease and insect resistance. *In* Mays, D. A. (ed.), *Forage Fertilization*. Am. Soc. Agron., Madison, WI, pp. 481–503.

Legay, J. M. 1958. Recent advances in silk-worm nutrition. *Annu. Rev. Entomol.* 3: 75–86.

Leigh, T. F., D. W. Grimes, H. Yamada, D. Bassett, and J. R. Stockton. 1970. Insects in cotton as affected by irrigation and fertilization practices. *California Agric.* 24(3): 12–14.

Lerer, M., and A. Bar-Akiva. 1976. Nitrogen constituents in manganese-deficient lemon leaves. *Physiol. Plant.* 38: 13–18.

LeRoux, E. J. 1954. Effects of various levels of nitrogen, phosphorus, and potassium in nutrient solution, on the fecundity of the two-spotted spider mite, *Tetranychus bimaculatus* Harvey (Acarina: Tetranychidae) reared on cucumber. *Can. J. Agric. Sci.* 34: 145–151.

Leuck, D. B. 1972. Induced fall armyworm resistance in pearl millet. *J. Econ. Entomol.* 65: 1608–1611.

Levitt, J. 1972. *Responses of Plants to Environmental Stresses*. 1st ed. Academic Press, New York.

Levitt, J. 1980. *Responses of Plants to Environmental Stresses. Water, Radiation, Salt and Other Stresses*, Vol. 2. 2d ed. Academic Press, New York.

Liener, I. E. 1973. Toxic factors associated with legume proteins. *Indian J. Nutr. Diet* 10: 303–322.

Liener, I. E. 1977. Protease inhibitors and hemagglutinins of legumes. *In* Bodwell, C. E. (ed.), *Evaluation of Proteins for Humans*. AVI, Westport, CT, pp. 284–303.

Lunderstädt, J., and I. M. Hoppe. 1975. Zur Nahrungsqualität von Fichtennadeln für forstliche Schädinsekten. *Z. Angew. Entomol.* 79: 177–193.

Lynch, R. E. 1984. Effects of "coastal" bermudagrass fertilization levels and age of regrowth on fall armyworm (Lepidoptera: Noctuidae): larval biology and adult fecundity. *J. Econ. Entomol.* 77: 948–953.

Macauley, B. J., and L. R. Fox. 1980. Variations in total phenols and condensed tannins in *Eucalyptus*, leaf phenology and insect grazing *Aust. J. Ecol.* 5: 31–35.

MacDonald, J. D. 1984. Salinity effects on the susceptibility of chrysanthemum roots to *Phytophthora cryptigea. Phytopathology* 74: 621–624.

Macleod, L. B., and M. Suzuki. 1967. Effect of potassium on the content of amino acids in alfalfa and orchard grass grown with NO_3^- and NH_4^+ nitrogen in nutrient solution culture. *Crop Science* 7: 599–605.

Maddox, D. M., and M. Rhyne. 1975. Effects of induced host-plant mineral deficiencies on attraction, feeding and fecundity of the alligator flea beetle. *Environ. Entomol.* 4: 682–686.

Mali, P. C., and S. L. Mentha. 1977. Effect of drought on enzymes and free proline in rice varieties. *Phytochemistry* 16: 1355–1357.

Manglitz, G. R., H. J. Gorz, F. A. Haskins, W. R. Akeson, and G. L. Beland. 1976. Interactions between insects and chemical components of sweetclover. *J. Environ. Qual.* 5: 347–352.

Markkula, M., and K. Tiittanen. 1969. Effect of fertilizers on the reproduction of *Tetranychus telarius* (L.), *Myzus persicae* (Sulz.) and *Acyrthosiphon pisum* Harris. *Ann. Agric. Fenn.* 8: 9–14.

Markkula, M., K. Tiittanen, and V. Kanervo. 1969. Growth substrate of plants and the reproduction rate of *Tetranychus telarius* (L.), *Acyrthosiphon pisum* Harris and *Myzus persicae* (Sulz.). *Ann. Agric. Fenn.* 8: 281–285.

Mathavan, S., and T. J. Pandian. 1975. Effect of temperature on food utilization in the monarch butterfly *Danaus chrysippus*. *Oikos* 26: 60–64.

Mathews, E. G. 1976. *Insect Ecology*. Univ. of Queensland P., Queensland, Australia.

Mathur, R. N. 1941. Certain observations on the nitrogen nutrition of the sugarcane plant in relation to susceptibility to attack of whitefly. *In Proc. 10th Annu. Conf. Sug. Tech. Assoc. India*, pp. 45–53.

Mattson, W. J. 1980. Herbivory in relation to plant nitrogen content. *Annu. Rev. Ecol.* 11: 119–161.

Mattson, W. J., and N. D. Addy. 1975. Phytophagous insects as regulators of forest primary production. *Science* 190: 515–522.

Maurizio, A. 1975. How bees make honey. *In* Crane, E. (ed.), *Honey: A Comprehensive Survey*. Heineman, London, pp. 77–105.

Maxwell, R. C., and R. F. Harwood. 1960. Increased reproduction of pea aphids on broad beans treated with 2,4-D. *Ann. Entomol. Soc. Am.* 53: 199–205.

Mayer, A. M., and E. Harel. 1979. Polyphenol oxidases in plants. *Phytochemistry* 18: 193–215.

McClure, M. S. 1977. Dispersal of the scale, *Fiorinia externa* (Homoptera: Diaspididae) and effects of edaphic factors on its establishment on hemlock. *Environ. Entomol.* 7: 539–544.

McClure, M. S. 1980. Foliar nitrogen: a basis for host suitability for elongate hemlock scale, *Fiorinia externa* (Homoptera: Diaspididae). *Ecology* 61: 72–79.

McDole, R. E., and G. M. McMaster. 1978. Effects of moisture stress and nitrogen fertilization on tuber nitrate-nitrogen content. *Am. Potato J.* 55: 611–619.

McGarr, R. L. 1942. Relation of fertilizers to the development of the cotton aphid. *J. Econ. Entomol.* 35: 482–483.

McKey, D. M. 1979. The distribution of secondary compounds within plants. *In* Rosenthal, G. A., and D. H. Janzen (eds.), *Herbivores: Their Interaction with Secondary Plant Metabolites*. Academic Press, New York, pp. 55–133.

McMurtry, J. A. 1962. Resistance of alfalfa to spotted alfalfa aphid in relation to environmental factors. *Hilgardia* 32: 501–539.

McNaughton, S. J. 1979. Grazing as an optimization process: grass–ungulate relationships in the Serengeti. *Am. Nat.* 113: 691–703.

McNeill, S. 1973. The dynamics of a population of *Leptopterna dolabrata* (Heteroptera: Miridae) in relation to its food resources. *J. Anim. Ecol.* 42: 495–507.

McNeill, S., and T. R. E. Southwood. 1978. The role of nitrogen in the development of insect–plant relationships. *In* Harborne, J. B. (ed.), *Biochemical Aspects of Plant and Animal Coevolution*. Academic Press, London, pp. 77–98.

Mehrotra, K. N., P. J. Rao, and T. N. A. Farooqi. 1972. The consumption, digestion, and utilization of food by locusts. *Entomol. Exp. Appl.* 15: 90–96.

Mehto, D. N., and B. S. Lall. 1981. A note on fertilization response against brinjal fruit and shoot borer. *Indian J. Entomol.* 43: 106–107.

Mellors, W. K., and S. E. Propts. 1983. Effects of fertilizer level, fertility balance, and soil moisture on the interaction of two-spotted spider mites (Acarina: Tetranychidae) with radish plants. *Environ. Entomol.* 12: 1239–1244.

Merino, L. G., and V. Vazquez. 1966. Influence of fertilizers on the damage caused by *Epitrix* sp. larvae in potato tubers in Ecuador. *Turrialba* 16: 84–85.

Merker, E. 1958. Die Schutzwirkung der Düngung im Walde gegen schädliche Insekten. *Forst-u.-Holzw.* 13: 316–319.

Merker, E. 1960. Die Bekämpfung von Waldschädlingen durch geeignete Düngung der Bestandesboden. *XI Intl. Entomologen—Kongress Wien*, Verhandl. II: 198–202, 4962.

Merker, E. 1961. Welche Ursachen hat die Schädigung der Insekten durch Dungung im Walde? *Allgem. Forst-u. J.-Ztg.* 132: 73–82.

Merker, E. 1962a. Augenblicklicher Stand der Untersuchungen über die schädigende Wirkingsweise von Düngestoffen auf Waldschädlinge. *Allgem. Forst-u. J.-Ztg.* 133: 81–83.

Merker, E. 1962b. Über die Ursachen der Vernichtung von Waldschädlingen durch Düngung. Mededel. Land bouwhogeschool Opzoekingssta. *Staat Gent.* 11: 821–828.

Merker, E. 1962c. Studien über unmittelbare und mittelbare verhängnisvolle Wirkungen der Bestandes-dügungen auf Waldverderber. *Anz. Schädlingskunde* 35: 133–140.

Merker, E. 1963. Die Bekämpfung der kleinen Fichtenblattwespe durch Düngung der Bestandesböden. *Allgem. Forst-u. J.-Ztg.* 134: 72–76.

Merker, E. 1969. Die Zuverlassigkeit der Bestandsdüngung gegen Waldschälinge. *Waldhygiene* 8: 1–100.

Merker, E., and J. Büttner. 1959. Die Wirkung von Mulldunger auf den Befall von Kiefernknospentriebwicklern. *Allg. Forstz.* 14: 792.

Metcalfe, J. R. 1970. Studies on the effect of the nutrient status of sugar cane on the fecundity of *Saccharosydne saccharivora* (Homoptera: Delphacidae). *Bull. Entomol. Res.* 60: 309–325.

Michael, E. 1963. Etude de la fecondite du puceron *Myzus persicae* en fonction de son alimentation sur tabac. I. Influence de la nutrition minerale de la plante. *Ann. Inst. Exp. du Tabac* 4: 421–434.

Michael, E., and J. Chouteau. 1963. Etude de la fecondite du puceron *Myzus persicae* en fonction de son alimentation sur tabac. II. Relation avec les teneurs en N. P. K. du parenchyne foliare. *Ann. Inst. Exp. du Tabac* 4: 435–441.

Michael Raj, S., and Y. B. Morachan. 1973. Effect of fertilization and diazinon application on the incidence of stem borer and leaf roller on rice. *Madras Agric. J.* 60: 431–435.

Miller, D. R., and M. Kosztarab. 1979. Recent advances in the study of scale insects. *Annu. Rev. Entomol.* 24: 1–27.

Mitchell, T. E., and R. F. Chandler. 1939. The nitrogen nutrition and growth of certain deciduous trees of northeastern United States. *Black Rock For. Bull. No. 11*. Cornwall-on-the-Hudson, NY.

Mitter, C., and D. J. Futuyma. 1983. An evolutionary-genetic view of host-plant utilization by insects. *In* Denno, R. F., and M. S. McClure (eds.). *Variable Plants and Herbivores in Natural and Managed Systems*. Academic Press, New York, pp. 427–459.

Mittler, T. E. 1970. Uptake rates of plant sap and synthetic diet by the aphid *Myzus persicae*. *Ann. Entomol. Soc. Am.* 63: 1701–1705.

Mochida, O., T. Suryana, and W. Sutarle. 1979. The effect of the application of nitrogen fertilizer to the rice plant and the irrigation water on the population growth of the brown planthopper, *Nilaparvata lugens* Stål. (Hom.: Delphacidae). *In Proc. Kongres Entomologi I.*, Jakarta, Indonesia, January 9–11, 1979.

Moore, D., and R. O. Clements. 1984. Stem-boring Diptera in perennial ryegrass in relation to fertilizer. I. Nitrogen level and form. *Ann. Appl. Biol.* 105: 1–6.

Morris, M. G. 1971. Differences between the invertebrate faunas of grazed and ungrazed chalk grassland. IV. Abundance and diversity of Homoptera: Auchenorrhyncha. *J. Appl. Ecol.* 8: 37–52.

Morris, O. N. 1961. The development of the clover mite *Bryobia praetiosa* (Acarina: Tetranychidae) in relation to nitrogen, phosphorus, and potassium nutrition of its plant host. *Ann. Entomol. Soc. Am.* 54: 551–557.

Morrow, P. A., and L. R. Fox. 1980. Effects of variation in *Eucalyptus* essential oil yield on insect growth and grazing damage. *Oecologia* 45: 209–219.

Mumford, E. P., and D. H. Hey. 1930. The water balance of plants as a factor in their resistance to insect pests. *Nature* 125: 411–412.

Myers, J. H., and B. J. Post. 1981. Plant nitrogen and fluctuations of insect populations: a test with the cinnabar moth-tansy ragwort system. *Oecologia* 48: 151–156.

Narayanan, K., J. Chandrasekaran, M. Meerzainudeen, and S. Jayaraj. 1973. Effect of graded levels of nitrogen on the incidence of rice gall midge. *Madras Agric. J.* 60: 572.

Nef, L. 1967. Comparison de populations de *Rhyacionia buoliana* Schiff. en response a une fumure minerale. *XIV IUFRO-Kongress Referate* V: 650–658.

Nowacki, E. M., M. Jurzysta, P. Gorski, D. Nowacka, and G. R. Waller. 1976. Effect of nitrogen nutrition on alkaloid metabolism in plants. *Biochem. Physiol. Pflanzen* 169: 231–240.

Odum, W. E., and E. J. Heald. 1972. Trophic analyses of an estuarine mangrove community. *Bull. Mar. Sci.* 22: 671–728.

Oka, I. N., and D. Pimentel. 1976. Herbicide (2,4-D) increases insect and pathogen pests on corn. *Science* 193: 239–240.

Okigbo, B. N., and G. G. Gyrisco. 1962. Effects of fertilizer on Hessian fly infestation. *J. Econ. Enotmol.* 55: 753–760.

Oldiges, H. 1959. Der Einfluss der Waldbodendüngung auf das Auftreten von Schadinsekten. *Z. Angew. Entomol.* 45: 49–59.

Onuf, C. P., J. M. Teal, and I. Valiela. 1977. Interactions of nutrients, plant growth, and herbivory in a mangrove ecosystem. *Ecology* 58: 514–526.

Painter, R. H. 1954. Some ecological aspects of the resistance of crop plants to insects. *J. Econ. Entomol.* 47: 1036–1040.

Painter, R. H. 1958. Resistance of plants to insects. *Annu. Rev. Entomol.* 3: 267–290.

Parisi, R. A., A. Ortega, and R. Reyna. 1973. El dano de *Diatraea saccharalis* Fabricius (Lepidoptera: Pyralidae) en relacion con la densidad de plantas, nivel de fertilidad e hibridos de maiz, en Argentina. *Agrociencia* 13: 43–63.

Parker, J., and R. C. Patton. 1975. Effects of drought and defoliation on some metabolites in roots of black oak seedlings. *Can. J. For. Res.* 5: 457–463.

Parnell, F. R. 1927. Report for season, 1925–1926. *In Empire Cotton Growing Crop Rpt. Exp. Sta.* South Africa Cotton Breeding Station, Barberton, pp. 37–39.

Patch, L. H. 1947. Manual infestation of dent corn to study resistance to European corn borer. *J. Econ. Entomol.* 40: 667–671.

Pate, J. S. 1980. Transport and partitioning of nitrogenous solutes. *Annu. Rev. Plant Physiol.* 31: 313–340.

Pate, J. S., and C. A. Atkins. 1983. Nitrogen uptake, transport, and utilization. *In* Broughton, W. J. (ed.), *Ecology of Nitrogen Fixation. Legumes*, Vol. 3. Oxford Univ. P., Oxford, pp. 245–297.

Pate, J. S., C. A. Atkins, S. T. White, R. M. Rainbird, and K. C. Woo. 1980. Nitrogen nutrition and xylem transport of nitrogen in ureide-producing grain legumes. *Plant Physiol.* 65: 961–965.

Pathak, M. D. 1975. Utilization of insect–plant interactions in pest control. *In* Pimentel, D. (ed.), *Insects, Science, and Society*. Academic Press, New York, pp. 121–148.

Pathak, P. K., and E. A. Heinrichs. 1982. Selection of biotype populations 2 and 3 of *Nilaparvata lugens* by exposure to resistant rice varieties. *Environ. Entomol.* 11: 85–90.

Perrenoud, S. 1976. Contribution to the discussion: The effect of K on insect and mite development. *In Fertilizer Use and Plant Health.* Intl. Potash Inst., CH-3048 Worblaufen-Bern, Switzerland, pp. 317–319.

Pfeiffer, D. G., and E. C. Burts. 1983. Effects of tree fertilization on numbers and development of pear psylla (Homoptera: Psyllidae) and on fruit damage. *Environ. Entomol.* 12: 895–901.

Phillips, R., and G. G. Henshaw. 1977. The regulation of synthesis of phenolics in stationary phase cell cultures of *Acer pseudoplatanus* L. *J. Exp. Bot.* 28: 785–794.

Phillipson, J. 1981. Bioenergetic options and phylogeny. *In* Townsend, C. R., and P. Calow (eds.), *Physiological Ecology: an Evolutionary Approach to Resource Use.* Sinauer Associates, Sunderland, MA, pp. 20–45.

Piene, H. 1978. Effect of increased spacing on carbon mineralization and temperature in young balsam fir. *Can. J. For. Res.* 8: 398–406.

Pillai, K. G., M. B. Kalode, and A. V. Rao. 1979. Effects of nitrogen levels, plant spacings, and row orientation on the incidence of the brown planthopper of rice. *Indian J. Agric. Sci.* 49: 125–129.

Pimentel, D., and C. A. Edwards. 1982. Pesticides and ecosystems. *BioScience* 32: 595–600.

Poljakoff-Mayber, A. 1975. Morphological and anatomical changes in plants as a response to salinity stress. *In* Poljakoff-Mayber, A., and J. Gale (eds.), *Plants in Saline Environments*. Springer-Verlag, New York, pp. 97–117.

Poljakoff-Mayber, A., and J. Gale. 1975. *Plants in Saline Environments*. Springer-Verlag, New York.

Prestidge, R. A. 1980. The influence of mineral fertilization on grassland leaf-hopper associations. Ph.D dissertation. University of London.

Prestidge, R. A. 1982. Instar duration, adult consumption, oviposition and nitrogen utilization efficiencies of leafhoppers feeding on different quality food (Auchenorryncha: Homoptera). *Ecol. Entomol.* 7: 91–101.

Prew, R. D., B. M. Church, A. M. Dewar, J. Lacey, A. Penny, R. T. Plumb, G. N. Thorne, A. D. Todd, and T. D. Williams. 1983. Effects of eight factors on the growth and nutrient uptake of winter wheat and on the incidence of pests and diseases. *J. Agri. Sci.* (Cambridge), 100: 363–382.

Pritchett, W. L., and W. H. Smith. 1972. Fertilizer responses in young pine plantations. *Soil Sci. Soc. Am. Proc.* 36: 660–663.

Quanjer, H. M. 1929. Der Einfluss der Düngung auf die Gesundheit der Kartoffel. *Ernährung der Pflanze* 25: 194.

Rahman, A. A., A. F. Shalaby, and M. O. Monayeri. 1971. Effects of moisture stress on metabolic products and ion accumulation. *Plant and Soil* 34: 65–90.

Rains, D. W. 1972. Salt transport by plants in relation to salinity. *Annu. Rev. Plant Physiol.* 23: 367–388.

Rajaratnam, J. A., and Law Ing Hock. 1975. Effect of boron nutrition on intensity of red spider mite attack on oil palm seedlings. *Exper. Agric.* 11: 59–63.

Rangarajan, A. V., N. R. Mahadevan, S. Ganapathy, and B. Mahalingam. 1974. Effect of N, P and K nutrition on the pest incidence in sunflower. *Potash Newsl.* 9(3): 27–29.

Reagan, T. E., R. L. Rabb, and W. K. Collins. 1978. Selected cultural practices affecting production of tobacco. *J. Econ. Entomol.* 71: 79–82.

Redak, R. A., and R. G. Cates. 1984. Douglas-fir (*Pseudotsuga menziesii*) spruce budworm (*Choristoneura occidentalis*) interactions: the effect of nutrition, chemical defenses, tissue phenology, and tree physical parameters on budworm success. *Oecologia* 62: 61–67.

Reese, J. C. 1978. Chronic effects of plant allelochemics on insect nutritional physiology. *Entomol. Exp. Appl.* 24: 625–631.

Regupathy, A., and A. Subramanian. 1972. Effect of different doses of fertilizers on the mineral metabolism of IR8 rice in relation to its susceptibility to gall fly, *Pachydiplosis oryzae* Wood-Mason and leaf roller *Cnaphalocrocis medinalis* Guenee. *Oryza* 9: 81–85.

Rekhi, R. S., J. Singh, and O. P. Meelu. 1985. Effect of green manure and nitrogen on mole rat damage and leaffolder (LF) incidence in rice. *Intl. Rice Res. Newsl.* 10(1): 26.

Rendig, V. V., C. Oputa, and E. A. McComb. 1976. Effects of sulfur deficiency on non-protein nitrogen, soluble sugars, and N/S ratios in young corn (*Zea mays* L.) plants. *Plant and Soil* 44: 423–437.

Rhoades, D. F., and R. G. Cates. 1976. Toward a general theory of plant antiherbivore chemistry. *In* Biochemical Interactions Between Plants and Insects. *Rev. Adv. Phytochem.* 10: 168–213.

Rice, E. L. 1974. *Allelopathy*. Academic Press, New York.

Richards, L. J., and J. H. Myers. 1980. Maternal influence on the size and emergence time of the cinnabar moth. *Can. J. Zool.* 58: 1452–1457.

Roberts, G. R., U. L. L. DeSilva, and D. T. Wetasinghe. 1976. The effect of cultural operations on the composition of the xylem sap of tea plants recovering from pruning. *Ann. Bot.* 40: 825–831.

Rodriguez, J. G. 1951. Mineral nutrition of the two-spotted spider mite, *Tetranychus bimaculatus* Harvey. *Ann. Entomol. Soc. Am.* 44: 511–526.

Rodriguez, J. G. 1958. The comparative NPK nutrition of *Panonychus ulmi* (Koch.) and *Tetranychus telarius* (L.) on apple trees. *J. Econ. Entomol.* 51: 369–373.

Rodriguez, J. G. 1960. Nutrition of the host and reaction to pest. *In* Reitz, L. P. (ed.), *Biological and Chemical Control of Plant and Animal Pests*. Am. Assoc. Adv. Sci., Washington, DC, 61: 149–167.

Rodriguez, J. G., and R. B. Neiswander. 1949. The effect of soil soluble salts and cultural practices on mite populations on hothouse tomatoes. *J. Econ. Entomol.* 42: 56–59.

Rodriguez, J. G., and L. D. Rodriguez. 1952. The relation between minerals, B-complex vitamins, and mite populations on tomato foliage. *Ann. Entomol. Soc. Am.* 45: 331–338.

Roehrig, N. E., and J. L. Capinera. 1983. Behavioral and developmental responses of range caterpillar larvae, *Hemileuca oliviae* to condensed tannin. *J. Insect Physiol.* 29: 901–906.

Room, P. M., and P. A. Thomas. 1985. Nitrogen and establishment of a beetle for biological control of the floating weed *Salvinia* in Papua New Guinea. *J. Appl. Ecol.* 22: 139–156.

Ross, A. F., J. A. Chucka, and A. Hawkins. 1948. The effect of fertilizer practice including the use of minor elements on stem-end browning, net necrosis, and spread of leafroll virus in the green mountain variety of potato. *Maine Agric. Exp. Sta. Bull.* 447: 97–142.

Russell, F. C. 1947. The chemical composition and digestibility of fodder shrubs and trees. *Jt. Publ. Commonw. Agric. Bur. Pastures Field Crops for Anim. Nutr.* 10: 185–231.

Ryan, C. A., and T. R. Green. 1974. Proteinase inhibitors in natural plant protection. *Rec. Adv. Phytochem.* 8: 123–140.

Salama, H. S., A. H. Amin, and M. Hawash. 1972. Effect of nutrients supplied to citrus seedlings on their susceptibility to infestation with the scale insects *Aonidiella aurantii* (Maskell) and *Lepidosaphes beckii* (Newman) (Coccoidea). *Z. Angew. Entomol.* 71: 395–405.

Salama, H. S., A. F. El-Sherif, and M. Megahed. 1985. Soil nutrients affecting the population density of *Parlatoria zizyphus* (Lucas) and *Icerya purchasi* Mask. (Homopt., Coccoidea) on citrus seedlings. *Z. Angew. Entomol.* 99: 471–476.

Salisbury, F. B., and C. W. Ross. 1978. *Plant Physiology*. 2d ed. Wadsworth, Belmont, CA.

Saradamma, K., K. Sasidharan Pillai, and N. M. Das. 1973. Relative susceptibility of the rice variety "Rohini" grown under different levels of nitrogen to the storage pests. *Agric. Res. J. Kerala* 11: 182–183.

Saroja, R., and N. Raju. 1981. Effect of potash levels on the incidence of rice whorl maggot. *Intl. Rice Res. Newsl.* 6(3): 15–16.

Saroja, R., S. N. Peeran, and N. Raju. 1981. Effects of method of nitrogen application on the incidence of rice leaffolder. *Intl. Rice Res. Newsl.* 6(4): 15–16.

Sarup, P., V. P. S. Panwar, K. K. Marwaha, and K. H. Siddiqui. 1978. Management of maize pests with particular reference to the stalk borer, *Chilo partellus* (Swinhoe) under resource constraints. *J. Entomol. Res.* (New Delhi) 2: 5–14.

Sathasivan, K., S. P. Palaniappan, and P. Balasubramaniyan. 1983. Comparative efficiency of carbofuran and its analogue FMC 35001 in combination with urea in lowland rice. *Pesticides* 17(12): 31–32.

Sathiyanandam, V. K. R., and A. Subramanian. 1981. Effect of potash on the incidence of whorl maggot. *Aduthurai Rep.* 5(1): 10.

Saxena, R. C., and M. D. Pathak. 1979. Factors governing susceptibility and resistance of certain rice varieties to the brown planthopper. *In Brown Planthopper: Threat to Rice Production in Asia*. Intl. Rice Res. Inst., Manila, Phillippines, pp. 304–317.

Scheurer, S. 1976. The influence of phytohormones and growth regulating substances on insect development processes. *Symp. Biologica Hungarica* 16: 255–259.

Schindler, U. 1967. Einfluss der Düngung auf Forstinsekten. *Bericht über das Kolloquium für Forstdungung Jyvaskylal Finland.* Intl. Kali-Inst., Bern, pp. 321-–327.

Schindler, U., and H. Baule. 1964. Forstliche Düngung und Kiefernknospentriebwicklerbefall. *Allg. Forstz.* 19: 534–537.

Schoonhoven, L. M. 1973. Plant recognition by lepidopterous larvae. *Symp. R. Entomol. Soc. Lond.* 6: 87–99.

Schroeder, L. A. 1976. Effect of food deprivation on the efficiency of utilization of dry matter, energy, and nitrogen by larvae of the cherry scallop moth *Calocalpe undulata. Ann. Entomol. Soc. Am.* 69: 55–58.

Schroeder, L. A. 1977. Energy, matter, and nitrogen utilization by larvae of the milkweed tiger moth, *Euchaetias egle. Oikos* 28: 27–31.

Schweig, C., and A. Grunberg. 1936. The problem of black scale, *Chrysomphalus ficus* Ashm., in Palestine. *Bull. Entomol. Res.* 27: 677–714.

Schweissing, F. C., and G. Wilde. 1979. Temperature and plant nutrient effects on resistance of seedling sorghum to the greenbug. *J. Econ. Entomol.* 72: 20–23.

Schwenke, W. von. 1960. Über die Wirkung der Walddüngung auf die Massenvermehrung der Kiefernbuschhornblattwespe (*Diprion pini* (L.)) 1959 in Mittelfranken und die hieraus ableitbaren gradologischen Folgerungen. *Z. Angew. Entomol.* 46: 371–378.

Schwenke, W. von. 1961. Walddüngung und Schadinsekten. *Anz. Schädlingskde* 34: 129–134.

Schwenke, W. von. 1962. Über die Wirkung der Düngung auf phytophage Milben und Insekten. *Mededel. Land bouwhogeschool Opzoekingsta. Staat Gent* 27: 817–820.

Schwerdtfeger, F. 1970. *Die Waldkrankheiten.* Paul Parey, Hamburg.

Scott, G. E., F. E. Dicke, and L. H. Feeny. 1965. Effects of first brood European corn borers on single crosses grown at different nitrogen and plant population levels. *Crop Sci.* 5: 261–263.

Scriber, J. M. 1977. Limiting effects of low leaf-water content on the nitrogen utilization, energy budget, and larval growth of *Hyalophora cecropia* (Lepidoptera: Saturniidae). *Oecologia* 28: 269–287.

Scriber, J. M. 1984. Nitrogen nutrition in plants and insect invasion. *In Nitrogen in Crop Production.* Am. Soc. Agron., Madison, WI, pp. 441–460.

Scriber, J. M., and P. Feeny. 1979. The growth of herbivorous caterpillars in relation to degree of feeding specialization and to growth form of their food plants (Lepidoptera: Papilionidae and Bombycoidea). *Ecology* 60: 829–850.

Semtner, P. S., M. Rasnake, and T. R. Terrill. 1980. Effect of host plant nutrition on the occurrence of tobacco hornworms and tobacco flea beetles on different types of tobacco. *J. Econ. Entomol.* 73: 221–224.

Sharma, M. L. 1970. Responses of phytophagous insects to treatments of nitrogen, potassium, and phosphorus applied to plants (general review). *Ann. Soc. Entomol. Que.* 15: 88–95.

Shasha'a, N. S., W. F. Campbell, and W. P. Nye. 1976. Effects of fertilizer and moisture on seed yield of onion. *HortScience* 11: 425–426.

Shepherd, R. F. 1959. Phytosociological and environmental characteristics of outbreak and non-outbreak areas of the two-year cycle spruce budworm, *Choristoneura fumiferana. Ecology* 40: 608–620.

Shuel, R. W. 1957. Some aspects of nectar secretion and nitrogen, phosphorus, and potassium nutrition. *Can. J. Plant Sci.* 37: 220–236.

Shull, A. F. 1930. Control of gamic and parthenogenetic reproduction in winged aphids by temperature and light. *Z. Abstamm. Vererbungslehre* 55: 108–126.

Simmons, I. G. 1981. *The Ecology of Natural Resources*. English Language Book Society and Edward Arnold, London.

Singh, P. 1970. Host-plant nutrition and composition: effects on agricultural pests. *Canada Dept. Agric. Res. Inst. Inf. Bull. No. 6*. Ottawa, Ontario.

Singh, R., and R. A. Agarwal. 1983. Fertilizers and pest incidence in India. *Potash Rev*. 23(11): 1–4.

Singh, S. R., and R. H. Painter. 1964. Reactions of four biotypes of corn leaf aphid, *Rhopalosiphum maidis* (Fitch) to differences in host plant nutrition. *In Proc. 12th Intl. Cong. Entomol*., p. 543.

Singh, T. P., and R. Singh. 1969. Incidence of stem borer, *Chilo zonellus* Swinhoe, and lodging in Jaunpuri variety of maize under different levels of nitrogen. *Indian J. Entomol*. 31: 158–160.

Singh, T. P., R. Singh, and L. B. Chaudhary. 1968. Interrelation of stem borer incidence and certain agronomic traits in Jaunpuri variety of maize under different levels of nitrogen. *Indian J. Entomol*. 30: 220–222.

Slansky, F., Jr. 1976. Phagism relationships among butterflies. *J. N. Y. Entomol. Soc*. 84: 91–105.

Slanksy, F., Jr., and P. Feeny. 1977. Stabilization of the rate of nitrogen accumulation by larvae of the cabbage butterfly on wild and cultivated food plants. *Ecol. Monogr*. 47: 209–228.

Slansky, F., Jr., and J. M. Scriber. 1985. Food consumption and utilization. *In* Kerkut, G. A., and L. I. Gilbert (eds.), *Comprehensive Insect Physiology, Biochemistry, and Pharmacology*, Vol. 4. Pergamon, Oxford, pp. 87–163.

Sloan, W. J. S. 1938. Cotton jassids or leafhoppers. *Queensland Agric. J*. 50: 450–455.

Smirnoff, W. A., and B. Bernier. 1973. Increased mortality of the Swaine jack-pine sawfly, and foliar nitrogen concentrations after urea-fertilization. *Can. J. For. Res*. 3: 112–121.

Smith, D. 1964. Chemical composition of herbage with advance in maturity of alfalfa, medium red clover, ladino clover, and birdsfoot trefoil. *Res. Report* 16, University of Wisconsin, Madison.

Smith, D. S., and F. E. Northcott. 1951. The effects on the grasshopper, *Melanoplus mexicanus mexicanus* (Sauss) (Orthoptera: Acrididae) of varying the nitrogen content in its food plant. *Can. J. Zool*. 29: 297–304.

Smith, E. H., and R. W. Harris. 1952. Influence of tree vigor and winter injury on the lesser peach tree borer. *J. Econ. Entomol*. 45: 607–610.

Sogawa, K. 1971. Effects of feeding of brown planthopper on the components in the leaf blade of rice plants. *Jap. J. Appl. Entomol. Zool*. 15: 175–179.

Stafford, H. A. 1974. The metabolism of aromatic compounds. *Annu. Rev. Plant Physiol*. 25: 459–486.

Stapel, C. 1958. Secondary effects of potassium deficiency. Some experiences in Denmark. *Potash Rev*. 23: 1–8.

Stark, R. W. 1965. Recent trends in forest entomology. *Annu. Rev. Entomol*. 10: 303–324.

Steward, F. C., F. Crane, K. Millar, R. M. Zacharias, R. Parson, and D. Margolis. 1959. Nutritional and environmental effects on the nitrogen metabolism of plants. *Symp. Soc. Exp. Biol*. 13: 148–176.

Stewart, G. R., and T. O. Orebamjo. 1983. Studies of nitrate utilization by the dominant species of regrowth vegetation of tropical West Africa: a Nigerian example. *In* Lee, J. A., S. McNeill, and I. H. Rorison (eds.), *Nitrogen as an Ecological Factor*. Blackwell Scientific, Oxford, pp. 167–188.

Stewart, W. D. P. 1977. Present-day nitrogen-fixing plants. *Ambio* 6: 166–173.

Storms, J. J. H. 1969. Observations on the relationship between mineral nutrition of apple root stocks in gravel culture and the reproduction rate of *Tetranychus urticae* (Acarina: Tetranychidae). *Entomol. Exp. Appl.* 12: 297–311.

Strong, D. R., J. H. Lawton, and T. R. E. Southwood. 1984. *Insects on Plants: Community Patterns and Mechanisms*. Harvard Univ. P., Cambridge, MA.

Subbaiah, K. K., and Y. B. Morachan. 1974. Effect of nitrogen nutrition and rice varieties on the incidence of leaf roller, *Cnaphalocrocis medinalis* Guen. *Madras Agric. J.* 61: 716.

Subramanian, A., M. Radhakrishnan, and V. K. R. Sathiyanandam. 1980. Incidence of leaf folder and brown planthopper in different manurial contents. *Aduthurai Rep.* 4: 79–80.

Subramanian, R., and M. Balasubramanian. 1976. Effect of potash nutrition on the incidence of certain insect pests of rice. *Madras Agric. J.* 63: 561–564.

Sutherland, O. R. W. 1967. Role of host plant in production of winged forms by a green strain of pea aphid, *Acyrthosiphon pisum* Harris. *Nature* 216: 387–388.

Swaminathan, K., R. Saroja, and N. Raju. 1985. Influence of source and level of nitrogen application on pest incidence. *Intl. Rice Res. Newsl.* 10(1): 24.

Taylor, C. E. 1955. Growth of the potato plant and aphid colonization. *Ann. Appl. Biol.* 43: 151–156.

Taylor, C. E. 1962. The population dynamics of aphids infesting the potato plant with particular reference to the susceptibility of certain varieties to infestation. *Eur. Potato. J.* 3: 204–219.

Taylor, L. F., J. W. Apple, and K. C. Berger. 1952. Response of certain insects to plants grown on varying fertility levels. *J. Econ. Entomol.* 45: 843–848.

Teal, J. M. 1962. Energy flow in the salt marsh ecosystems of Georgia. *Ecology* 43: 614–624.

Terman, G. L., J. C. Noggle, and C. M. Hunt. 1976. Nitrate-N and total N concentration relationships in several plant species. *Agron. J.* 68: 556–560.

Terry, N., and L. J. Waldron. 1984. Salinity, photosynthesis, and leaf growth. *Calif. Agric.* 38(10): 38–39.

Thalenhorst, W. 1972. Düngung, Wuchsmerkmale der Fichte und Arthropodenbefall. *Aus dem Walde* 18: 1–248.

Thompson, W. L. 1942. The effect of magnesium deficiency on infestation of purple scale on citrus. *J. Econ. Entomol.* 35: 351–354.

Thornhill, R., and J. Alcock. 1983. *The Evolution of Insect Mating Systems*. Harvard Univ. P., Cambridge, MA.

Tingey, W. M., and S. R. Singh. 1980. Environmental factors influencing the magnitude and expression of resistance. *In* Maxwell, F. G., and P. R. Jennings (eds.), *Breeding Plants Resistant to Insects*. Wiley, New York, pp. 89–113.

Tisdale, S. L., and W. L. Nelson. 1975. *Soil Fertility and Fertilizers*. 3d ed. Macmillan, New York.

Todd, J. W., M. B. Parker, and T. P. Gaines. 1972. Populations of Mexican bean beetles in relation to leaf protein of nodulating and non-nodulating soybeans. *J. Econ. Entomol.* 65: 729–731.

Torrey, J. G. 1978. Nitrogen fixation by actinomycete-nodulated angiosperms. *BioScience* 28: 586–592.

Trolldenier, G., and E. Zehler. 1976. Relationships between plant nutrition and rice diseases. *In Fertilizer Use and Plant Health*, Intl. Potash Inst., CH-3048 Worblaufen-Bern, Switzerland, pp. 85–93.

Tulisalo, U. 1971. Free and bound amino acids of three host plant species and various fertilizing treatments affecting the fecundity of the two-spotted spider mite, *Tetranychus urticae* Koch (Acarina: Tetranychidae). *Ann. Entomol. Fennici* 37: 155–163.

Tulisalo, U., and M. Markkula. 1970. Resistance of pea to the pea weevil *Sitona lineatus* (L.). *Ann. Agric. Fenn.* 9: 139–141.

Upadhyay, V. R., A. H. Shah, and N. D. Desai. 1981. Influence of levels of nitrogen and rice varieties on the incidence of rice leaffolder *Cnaphalocrocis medinalis* Guenee. *Gujarat Agric. Univ. Res. J.* 6: 115–117.

Uthamasamy, S., V. Velu, M. Gopalan, and K. M. Ramanathan. 1983. Incidence of brown planthopper *Nilaparvata lugens* Stål on IR50 at graded levels of fertilization at Aduthurai. *Intl. Rice Res. Newsl.* 8(5): 13.

Vaithilingam, C. 1975. Effect of potash nutrient on the incidence and severity of different insect pests of rice. M.S. (Ag.) thesis. Annamalai University, Annamalai, India.

Vaithilingam, C., M. Balasubramanian, and R. Subramanian. 1976. Effect of potash nutrition on the feeding and excretion of brown planthopper in three rice varieties. *Madras Agric. J.* 63: 571–572.

van de Vrie, M. 1972. Potential of *Typhlodromus potenlillae* (Garman) in reducing *Panonychus ulmi* (Koch) on apple with various levels of nitrogen fertilizing. *Zesyly Problemoue Postepow. Nauk. Roslniczych* 129: 235–242.

van de Vrie, M., and P. Delver. 1979. Nitrogen fertilization of fruit trees and its consequence for the development of *Panonychus ulmi* populations and the growth of fruit trees. *Rec. Adv. Acarol.* 1: 23–30.

van Emden, H. F. 1966. Studies on the relations of insect and host plant. III. A comparison of the reproduction of *Brevicoryne brassicae* L. and *Myzus persicae* Sulz. (Hemiptera: Aphididae) on brussels sprout plants supplied with different rates of nitrogen and potassium. *Entomol. Exp. Appl.* 9: 444–460.

van Emden, H. F. 1973. Aphid host plant relationships. *In* Lowe, A. D. (ed.), *Phytochem. Ecol. Bull.*, Entomol. Soc. New Zealand, no. 2, pp. 54–64.

van Emden, H. F., V. F. Eastop, R. D. Hughes, and M. J. Way. 1969. The ecology of *Myzus persicae*. *Annu. Rev. Entomol.* 14: 197–270.

Varadharajan, G., and P. R. Nagaraja Rao. 1965. The influence of nitrogen and phosphoric acid on the incidence of rice stem borer, *Tryporyza incertulas* Wlk. *Rice Newsl.* 13: 105–107.

Vereijken, P. H. 1979. Feeding and multiplication of three cereal aphid species and their effect on yield of winter wheat. *Agric. Res. Dept., Netherlands* 888: 1–58.

Viets, F. G. 1972. Water deficits and nutrient availability. *In* Kozlowski, T. T. (ed.), *Water Deficits and Plant Growth*, vol. 3. *Plant Response and Control of Water Balance*. Academic Press, New York, pp. 217–239.

Vijverberg, A. J. 1965. De invloed van de fysiologische taestand van de voedselplant op het populatieverloop van zuigende insekten met name bladivizen. *Inst. Plziekt. Onderzoek, Wageninen, Jaorverslag* 1964: 56–59.

Vince, S. W., I. Valiela, and J. M. Teal. 1981. An experimental study of the structure of herbivorous insect communities in a salt marsh. *Ecology* 62: 1662–1678.

Vladimov, A. D. 1945. Influence of nitrogen sources in the formation of oxidised and reduced organic compounds in plants. *Soil Sci.* 60: 266–275.

Volk, J., and O. Bode. 1954. Weitere Untersuchungen zur Frage eines Zusammenhanges zwischen Düngung, Blattlausbesatz und Krankheitsausbreitung in Kartoffelbestanden. II. *Mitteilung Z. Pflkrankh.* 61: 49–70.

Volk, J., O. Bode, and I. Hanschild. 1952. Untersuchungen zur Frage eines Zusammenhanges zwischen Düngung, Blattlausbesatz und Krankheitsausbreitung in Kartoffelbestanden. I. *Mitteilung Z. Pflkrankh.* 59: 97–110.

von Rudloff, E. 1975. Volatile leaf oil analysis in chemosystematic studies of North American conifers. *Biochem. Syst. Ecol.* 2: 131–167.

Waghray, R. N., and D. R. Singh. 1965. Influence of host plant nutrition on *Aphis craccivora* Koch. *Indian J. Entomol.* 29: 196–201.

Wagner, M. R., and E. A. Blake. 1983. Western spruce budworm: consumption effect of host species and foliage chemistry. Proc. Forest-Defoliator Interactions: A Comparison between Gypsy Moth and Spruce Budworms. *U.S. Dept. Agric. For. Serv. Tech. Rep. NE.* 85 p.

Waldbauer, G. 1968. The consumption and utilization of food by insects. *Adv. Insect Physiol.* 5: 229–288.

Waller, G. D., E. W. Carpenter, and O. A. Ziehl. 1972. Potassium in onion nectar and its probable effect on attractiveness of onion flowers to honey bees. *J. Am. Soc. Hortic. Sci.* 97: 535–539.

Waloff, N. 1980. Studies on grassland leafhoppers (Auchenorrhyncha: Homoptera) and their natural enemies. *Adv. Ecol. Res.* 11: 81–215.

Waloff, N., and M. G. Solomon. 1973. Leafhoppers (Auchenorryncha: Homoptera) on acidic grassland. *J. Appl. Ecol.* 10: 189–212.

Watters, R. F. 1960. The nature of shifting cultivation: a review of recent research. *Pacific Viewpoint* 1: 59–99.

Webster, G. C. 1959. *Nitrogen Metabolism in Plants.* Row, Peterson, Evanston, IL.

Wellenstein, G. 1973. Der Forstschutz im Zeitalter des Umweltschutzes. *Allg. Forst-u. J.-Ztg.* 144: 69–75.

Wermelinger, B., J. J. Oertli, and V. Delucchi. 1985. Effect of host plant nitrogen fertilization on the biology of the two-spotted spider mite, *Tetranychus urticae. Entomol. Exp. Appl.* 38: 23–28.

White, T. C. R. 1969. An index to measure weather-induced stress of trees associated with outbreaks of psyllids in Australia. *Ecology* 50: 905–909.

White, T. C. R. 1974. A hypothesis to explain outbreaks of looper caterpillars with special reference to populations of *Selidosema suavis* in a plantation of *Pinus radiata* in New Zealand. *Oecologia* 16: 279–301.

White, T. C. R. 1976. Weather, food, and plagues of locusts. *Oecologia* 22: 119–134.

White, T. C. R. 1978. The importance of a relative shortage of food in animal ecology. *Oecologia* 33: 77–86.

White, T. C. R. 1984. The abundance of invertebrate herbivores in relation to the availability of nitrogen in stressed food plants. *Oecologia* 63: 90–105.

Whitham, T. G. 1978. Habitat selection by *Pemphigus* aphids in response to resource limitation and competition. *Ecology* 59: 1164–1176.

Whitham, T. G. 1981. Individual trees as heterogeneous environments: adaptation to herbivory or epigenetic noise? *In* Denno, R. F., and H. Dingle (eds.), *Insect Life History Patterns: Habitat and Geographic Variation.* Springer-Verlag, New York, pp. 9–27.

Willmer, P. G., and D. M. Unwin. 1981. Field analysis of insect heat budgets: reflectance, size, and heating rates. *Oecologia* 50: 250–255.

Wint, G. R. W. 1983. The effect of foliar nutrients upon the growth and feeding of a lepidopteran larva. *In* Lee, J. A., S. McNeill, and I. H. Rorison (eds.), *Nitrogen as an Ecological Factor.* Blackwell Scientific, Oxford, pp. 301–320.

Wittwer, S. H., and L. Haseman. 1945. Soil nitrogen and thrips injury on spinach. *J. Econ. Entomol.* 38: 615–617.

Wolfson, J. L. 1980. Oviposition response of *Pieris rapae* to environmentally induced variation in *Brassica nigra. Entomol. Exp. Appl.* 27: 223–232.

Woltz, S. S., and E. G. Kelsheimer. 1959. Effect of variation in nitrogen nutrition of chrysanthemums on attack by serpentine leaf miner. *Proc. Fla. St. Hortic. Soc.* 71: 404–406.

Wooldridge, A. W., and F. P. Harrison. 1968. Effects of soil fertility on abundance of green peach aphids on Maryland tobacco. *J. Econ. Entomol.* 61: 387–391.

Yein, B. R., and H. Singh. 1982. Effect of pesticides and fertilizer on the population of whitefly and incidence of yellow-mosaic virus in greengram. *Indian J. Agric. Sci.* 52: 852–855.

Yoshihara, T., K. Sogawa, M. D. Pathak, B. O. Juliano, and S. Sakamura. 1980. Oxalic acid as a sucking inhibitor of the brown planthopper in rice (Delphacidae: Homoptera). *Entomol. Exp. Appl.* 27: 149–155.

Zeng, Y. L., P. Y. Gong, L. R. Jiang, and M. L. Zhang. 1982. Effects of nitrogen fertilizer application on the cotton plant and the bollworm (in Chinese). *Acta Entomol. Sin.* 25: 16–23.

Zitzman, A., Jr. 1984. Feeding ecology and nitrogen and lipid composition of the Colorado potato beetle, *Leptinotarsa decemlineata* (Say), as a function of the nitrogen supply of its host plant, the potato plant. M.S. thesis. Rutgers, The State University, New Brunswick, NJ.

Zwolfer, W. 1957. Studien zur Waldbodenkleinfauna. *Forstwiss. Zentr.* 76: 65–128.

3

HOST PLANT SUITABILITY IN RELATION TO WATER STRESS

Thomas O. Holtzer

Department of Entomology
University of Nebraska
Lincoln, Nebraska

Thomas L. Archer

Texas A & M University System
Agricultural Research and Extension Center
Lubbock, Texas

John M. Norman

Department of Agronomy and
Center for Agricultural Meteorology and Climatology
University of Nebraska
Lincoln, Nebraska

1. INTRODUCTION

Water, available in the necessary amounts at the appropriate times, is crucial to the very lives of plants. All aspects of growth, development, and reproduction are directly or indirectly dependent on suitable water relationships. The influence of water status is through biochemical, physiological, and physical routes within the plant. In addition, through these same pathways, the plant's water status can influence its immediate external environment. Thus lack of sufficient water has the potential to alter insect–plant relationships in myriad interacting, dynamic ways. As a result, predicting the effects of plant water deficits on insect population growth is far from straightforward.

At the trivial extreme, plant death from water deficit usually means the elimination of local populations of insect herbivores. More interesting are of course the subtle influences that less severe water stress, occurring on mesophytes, has on the insects associated with the plants. There are many studies (e.g., Slosser 1980, Mellors and Propts 1983, Mellors et al. 1984, Santos, 1984, Stanton and Cook 1984, Perring et al. 1986) indicating relationships between indicators of plant water status and an insect response. Unfortunately, the causal relationships leading from water stress to the insect response are very difficult to disentangle and frequently remain obscure.

Listed in Table 1 are some factors that are important influences on insect–plant relationships. In subsequent sections we will focus on water stress in relation to these factors and attempt, where possible, to examine the mechanisms through which plant water stress influences insect herbivores.

2. PLANT ENVIRONMENTS AND THE PROCESS OF WATER STRESS DEVELOPMENT

The effects of water stress are complex, in part because the development of water stress is itself a complex phenomenon. Hsiao et al. (1976) stated

Table 1. Some Plant Factors That Affect Associated Insects

Nutritional quality and quantity
Allelochemical availability
Cell division and expansion
Canopy and root system structure
Partitioning of resources between vegetative growth and reproduction
Timing of senescence
Water potential
Microenvironment
Third trophic level interactions

that terrestrial plants in natural environments are rarely free of water stress for more than a few days. This pattern can occur because water loss due to transpiration exceeds absorption during some part of nearly every day (Kramer 1983). Greater deficits develop under conditions that favor transpiration (high air temperatures, bright sun, and strong winds) and/or reduced root water uptake (dry soils, low soil temperatures, and poor soil aeration). Typically, plants experience diurnal cycles in which water deficits develop in the late morning and afternoon and may be relieved during the evening and at night when transpirational losses decrease. These transient midday deficits frequently occur even when there is no adverse depletion of soil moisture. However, if soil moisture is not replenished in a timely manner, a pattern of long-term water deficit develops (Fig. 1). Characteristically the

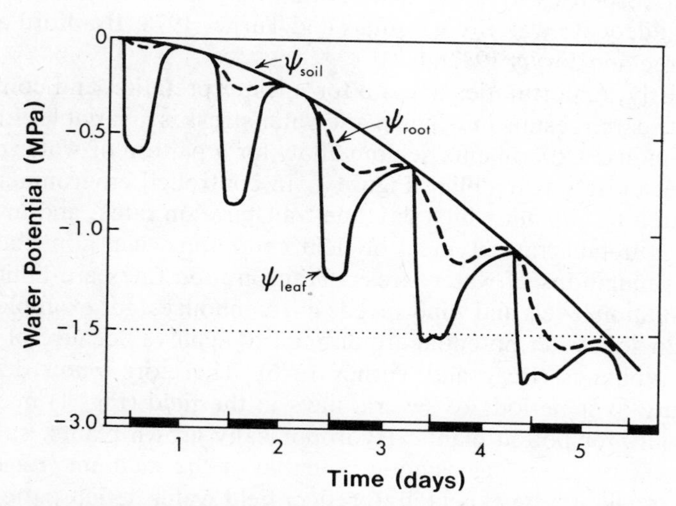

Figure 1. Diagram showing probable changes in leaf water potential (ψ_{leaf}) and root water potential (ψ_{root}) of a transpiring plant rooted in soil allowed to dry from a near-zero water potential to a water potential at which wilting occurs. The dark bars indicate night. (From Slatyer 1967, Kramer 1983)

water potential in the leaves (ψ_{leaf}) is low during the day but recovers to the potential of the soil (ψ_{soil}) at night. At some point the daytime leaf water potential may fall below the wilting point (day 3 in Fig. 1). Permanent wilting eventually occurs when the soil water potential goes below the wilting point (day 5) (Kramer 1983, pp. 344–347).

Although Figure 1 illustrates a reasonable pattern of water stress development, field conditions are frequently even more complex. For example, the five-day drying cycle shown in Figure 1 could easily be extended by light rainfall on the third day. If soil moisture depletion plateaued at a water potential of about −1 MPa (megapascal) for several days, the relationships illustrated doubtless would change. For example, internal compensation may permit ψ_{leaf} to fall below −1.5 MPa before wilting occurs, and root growth stimulated by more gradual water stress development may permit both ψ_{root} and ψ_{leaf} to remain closer to ψ_{soil} during diurnal cycles.

The effects of differences in the patterns of water stress development on the factors in Table 1 remain largely unresolved. This is in part because the problems associated with research in this area are formidable. Even the effects of recurring, transient, midday deficits on a variable as well studied as crop yield are not clearly established (however, losses are thought to be substantial). Nevertheless, some responses to water stresses, such as changes in the availability of allelochemicals and nutrients, are very likely affected by the pattern of the development of stress (e.g., proline concentrations) (Hanson and Hitz 1982). Additionally, available data suggest that plants experiencing a period of water deficit acclimate in complex ways that alter their responses to water deficits that may occur after an intervening period of adequate water (e.g., Jones and Turner 1978, Bradford and Hsiao 1982, Tyree and Jarvis 1982).

Obviously, opportunities abound for misinterpretation and confusion regarding research results in which plant water stress is a variable. Laboratory and greenhouse experiments seldom allow for a pattern of water deficit development closely resembling Figure 1. In controlled environments, problems related to obtaining high daytime transpiration rates, and to obtaining root water uptake rates typical of field conditions, can contribute to unrealistic simulations of water stress. Transpiration rates are limited by reduced radiation levels and wind speeds in greenhouses, for example. Gradual changes in soil water potential are difficult to achieve because of small soil and root volumes (Begg and Turner 1976). Therefore water deficits that accumulate over periods of several days in the field (Fig. 1) may develop within hours for potted plants. Hydroponically grown plants, subjected to water stress by adjusting osmotic potential of the medium (Lagerwerff et al. 1961) would not be expected to reflect field water deficit patterns. Also, minor differences in protocols for handling plants can result in substantial differences in plant growth and development that may alter the effects of water stress (Jaffe and Biro 1979, Jaffe and Telewski 1984).

On the other hand, field experiments present difficulties related to precise manipulation or even precise knowledge of the relevant variables.

3. PLANT WATER POTENTIAL AND ITS DIRECT EFFECTS ON INSECT HERBIVORES

Total water potential of a plant ψ_T can be approximated as

$$\psi_T = \psi_P + \psi_S \tag{1}$$

where ψ_P = hydrostatic or pressure potential inside cells or tissues (turgor potential), and ψ_S = potential due to solutes (osmotic potential). Negative water potential indicates that work must be done to extract water from the plant.

Typically, ψ_S is negative and ψ_P is positive when the plant is under turgor. In a fully turgid plant ψ_T is zero and may fall to -2 MPa at wilting (Fig. 2) (Nobel 1983, Kramer 1983). As ψ_T decreases, its components change relative to each other. Two factors are responsible: (1) decreasing water volume and decreasing plant cell volume passively increase the concentration of solutes, and (2) many plants actively add solutes (osmoregulate) as ψ_T decreases—perhaps because growth, stomatal opening, and other crucial plant processes are critically dependent on ψ_P. Osmotic adjustment drives ψ_S further neg-

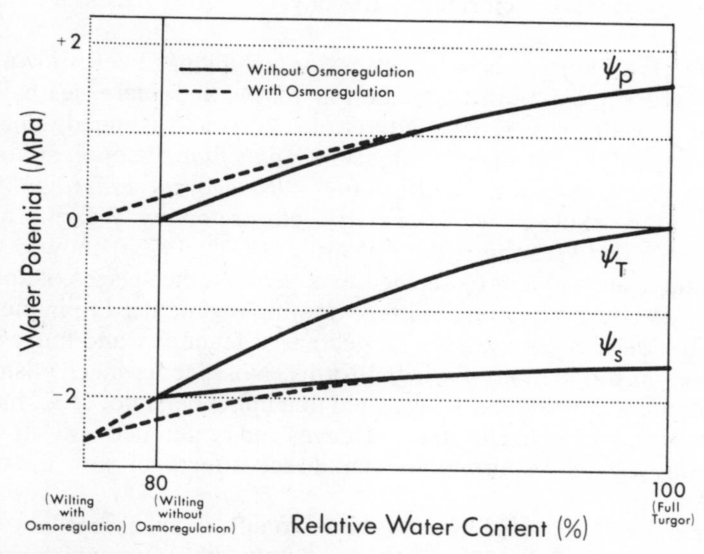

Figure 2. Diagram showing probable changes in pressure potential (ψ_P), solute potential (ψ_S), and total potential (ψ_T) for plants with and without osmoregulation.

ative and tends to maintain ψ_P despite decreasing ψ_T (dashed lines in Fig. 2).

The components of water potential vary from cell to cell, so bulk tissue estimates made with psychrometers or pressure chambers may not represent all individual cells. In particular, component potentials in phloem and xylem cells can be much different from bulk tissue. In xylem, ψ_S is very small (typically < 0.01 MPa) and ψ_P is negative (i.e., near ψ_T). On the other hand, ψ_S for phloem can be very strongly negative, and ψ_P is positive. These striking differences in the components of water potential among phloem, xylem, and bulk cells can occur despite their close physical relationships (Nobel 1983).

Although a variety of solutes are more abundant in water-stressed plants, the specific role and importance in osmotic regulation of the individual solutes and the mechanisms leading to their increase are not well understood (Begg and Turner 1976, Stewart and Larher 1980). Among the solutes that have been shown to increase are inorganic ions, organic acids, soluble carbohydrates, soluble protein, various amino acids and other nitrogen-containing compounds (especially proline and glycinebetaine), and their sulfur analogues (Kramer 1983, Wyn Jones 1984). Clearly, changes in the amounts of some of these solutes could have powerful influences on insect herbivores (see Section 4 for discussion).

Despite the ameliorating influence of osmoregulation, ψ_P in phloem cells decreases to zero under severe water stress. The direct effects of lowered ψ_P are likely to be most consequential to aphids, leafhoppers, and other "sucking" insects. For them, hydrostatic pressure potential must be considered a key variable determining the physical effort necessary to obtain food.

The effects of host plant water stress on aphids has been shown empirically to be complex. Both population increases and decreases have been attributed to water stress. Kennedy et al. (1958) and Kennedy and Booth (1959) attributed the detrimental effects on *Aphis fabae* Scopoli of host plant water stress in part to low ψ_P. Results for *Brevicoryne brassicae* (L.) and *Myzus persicae* (Sulzer) were similarly interpreted by Wearing and van Emden (1967) and Wearing (1967). Wearing (1972a, 1972b) showed that the relative importance of low ψ_P seemed to depend on the species of aphid, the age of leaves being fed upon, and the watering regime used to induce water stress. Low ψ_P was associated with decreased fecundity and longevity and an increase in production of winged forms. No specific mechanisms were identified, but the researchers suggested that aphid behavior (e.g., increased restlessness when feeding on stressed leaves and preference for well-watered leaves) may be as important as the direct effects of low ψ_P on food acquisition.

Host plant water stress has been shown to influence susceptibility of pine to bark beetles (Coulson 1979). One factor is oleoresin flow which is reduced by low ψ_P of the epithelial cells lining the resin ducts. Low oleoresin pressures reduce the resistance of trees to the early establishment phase of bark beetle attack (Vite 1961).

Recent advances in methodologies for measuring ψ_T, ψ_S, and ψ_P (reviewed by Kirkham 1985) and for manipulating water stress have provided new opportunities for research. For example, Sumner et al. (1983) developed a method for studying the interaction of caged greenbugs, *Schizaphis graminum* (Rondani), with hydroponically grown wheat subjected to various levels of water stress. They simulated drought stress by decreasing ψ_S of the hydroponic medium by adding polyethylene glycol (a somewhat controversial technique (Lawlor 1970) developed for use in plant water stress research (Lagerwerff et al. 1961, Jackson 1962, Begg and Turner 1976) but not previously adapted to insect – plant research). Aphid longevity and fecundity showed a nonlinear response, declining sharply to polyethylene glycol concentrations sufficient to cause ψ_S of the hydroponic medium to be -0.75 MPa or less.

Although techniques for measuring and manipulating plant water stress continue to be developed, many are subject to considerable experimental error, and interpreting results can be challenging. These limitations underscore the need for careful design of experiments and for making use of insights derived from multiple arenas (e.g., laboratory, greenhouse, field, and computer simulations) in developing theoretical explanations of the effects of water stress.

4. EFFECTS ON INSECT HERBIVORES OF CHANGES IN PLANT METABOLISM INDUCED BY WATER STRESS

From the standpoint of insect herbivores, water stress induces a number of changes in plant metabolism that have large potential impacts. Increases in plant metabolites may affect insects at the nutritional–physiological level, or they may alter behavioral responses such as attraction to plants, length of time spent feeding, and preference for specific plants or locations on plants (Bernays and Chapman 1978, Rhoades 1979).

In this section we discuss metabolites generally regarded as insect nutrients, plant defense chemicals, and plant growth regulators.

4.1 Insect Nutrients

4.1.1 Nitrogen

Organic nitrogen (proteins or amino acids and related compounds) availability has been shown to be a critical, if not limiting, factor in the population growth of many insect hervibores (see reviews by McNeill and Southwood 1978, Mattson 1980, White 1978 and 1984, Scriber 1984). However, the situation is complicated by other studies that show insect populations remaining unchanged or decreasing in response to increased N (see Scriber 1984). One possible explanation for lack of responses is that in some plants the normal N level is so high that associated insects are insensitive to even substantial shifts. Miles et al. (1982a, b) present convincing evidence that this is the case for some sucking and chewing insects.

In addition to nutritional effects, nitrogen availability can be important at the behavioral level. The increase in attractiveness of water-stressed plants for grasshoppers was reported to be related to the phagostimulatory effect of increased levels of proline and valine (Haglund 1980). However, Bright et al. (1982) were unable to show a relationship between attractiveness and proline increase in barley. Changes in the availability of nitrogen in various plant parts would be expected to have a profound influence on within plant distribution of insect herbivores. For example, White (1984) observed that young lepidoptera larvae move to younger tissues on water-stressed plants.

For plants, nitrogen metabolism is of course an integral part of growth, development, and reproduction, and it influences and is influenced by many other factors. Nitrogen metabolism is among the processes most sensitive to water deficits (Hsiao et al. 1976). The effects of water deficit on nitrogen metabolism and the nitrogen economy of plants depends on the nature of the nitrogen economy of the plant species and on the timing of water stress relative to other events in the plant's life history (Hanson and Hitz 1983). The effects of water deficit on the nitrogen economy of annual crop plants is perhaps better known than that of other subsets of plants. In their review Hanson and Hitz (1983) recognized two phases of the nitrogen economy of annual crops: phase one dominated by acquisition of reduced N, and phase two dominated by reallocation of reduced N. Figure 3 illustrates the pattern of reduced N in a representative cereal and legume.

Water deficit during the acquisition phase in cereal crops is often accompanied by increased N concentrations in dry matter, provided enough soil moisture is present for roots to take up NO_3^-. Increase in N concentrations may result from N acquisition being less curtailed by water stress than is cell growth. In contrast to plants receiving N from soil NO_3^-, plants dependent on N_2 fixation by symbionts often show reductions in dry matter N concentration when they are water stressed during the acquisition phase. This decrease may result from direct or indirect inhibition of the symbionts. During the acquisition phase much of the reduced N is incorporated into the enzyme ribulose-1,5-bisphosphate carboxylase/oxygenase (Rubisco) which is central to carbon fixation (Hanson and Hitz 1982, 1983).

In the reallocation phase there is a net breakdown of leaf and stem proteins as leaves senesce. Within the leaves amino acid metabolism alters the relative concentration of various amino acids and the resulting amino and amino-related N is transported through the phloem and is subsequently incorporated into developing seeds. Water deficits during the reallocation phase seem to increase the net breakdown of protein, particularly of Rubisco, and

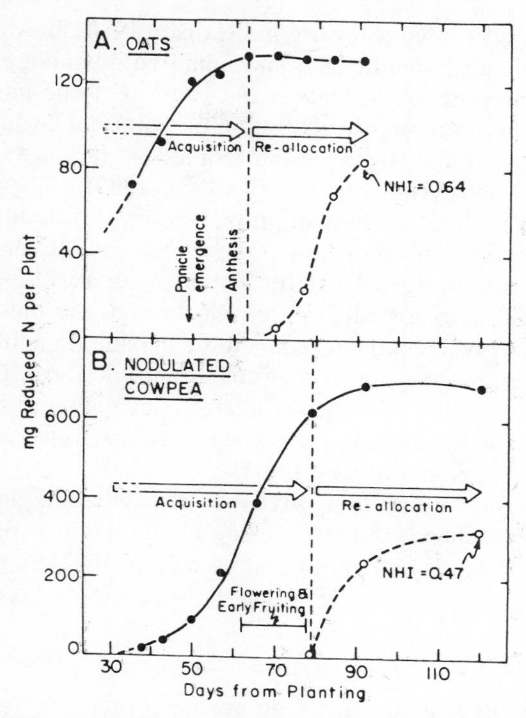

Figure 3. Accumulation of reduced N in the aboveground parts of (*A*) oats grown in soil and supplied NO_3–N and (*B*) nodulated cowpeas grown in N-free mineral nutrient medium. O—O: total reduced-N in aboveground parts; O- - - - -O: reduced-N in seeds. The broken vertical lines separate the phases of N-acquisition and reallocation. (From Hanson and Hitz 1983; reproduced from *Limitations to Efficient Water Use in Crop Production*, pp. 331–343, 1983 by permission of ASA-CSSA-SSSA)

the export of amino N out of the leaf via the phloem. Under sustained water stress net protein breakdown and amino acid metabolism continue, but phloem transport of N compounds out of the leaf diminishes. The result is higher concentrations of free amino acids (particularly proline) in water-stressed leaves as they approach the final stages of senescence (Hanson and Hitz 1983).

Proline accumulation is a widely observed response of plants exposed to a variety of environmental stresses (Stewart and Larher 1980, Hanson and Hitz 1982). Typically free proline accumulation in water-stressed plants accompanies net breakdown of protein. In addition to the free proline coming directly from the proline in protein that is being degraded, other amino acids are converted metabolically to proline (Stewart 1981). A variety of functions (e.g., osmotic adjustment; storage of N in a nontoxic, readily available state for use when water stress is relieved; and protein stabilization) have been identified for proline (Aspinall and Paleg 1981, Wyn Jones 1984), and its biochemistry is well known (Stewart and Hanson 1980, Stewart 1981). However, the physiological significance and adaptive advantage of proline accumulation in water-stressed plants is uncertain. On one hand, proline accumulation may be primarily the symptom of water deficit injury, or it may be an adaptive metabolic response that confers a measure of water stress resistance or tolerance (Hanson and Hitz 1982, 1983).

In contrast to proline, accumulation of betaines (particularly glycinebetaine) in water-stressed plants is much more restricted taxonomically. Metabolically, it can be derived from the amino acids resulting from protein breakdown, and it is exported from leaves through the phloem. The most likely function for glycinebetaine seems to be in osmotic adjustment and for storage of N (from protein breakdown) in a nontoxic form. However, there is little evidence that the N in glycinebetaine is available to plants (as is the N in proline) for metabolic use once water stress is relieved (Hanson and Hitz 1982, Wyn Jones and Storey 1981).

Little is known regarding the activity of glycinebetaine in insects, but it likely can be metabolized, thus increasing the amino acid pool (McGilvery 1979). However, there is evidence (Corcuera et al. 1987) that glycinebetaine incorporated into defined, artificial diets is beneficial to greenbugs.

4.1.2 Carbohydrates

Water stress has important effects on carbohydrate metabolism in plants. These include changes in photosynthesis, translocation of photosynthate, partitioning of photosynthate between sugars and starches, rate of hydrolysis of starches, and respiration (Kramer 1983). The increase of soluble sugars is thought to play a role in osmotic adjustment in plants (see Section 3). However, there is little evidence that effects of water stress on carbohydrate availability play an appreciable direct role in insect population growth. In fact, carbohydrate availability typically is not dealt with in the insect–plant relationship and insect nutritional ecology literature (Southwood 1973, Harborne 1978, Scriber and Slansky 1981, Scriber 1984). However, there may be phagostimulatory (Bernays and Chapman 1978) or perhaps nutritional effects of increases in soluble sugars sometimes associated with the early stages of water stress.

The indirect role of changes in plant carbohydrate metabolism induced by water stress may be substantial. For example, nitrogen metabolism is closely tied to carbohydrate metabolism as is size and quantity of fruit and seeds (Kramer 1983). One consequence of these indirect effects may be that while the quality of available food may be enhanced for some insect herbivores because of the increased concentration of nitrogen, the quantity of food (biomass) available to the same or other herbivores is reduced (Hanson and Hitz 1982). The ecological implications of this trade-off have not received much attention.

4.1.3 Water Content

Water content of leaves is frequently used as an indicator of plant water stress and is related to leaf water potential (see Section 3). Potentially, low leaf water content also could limit insect growth through its direct nutritional role, particularly for chewing feeders. Scriber (1977) showed relationships between leaf water content and several aspects of larval growth of *Hyalophora cecropia* (L.) including nitrogen utilization. Reese and Beck (1978) and Schmidt and Reese (1987) showed that dietary moisture strongly influenced three nutritional indexes (approximate digestibility, efficiency of conversion of assimilated food, and efficiency of conversion of ingested food) for black cutworm fed artificial diets. Others (e.g., Soo Hoo and Fraenkel 1966, Waldbauer 1968, McClure 1980) also have reported reduced growth (not related to reduced consumption) of insects reared on low water content food. However, Tabashnik (1982) did not demonstrate a consistent, strong relationship between leaf water content and larval growth with the *Colias* butterfly complex. Chiang and Norris (1983) found a negative correlation between leaf water content (obtained in the lab) of several soybean varieties and population density of a leaf miner (in the field). Also Lewis (1984) found that the grasshopper *Melanoplus differentialis* (Thomas) preferred to feed on and performed better on wilted sunflower. These studies point out that although water may be a critically important nutrient, the effects of variation in leaf water content are situation dependent.

Bernays and Chapman (1978) reviewed the phagostimulatory aspects of water for acridoids. In general, water content affected food selection and amount eaten, and the water balance of the insect affected the degree of phagostimulation provided by food water content. Wolfson (1980) showed that leaf water content influenced oviposition choice in *Pieris rapae* (L.).

For aphids feeding on phloem sap, ψ_P (discussed in Section 3) is probably more important than leaf water content.

4.2 Defensive Chemicals

A wide variety of plant secondary metabolites have been shown to function in the chemical defense of plants against insect herbivores. Rosenthal and Janzen (1979) included chapters dealing with toxic amino acids, cyanogenic glycosides, alkaloids, toxic liquids, glucosinolates, terpenoids, saponins, phytohemagglutinins, proteinase inhibitors, flavonoids, tannins and lignins, and insect hormones and antihormones. Theoretical treatments of the defensive chemistry of plants against herbivory have been presented by Feeny (1976), Rhoades and Cates (1976), several authors in Rosenthal and Janzen (1979), Rhoades (1983), and Coley et al. (1985).

Plant defensive chemicals can be toxic to herbivores, or they can reduce the plant's digestibility (Rhoades and Cates 1976). Reese (1979, 1981) has stated that interactions between some defensive chemicals and nutrients can make nutrients unavailable even if they are present in large quantities. In addition to direct effects on herbivores, defensive chemicals have been shown to play a role in the effectiveness of natural enemies (Schultz 1983, Barbosa and Saunders 1985).

Gershenzon (1984) thoroughly reviewed the literature dealing with the effects of water stress (and nutrient stress) on the levels of plant secondary metabolites. Although he found relatively few studies dealing with the effects of water stress, he reached the conclusion that it does significantly alter the amounts of defensive compounds present in plants. In general, he found N and S containing defensive chemicals (cyanogenic glycosides, glucosinolates, and alkaloids) increased under water stress, while terpenoids increased in water-stressed herbs and shrubs and decreased in trees. No pattern emerged for phenolic compounds. Depending on the plant species, the compound, and the situation, responses to water stress included increased levels, decreased levels, and no change. Other data (Miles et al. 1982a, Cates et al. 1983, Louda et al. 1987) are consistent with these general conclusions. However, Gershenzon (1984) correctly drew attention to the large gaps in the knowledge base used to make generalizations about the effects of water stress on levels of secondary compounds. The lack of data is particularly evident for herbs and grasses.

Increases in certain defensive chemicals in water-stressed plants have several possible explanations. First, increased accumulations may be the incidental result of other changes induced by water stress. For example, low ψ_P leads to very rapid cessation of cell expansion. If the metabolism of secondary compounds remained relatively unaffected, concentrations would increase (Gershenzon 1984). This argument is similar to that summarized by Hanson and Hitz (1983) for accumulation of proline. That is, the data do not clearly indicate whether the accumulation of a plant metabolite during stress is merely the result of injury to some aspect of the metabolic system or whether it is an adaptive response to the stress.

Second, under water stress, the accumulation of less costly defensive compounds may be favored over the accumulation of more costly compounds in plants that utilize more than one chemical defense (Rhoades 1979). Thus within individuals, some defensive chemicals would show increased concentrations.

Third, changes induced by water stress may alter the relative costs and benefits of chemical defenses in a variety of ways that combine to favor the accumulation of some kinds of compounds. For example, the accumulation of proline, other amino acids, and other N and S containing small molecules in water-stressed plants (Section 4.1) may lower the cost of producing N and S containing defensive chemicals. In other words, the total cost of producing defensive chemicals is lowered because inputs that normally have a high cost are accumulating as a result of water stress. A second example is related to the argument of Seigler (1977) and Seigler and Price (1976) that the cost of synthesizing N and S containing secondary chemicals, attributable to defense, is less than the total cost because the chemicals also function as storage sites for N and S. In water-stressed plants, where N and S containing small molecules are accumulating, the need for additional storage sites is increasing. Thus the cost of synthesis attributable to defense may be still further diminished.

At the same time the benefit of defending leaves with high accumulations of proline and related compounds would be relatively greater. Not only is proline a valuable resource that could be utilized by the plant when stress is relieved; but left undefended, accumulations of proline could serve as a resource for insect herbivores. Such resources may permit herbivore populations to grow to levels that would continue to be devastating to the plant even after water stress was relieved.

4.3 Plant Growth Regulators

There is increasing evidence for the importance of plant growth regulators (abscisic acid, indoleacetic acid, cytokinins, ethylene, and gibberellins) in the response of plants to water stress (Aspinall 1980, Milborrow 1981, Hsiao and Bradford 1983, Kramer 1983). For example, abscisic acid increases in water-stressed plants alter stomatal behavior, root permeability to water, ion transport, and growth. Work by Visscher-Neumann (1980, 1982) and Chrominski et al. (1982) indicates that several species of grasshoppers and crickets respond to plant growth regulators applied to their food plants. Depending on the growth regulator, increased concentrations either increased or decreased longevity and reproduction. Although neither the details of the sensitivity of insects to plant growth regulators nor the taxonomic ubiquity of the phenomenon are known, these relationships may turn out to play an important role in the overall response of insect herbivores to water stressed plants.

4.4 Additional Considerations

The effects of water stress on plant metabolic processes clearly can have important consequences for associated insects. Although certain trends in the nature of these relationships seem recognizable, the role of metabolic changes escapes firm generalization. Examples that run counter to the general tendencies are abundant. More information from studies of insect nutrition on well-defined diets where levels of constituents can be varied independently would be especially valuable in establishing a more complete understanding of the role of plant metabolites in the relationship between plant water stress and insect performance.

5. THE MICROENVIRONMENT OF INSECT HERBIVORES AND THE INFLUENCE OF PLANT WATER STRESS

Population growth of insect herbivores is critically dependent, in numerous interacting ways, on environmental conditions such as temperature, humidity, availability of free water, and wind (Wellington and Trimble 1984). Of course, it is the microenvironment, adjacent to individual organisms, that is most relevant (Willmer 1982). For example, the important environment for small arthropods such as spider mites may be within a few hundred microns of the leaf surface. In this microenvironment the relevant temperature may be the leaf temperature. The humidity in this microenvironment falls between 100% (the humidity at the leaf surface) and canopy humidity (the humidity a few millimeters away from the leaf). If the mites are on webbing, then the canopy air conditions may be most appropriate. The soil surface conditions may be most important for a carabid beetle, whereas for a large European corn borer larva, the temperature inside the corn stalk may be most important. Corn rootworm larvae experience the temperature conditions in the upper layers of the soil, and the temperature experienced by a leaf miner is that of the host plant leaf. A newly hatched caterpillar like *Papilio glaucus* L. may be near leaf temperature because of its small size and close proximity to the leaf, whereas a full-grown larva may be at canopy temperature.

Clearly, water stress experienced by a host plant can have considerable impact on the microenvironment experienced by herbivores. For example, long-term water stress can produce permanent changes in canopy structure, whereas transient water stress affects leaf temperature and humidity through changes in stomatal behavior. As with changes in metabolites (Section 4) microenvironmental influences may operate at several levels, including the physiological and behavioral (e.g., see Perring et al. 1984b, Myers 1985). Unfortunately, the microenvironment of many insect herbivores can be exceedingly difficult to measure. As a result we are often faced with two alternatives: (1) assume that a weather station near the field location, or perhaps miles away, represents the microenvironment of interest or (2) assert that even if the microenvironment of the organism differs from the weather station instantaneously, on the average, over 24 hours the two environments are similar. These assumptions obscure the importance of the host plant (and its water status) in determining the microenvironment of herbivores. In addition, because of the difficulty of testing these assumptions, results of their failure may be assigned to other causes, such as the inappropriateness of laboratory response functions for a particular organism when it is under field conditions. Information from micrometeorological studies can be most helpful in assessing whether or not the preceding assumptions are reasonable for a given situation such as a herbivore feeding on the leaves of a water-stressed plant.

5.1 Characterizing the Microenvironment of Insect Herbivores in the Soil–Plant–Atmosphere Continuum

Small herbivores that feed in or on leaves can experience a microenvironment dominated by the leaf. The microenvironment of larger insects is less dominated by individual leaves and, instead, is more closely related to canopy conditions. The leaf environment is determined by the energy exchanges of the leaf, and the canopy environment derives from the collective effect of all the leaves combined with atmospheric and soil exchanges. The energy exchanges of a leaf can be represented by the following energy budget equation:

$$R_n = T + Q \tag{2}$$

where R_n is the net radiation absorbed by the leaf, T is the transpiration, and Q is the sensible heat exchange (convective heat flux resulting from temperature differences) (Rosenberg et al. 1983). Net radiation depends strongly on the amount of sunlight; thus with more solar radiation, sensible heat and transpiration fluxes will be larger. In this context transpiration is a cooling process that depends on stomatal resistance, leaf size, and air temperature, wind speed, and the vapor pressure of water near the leaf. Sensible heat exchange depends on leaf size and on wind speed and air temperature within the canopy, near the leaf. Conditions within the canopy depend on wind speed, temperature, humidity, and radiation above the canopy as well as canopy structure and exchanges of energy and water at the soil surface. Leaf area per unit of ground area, or leaf area index (LAI), has a strong influence on canopy environment because of the attenuation of radiation and wind by the leaves. Further, leaf size affects leaf energy exchanges of water vapor and heat (Kramer 1983, Nobel 1983).

As a result of the processes occurring within the canopy and the soil, air temperature within the canopy can differ from air temperature a few meters above the canopy by several degrees or more. The vapor pressure of water in canopy air may differ from above-canopy air by a few tenths of kilopascals or even a kilopascal if the soil surface is wet and the air is dry. Leaf temperature may vary from 5° below to 10°C or more above canopy air temperature for larger leaves such as maize (Fig. 4). Leaves are hottest when the plant is water stressed, and transpiration is therefore minimal because of increased stomatal resistance. Relative humidity near the leaf is important because of its effect on evaporation. Very small leaves such as pine needles may remain very near to canopy air temperature under a wide range of wind, radiation, and humidity conditions. Within a canopy, leaves may vary widely in their environmental condition. Clearly, some will be sunlit and some shaded. Insects that prefer one environment over another may behave so as to develop in the most favorable environment. When this occurs, even if the average daily leaf temperature is near the average daily weather-shelter

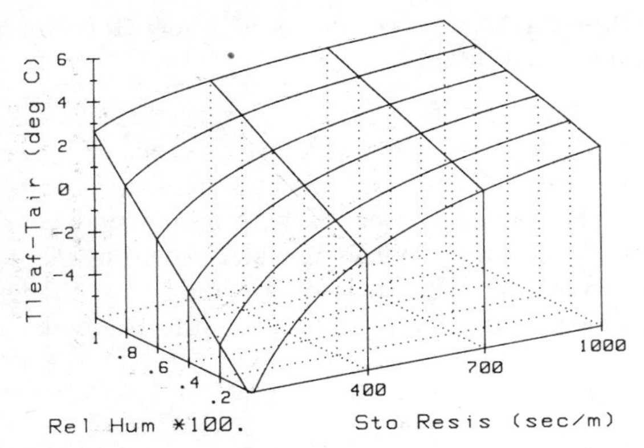

Figure 4. Response surface for $T_{leaf} - T_{air}$ as a function of stomatal resistance and ambient RH. (From Toole et al. 1984; reproduced by permission of the Entomological Society of America)

temperature, such a temperature may not be appropriate. Perhaps the mean temperature of only the sunlit leaves is a better estimate of the temperature actually experienced.

The departure of canopy conditions from weather-shelter conditions depends very much on wind speed. When wind speeds are low (less than 1 m/s) larger leaves can become hot in the daytime (particularly if water stress is present) and cool at night. If water stress is severe, canopy architecture may be altered by wilting or leaf rolling. This results in less solar radiation being intercepted by leaves. However, more radiation will penetrate deeper in the canopy, or even to the soil surface, where wind speeds are low. Thus temperatures deep in the canopy may be raised considerably. In addition, if the soil surface is dry, very high soil temperatures may occur.

The soil is of interest both because it can strongly influence the canopy environment and because some insects may spend some part of their life cycle there. The soil serves as a water storage reservoir for the plant. The deeper regions of this reservoir are tapped by the plant roots; the amount of water available to the plant depends on rooting depth and soil water-holding capacity, which depends on soil structure and texture. How long this reservoir of water can supply the plant depends on precipitation, evaporative demand, plant properties, and drainage out of the reach of roots. All of these factors affect the rate at which water stress might develop.

The top region of this soil reservoir can alter the temperature and humidity of the canopy space, depending on whether the surface regions are wet or dry. The soil surface energy exchanges, which affect the state of both the canopy space and the soil medium, are influenced by water content, soil texture, roughness, and surface reflectance.

Free water plays a key role in the soil–plant–atmosphere–herbivore system. Obviously, the plant uses water from the soil, and insect herbivores may get water from the soil or from plant tissues. However, another source of water is condensate on plant surfaces. This condensate may originate from the atmosphere as dew on leaves that are radiatively cooled at night, or it may come from a moist soil by distillation when the atmosphere is moderately dry (Monteith and Butler 1979, Norman 1982). Distillation occurs when convection transports moisture from the warm, moist soil (warmed from the daytime heat input) to cooler leaves in the upper canopy. Plant surfaces may also experience wetness from intercepted rainfall or irrigation water (Norman and Campbell 1983). Regardless of the source, the presence of water on leaves can produce two important effects: (1) it can delay the onset of transpiration and slightly reduce plant water use and plant temperatures, and (2) it can drastically alter the microenvironment where the free water has collected.

Many factors affect leaf wetness duration, and it remains one of the most difficult quantities to predict. If leaf wetness occurs because of irrigation or precipitation, then wetness duration depends mainly on the duration of the event, wind speed, temperature, relative humidity, incident radiation, leaf size, canopy structure, and whether water sits on the leaf as droplets or a thin film. If leaf wetness arises from dew at night, then radiative cooling, wind speed, relative humidity, temperature, and soil surface wetness are most important. Because dew usually occurs at lower temperatures than rain, the microenvironment of leaves wetted by dew is often cooler than rain-wetted leaves. The interactions of many environmental, soil, and plant factors makes predicting dew very difficult. For example, a high daytime wind speed could dry the soil surface faster than a low daytime wind speed so that even though nighttime conditions following the two days are identical, no distillation (or leaf wetness) would occur following the windy day.

Clearly, many factors influence the microenvironment immediately surrounding a particular insect, and broad generalizations are difficult to find. However, plant water stress has the potential to play a very significant role in determining microenvironmental conditions.

5.2 Comprehensive Plant–Environment Models

5.2.1 Development of a Comprehensive Model

In situations where the microenvironment can be expected to differ substantially from weather station conditions, comprehensive plant–environment models can be very useful for assessing the role of microenvironment in the population growth of herbivores. A special case of interest involves pests of mesophytic crops grown in semiarid regions. Despite irrigation these crops frequently experience transient water stress and may be subjected to moderate long-term stress, particularly when there is economic pressure to reduce production costs. Comprehensive models, although susceptible to error like any other research tool, can address a specific question using a broad spectrum of knowledge from disciplines such as meteorology, plant physiology, biochemistry, entomology, plant pathology, soil science, and physiological ecology. The use of the plant–environment model Cupid (Norman 1979 and 1982, Norman and Campbell 1983) to predict spider mite infestations is an example of a specific application (Toole et al. 1984). Cupid uses inputs including above-canopy weather conditions, initial soil conditions, and site variables and predicts the detailed microenvironment of leaves making various angles to the sun, at various depths in the canopy. The energy balance of individual mites can then be solved to estimate their body temperature regardless of whether they are on the upper or lower surfaces of leaves or whether they are in the sunlight or shade (Norman et al. 1984). Using these mite temperature distributions, a distributed delay, dynamic, mite population submodel is run to estimate mite development rates, fecundity, and longevity. Each developmental stage (egg, protonymph, deutonymph, preovipositional adult, and reproductive adult) has its own temperature and humidity response functions determined in the lab (Perring et al. 1984a, b). Mite population density is then estimated from the distributed delay model. The mites, in turn, damage the leaves and, when they reach some density, move to adjacent leaves. Stomatal and cuticular resistances of the leaves are adjusted in response to mite damage. In general, cuticular resistance is reduced and stomatal resistance increased, perhaps by as much as an order of magnitude. Increases in stomatal resistance cause leaf temperatures to rise and this raises the temperature that the mites are exposed to.

Because of the comprehensiveness of Cupid and the mite submodel, it is possible to account for the effects of plant water stress on the microenvironment of the mites. Thus the models are useful for studying spider mite population growth on plants subjected to conditions that induce plant water stress. Figure 5 is an example of predicted versus measured mite populations for well-watered and water-stressed maize.

5.2.2 Simulation Studies

Irrigation can relieve water stress and can directly change the environment of the canopy. As a result irrigation generally cools the environment within the canopy. If air temperatures are high, this cooling can reduce mite development rates and retard damage. A special high speed center pivot irrigation system, which disperses 1 mm of water over 130 acres in three hours, was simulated with Cupid (Barfield and Norman 1983). Results were compared to a simulation in which the same total amount of irrigation water was applied but was concentrated during one-hour periods every three days (Fig. 6). The simulations predict that mite populations would be significantly lower when small irrigation amounts were applied frequently.

Simulation studies with Cupid and comparisons to field data described earlier clearly indicate that the microenvironmental effects of plant water

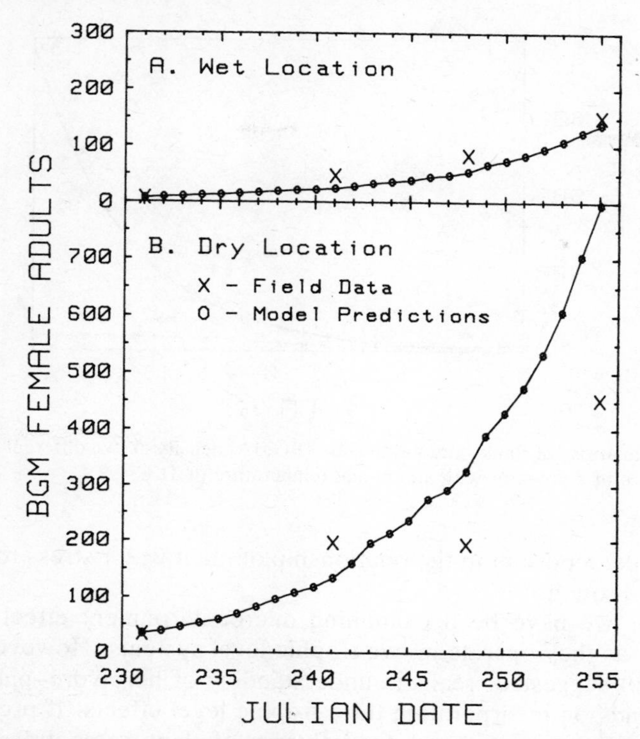

Figure 5. Comparison of model simulation to field data for the Banks grass mite (BGM) on maize in (*A*) wet and (*B*) dry locations. (From Toole et al. 1984; reproduced by permission of the Entomological Society of America).

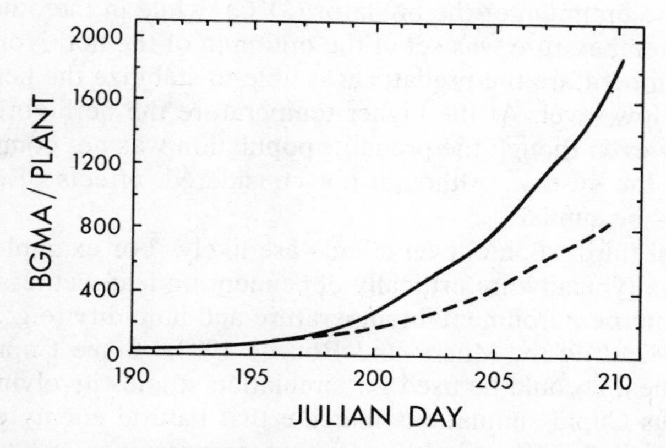

Figure 6. Simulations of Banks grass mite adult density (BGMA) in which the same 21-day total irrigation water was applied at either 3-hour (---) or 3-day (——) intervals.

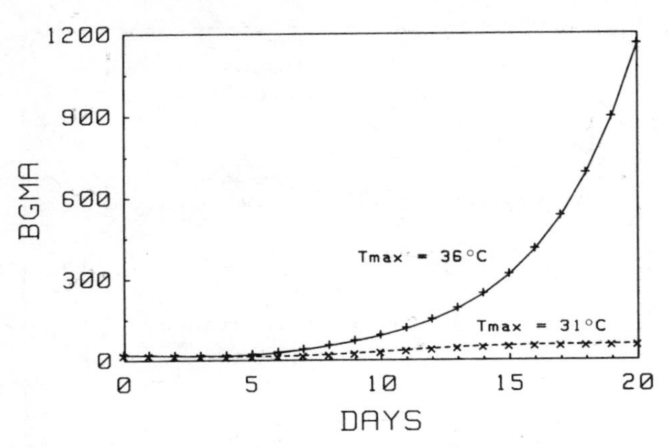

Figure 7. Simulations of Banks grass mite adult (BGMA) density at two different temperatures in the presence of a predator with an optimal temperature of 31°C.

stress can be important in the relationship of plant water stress to herbivore population growth.

Thus far we have been examining microenvironment effects of water stress only as they operate in two trophic level systems. However, as Price et al. (1980) suggested, realistic understanding of herbivore–plant interactions depends on incorporating third trophic level effects. If predators and pathogens also are considered, the influence of plant water stress, operating through effects on microenvironment, may be even more pronounced. For example, predators and herbivores may have different response functions to temperature. This may make the predators more effective at some temperatures than at others (Wollkind and Logan 1978, Logan 1982). Figure 7 shows trends in number of herbivorous mites for simulations of a system that includes a predatory mite. In one simulation maximum daily temperature was set at the optimum of the predator (31°C), while in the other, the maximum daily temperature was set at the optimum of the herbivore (36°C). At the lower temperature the predator was able to stabilize the herbivore population at a low level. At the higher temperature the herbivore population grew rapidly even though the predator population was not completely eliminated from the system. Although not considered, effects of different humidities may be similar.

Additional third trophic level effects are likely. For example, pathogens of herbivores typically are critically dependent on leaf wetness duration as well as on microenvironmental temperature and humidity (e.g., Millstein et al. 1982, 1983a, 1983b; Milner and Bourne 1983). Since Cupid calculates these variables, it could be used for simulation studies involving pathogens as well. Thus Cupid simulations indicate that natural enemy effectiveness could be decisively influenced by plant water stress.

Simulation studies of the corn plant–herbivore mite–predator mite have been encouraging. Without this experience, we might have doubted that a model with so many equations and variables would work. The reason that such a complex model functions as well as it does is not so much an accomplishment in model building as it is a testimony to the validity of the wealth of basic research information that has been accumulated by many scientists over many years.

6. EPILOGUE

Quite clearly, a complete understanding of the effects of plant water stress on insect herbivores is beyond the current level of our knowledge. There are many areas where we know little, and there are some areas where the available evidence seems inconsistent. Often cause and effect relationships are blurred by complexity and intercorrelations among factors. What we can say with certainty is that an unobscured view will continue to elude us if our research is aimed primarily at individual processes and factors. The influences of water potential, nutrients, chemical defenses, other metabolites, and microenvironment need to be considered simultaneously at various levels of integration. Even though additional research on each component would be welcomed, the main research emphasis must be placed on understanding the components as they interact to form a whole system.

In addition the focus must be expanded to include third trophic level considerations, interactions with other environmental stresses, feedback of insect herbivory on water stress effects, and interactions of water stress and herbivory as they affect plant dynamics at the population and community level.

Studies by Dorschner et al. (1986), Louda (1986), Louda et al. (1987), and Parker (1985) provide recent examples of the importance of the interacting synergistic effects of water stress with other stresses on plants at the physiological, population, and community levels. Their work also points out that progress toward a more complete understanding of the role of plant water stress will be accelerated by insights gained from the perspectives of scientists working in managed as well as more natural systems. Also obvious is the fact that interdisciplinary efforts and cooperative exchanges among research groups have great potential for hastening the development of a comprehensive understanding of water stress effects on insect herbivores.

ACKNOWLEDGMENTS

We wish to thank S. Chaffin, J. Kalisch, R. Newton, S. Spomer, M. Weidner, and M. Worth for assistance during manuscript preparation. We also wish to thank the following for their many helpful comments on earlier drafts: A. Joern, Z B Mayo, J. C. Reese, J. M. Scriber, and F. T. Wilhelms. The research of T. O. Holtzer and J. M. Norman was supported in part by the U.S. Department of Agriculture under grant 5901-0410-9-0345-0 from the Competitive Research Grants Office and Special Grant Agreement 84-CSRS-2-2515.

REFERENCES

Aspinall, D. 1980. Role of abscisic acid and other hormones in adaptation to water stress. *In* Turner, N. C., and P. J. Kramer (eds.), *Adaptation of Plants to Water and High Temperature Stress*. Wiley-Interscience, New York, pp. 155–172.

Aspinall, D., and L. G. Paleg. 1981. Proline accumulation: physiological aspects. *In* Paleg, L. G., and D. Aspinall (eds.), *The Physiology and Biochemistry of Drought Resistance in Plants*. Academic Press, Sydney, Australia, pp. 206–241.

Barbosa, P., and J. A. Saunders. 1985. Plant allelochemicals: linkages between herbivores and their natural enemies. *In* Cooper-Driver, G. A., T. Swain, and E. E. Conn (eds.), *Chemically Mediated Interactions Between Plants and Other Organisms. Rec. Adv. Phytochem.*, Vol. 19. Plenum, New York, pp. 107–137.

Barfield, B. J., and J. M. Norman. 1983. Potential for plant environment modification. *Agric. Water Manage.* 7: 73–88.

Begg, J. E., and N. C. Turner. 1976. Crop water deficits. *Adv. Agron.* 28: 161–217.

Bernays, E. A., and R. F. Chapman. 1978. Plant chemistry and acridoid feeding behavior. *In* Harborne, J. B. (ed.), *Biochemical Aspects of Plant and Animal Coevolution*. Academic Press, London, pp. 99–141.

Bradford, K. J., and T. C. Hsiao. 1982. Physiological responses to moderate water stress. *In* Lange, O. L., P. S. Nobel, C. B. Osmond, and H. Ziegler (eds.), *Encyclopedia of Plant Physiology. New Series*, Vol. 12. Springer-Verlag, New York, pp. 263–324.

Bright, S. W. J., P. J. Lea, J. S. H. Kueh, C. Woodcock, C. W. Hollomon, and G. C. Scott. 1982. Proline content does not influence pest and disease susceptibility of barley. *Nature* 295: 592–593.

Cates, R. G., R. A. Redak, and C. B. Henderson. 1983. Patterns in defensive natural product chemistry: Douglas fir and western spruce budworm interactions. *In* Hedin, P. A. (ed.), *Plant Resistance to Insects. ACS Symp. Ser.* 208. American Chemical Society, Washington, DC, pp. 3–19.

Chiang, H. S., and D. M. Norris. 1983. Morphological and physiological parameters of soybean resistance to agromyzid beanflies. *Environ. Entomol.* 12: 260–265.

Chrominski, A., S. Visscher-Neumann, and R. Jurenka. 1982. Exposure to ethylene changes nymphal growth rate and female longevity in the grasshopper. *Naturwissenschaften* 69: 45–46.

Coley, P. D., J. P. Bryan, and F. S. Chapin, III. 1985. Resource availability and plant anti-herbivore defense. *Science* 230: 895–899.

Corcuera, L. J., V. H. Argandona, and G. E. Zuniga. 1987. Resistance of cereal crops to aphids: role of allelochemicals. *In* Walker, G. R. (ed.). *Allelochemicals: Role in Agriculture and Forestry. ACS Symp. Ser.* 330. American Chemical Society, Washington, DC, pp. 129–135.

Coulson, R. N. 1979. Population dynamics of bark beetles. *Annu. Rev. Entomol.* 24: 217–246.

Dorschner, K. W., R. C. Johnson, R. D. Eikenbary, and J. D. Ryan. 1986. Insect–plant interactions: greenbug (Homoptera: Aphididae) disrupt acclimation of winter wheat to drought stress. *Environ. Entomol.* 15: 118–121.

Feeny, P. P. 1976. Plant apparency and chemical defense. *In* Wallace, J., and R. Mansell (eds.),

Biochemical Interactions Between Plants and Insects. Rec. Adv. Phytochem., Vol. 10. Plenum, New York, pp. 1–40.

Gershenzon, J. 1984. Changes in the levels of plant secondary metabolites under water and nutrient stress. *In* Timmermann, B. N., C. Steelink, and F. A. Loewus (eds.), *Phytochemical Adaptations to Stress. Rec. Adv. Phytochem.*, Vol. 18. Plenum, New York, pp. 273–320.

Haglund, B. M. 1980. Proline and valine—cues which stimulate grasshopper herbivory during drought stress? *Nature* 28: 697–698.

Hanson, A. D., and W. D. Hitz. 1982. Metabolic responses of mesophytes to plant water deficits. *Annu. Rev. Plant Physiol.* 33: 163–203.

Hanson, A. D., and W. D. Hitz. 1983. Whole-plant response to water deficits: water deficits and the nitrogen economy. *In* Taylor, H. M., W. R. Jordan, and T. R. Sinclair (eds.), *Limitations to Efficient Water Use in Crop Production.* American Society of Agronomy, Crop Science Society of America, and Soil Science Society of America, Madison, WI, pp. 331–343.

Harborne, J. B., (ed.). 1978. *Biochemical Aspects of Plant and Animal Coevolution.* Academic Press, London.

Hsiao, T. C., and K. J. Bradford. 1983. Physiological consequences of cellular water deficits. *In* Taylor, H. M., W. R. Jordan, and T. R. Sinclair (eds.), *Limitations to Efficient Water Use in Crop Production.* American Society of Agronomy, Crop Science Society of America, and Soil Science Society of America, Madison, WI, pp. 227–265.

Hsiao, T. C., E. Acevedo, E. Fereres, and D. W. Henderson. 1976. Stress metabolism: water stress, growth, and osmotic adjustment. *Phil. Trans. R. Soc. Lond.* 273: 479–500.

Jackson, W. T. 1962. Use of carbowaxes (polyethylene glycols) as osmotic agents. *Plant Physiol.* 37: 513–519.

Jaffe, M. J., and R. Biro. 1979. Thigmomorphogenesis: the effect of mechanical perturbation on the growth of plants, with special reference to anatomical changes, the role of ethylene, and interaction with other environemtnal stresses. *In* Mussell, H. W., and R. C. Staples, (eds), *Stress Physiology in Crop Plants.* Wiley-Interscience, New York, pp. 25–59.

Jaffe, M. J., and F. W. Telewski. 1984. Thigmomorphogenesis: callose and ethylene in the hardening of mechanically stressed plants. *In* Timmermann, B. N., C. Steelink, and F. A. Loewus (eds.), *Phytochemical Adaptations to Stress. Rec. Adv. Phytochem.*, Vol. 18. Plenum, New York, pp. 79–95.

Jones, M. M., and N. C. Turner. 1978. Osmotic adjustment in leaves of sorghum in response to water deficits. *Plant Physiol.* 61: 122–126.

Kennedy, J. S., and C. O. Booth. 1959. Responses of *Aphis fabae* Scop. to water shortage in host plants in the field. *Entomol. Exp. Appl.* 2: 1–11.

Kennedy, J. S., K. P. Lamb, and C. O. Booth. 1958. Responses of *Aphis fabae* Scop. to water shortage in plants in pots. *Entomol. Exp. Appl.* 1: 274–291.

Kirkham, M. B. 1985. Techniques for water use measurements of crop plants. *HortScience* 20: 993–1001.

Kramer, P. J. 1983. *Water Relations of Plants.* Academic Press, New York.

Lagerwerff, J. V., G. Ogata, and H. E. Eagle. 1961. Control of osmotic pressure of culture solutions with polyethylene glycol. *Science* 133: 1486–1487.

Lawlor, D. W. 1970. Absorption of polyethylene glycols by plants and their effects on plant growth. *New Phytol.* 69: 501–513.

Lewis, A. C. 1984. Plant quality and grasshopper feeding: effects of sunflower condition on preference and performance in *Melanoplus differentialis. Ecology* 65: 836–843.

Logan, J. A. 1982. Recent advances and new directions in phytoseiid population models. *In* Hoy, M. A. (ed.), *Recent Advances in Knowledge of the Phytoseiidae.* Publication 3284, Agricultural Science Publications, University of California, Berkeley, pp. 49–71.

Louda, S. M. 1986. Insect herbivory in response to root-cutting and flooding stress on a native crucifer under field conditions. *Acta Oecologica/Oecologia Generalis* 7: 37–53.

Louda, S. M., M. A. Farris, and M. J. Blua. 1987. Variation in methylglucosinolate and insect damage to *Cleome serrulata* (Capparaceae) along a natural soil moisture gradient. *J. Chem. Ecol.* 13: 569–581.

Mattson, W. J., Jr. 1980. Herbivory in relation to plant nitrogen content. *Annu. Rev. Ecol. Syst.* 11: 119–161.

McClure, M. S. 1980. Foliar nitrogen: a basis for host suitability for elongate hemlock scale, *Fiorinia externa* (Homoptera: Diaspididae). *Ecology* 61: 72–79.

McGilvery, R. W. 1979. *Biochemistry, A Functional Approach*. 2d ed. Saunders, Philadelphia.

McNeill, S., and T. R. E. Southwood. 1978. The role of nitrogen in the development of insect plant relationships. *In* Harborne, J. B. (ed.), *Biochemical Aspects of Plant and Animal Coevolution*. Academic Press, London, pp. 77–98.

Mellors, W. K., and S. E. Propts. 1983. Effects of fertilizer level, fertility balance, and soil moisture on the interaction of two-spotted spider mites (Acarina: Tetranychidae) with radish plants. *Environ. Entomol.* 12: 1239–1244.

Mellors, W. K., A. Allergo, and A. N. Hsu. 1984. Effects of carbofuran and water stress on growth of soybean plants and two-spotted spider mite (Acari: Tetranychidae) populations under greenhouse conditions. *Environ. Entomol.* 13: 561–567.

Milborrow, B. V. 1981. Abscisic acid and other hormones. *In* Paleg, L. G., and D. Aspinall (eds.), *The Physiology and Biochemistry of Drought Resistance in Plants*. Academic Press, Sydney, Australia, pp. 348–388.

Miles, P. W., D. Aspinall, and A. T. Correll. 1982a. The performance of two chewing insects on water-stressed food plants in relation to changes in their chemical composition. *Aust. J. Zool.* 30: 347–355.

Miles, P. W., D. Aspinall, and L. Rosenberg. 1982b. Performance of the cabbage aphid, *Brevicoryne brassicae* (L.), on water-stressed rape plants, in relation to changes in their chemical composition. *Aust. J. Zool.* 30: 337–345.

Millstein, J. A., G. C. Brown, and G. L. Nordin. 1982. Microclimatic humidity influence on conidial discharge in *Erynia* sp. (Entomophthorales: Entomophthoraceae), an entomopathogenic fungus of the alfalfa weevil (Coleoptera: Curculionidae). *Environ. Entomol.* 11: 1166–1169.

Millstein J. A., G. C. Brown, and G. L. Nordin. 1983a. Microclimatic moisture and conidial production in *Erynia* sp. (Entomophthorales: Entomophthoraceae), in vivo moisture balance and conidiation phenology. *Environ. Entomol.* 12: 1339–1343.

Millstein J. A., G. C. Brown, and G. L. Nordin. 1983b. Microclimatic moisture and conidial production in *Erynia* sp. (Entomophthorales: Entomophthoraceae): in vivo production rate and duration under constant and fluctuating moisture regimes. *Environ. Entomol.* 12: 1344–1349.

Milner, R. J., and J. Bourne. 1983. Influence of temperature and duration of leaf wetness on infection of *Acyrthosiphon kondoi* with *Erynia neoaphidis*. *Ann. Appl. Biol.* 102: 19–27.

Monteith, J. L., and D. R. Butler. 1979. Dew and thermal lag: a model for cocoa pods. *Quart. J. R. Meteor. Soc.* 105: 207–215.

Myers, J. H. 1985. Effect of physiological condition of the host plant on the ovipositional choice of the cabbage white butterfly, *Pieris rapae*. *J. Animal Ecol.* 54: 193–204.

Nobel, P. S. 1983. *Biophysical Plant Physiology and Ecology*. W. H. Freeman, San Francisco.

Norman, J. M. 1979. Modeling the complete crop canopy. *In* Barfield, B., and J. Geiber (eds.), *Modification of the Aerial Environment of Crops. Monogr. No. 2 ASAE*, American Society of Agricultural Engineering, St. Joseph, MI, pp. 249–277.

Norman, J. M. 1982. Simulation of microclimates. *In* Hatfield, J. L., and I. J. Thomason (eds.), *Biometeorology in Integrated Pest Management*. Academic Press, London, pp. 65–99.

Norman, J. M., and G. S. Campbell. 1983. Application of a plant–environment model to problems in irrigation. *In* Hill, D. (eds.), *Advances in Irrigation*. Academic Press, London, pp. 155–188.

Norman, J. M., J. L. Toole, T. O. Holtzer, and T. M. Perring. 1984. Energy budget for the Banks grass mite (Acari: Tetranychidae) and its use in deriving mite body temperatures. *Environ. Entomol.* 13: 344–347.

Parker, M. A. 1985. Grasshopper attack and water balance of the shrub *Gutierrezia microcephala*. *Amer. Midland Nat.* 113: 193–195.

Perring, T. M., T. O. Holtzer, J. L. Toole, J. M. Norman, and G. L. Myers. 1984a. Influences of temperature and humidity on pre-adult development of the Banks grass mite (Acari: Tetranychidae). *Environ. Entomol.* 13: 338–343.

Perring, T. M., T. O. Holtzer, J. A. Kalisch, and J. M. Norman. 1984b. Temperature and humidity effects on ovipositional rates, fecundity, and longevity of adult female Banks grass mites (Acari: Tetranychidae). *Ann. Entomol. Soc. Am.* 77: 581–586.

Perring, T. M., T. O. Holtzer, J. L. Toole, and J. M. Norman. 1986. Relationships between corn-canopy microenvironments and Banks grass mite (Acari: Tetranychidae) abundance. *Environ. Entomol.* 15: 79–83.

Price, P. W., C. E. Bouton, P. Gross, B. A. McPheron, J. N. Thompson, and A. E. Weis. 1980. Interactions among three trophic levels: influence of plants on interactions between insect herbivores and natural enemies. *Annu. Rev. Ecol. Syst.* 11: 41–65.

Reese, J. C. 1979. Interactions of allelochemicals with nutrients in herbivore food. *In* Rosenthal, G. A., and D. H. Janzen (eds.), *Herbivores: Their Interaction with Secondary Plant Metabolites*. Academic Press, New York, pp. 309–330.

Reese, J. C. 1981. Insect dietetics: complexities of plant–insect interactions. *In* Bhaskaran, G., S. Friedman, and J. G. Rodriguez (eds.), *Current Topics in Insect Endocrinology and Nutrition*. Plenum, New York, pp. 317–335.

Reese, J. C., and S. D. Beck. 1978. Interrelationships of nutritional indices and dietary moisture in the black cutworm (*Agrotis ipsilon*) digestive efficiency. *J. Insect Physiol.* 24: 473–479.

Rhoades, D. F. 1979. Evolution of plant chemical defense against herbivores. *In* Rosenthal, G. A., and D. H. Janzen (eds.), *Herbivores: Their Interaction with Secondary Plant Metabolites*. Academic Press, New York, pp. 4–54.

Rhoades, D. F. 1983. Herbivore population dynamics and plant chemistry. *In* Denno, R. F. and M. S. McClure (eds.), *Variable Plants and Herbivores in Natural and Managed Systems*. Academic Press, New York, pp. 155–220.

Rhoades, D. F., and R. G. Cates. 1976. Toward a general theory of plant antiherbivore chemistry. *In* Wallace, J., and R. Mansell (eds.), *Biochemical Interactions between Plants and Insects. Rec. Adv. Phytochem.*, Vol. 10. Plenum, New York, pp. 168–213.

Rosenberg, N. J., B. L. Blad, and S. B. Verma. 1983. *Microclimate, The Biological Environment*. 2d ed. Wiley-Interscience, New York.

Rosenthal, G. A., and D. H. Janzen (eds.). 1979. *Herbivores, Their Interaction with Secondary Plant Metabolites*. Academic Press, New York.

Santos, M. A. 1984. Effects of host plant on the predator–prey cycle of *Zetzellia mali* (Acari: Stigmaeidae) and its prey. *Environ. Entomol.* 13: 65–69.

Schmidt, D. J., and J. C. Reese. 1987. The effects of physiological stress on black cutworm (*Agrotis ipsilon*) larval growth and food utilization. *J. Insect Physiol.* 34: 5–10.

Schultz, J. C. 1983. Impact of variable plant defensive chemistry on susceptibility of insects to natural enemies. *In* Hedin, P. S. (ed.), *Plant Resistance to Insects. ACS Symp. Ser.* 208. American Chemical Society, Washington, DC, pp. 37–54.

Scriber, J. M. 1977. Limiting effects of low leaf–water content on the nitrogen utilization, energy budget, and larval growth of *Hyalophora cecropia* (Lepidoptera: Saturniidae). *Oecologia* 28: 269–287.

Scriber, J. M. 1984. Host-plant suitability. *In* Bell, W. J., and R. T. Carde (eds.), *Chemical Ecology of Insects*. Chapman and Hall, London, pp. 159–202.

Scriber, J. M. and F. Slansky, Jr. 1981. The nutritional ecology of immature insects. *Annu. Rev. Entomol.* 26: 183–211.

Seigler, D. S. 1977. Primary roles for secondary compounds. *Biochem. Sys. Ecol.* 5: 195–199.

Seigler, D., and P. W. Price. 1976. Secondary compounds in plants: primary functions. *Am. Nat.* 110: 101–105.

Slatyer, R. O. 1967. *Plant Water Relationships*. Academic Press, New York.

Slosser, J. E. 1980. Irrigation timing for bollworm management in cotton. *J. Econ. Entomol.* 73: 346–349.

Soo Hoo, C. F., and G. Fraenkel. 1966. The consumptions, digestion, and utilization of food plants by a polyphagous insect, *Prodenia eridania* (Cramer). *J. Insect Physiol.* 12: 711–730.

Southwood, T. R. E. 1973. The insect/plant relationship—an evolutionary perspective. *In* van Emden, H. F. (eds.), *Insect/Plant Relationships*. Wiley, New York, pp. 3–30.

Stanton, M. L., and R. E. Cook. 1984. Sources of intraspecific variation in the host-plant seeking behavior of *Colias* butterflies. *Oecologia* 61: 265–270.

Stewart, C. R. 1981. Proline accumulation: biochemical aspects. *In* Paleg, L. G., and D. Aspinall (eds.), *The Physiology and Biochemistry of Drought Resistance in Plants*. Academic press, Sydney, Australia, pp. 243–259.

Stewart, C. R., and A. D. Hanson. 1980. Proline accumulations as a metabolic response to water stress. *In* Turner, N. C., and P. J. Kramer (eds.), *Adpatation of Plants to Water and High Temperature Stress*. Wiley-Interscience, New York, pp. 173–189.

Stewart, C. R., and F. Larher. 1980. Accumulation of amino acids and related compounds in relation to environmental stress. *In* Miflin, B. J. (ed.), *The Biochemistry of Plants, A Comprehensive Treatise*, Vol. 5. *Amino Acids and Derivatives*. Academic Press, New York, pp. 609–635.

Sumner, L. C., J. T. Need, R. W. McNew, K. W. Dorschner, R. D. Eikenbary, and R. C. Johnson. 1983. Response of *Schizaphis graminum* (Homoptera: Aphididae) to drought-stressed wheat, using polyethylene glycol as a matricum. *Environ. Entomol.* 12: 919–922.

Tabashnik, B. E. 1982. Responses of pest and non-pest *Colias* butterfly larvae to interspecific variation in leaf nitrogen and water content. *Oecologia* 55: 389–394.

Toole, J. L., J. M. Norman, T. O. Holtzer, and T. M. Perring. 1984. Simulating Banks grass mite (Acari: Tetranychidae) population dynamics as a subsystem of a crop canopy-microenvironment model. *Environ. Entomol.* 13: 329–337.

Tyree, M. T., and P. G. Jarvis. 1982. Water in tissues and cells. *In* Lange, O. L., P. S. Nobel, C. B. Osmond, and H. Ziegler (eds.), *Encyclopedia of Plant Physiology. New Series*, Vol. 12B. Springer-Verlag, New York, pp. 35–77.

Visscher-Neumann, S. 1980. Regulation of grasshopper fecundity, longevity and egg viability by plant growth hormones. *Experientia* 36: 130–131.

Visscher-Neumann, S. 1982. Plant growth hormones affect grasshopper growth and reproduction. *In* Visser, J. H. and A. K. Minks (eds.), *Proc. 5th Int. Symp. on Insect–Plant Relationships*. Centre for Agricultural Publishing and Documentation, Wageningen, The Netherlands, pp. 57–62.

Vité, J. P. 1961. The influence of water supply on oleoresin exudation pressure and resistance to bark beetle attack in *Pinus ponderosa*. *Contrib. Boyce Thompson Inst. Pl. Res.* 21: 37–66.

Waldbauer, G. P. 1968. The consumption and utilization of food by insects. *In Adv. Insect Physiol.*, Vol. 5. Academic Press, New York, pp. 229–288.

Wearing, C. H. 1967. Studies on the relations of insect and host plant: II. Effects of water stress in host plants on the fecundity of *Myzus persicae* (Sulz.) and *Brevicoryne brassicae* (L.). *Nature* 213: 1052–1053.

Wearing, C. H. 1972a. Responses of *Myzus persicae* and *Brevicoryne brassicae* to leaf age and water stress in brussels sprouts grown in pots. *Entomol. Exp. Appl.* 15: 61–80.

Wearing, C. H. 1972b. Selection of brussels sprouts of different water status by apterous and alate *Myzus persicae* and *Brevicoryne brassicae* in relation to the age of leaves. *Entomol. Exp. Appl.* 15: 139–154.

Wearing, C. H., and H. F. van Emden. 1967. Studies on the relations of insect and host plant: I. Effects of water stress in host plants on infestation by *Aphis fabae* Scop., *Myzus persicae* (Sulz.) and *Brevicoryne brassicae* (L.). *Nature* 213: 1051–1052.

Wellington, W. G., and R. M. Trimble. 1984. Weather. *In* Huffaker, C. B., and R. L. Rabb (eds.), *Ecological Entomology*. Wiley-Interscience, New York, pp. 399–425.

White, T. C. R. 1978. The importance of a relative shortage of food in animal ecology. *Oecologia* 33: 71–86.

White, T. C. R. 1984. The abundance of invertebrate herbivores in relation to the availability of nitrogen in stressed food plants. *Oecologia* 63: 90–105.

Willmer, P. G. 1982. Microclimate and the environmental physiology of insects. *In* Berridge, M. J., J. E. Treherne, and V. B. Wigglesworth (eds.), *Adv. Insect. Physiol.*, Vol. 16. Academic Press, New York, pp. 1–57.

Wolfson, J. L. 1980. Oviposition response of *Pieris rapae* to environmentally induced variation in *Brassica nigra*. *Entomol. Exp. Appl.* 27: 223–232.

Wollkind, D. J., and J. A. Logan. 1978. Temperature-dependent predator–prey mite ecosystem on apple tree foliage. *J. Math. Biol.* 6: 265–283.

Wyn Jones, R. G. 1984. Phytochemical aspects of osmotic adaptation. *In* Timmermann, B. N., C. Steelink, and F. A. Loewus (eds.), *Phytochemical Adaptations to Stress. Rec. Adv. Phytochem.*, Vol. 18. Plenum, New York, pp. 55–78.

Wyn Jones, R. G., and R. Storey. 1981. Betaines. *In* Paleg, L. G., and D. Aspinall (eds.), *The Physiology and Biochemistry of Drought Resistance in Plants*. Academic Press, Sydney, Australia, pp. 172–204.

4

INFLUENCE OF TEMPERATURE-INDUCED STRESS ON HOST PLANT SUITABILITY TO INSECTS

J. H. Benedict

Texas A & M University
Corpus Christi, Texas

J. L. Hatfield

Cropping Systems Research
USDA-ARS
Lubbock, Texas

1. INTRODUCTION

Temperature is one of the most important physical factors of the environment, affecting the physiological and behavioral interactions of insects and plants. In the context of this chapter, "temperature-induced stress" is defined as a physical environmental factor that limits survival, growth, or reproduction of herbivores and their hosts. In relation to host plant suitability, temperature-induced stress is further defined as an external constraint to full genetic expression of plant morphological or biochemical defense mechanisms against herbivores (DiCosmo and Towers 1984). The study of how temperature-induced stress affects host plant suitability for insect herbivores is plagued with the nearly impossible task of separating temperature effects from free moisture and humidity effects. Water constitutes the major portion of living tissues and is the principal medium for biochemical reactions. The survival of plants and insect herbivores is dependent on their ability to maintain a suitable water balance.

Temperature-induced stress is a relative phenomenon, similar to the phenomenon of host plant resistance to insects, in that a given environmental temperature may be stressful to one organism's growth, reproduction, or defense, and near optimum for another organism. We will focus on investigations where temperature induced a change in the host plant and/or its insect herbivore. We will discuss possible mechanisms whereby temperature-induced stress can or does effect host plant suitability, and will identify meaningful areas for future research effort. Further we will indicate the relationship of temperature-induced stress to the interactions between ecosystem components, such as plant, insect herbivore, light, moisture, and pathogens.

Tingey and Singh (1980) proposed three mechanisms whereby temperature can induce changes in host plant suitability to insect herbivores. First, temperature-induced stress can cause changes in plant physiology that affect the expression of genetic resistance, resulting in changes in the levels of allelochemical and/or morphological defenses and/or nutritional quality of the host. Insect herbivores feeding on such temperature-stressed plants would have altered growth, development, reproduction, survival and/or behavior (i.e., increased or decreased antibiosis, and/or nonpreference effects). Second, temperature-induced stress can directly affect plant physiology, resulting in altered plant growth and development and thus changing plant response to insect injury. Therefore only the plant's response to insect damage is changed (i.e., increased or decreased tolerance). Third, temperature-induced stress can directly affect insect behavior and physiology, and thus change herbivore, growth, reproductive biology, and population dynamics. These three mechanisms of temperature-induced change in the host–plant–insect interrelationship should be thought of as factors that modify (increase or decrease) directly or indirectly, herbivore "capability" or "fitness" to utilize a food resource.

2. TEMPERATURE-INDUCED STRESS EFFECTS ON PLANT NUTRITION AND CHEMICAL DEFENSES

Plants are exposed to a wide range of temperatures diurnally and seasonally, however temperatures of foliage and/or other plant parts are not necessarily the same as the air temperature. Jackson (1982) reviewed the measurement of plant temperatures and reported that foliage temperature ranges from 10°C below ambient air temperature to as much as 5°C above it. There are also differences in foliage temperatures within the canopy. These differences between foliage temperatures and ambient air temperature are dependent on both net radiation and soil water availability. Foliage temperature is then a function of the size of the canopy, the distribution of radiation within the canopy, and the soil water status (Hatfield 1979). It is not possible to make generalizations about temperature effects at various canopy positions without an understanding of the temperature of that canopy strata. For example, Gamble and Burke (1984) showed that glutathione reductase activity was dependent on canopy position and canopy temperature distributions. These temperature relationships must be understood in order to develop a comprehensive understanding of the role of plant metabolites and temperature on insect response.

Levitt (1980) in reviewing the effects of cool and warm temperatures on plant physiology suggested three classifications of response: direct, indirect, and secondary stress injury. He found changes in concentration of secondary plant metabolites would occur as a result of increased plant anaerobic respiration under temperature stress. This effect he classified as indirect. Most of these secondary plant compounds are derived from the metabolic processes associated with sugar and amino acid metabolism. Klun (1974) described how certain secondary metabolites would affect insect response but did not provide an analysis of how these compounds were formed. He also suggested that these compounds would act as allomones (i.e., toxins) (Nordlund 1981) because they would cause the plant to be poisonous and distasteful to the insect. As shown by Swain (1977) and Schoonhoven (1981), those plant biosynthetic products involved as allomones and kairomones (Nordlund 1981) are secondary products. Thus their presence and concentration are dependent on the kinetics of enzymes within the plant tissue and as such are affected by temperature.

2.1 Low Temperature

Since secondary products are a result of amino acid and carbohydrate synthesis, it is important to realize how the biosynthesis of these products is affected by low temperatures. Graham and Patterson (1982) suggested that when plants are exposed to cool temperatures energy levels decrease as a result of decreases in ATP levels. This decrease in ATP leads to an increase in carbohydrate and amino acid levels due to the lack of energy available for the conversion of these products to secondary products. The two enzymatic reaction parameters which are temperature dependent are the reaction rate and the affinity constant, Km (Dixon et al. 1979). Both of these parameters respond to temperature changes to enable the plant to continue primary metabolic synthesis throughout a range of temperatures. The affinity constant, Km, of sweet potato, phenylalanine ammonium lyase for phenylalanine, has been shown to decrease at low temperatures, and it is coupled with an increase in concentration of phenolics around 0°C (Rhodes et al. 1979). These changes in phenolics could have an impact on the plant desirability as forage for insects. Graham and Patterson (1982) in reviewing the effect of chilling temperatures on proteins suggested that there may be changes in enzyme stability which may increase the affinity at low temperature, at the expense of decreased affinity at high temperatures. Thus we suggest that the biosynthetic response of plants may be, in part, temperature dependent, and that defensive products and their concentrations produced at one temperature, may not be produced at higher or lower temperatures.

Levitt (1980) showed that sugars accumulate in plant tissues when they are exposed to low temperatures. This accumulation is not linked to decreases in plant respiration since anaerobic respiration has been found to increase in cold tissue (Levitt 1980). Lyons (1973) found that as plants were chilled, the membrane lipids became more solidified, suggesting that there are complex interactions between the cell membrane and available water at chilling temperatures.

In summary, we can say that as the plant is exposed to increasingly lower temperatures, changes in metabolism occur that can impact on the synthesis of secondary products. Thus the biosynthesis of these secondary products may change through time in response to a decrease in temperature rather than solely in response to insect wounding as is sometimes assumed. It is difficult to make unique generalizations about plant metabolism across plant species as it has been shown that starch production response varies between species subjected to decreased temperatures (Downton and Hawker 1975). Studies on the effect of exposing plant canopies to low ambient temperatures show that it is difficult to derive unique biosynthetic curves for responses and concentrations of secondary compounds on the basis of temperature. Further the response of plant biosynthesis to low temperature is not always a single factor response (e.g., in maize leaves, irradiance interacts with low temperature to cause a loss in NADP-malate dehydrogenase pyruvate Pi dikinase and catalase activities) (Taylor et al. 1974). We conclude that environmental factors may interact in complex, little understood ways, to affect biosynthesis of secondary metabolites.

2.2 High Temperature

High temperatures cause several metabolic disruptions affecting the plant. Unfortunately, it is difficult to separate the effects of high temperatures from drought stress. If secondary compounds are linked to protein synthesis, then changes in secondary compounds may be related to changes in DNA and/or RNA expression resulting from temperature variation. Bernstam (1978), in a review of temperature effects, concluded that high temperature damage to DNA could be repaired, depending on the physiological condition of the cell. Levitt (1980) suggested that protein denaturation could occur in plants exposed to extremely high temperatures. Such denaturation could affect proteins and lipids that are associated with the cell membrane and involved with synthesis of allomones and kairomones. Bjorkman et al. (1980), in comparing a cool-adapted C_4 species (*Atriplex sabulosa*) and a warm-adapted C_4 species (*Tidestromia oblongifolia*), found that the photosynthetic rates of the two species differed. They suggested that the difference was due to several rate-limiting, temperature-dependent steps in the photosynthetic process. Enzymes that may become inactivated at high temperatures include NADP glyceraldehyde-3P-dehydrogenase, RUBP carboxylase, PEP carboxylase, RUSP kinase, and NADP malate dehydrogenase. Bernstam (1978) suggested that plants exposed to higher temperatures undergo changes in cell membranes resulting from changes in RNA and DNA.

Increased rates of metabolism occur in plants exposed to warm temperatures due to an increase in enzyme reaction rates. This increase leads to decreases in any temporary storage compounds (e.g., starches), and if the high temperatures are prolonged, increases may occur in amino acid concentrations due to further disruption of cellular function and cellular membrane function.

There are many changes that occur within the plant when it is subjected to abnormally high temperatures. It has been shown that exposure to unusually high heat reduces or totally blocks the expression of some genes and stimulates others. The gene products, which are present in heat-stressed plants, are termed HSP and are assumed to be related to plant mechanisms for survival or adaption. In a recent field study on heat shock proteins of leaves by Burke et al. (1985), it was found that HSP's accumulated in cotton leaves when leaf temperatures exceeded 37°C. The HSP polypeptides they found in field plants had weights of 21, 37, 58, 60, 75, 89, 94, and 100 KD. These proteins have not been investigated sufficiently to determine what, if any, role they play as secondary metabolites that protect plants against insects.

Research on lipids has shown that as temperatures increase, the saturation increases, with the primary change being a decrease in linolenic acid (Pearcy 1978). Utkhede and Jain (1976) have shown that concentrations of nucleic acids in wheat do not respond to changes in temperature. However, amounts of triglycerides and fatty acids in plants rise when temperatures increase. In general, there appears to be a large diversity in responses of secondary compounds to high temperatures, and further there tends to be increases in the plant products associated with these secondary metabolites. No direct evidence linking the biochemical pathways of the secondary compounds, and their concentrations, to temperature conditions in the field has been found.

2.3 Temperature Effects on Plant Nitrogen Status

Scriber (1984) in reviewing factors which affect host plant suitability suggested that nitrogen content of the plant is the primary factor determining insect response to plants. He stated that growth and reproductive success of insects depended primarily on the insect's ability to ingest and convert plant nitrogen efficiently and rapidly. Thus the suitability of plants as food sources would be directly linked to the environmental factors related to plant nitrogen status (see Chapter 2 for further discussion). Many allomones and kairomones are nitrogenous compounds or compounds with a nitrogen subunit. Thus any environmental factor affecting plant nitrogen status may affect the quantity and/or quality of secondary plant metabolites.

The physiological response of plants to nitrogen is dependent on whether the plant utilizes soil nitrogen or symbiotically fixed nitrogen. Temperature and water status of the leaves are often closely linked, and studies have shown that nitrate reductase activity declines with even moderate water stress. Hsiao (1973) has shown nitrate reductase activity to be similar to protein synthesis; thus we could expect that overall nitrogen metabolism would behave similarly to protein synthesis under cool or hot temperatures. Nitrogen metabolism is also tied to the activity of RUBP carboxylase which changes with plant age. Normal senescence patterns lead to a decrease of available nitrogen as the plant matures (Peterson and Huffaker 1975).

3. TEMPERATURE-INDUCED STRESS EFFECTS ON PLANT MORPHOLOGICAL DEFENSES

Morphological characteristics of a plant range from the aspects of a single leaf, to the presentation of the total canopy foliage to sunlight (i.e., the foliage arrangement). Changes in plant morphology can be due to a number of factors. We suggest two separate hypotheses to illustrate this concept. The first is that temperature-induced changes of the leaf surface may cause the plant to produce a less desirable forage for phytophagus insects, thereby reducing the injury due to insects through reduced feeding or survival. Second, temperature-induced morphological responses such as leaf rolling or wilting may change the microclimate within the canopy causing a shift toward a less favorable environment for insects, in terms of temperature or humidity. Hatfield (1982) has delineated ways in which the microclimate can be modified by agronomic management to control insects and diseases. However, it is difficult to separate temperature-induced changes in morphology from those induced by drought stress.

3.1 Leaf Surface Changes

The leaf surface is covered with a cuticle, composed of waxes, and frequently pubescence (Jeffree 1986). Leaf reflective properties are associated with these morphological features of the leaf. Bunnick (1978) reviewed the changes in light reflective properties and reported that leaf reflectance was dependent on both the plant water status and characteristics of the cuticular layer. Gausman and Cardenas (1973) showed that in the visible and near-infrared radiation wavelengths (0.5–2.5 μm) the difference in reflective properties between glabrous and pubescent soybean lines was due to the trapping of the near-infrared radiation by the leaf hairs of the pubescent lines. Leaf reflectance also increases as the leaf ages and/or undergoes water stress (Gausman et al. 1970). Syvertsen and Levy (1982) have shown that the thickness of the cuticle of the citrus leaf changes throughout the day in response to leaf water potential, with cuticle thickness increasing as the leaf water potential decreases. Plants exposed to cold temperature also exhibit increases in leaf thickness. Parsons (1982) in reviewing the changes in leaf waxes to water stress found that the amount of leaf waxes increased in response to water stress.

The epicuticular layer of higher plants is composed mainly of waxes resulting from lipid and other fatty acid syntheses (Jeffree 1986). The epicuticule also contains minor consitutents that are commonly secondary plant compounds. Certain of the waxes and secondary plant compounds are biologically active to insects (Städler 1986, Woodhead and Chapman 1986). Wall et al. (1981, 1983) have found lipophilic insect pheromone becomes attached to vegetation surrounding the site of pheromone release, presumably to the epicuticular waxes. We suggest that the concentration of lipophilic insect pheromones and plant volatiles in the epicuticular waxes and the surrounding atmosphere of an infested plant exist in a dynamic state in response to ambient temperatures.

3.2 Plant Morphology Changes

Morphological changes which a temperature-stressed plant may commonly exhibit are leaf rolling or wilting. Both responses reduce the radiation absorbed by the plant hence, any effect of these leaf responses on insect behavior could be considered secondary. Gerik (1985) in studying *Heliothis* behavior in cotton fields found that there were subtle changes in cotton growth, culture, and microclimate that modified insect densities. When he used irrigation and double row plantings, there were significant increases in infestation that resulted in increased damage. In the canopy of the irrigated, double row plantings, relative humidity was increased and temperatures decreased, providing conditions known to be favorable for survival and development of *Heliothis*.

As discussed, changes in plant morphology will affect the microclimate, and even subtle temperature changes (less than 3°C) can be expected to have an impact on insect population dynamics. Unfortunately, little is known about the interactions of temperature with plant morphological changes, insect growth, and survival or foraging habits. New research is required to quantify effects of temperature extremes on the interactive changes of plants and herbivores in the field environment.

4. EFFECTS OF TEMPERATURE-INDUCED STRESS ON PLANT TOLERANCE TO INSECT INJURY

Although little experimental evidence can be found, we would expect that certain plant phenotypes or species can better tolerate insect damage within a particular temperature range than other phenotypes or species (Tingey and Singh 1980). Plant physiological mechanisms for increased tolerance could include lowered sensitivity to a particular stress ethylene concentration (Morgan 1985), causing increased fruit set, or increased growth rates, resulting in more fruit produced per unit of time. Sosa and Foster (1976), in studying the effect of high temperature on the level of resistance to Hessian fly in wheat, found that the level of host plant suitability increased and/or virulence of the insect increased at higher temperatures. However, increased tiller production of some wheat phenotypes at high temperatures was suggested to be an important form of tolerance that offset part of the seed loss caused by the decrease in level of antibiosis and/or nonpreference to the Hessian fly.

5. EFFECTS OF TEMPERATURE-INDUCED STRESS ON INSECT BEHAVIOR, PHYSIOLOGY, AND REPRODUCTIVE SUCCESS

Although most insects have little or no physiological control over body temperature, they can regulate temperature through behavioral actions, such as moving into suitable temperature environments. Some insects may also regulate temperature by evaporative cooling, which results from the release of water vapor through the spiracles. Insects can survive a wide range of temperatures (below 0° to 50°C), though most species have a relatively narrow range of temperatures for optimum growth, fecundity, and survival. Insect tolerance to temperature-induced stress is a result of the effect of temperature on insect metabolic functions, such as digestion, assimilation, and excretion, which are regulated in part by enzymatic reactions. Most enzymes function efficiently only within a specific temperature range. Fluctuating diurnal temperatures tend to provide conditions, at some time during each day, when the enzyme systems of an insect will operate most efficiently.

Bogus and Cymborowski (1977) found that cooling temperatures caused an increase in the number of wax-moth larval molts and reduced the number of larvae retaining the ability to secrete silk. The time of day at which chilling stress took place, differentially affected the number of extra larval molts. They suggested that this differential response is due to diurnal cold-sensitive rhythms in juvenile hormone secretion or in hormone-sensitive tissues.

In some insects moderately high temperatures induce production of heat shock proteins (HSP) which protect against death due to subsequent exposure to higher temperatures (Stephanou et al. 1983). Heat and cold stress also induce changes in insect hemolymph. At high temperatures amino acid concentration and hemolymph volume are increased in beet armyworm (Cohen and Patana 1982). Although stress effects differ with stage of insect development, heat- and cold-induced stress are known to cause a reduction in growth rate and to alter metabolic rate and fat consumption for many insects (Wigglesworth 1965).

Adaptation or acclimation to temperature changes are thought to be common in insects. Physiological responses to increasing temperature can be altered with previous exposure (Chapman 1971) (Fig. 1). Temperature preconditioning may be a major factor in subsequent physiological responses. Gonen (1977) reared *Sitophilus granarius* (L.) at high "supraoptimal" temperatures and found that tolerance of *S. granarius* to a short-term exposure (0 to 30 hr) to the lethal temperature of 40°C was increased. Insect mortality was found to increase as the length of exposure to the lethal temperature was increased, regardless of preconditioning.

Temperature thresholds, at which growth is initiated, have been identified for some insects (e.g., European pea aphid) (Fig. 2). The temperature threshold may differ for each insect growth stage. Optimum temperatures for food assimilation, conversion, and growth of many insect pests are in the range from 15° to 30°C. It would be interesting to know whether tem-

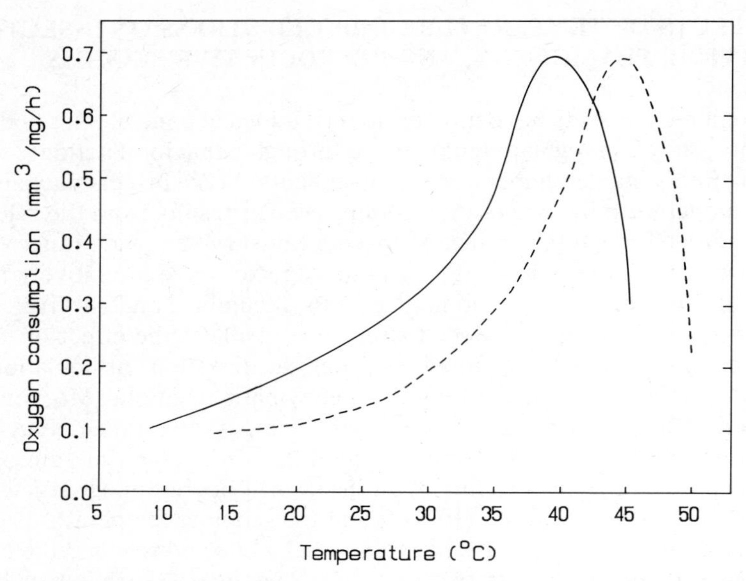

Figure 1. The effect of preconditioning at 12°C (——) and 25°C (– – –) on oxygen consumption of *Melasoma*. (Modified after Chapman 1971)

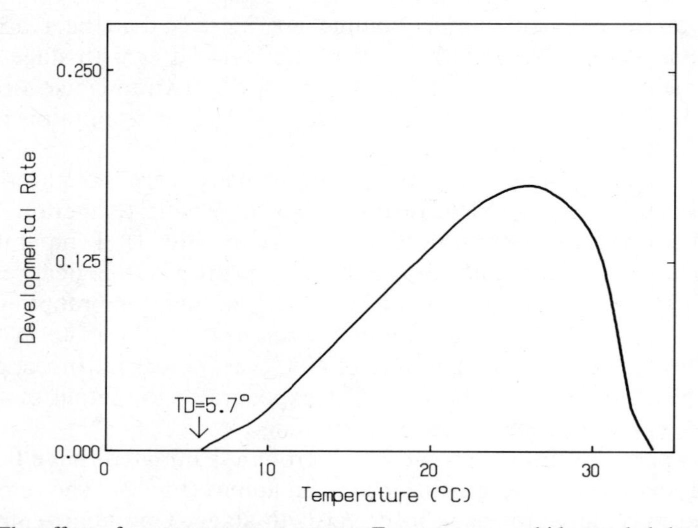

Figure 2. The effect of constant temperature on European pea aphid nymphal development. TD is the temperature threshold (°C) for development (the minimum temperature at which growth can proceed). (Modified after Bieri et al. 1983)

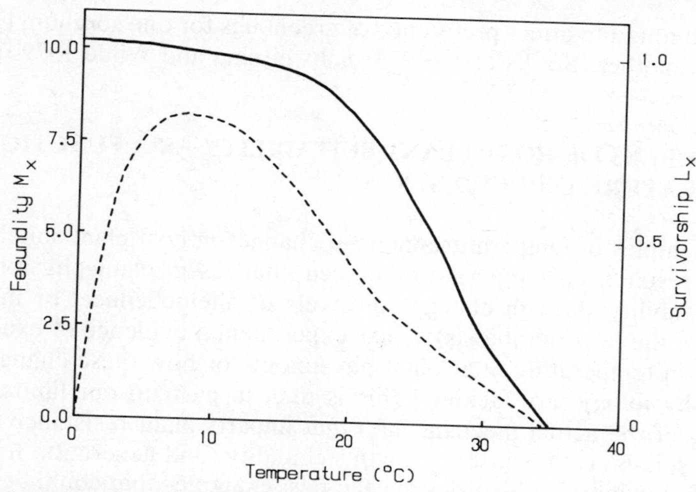

Figure 3. The effect of constant temperature on fecundity (M_x) (---) and survivorship (L_x) (——) of European pea aphid. (Modified after Bieri et al. 1983)

perature thresholds for initiation of growth could be changed with a change in host plant suitability.

Temperature influences herbivore population growth by affecting survival, fecundity, developmental rate, and generation time. Bieri et al. (1983) determined that European pea aphid fecundity was maximum at about 7°C (Fig. 3) and that the maximum developmental rate for nymphs was at 25°C (Fig. 2). Thus European pea aphid fecundity was determined to be more sensitive to temperature extremes than nymphal survival. In a study of pea aphid population growth at two temperatures on resistant or susceptible alfalfa cultivars, Dahms and Painter (1940) found that aphids on the resistant plants produced the most progeny at 7°C (the lower temperature). These scientists suggested that aphids feeding on the less suitable, resistant plants appeared to have a lower optimum temperature for growth and fecundity than did aphids on susceptible plants. We suggest an alternative explanation, that host plant suitability was increased on the "resistant" plants, or decreased in the "susceptible" plants, at the lower temperature due to physiological changes in the plant.

Sensory receptors that are sensitive to temperature have been identified in different insect species (Staedler 1977). Presently no evidence exists that demonstrates that temperature receptors influence feeding behavior, or are involved in determining host suitability. However, feeding behavior and insect movements have been observed to change with changes in temperature (Chapman 1971, Logan et al. 1985, Willmer 1986). Rate of feeding tends to be maximum at a temperature that is specific for a species, developmental stage, or biotype (Starks et al. 1973). Ambient temperature extremes have also been found to affect preference of greenbugs for one sorghum line compared to another (Starks et al. 1973, Schweissing and Wilde 1979b).

6. EXAMPLES OF HOST PLANT SUITABILITY AS A FUNCTION OF TEMPERATURE-INDUCED STRESS

Most examples of temperature-induced changes in host plant suitability appear to result from temperature-induced changes in plant physiology that alter suitability through changes in levels of allelochemical or nutritional quality of the host (antibiosis). Clear experimental evidence of exactly how changes in temperature alter plant physiology, or how these changes affect insect physiology, are lacking. This is due, in part, to our limited understanding of the actual mechanism(s) that imparts plant resistance (reduced level of suitability) for insect growth, fecundity, and generation time.

The following case histories are the best examples that could be found in the literature, of the general effects of fluctuating, constant and extreme temperatures on host suitability to insects. Most of these examples are from cool season aphids. Information on temperature effects for other insect pest species is limited.

Tingey and Singh (1980) provide an excellent review of studies demonstrating the effects of temperature on aphid resistance in alfalfa and sorghum. They suggest that the interactions between temperature, pest developmental biology (i.e., growth, survival and fecundity), plant physiology (i.e., wounding response, growth, and fruit production), and other environmental factors are complex.

6.1 Aphids on Alfalfa

Possibly the first study to suggest that temperature modifies host plant suitability was that by Dahms and Painter (1940) who found that when field temperatures went from about 16° to 7°C, resistance to pea aphid population growth was lost compared to the aphid population growth on susceptible alfalfa plants. Utilizing constant temperatures (13°, 21°, and 30°C), Isaak et al. (1963) reported results similar to Dahms and Painter for pea aphid and spotted alfalfa aphid in that resistant and susceptible alfalfa clones could not be separated at low temperatures. The clones were equally suitable as hosts for these aphids. In investigations of spotted alfalfa aphid utilizing constant temperatures (16°, 21°, 27°, and 32°C), Hackerott and Harvey (1959) found that the temperature for optimum length of reproductive period, fecundity of adults, and numbers of nymphs that matured and reproduced, decreased as resistance of the plants increased. McMurtry (1962) also found that decreasing temperature caused the resistance to population growth of spotted alfalfa aphid to be reduced differentially, in two resistant clones. In fact, one clone, "C-902," became totally susceptible at 10° and 16°C. Further, if the plants of "C-902" were moved from 10°C to temperatures of 20°C or higher for a few days, they again became resistant. Other researchers (Isaak et al. 1965) have observed the identical phenomenon with pea aphid and spotted alfalfa aphid. They found that resistant clones showed distinct rates of change in suitability for aphid population growth when moved from high temperatures to low and then back to high. Although spotted alfalfa aphid populations responded differently than pea aphid populations to the 10 alfalfa clones in the Isaak et al. (1965) study, the loss or gain in host plant suitability in response to temperature was consistent for all resistant alfalfa clones. Thus McMurtry (1962) and Isaak et al. (1965) concluded that the expression of some unknown aphid resistance factor(s) in the resistant clones is modified by temperature. This could be due to improved nutritional quality as a result of the accumulation of amino acids and carbohydrates at the expense of allelochemicals, when plants are exposed to cool temperatures. McMurtry (1962) also suggests that under conditions of fluctuating high and low temperatures, the mean temperature rather than the high or low temperature, appears to be most important in controlling the level of host plant suitability. Current ecological theory states that tropical plants have higher levels of secondary metabolites because they have been subjected over evolutionary time to more intense grazing pressure from herbivores than temperate plant species (Rhoades 1983). We suggest that higher mean temperatures throughout the life of plants in the tropics may be responsible, in part, for the higher levels of defensive compounds.

The influence of fluctuating versus constant temperatures was studied using spotted alfalfa aphid on alfalfa in no-choice and free-choice studies by Kindler and Staples (1970). In the no-choice study with susceptible alfalfa plants, they found that fluctuating temperatures stimulated more rapid aphid development, and enhanced fecundity and longevity compared with aphids at constant temperatures. However, when aphid developmental biology on resistant clones was compared at constant and fluctuating temperatures there was no effect at any temperature. In their free-choice study aphids maintained at two constant temperatures (10° and 27°C) migrated to the susceptible clone, demonstrating that resistance was still retained as nonpreference for the resistant clone. Schalk et al. (1969) found a similar preference exhibited for susceptible clones at 27°C, but at 10°C the aphids responded to some susceptible clones as though they they had become resistant (Fig. 4). Also aphid preference for one clone changed, depending on whether excised leaves or whole plants were offered. At 10°C the aphids on excised leaves of the resistant clone behaved as though it was susceptible. The results of Schalk et al. (1969) and McMurtry (1962) suggest that the level of expression and ease of identification of aphid resistance in alfalfa are dependent on temperature and genetic sources of resistance.

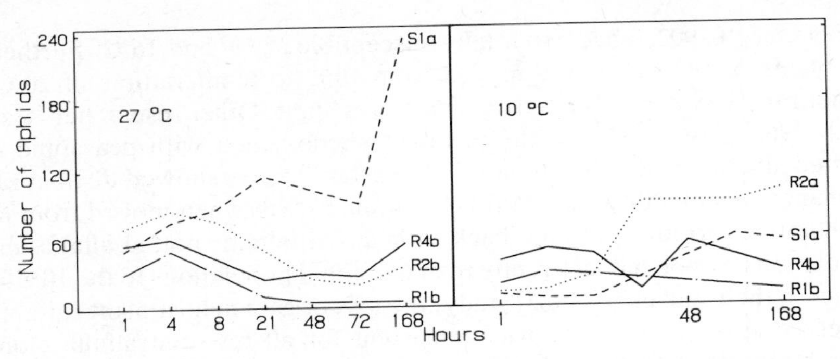

Figure 4. The effect of two constant temperatures (27° or 10°C) on the preference of the spotted alfalfa aphid for whole alfalfa plants. Clones followed by the same letter within each temperature are not significantly different at the 10% level of probability, as determined by *t*-tests for heterogeneous variances at the 168-hour reading (R = resistant clones, S = susceptible clones). (Modified after Schalk et al. 1969)

6.2 Greenbugs on Barley, Oats, Rye, and Sorghum

The number of greenbug nymphs produced at constant temperatures (10° to 32°C) on resistant sorghums increased as the temperature decreased (Schweissing and Wilde 1979a, Wood and Starks 1972), as with pea and spotted alfalfa aphids on alfalfa. It is interesting that greenbug biotypes A, B, and C were found to differ in response to constant temperatures on sorghum. Based on reproductive rate, biotype C was the most adapted to temperature extremes (Wood and Starks 1972). Schweissing and Wilde (1979a) found that tolerance to damage by biotype C increased in susceptible sorghum as temperature decreased, whereas tolerance to leaf damage was unchanged in the resistant sorghum, KS 30.

Greenbug resistance in barley, oats, and rye did not decrease with decreasing temperature (Schweissing and Wilde 1978, Wood and Starks 1972) as with sorghum. Further, tolerance to leaf damage per aphid tended to remain the same or increase slightly with decreasing temperature on barley, oats, and rye. Schweissing and Wilde (1978) suggest that greenbug resistance is associated with optimum temperatures for plant growth—that is, greater resistance is seen in cool season crops (barley, oats, and rye) cultivated under cool temperatures, and in the warm season crops (sorghum) cultivated under warm temperatures.

6.3 Hessian Fly on Wheat

Wheat possessing certain genes for resistance was observed in the field to lose expression of resistance to Hessian fly as temperature increased. In laboratory studies Cartwright et al. (1946) showed that under constant temperatures, the higher temperatures (24° to 27°C) caused a pronounced increase in percentage infested plants in resistant and susceptible lines. The magnitude of effect was greater in some of the resistant lines than in the susceptible lines. However, resistance in a nearly immune line, PI94587, was not altered with high temperatures. Sosa and Foster (1976) confirmed the Cartwright et al. finding, that wheat lines with different specific genes for Hessian fly resistance responded differently to insect attack at high than low temperatures. They also proved that when these resistant wheat lines were infested by Hessian fly races B, C, D, and Great Plains, the cultivars had different tillering rates and population densities of Hessian fly, depending on the race, cultivar, and constant temperature used. Again we do not know whether temperature-induced changes in the host plant or insect pest are responsible for changes observed in host suitability at higher temperatures.

In none of these examples, or elsewhere in the literature, could we find evidence that would explain the changes observed in host plant suitability, such as a specific change in insect or plant physiology resulting from temperature-induced stress. We were surprised at how little is known concerning the actual nature of plant resistance to insects. We are convinced that increased research effort in this area will prove very fruitful to plant physiologists, chemists, applied entomologists, and ecologists in expanding knowledge of the basic physiological and behavioral mechanisms governing the interactions of host plants and their insect herbivores. A thorough understanding of the nature of resistance is essential to the development of new concepts in plant–insect interactions, and in making real advances in breeding and genetic engineering of plants to resist insects.

7. INTERACTIONS BETWEEN TEMPERATURE, PLANTS, HERBIVORES, AND ENVIRONMENTAL FACTORS

Abiotic and biotic environmental factors interact to affect suitability of plants to herbivores. The feeding of the herbivore itself causes changes in plant suitability (see Chapters 9 and 10). The plant–insect interactions are a continuum of stimuli and responses from living and nonliving factors, such as pathogens, predators, beneficial organisms, temperature, moisture, day length, light intensity, and soil. An excellent review of these interactions is presented by Denno and McClure (1983). The interactions of temperature with other factors may directly affect the host plant (e.g., humidity, light intensity, and nutrients) or indirectly (e.g., plant pathogens and beneficial microorganisms). These interactions can also affect herbivore ability to damage the plant, and herbivore developmental biology and behavior. Any factor(s) that reduces herbivore colonization rate (i.e. success) will have the "apparent function" of decreasing host plant suitability.

7.1 Temperature and Light

The general effects of temperature and photoperiod on diapause initiation are well known (Chapman 1971), but less well known is the interaction of photoperiod and temperature on insect development and adult fecundity (Druzhelyubova 1976, Ali and Ewiess 1977). Low temperatures and shortening day length induce reproductive diapause and hibernation in most insects. However, studies of host plant suitability with alfalfa and aphids (Kindler and Staples 1970, McMurtry 1962) did not show an interaction of temperature and photoperiod. In the interpretation of the effect of decreasing or increasing temperatures on host plant suitability, we should be careful not to ignore the possible effects of photoperiod.

7.2 Temperature and Humidity

Both humidity and temperature continuously interact to produce conditions within the plant canopy that play a key role in regulating insect growth, survival, and fecundity. Isaak et al. (1963) found that humidity conditioned the effects of temperature on pea aphid and spotted alfalfa aphid nymphal production per day. Nymphal production appeared to be depressed at high temperature (30°C) by high relative humidities (50 to 90%). In studies of Mexican bean beetles as affected by temperature, relative humidity and stage of soybean development, Wilson et al. (1982) found that larval survival was decreased as temperature increased, especially when it exceeded 27°C, or when relative humidity was decreased. These stresses increased developmental time and reduced pupal weight. Further, when larvae were reared on postbloom soybean, they were more sensitive to low relative humidity (i.e., reduced survival) than larvae reared on prebloom soybean. These studies indicate that when host quality is low, due to inferior host species or genotype, or inferior plant growth stage, herbivores may be more sensitive to physical stresses such as extreme temperature or humidity.

We conclude that a myriad of behavioral, biochemical, and physical factors influence an insect's survival on a given host plant and that the relative importance of each is likely to change as the insect develops. Further we suggest that behavior, temperature, and humidity interact to affect plant growth, expression of resistance, and plant attractiveness to herbivores and beneficial insects through plant produced allomones, synomones, and kairomones (Nordlund 1981).

7.3 Temperature and Plant Nutrients

Greenbug numbers and development, and leaf damage of resistant and susceptible sorghums can be affected by temperature and nutrients. Schweissing and Wilde (1979a) found that a greenbug resistant line showed interactions between temperature and six nutrient regimes containing combinations of

N, P, and K. Resistant plants grown with excess potassium had the lowest host plant quality, as measured by aphid population growth and foliage loss, over three fluctuating temperature regimes (L:D) (21.1°:10°C, 26.7°:14.6°C, and 32.2°:21.1°C). High temperatures and excess nitrogen increased host plant quality as measured by increased numbers of aphids produced, and greater leaf damage. Most treatments had significant effects on these damage parameters. In general, host plant suitability of these sorghums tended to be similar but as temperature increased, the differences in quality were magnified. These scientists suggest that nutrient deficiencies in the field, especially potassium, could be responsible for the erratic performance of resistant sorghum at higher temperatures. At this time, we do not know if nutrient-temperature-induced changes in host plant quality are due to changes in plant nutritional quality or resistance factors such as allelochemicals, or both.

7.4 Temperature and Drought

Drought is caused by reduced soil moisture which results in increased ambient and leaf temperatures and reduced relative humidity. Consequently drought is a three-way interaction of temperature, humidity, and soil moisture that directly affects herbivore and plant physiology, and thus apparent host plant suitability. A key response of the plant to "drought stress" is increased temperature of plant tissues as a result of reduced transpiration. Hoffman et al. (1975) suggested that populations of a psyllid, *Acizzia russellae* Webb and Moran, which suffer major reductions in density during droughty summer months, were dying as a result of exposure to lethal temperatures (in excess of 45°C) and low humidity. In contrast, mild drought improves host plant suitability for some aphid species, resulting in increased survival, growth, and fecundity (Tingey and Singh 1980). Mild drought appears generally to increase levels of soluble nitrogen, sucrose, and aphid performance. Kennedy and Booth (1959) have observed that aphid outbreaks in natural environments frequently follow mild droughts. Rhoades (1983) theorized that physical stress on plants such as drought could reduce nutrient budgets, which would result in lower plant commitment to defense and thus increase host plant suitability to some insects. White (1974, 1976) hypothesized that the major factor regulating density of insect herbivores is a shortage of nitrogen. He argues that mild or intermittent drought is responsible for increasing nitrogen content of some plant species, and presents two excellent field examples in support of his argument. Visscher-Neumann (1980) suggests that plant hormones, such as abscisic acid and gibberellin, serve as biochemical signals to herbivores, synchronizing insect reproductive biology and population dynamics with host suitability and environmental conditions. Rhoades (1983) states that our inability to develop population models that accurately represent the large fluctuations in herbivore numbers is due to our failure to include changes in suitability of food resources. Host plant suitability at any point in time is a function of nutritional quality, plant hormones, and defensive chemistry and morphology. Future population models should provide a causal link between physical factors, such as temperature, and biological factors, such as host plant suitability, to accurately predict fluctuations in herbivore density.

7.5 Temperature and Pathogens

Temperature can directly affect the developmental biology and virulence of pathogens that attack plants or insects, resulting in what appears to the observer as changes in host plant suitability with temperature. Bell (1982) states that any stress, especially temperature, that decreases the plant's ability to defend itself actively, or constitutively, from pathogens will reduce plant resistance to disease. In plants the speed of expression of active defenses in response to pathogen attack is an important determinant of defense effectiveness (Bell 1984, Baily and Mansfield 1982). Diseased plants are frequently less suitable hosts for insects due to increased allelochemical concentrations induced by plant pathogens (see Chapter 12). Accumulated evidence shows that many disease-induced allelochemicals, including some phytoalexins, are active against plant diseases and insect herbivores (Bell 1981, 1986; Kogan and Paxton 1983). Woodhead (1981) found in the field that levels of phenolic acids increased in sorghum in response to attack from pathogenic fungi and insect herbivores and that the phenolic acid response was apparently enhanced with higher field temperatures.

Temperature-induced stress can also directly affect insect herbivore susceptibility to disease-causing organisms, independent of the host plant. Biever and Wilkinson (1978) fed larvae of *Pieris rapae* (L.) a diet medium containing granulosis virus, exposed them to 4°C, for four to 10 days, and then returned them to 25°C. They found that the low temperature greatly increased the number of infected larvae and the number of dead larvae. Diet medium low in moisture also increased mortality. Interestingly, low temperature stress and low moisture diet (without virus) together induced a lethal virus infection in normally healthy larvae.

Indirectly, temperature can affect insect susceptibility to pathogens by modifying host plant defensive chemistry. Ramoska and Todd (1985) have found evidence that host plant antimicrobial compounds can be transferred to the insect herbivore during feeding, and thus provide the herbivore with protection against insect pathogens.

7.6 Temperature and Toxicity of Allelochemics

Herbivores are known to utilize three methods for overcoming plant toxins. Insects may sequester and store the toxins in their body unchanged, or excrete them rapidly in unaltered form, or alter the toxin biochemically to nontoxic products that can be metabolized and/or excreted (Dowd et al.

1983). Many plant toxins and pesticides are similar in their lipophilic nature and are detoxified in the same manner (Brattsten 1979, Brattsten et al. 1986a, b). Most commonly, toxins are metabolized by the microsomal mixed-function oxidases (MFOs), or conjugating enzymes (esterases, reductases, group transfer enzymes) or epoxide hydrases. Little is known about the influence of temperature or humidity on the ability of these systems to detoxify plant toxins.

We do have good data for the influence of temperature on herbivore detoxification of insecticides (Brattsten et al. 1986a). Some pesticides are allelochemicals, whereas others are organophosphates, carbamates, and chlorinated hydrocarbons. Current thought is that these insecticides are metabolized in the same manner as plant toxins. Kansouh and coworkers (1981) studied the effect of constant temperatures (10°, 20°, and 30°C), on the activity of three enzymes involved in detoxification of lipophilic toxins. They found the level of activity of the enzymes (which is a measure of their ability to protect insects from toxins) was differentially affected by increasing temperature and varied with genetic strains of the cotton leafworm. Aliesterase was the most sensitive to temperature, showing a significant increase in activity with increasing temperature to 30°C. They also looked at the effects of humidity and found that increasing relative humidity from 65 to 80% caused a significant reduction in aliesterase activity.

A recent laboratory study by Brattsten et al. (1986a) demonstrated that southern armyworms reared at 15°C had a higher level of activity of detoxifying enzymes (cytochrome P-450, P-450 reductase, *p*-chloro *N*-methylaniline *N*-demethylation, methoxyresourufin *O*-dimethylation, and aldrin epoxidation) in their midguts than those larvae reared at 30°C. Further, toxicity of carbaryl, an insecticide, was decreased threefold at the lower temperature compared to 30°C.

A number of studies have been conducted on the effect of temperature on efficacy of insecticides (Barson 1983; Hanker 1973; Riskallah 1984; Sparks et al. 1982, 1983; Tyler and Binns 1982). The results are difficult to interpret, since temperature may affect insecticide absorption, distribution, and metabolism (Wilkinson 1976). However, we can generalize that the toxicity of organophosphates increases with increasing temperatures (5° to 35°C) and the toxicity of pyrethroids decreases with increasing temperature. However, there were exceptions, depending on the specific compound and insect species studied. It would be valuable to know how temperature affects the biological activity of allelochemicals.

8. OVERVIEW AND PROSPECTS

The complex and frequently unclear relationships between temperature-induced stress on insect herbivores and their host plants have been documented to a limited extent in the scientific literature. We find that temperature-induced changes in plant physiology can result in changes in carbon-nitrogen ratios, concentrations of amino acids, sugar ratios in sap, and quality and quantity of secondary plant compounds in vegetative and reproductive structures. These changes have, in some cases, altered host plant suitability for survival, growth, development, reproduction, and population development of the insect herbivore. Unfortunately, we have no evidence of the actual temperature-induced changes in plant chemistry or herbivore physiology that explain the observed changes in host plant suitability discussed previously. Evidence suggests that a loss in host plant suitability may occur as a result of either high or low temperature stress depending on herbivore, plant species, and genotype.

The best case histories of temperature-induced changes in host plant suitability are from studies of plant resistance to aphids. We find that low temperatures tend to improve host plant suitability to aphids, whereas plant suitability tends to decrease when temperatures are increased. Further, fluctuating temperatures may have different effects on host plant or herbivore fitness than constant temperature. Possibly the most interesting discovery is that temperature can interact separately with humidity and soil fertility to affect the expression of host plant suitability to herbivores.

The future prospects for elucidating the effect of temperature on physiology, biology, and behavior of insect herbivores and their host plants are very promising. Many scientists (Bell 1984, Kogan and Paxton 1983, Tingey and Singh 1980) believe that we have the tools and techniques to investigate and understand this area. They also believe that in the future we may control temperature to reduce host plant suitability or improve plant wound response and tolerance to insect herbivores. DiCosmo and Towers (1984) in studies of temperature stress in plant callus tissue show how phytochemical production can be induced with changes in nutritional or hormonal levels. Exciting opportunities exist for the use of plant growth regulators, plant nutrition, and plant genetics to alter host plant suitability. Temperature-induced changes in allelochemicals may have far more complex and wide ranging effects than are presently apparent. For example, not only do plants need to be nutritionally and allelochemically satisfactory (Reese 1983) for insect growth, survival, and reproduction, but they must also meet requirements of smell and taste to which the herbivores are genetically and conditionally prepared to respond. We find host plant allelochemicals, in some cases, appear to determine both the initiation and levels of insect pheromone production (Hughes 1973, Chang 1986, Chang et al. 1986). Phytochemical volatiles may also play a role in communicating plant defense responses to other plants and in communicating information necessary for predators and parasites (which are components of plant defense) to find and attack insect herbivores (Price 1981).

A poorly understood relationship exists between temperature, plant hormonal–nutritional–phytochemical balances and insect herbivore hormonal balances. For examples, Visscher-Neumann et al. (1979) studied the interacting effects of temperature on the host plant, western wheatgrass, and rearing temperature on a herbivore, rangeland grasshopper. They found that grass grown at cool temperatures caused increased egg production. Further Visscher-Neumann (1971) found that grasshopper egg viability and rate of embryonic development were determined largely by the physiological state of the female parent. When Visscher-Neumann (1980) examined the effect of plant hormones on female grasshopper reproduction, she found significant changes in fecundity rate. Abscisic acid (ABA) and gibberellin (GA) were active in reducing fecundity. Interestingly, these compounds are all terpenoids, similar to insect juvenile hormone (JHIII). From these data, and others, we suggest that plant growth hormones, insect growth hormones, and temperature interact in various ways to affect host plant suitability. Visscher-Neumann (1980) hypothesized that plant hormones serve as biochemical signals to synchronize insect herbivore reproductive biology and population dynamics with weather and host plant suitability. The relationship between plant and insect hormones in regulating insect population dynamics is a potentially fertile field for research.

Another potentially rewarding research area is the utilization of bioactive chemicals. Bioactive chemicals such as plant hormones have been incorporated into seeds to alleviate temperature-induced stress during seed germination and growth (Khan 1978). Khan found bioactive chemicals alleviated high temperature stress in lettuce seed but had no effect on low temperature stress. The effect of seed infusions (with plant hormones) on host plant suitability to insects is not known.

In the future, crop plants may have major alterations in plant architecture brought about through breeding, planting patterns, hormonal, morphology, and/or mechanical manipulations. Coyne (1980) suggests that plant architecture (pubescence, leaf shape, plant color, fruiting form, shape and location, and general plant branching and canopy pattern) can be utilized to change canopy temperatures and host plant suitability for insects. Once we more clearly identify factors affecting host plant suitability to temperature-induced stress, we may be able to forecast population outbreak conditions in agroecosystems (Podolsky 1984). The ability to make such predictions would allow us to modify economic thresholds and insect control strategies based on environmental factors.

Clearly, we have more to learn about the nature of resistance mechanisms in plants and how temperature interacts with other environmental factors to affect insect and host plant, and their reproductive success through time. Additonal knowledge in these areas is essential to basic, and applied, animal and plant sciences. Future advances in genetic engineering of crops, host plant resistance to insects, and pest management technology are dependent on our gaining a better understanding of environmental effects on crop–herbivore interactions. The basic concepts of population regulation of plant and herbivore, optimal foraging of herbivores, and the expression of defense mechanisms in plants may be more clearly understood when the role of temperature is better known.

We conclude that host plant suitability is a constellation of interacting factors of which temperature plays a central role, affecting herbivore and plant reproductive success in time and space. We can generalize about this phenomenon only so far as to state that each distinct plant–herbivore environment has a unique set of responses to temperature induced stress and that temperature-induced stress will affect to some degree all biotic organisms in the relationship.

ACKNOWLEDGMENTS

Special appreciation is extended to Drs. J. Mahan, P. Morgan, D. Rummel, J. Schneider, G. Teetes, and G. Wilde for their valuable reviews of this chapter. We also appreciate the patient assistance of Carolyn Villanueva and Mike Treacy in various aspects of the development of this chapter. A special thanks to Kristine Schmidt for preparation of the illustrations, and to Sharon, Kim, and Kelly Benedict for their love and encouragement.

Technical Article 28144 of the Texas Agricultural Experiment Station.

REFERENCES

Ali, M., and M. A. Ewiess. 1977. Photoperiodic and temperature effects on rate of development and diapause in the green stink bug, *Nezara viridula* L. (Heteroptera: Pentatomidae). *Acta Phytopathol.* 12: 337–347.

Bailey, J. A., and J. W. Mansfield (eds.). 1982. *Phytoalexins*. Wiley, New York.

Barson, G. 1983. The effects of temperature and humidity on the toxicity of three organophosphorus insecticides to adult *Oryzaephilus surinamensis* (L.). *Pesticide Science* 14: 145–152.

Bell, A. A. 1981. Biochemical mechanisms of disease resistance. *Annu. Rev. Plant Physiol.* 32: 21–81.

Bell, A. A. 1982. Plant pest interaction with environmental stress and breeding for pest resistance: plant diseases. *In* Christiansen, M. N., and C. F. Lewis (eds.), *Breeding Plants for Less Favorable Environments*. Wiley, New York, pp. 335–363.

Bell, A. A. 1984. Morphology, chemistry, and genetics of *Gossypium* adaptations to pests. *In* Timmerman, B. N., C. Steelink, and F. A. Loewus (eds.), *Phytochemical Adaptations to Stress*. Plenum, New York, pp. 197–230.

Bell, A. A. 1986. Physiology of secondary products in cotton. *In* Stewart, J. M. and J. R. Mauney (eds.), *Cotton Physiology—A Treatise*. National Cotton Council of America, Memphis.

Bernstam, V. A. 1978. Heat effects on protein biosynthesis. *Annu. Rev. Plant Physiol.* 29: 25–46.

Bieri, M., J. Baumgartner, G. Bianchi, V. Delucchi, and R. Von Arx. 1983. Development and fecundity of pea aphid (*Acyrthosiphon pisum* Harris) as affected by constant temperatures and by pea varieties. *Mitteilungen der Schweizerischen Entomologischen Gesellschaft* 56, 1/2: 163–171.

Biever, K. D., and J. D. Wilkinson. 1978. A stress-induced granulosis virus of *Pieris rapae*. *Environ. Entomol.* 7: 572–573.

Bjorkman, O., M. R. Badger, and P. A. Arnold. 1980. Response and adaptation of photosynthesis to high temperatures. *In* Turner, N. C., and P. J. Kramer (eds.), *Adaptation of Plants to Water and High Temperature Stress*. Wiley, New York, pp. 233–249.

Bogus, M. J., and B. Cymborowski. 1977. Daily changes in the sensitivity of wax-moth larvae, *Galleria mellonella*, to cooling stress. *Physiol. Entomol.* 2: 103–107.

Brattsten, L. B. 1979. Biochemical defense mechanisms in herbivores against plant allelochemicals. *In* Rosenthal, G. A., and D. H. Janzen (eds.), *Herbivores, Their Interaction with Secondary Plant Metabolites*. Academic Press, New York, pp. 199–270.

Brattsten, L. B., C. A. Gunderson, J. T. Fleming, and K. N. Nikbahkt. 1986a. Temperature and diet modulate cytochrome P-450 activities in southern armyworm, *Spodoptera eridania* (Cramer), caterpillars. *Pestic. Biochem. Physiol.* 25: 346–357.

Brattsten, L. B., C. W. Holyoke, Jr., J. R. Leeper, and K. F. Raffa. 1986b. Insecticide resistance: challenge to pest management and basic research. *Science* 231: 1255–1260.

Bunnick, N. J. T. 1978. The multispectral reflectance of shortwave radiation by agricultural crops in relation with their morphological and optical properties. *Mededelingen Landbouwhogeschool Wageningen Nederlund* (78-1). 175 p.

Burke, J. J., J. L. Hatfield, R. R. Klein, and J. E. Mullet. 1985. Accumulation of heat shock proteins in field-grown cotton. *Plant Physiol.* 78: 394–398.

Cartwright, W. B, R. M. Caldwell, and L. E. Compton. 1946. Relation of temperature to the expression of resistance in wheats to Hessian fly. *J. Am. Soc. Agron.* 38: 259–263.

Chang, J. F. 1986. Influence of cotton cultivars on boll weevil (Coleoptera: Curculionidae) behavior and pheromone production. Ph.D. dissertation. Texas A&M University, College Station.

Chang, J. F., J. H. Benedict, T. L. Payne, B. J. Camp, W. L. McGovern, and G. H. McKibben. 1986. Boll weevil pheromone production in response to selected cotton cultivars. In Nelson, T. C. (ed.), *Proc. Beltwide Cotton Prod. Res. Conf.* National Cotton Council of America, Memphis, pp. 489–492.

Chapman, R. F. 1971. *The Insects, Structure, and Function*. American Elsevier, New York.

Cohen, A. C., and R. Patana. 1982. Ontogenetic and stress related changes in hemolymph chemistry of beet armyworms *Spodoptera exigua*. *Comp. Biochem. Physiol. A. Comp. Physiol.* 71: 193–198.

Coyne, D. P. 1980. Modification of plant architecture and crop yield by breeding. *HortScience* 15: 244–247.

Dahms, R. G., and R. H. Painter. 1940. Rate of reproduction of the pea aphid on different alfalfa plants. *J. Econ. Entomol.* 33: 482–485.

Denno, R. F., and M. S. McClure (eds.). 1983. *Variable Plants and Herbivores in Natural and Managed Systems*. Academic Press, New York.

DiCosmo, F., and G. H. N. Towers. 1984. Stress and secondary metabolism in cultured plant cells. *In*. Timmermann, B. N., C. Steelink, and F. A. Loewus (eds.), *Phytochemical Adaptations to Stress*. *Rec. Adv. Phytochem.* Vol. 18. Plenum, New York, pp. 97–175.

Dixon, M., E. C. Webb, C. J. R. Thorne, and K. F. Tipton. 1979. *Enzymes*. 3d ed. Academic Press, New York.

Dowd, P. F., C. M. Smith, and T. C. Sparks. 1983. Detoxification of plant toxins by insects. *Insect Biochem.* 13: 453–468.

Downton, W. J. S., and J. S. Hawker. 1975. Evidence for lipid–enzyme interaction in starch synthesis in chilling-sensitive plants. *Phytochemistry* 14: 1259–1263.

Druzhelyubova, T. S. 1976. Temperature and light as factors affecting development and behavior in geographic populations of *Agrotis ypsilon* Rott. (Lepidoptera, Noctuidae) (field crop pests). *Entomol. Rev.* 55: 9–14. (Translated from *Entomologicheskoe Obozrenie* 55: 277–285.)

Gamble, P. E., and J. J. Burke. 1984. Effect of water stress on the chloroplast antioxidant system. I. Alterations in glutathione reductase activity. *Plant Physiol.* 76: 615–621.

Gausman, H. W., W. A. Allen, R. Cardenas, and A. J. Richardson. 1970. Relation of light reflectance to histological and physical evaluations of cotton leaf maturity. *Applied Optics* 9: 545–552.

Gausman, H. W., and R. Cardenas. 1973. Light reflectance by leaflets of pubescent, normal, and glabrous soybean lines. *Agron. J.* 65: 837–838.

Gerik, J. C. 1985. Cultural modifications on cotton development and *Heliothis* ecology. M.S. thesis. Texas Tech University, Lubbock, 112 p.

Gonen, M. 1977. Susceptibility of *Sitophilus granarius* and *S. oryzae* (Coleoptera: Curculionidae) to high temperature after exposure to supra-optimal temperature. *Entomol. Exp. Appl.* 21: 243–248.

Graham, D., and B. D. Patterson. 1982. Responses of plants to low, nonfreezing temperatures: proteins, metabolism, and acclimation. *Annu. Rev. Plant Physiol.* 33: 347–372.

Hackerott, H. L., and T. L. Harvey. 1959. Effect of temperature on spotted alfalfa aphid reaction to resistance in alfalfa. *J. Econ. Entomol.* 52: 949–953.

Hanker, I. 1973. Penetration, excretion, and metabolism of DDT-14c (carbon) and the effects of temperature in susceptible and resistant larvae of *Leptinotarsa decemlineata* Say (Coleoptera). *Acta Entomol. Bohemoslav.* 70: 243–253.

Hatfield, J. L. 1979. Canopy temperatures: The usefulness and reliability of remote measurements. *Agron. J.* 71: 889–892.

Hatfield, J. L. 1982. Modification of the microclimate via management. *In* Hatfield, J. L., and I. J. Thomason (eds.), *Biometeorology in Integrated Pest Management*. Academic Press, New York, pp. 147–170.

Hoffmann, J. H., V. C. Moran, and J. W. Webb. 1975. The influence of the host plant and saturation deficit on the temperature tolerance of a psyllid. *Entomol. Exp. Appl.* 18: 55–67.

Hsiao, T. C. 1973. Plant responses to water stress. *Annu Rev. Plant Physiol.* 24: 519–570.

Hughes, P. R. 1973. *Dendroctonus*: production of pheromones and related compounds in response to host monoterpenes. *Z. Angew. Entomol.* 73: 294–312.

Isaak, A., E. L. Sorensen, and E. E. Ortman. 1963. Influence of temperature and humidity on resistance in alfalfa to the spotted aphid and pea aphid. *J. Econ. Entomol.* 56: 53–57.

Isaak, A., E. L. Sorensen, and R. H. Painter. 1965. Stability of resistance to pea aphid and spotted alfalfa aphid in several alfalfa clones under various temperature regimes. *J. Econ. Entomol.* 58: 140–143.

Jackson, R. D. 1982. Canopy temperature and crop water stress. *In* Hillel, D. (ed.), *Advances in Irrigation*. Academic Press, New York, pp. 43–85.

Jeffree, C. E. 1986. The cuticle, epicuticular waxes and trichomes of plants, with reference to their structures, functions and evolution. *In* Juniper, B., and R. Southwood (eds.), *Insects and the Plant Surface*. Edward Arnold, London, pp. 23–64.

Kansouh, A. S. H., A. M. Ali, M. M. Hosny, and S. M. Madi. 1981. Activity of three enzyme systems in susceptible and resistant strains of *Spodoptera littoralis* (Boisd.) larvae with special reference to the influence of rearing temperature and relative humidity. *Bull. Entomol. Soc. Egypt, Economic Series*, 1976/77, publ. 1981, no. 10, 247–256.

Kennedy, J. S., and C. O. Booth. 1959. Responses of *Aphis fabae* Scop. to water shortage in host plants in the field. *Entomol. Exp. Appl.* 2: 1–11.

Khan, A. A. 1978. Incorporation of bioactive chemicals into seeds to alleviate environmental stress. *Acta Hortic.* (The Hague) 83: 225–234.

Kindler, S. D., and R. Staples. 1970. The influence of fluctuating and constant temperatures, photoperiod, and soil moisture on the resistance of alfalfa to the spotted alfalfa aphid. *J. Econ. Entomol.* 63: 1198–1201.

Klun, J. A. 1974. Biochemical basis of resistance of plants to pathogens and insects: insect hormone mimics and selected examples of other biologically active chemicals derived from plants. *In* Maxwell, F. G., and F. A. Harris (eds.), *Proc. Summer Institute on Biological Control of Plant Insects and Diseases*. Univ. of Mississippi P. Jackson, pp. 463–484.

Kogan, M., and J. Paxton. 1983. Natural inducers of plant resistance to insects. *In* Hedin, P. A. (ed.), *Plant Resistance to Insects. ACS Symp. Ser.* 208, pp. 153–171.

Levitt, J. 1980. *Responses of Plants to Environmental Stresses*, Vol. 1. Academic Press, New York.

Logan, P. A., R. A. Casagrande, H. H. Faubert, and F. A. Drummond. 1985. Temperature-dependent development and feeding of immature Colorado potato beetles, *Leptinotarsa decemlineata* (Say) (Coleoptera: Chrysomelidae). *Environ. Entomol.* 14: 275–283.

Lyons, J. M. 1973. Chilling injury to plants. *Annu. Rev. Plant Physiol.* 24: 445–446.

McMurtry, J. A. 1962. Resistance of alfalfa to spotted alfalfa aphid in relation to environmental factors. *Hilgardia* 32: 501–539.

Morgan, P. W. 1985. Ethylene as an indicator and regulator in the development of field crops. *In* Bopp, M. (ed.), *Plant Growth Substances*. Springer-Verlag, Berlin, pp. 375–379.

Nordlund, D. A. 1981. Semiochemicals: A review of the terminology. *In* Nordlund, D. A., R. L. Jones, and W. L. Lewis (eds.) *Semiochemicals, Their Role in Pest Control.* Wiley, New York, pp. 13–28.

Parsons, L. R. 1982. Plant responses to water stress. *In* Christiansen, M. N., and C. F. Lewis (eds.), *Breeding Plants for Less Favorable Environments*. Wiley, New York, pp. 175–192.

Pearcy, R. W. 1978. Effect of growth temperature on the fatty acid composition of leaf lipids in *Atroplex lentiforms* (Torr.) Wats. *Plant Physiol.* 61: 484–486.

Peterson, L. W., and R. C. Huffaker. 1975. Loss of ribulose 1,5-diphosphate carboxylase and increase in proteolytic activity during senescence of detailed primary barley leaves. *Plant Physiol.* 55: 1009–1015.

Podolsky, A. S. 1984. Combined phenological forecasting as a method for the study and management of agroecosystems. *In* Podolsky, A. S. (ed.), *New Phenology: Elements of Mathematical Forecasting in Ecology*. Wiley, New York, pp. 337–350.

Price, P. W. 1981. Semiochemicals in evolutionary time. *In* Nordlund, D. A., R. L. Jones, and W. L. Lewis (eds.), *Semiochemicals, Their Role in Pest Control*. Wiley, New York, pp. 251–279.

Ramoska, W. A. and T. Todd. 1985. Variation in efficacy and viability of *Beauveria bassiana* in the chinch bug (Hemiptera: Lygaeidae) as a result of feeding activity on selected host plants. *Environ. Entomol.* 14: 146–148.

Reese, J. C. 1983. Nutrient–allelochemical interactions in host plant resistance. *In* Hedin, P. A. (ed.), *Plant Resistance to Insects. ACS Symp. Ser.* 208, pp. 231–264.

Rhoades, D. F. 1983. Herbivore population dynamics and plant chemistry. *In* Denno, R. F., and M. S. McClure (eds.), *Variable Plants and Herbivores in Natural and Managed Systems*. Academic Press, New York, pp. 155–220.

Rhodes, M. J. C., L. S. C. Wooltorton, and E. J. Lorenco. 1979. Purification and properties of hydroxycinnamoyl Co. A quinate hydroxycinnamoyl transferase from potatoes. *Phytochemistry* 18: 1125–1129.

Riskallah, M. R. 1984. Influence of posttreatment temperature on the toxicity of pyrethroid insecticides to susceptible and resistant larvae of the Egyptian cotton leafworm, *Spodoptera littoralis* (Boisd.). *Experientia* 40: 188–190.

Schalk, J. M., S. D. Kindler, and G. D. Manglitz. 1969. Temperature and preference of the spotted alfalfa aphid for resistant and susceptible alfalfa plants. *J. Econ. Entomol.* 62: 1000–1003.

Schoonhoven, L. M. 1981. Chemical mediators between plants and phytophagous insects. *In* Nordlund, D. A., R. L. Jones, and W. J. Lewis (eds.), *Semiochemicals: Their Role in Insect Control*. Wiley, New York, pp. 31–50.

Schweissing, F. C., and G. Wilde. 1978. Temperature influence on greenbug resistance of crops in the seedling state. *Environ. Entomol.* 7: 831–834.

Schweissing, F. C., and G. Wilde. 1979a. Temperature and plant nutrient effects of resistance of seedling sorghum to the greenbug. *J. Econ. Entomol.* 72: 20–23.

Schweissing, F. C., and G. Wilde. 1979b. Predisposition and nonpreference of greenbug for certain host cultivars. *Environ. Entomol.* 8: 1070–1072.

Scriber, J. M. 1984. Host-plant suitability. *In* Bell, W. J., and R. J. Carde (eds.), *Chemical Ecology of Insects*. Sinauer Assoc., Sunderland, MD, pp. 159–202.

Sosa, O., Jr., and J. E. Foster. 1976. Temperature and the expression of resistance in wheat to the Hessian fly. *Environ. Entomol.* 5: 333–336.

Sparks, T. C., M. H. Shour, and E. G. Wellemeyer. 1982. Temperature–toxicity relationships of pyrethroids on three lepidopterans. *J. Econ. Entomol.* 75: 643–646.

Sparks, T. C., A. M. Pavloff, R. L. Rose, and D. F. Clower. 1983. Temperature–toxicity relationships of pyrethroids on *Heliothis virescens* (F.) (Lepidoptera: Noctuidae) and *Anthonomus grandis grandis* Boheman (Coleoptera: Curculionidae). *J. Econ. Entomol.* 76: 243–246.

Städler, E. 1986. Oviposition and feeding stimuli in leaf surface waxes. *In* Juniper, B., and R. Southwood (eds.), *Insects and the Plant Surface*. Edward Arnold, London, pp. 105–121.

Staedler, J. P. 1977. Sensory aspects of insect–plant interactions. In *Proc. 15th Int. Cong. Entomol.*, Washington, DC, pp. 228–248.

Starks, K. J., E. A. Wood, Jr., and G. L. Teetes. 1973. Effects of temperature on preference of two greenbug biotypes for sorghum selections. *Environ. Entomol.* 2: 351–354.

Stephanou, G. S., N. Alahiotis, V. J. Marmaras, and C. Christodoulou. 1983. Heat shock response in *Ceratitis capitata*. *Comp. Biochem. and Physiol.* 74:425–432.

Swain, T. 1977. Secondary compounds as protective agents. *Annu. Rev. Plant Physiol.* 28: 479–501.

Syvertsen, J. P., and Y. Levy. 1982. Diurnal changes in citrus leaf thickness, leaf water potential and leaf to air temperature differences. *J. Exp. Bot.* 33: 783–789.

Taylor, A. O., C. R. Slack, and H. G. McPherson. 1974. Plants under climatic stress. IV. Chilling and light effects on photosynthetic enzymes of *Sorghum* and maize. *Plant Physiol.* 54: 696–701.

Tingey, W. M. and S. R. Singh. 1980. Environmental factors influencing the magnitude and expression of resistance. *In* Maxwell, F. G., and P. R. Jennings (eds.), *Breeding Plants Resistant to Insects*. Wiley, New York, pp. 87–113.

Tyler, P. S., and T. J. Binns. 1982. The influence of temperature on the susceptibility to eight organophosphorus insecticides of susceptible and resistant strains of *Tribolium castaneum*, *Oryzaephilus surinamensis*, and *Stiophilus granarius*. *J. Stored Products Res.* 18:13–19.

Utkhede, R. S., and H. K. Jain. 1976. Ribonucleic acid and high temperature sensitivity in wheat. *Cytologia* 41: 1–4.

Visscher-Neumann, S. 1971. Studies on the embryogenesis of *Aulocara elliotti* (Orthoptera: Acrididae). III. Influence of maternal environment and aging on development of progeny. *Ann. Entomol. Soc. Am.* 64: 1057–1074.

Visscher-Neumann, S. 1980. Regulation of grasshopper fecundity, longevity and egg viability by plant growth hormones. *Experientia* 36: 130–131.

Visscher-Neumann, S., R. Lund, and W. Whitmore. 1979. Host plant growth temperatures and insect rearing temperatures influence reproduction and longevity in the grasshopper, *Aulocara elliotti* (Orthoptera: Acrididae). *Environ. Entomol.* 8: 253–258.

Wall, C. and J. N. Perry. 1983. Further observations on the responses of male pea moth, *Cydia nigricana* to vegetation previously exposed to sex attractant. *Entomol. Exp. Appl.* 33: 112–116.

Wall, C., D. M. Sturgeon, A. R. Greenway, and J. N. Perry. 1981. Contamination of vegetation with synthetic sex attractant released from traps for the pea moth *Cydia nigricana*. *Entomol. Exp. Appl.* 30: 111–115.

White, T. C. R. 1974. A hypothesis to explain outbreaks of looper caterpillars with special reference to populations of *Selidosema suavis* in a plantation of *Pinus radiata* in New Zealand. *Oceologia* 16: 279–301.

White, T. C. R. 1976. Weather, food, and plagues of locusts. *Oecologia* 22: 119–134.

Wigglesworth, V. B. 1965. Water and temperature. *In* Wigglesworth, V. B., *The Principles of Insect Physiology*. Methuen, London, pp. 594–628.

Wilkinson, C. F. 1976. Insecticide interactions. *In* Wilkinson, C. F., (ed.), *Insecticide Biochemistry and Physiology*. Plenum, New York, pp. 605–647.

Willmer, P. 1986. Microclimatic effects on insects at the plant surface. *In* Juniper, B., and R. Southwood (eds.), *Insects and the Plant Surface*. Edward Arnold, London, pp. 65–80.

Wilson, K. G., R. E. Stinner, and R. L. Rabb. 1982. Effects of temperature, relative humidity, and host plant on larval survival of the Mexican bean beetle *Epilachna varivestis* Mulsant. *Environ. Entomol.* 11: 121–126.

Wood, E. A., and K. J. Starks. 1972. Effect of temperature and host plant interaction on the biology of three biotypes of the greenbug. *Environ. Entomol.* 1: 230–234.

Woodhead, S. 1981. Environmental and biotic factors affecting the phenolic content of different cultivars of *Sorghum bicolor*. *J. Chem. Ecol.* 7: 1035–1047.

Woodhead, S. and R. F. Chapman. 1986. Insect behavior and the chemistry of plant surface waxes. *In* Juniper, B., and R. Southwood (eds.), *Insects and the Plant Surface*. Edward Arnold, London, pp. 123–135.

5

EFFECTS OF ELECTROMAGNETIC RADIATION ON INSECT–PLANT INTERACTIONS

May Berenbaum

Department of Entomology
University of Illinois
Urbana, Illinois

1. INTRODUCTION

Electromagnetic radiation (EMR) is energy that is propagated in wave motion through space via variation in electric and magnetic fields. Its most familiar manifestation is as visible light, whose behavior is typical of the behavior of other forms of EMR. All EM waves in a vacuum, for example,

Table 1. Electromagnetic Spectrum

Type of Radiation	Natural Source	Wavelength (nm)
Gamma	Radioactive minerals; cosmic rays	0.0001–0.14
X-rays	Sun–low intensity	0.0005–20
Hard		0.0005–0.1
Soft		0.1–20
Ultraviolet	Sun	40–390
UV-C		40–286
UV-B		286–320
UV-A		320–390
Visible	Sun	390–780
Violet		390–430
Blue		430–470
Blue-green		470–500
Green		500–530
Yellow-green		530–560
Yellow		560–590
Orange		590–620
Red		620–780
Infrared	Sun	$780–4 \times 10^5$
Near		$780–2 \times 10^3$
Far		$2 \times 10^3–4 \times 10^5$
Hertzian waves	Sun–low intensity	$10^5–3 \times 10^{13}$
Space heating		$10^5–10^6$
Radio		$10^6–10^{12}$
Radar		$10^6–10^9$
Television		$10^9–10^{11}$
Power, ac		10^{15}

From Giese 1976.

travel at the speed of light (300,000 km/sec), and all forms of EMR share with light the properties of wave motion, namely, reflection, refraction, diffraction, interference, and polarization.

Waves can be classified by the frequency at which they oscillate (measured in hertz) or by wavelength, the distance between peaks of oscillation (measured in variants of a meter, ranging from nanometers to kilometers). The orderly arrangement of waves by frequency and wavelength makes up the electromagnetic spectrum (Table 1). The different regions of the spectrum vary in their importance to life on earth, in general, and to plant life, in particular. By far the major source of EMR for life on earth is solar radiation. Approximately half of the solar spectrum is visible light, and much of the remainder is infrared (IR) or heat (Richardson 1977). Once solar radiation encounters the atmosphere, it is substantially modified. Ultraviolet light (UV) is absorbed by ozone, oxygen, and nitrogen, while infrared is

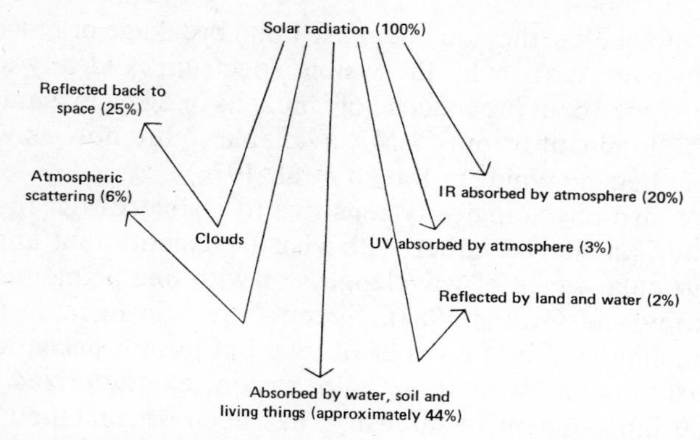

Figure 1. Fate of incoming solar radiation. (Giese 1976, Richardson 1977)

absorbed by carbon dioxide and water. Blue and ultraviolet wavelengths are particularly prone to scattering (and thereby account for the perceived color of clear sky). Heavy cloud cover can modify incoming solar radiation even further by attenuating it 80% or more.

Overall, an accounting of the fate of incoming solar radiation reveals that, on an average, a quarter to a third is reflected by the cloud layer, 15 to 20% absorbed by the atmosphere, and 5 to 10% is diffuse radiation scattered back to space by particulate matter. Thus only half or less of solar radiation reaches the earth's surface to interact with living things (Giese 1976, Richardson 1977) (Fig. 1).

In that visible light determines photosynthetic activity of plants, it directly affects energy relations at all higher trophic levels in plant-based food webs. Photosynthesis is the process by which radiant energy from the sun is made available to living things, primarily by green plants, although there also exist photosynthetic bacteria. Different regions of EMR affect photosynthesis in characteristic ways; only about 50% of the light actually reaching the earth's surface is "photosynthetically active" (Ricklefs 1973). Plant chloroplasts contain various forms of the pigment chlorophyll, the chief harvester of photosynthetically active radiation. Chlorophylls absorb well in red and blue and poorly in green and yellow (and their tendency to reflect light in this region accounts for the green color of the majority of plants on earth). Accessory pigments include structural variants of chlorophyll (known as chlorophyll *b*, *c*, *c*2, or *d*), carotenoids (primarily yellow pigments), and phycobilins (the red or blue protein pigments of some algae).

Visible wavelengths may be important to life on earth because they are energy rich without being damaging to hydrogen bonds, which are vital to basic life processes. Wavelengths less than 200 nm are ionizing in that they propel electrons out of atoms. Wavelengths greater than the range of the visible spectrum are absorbed by water, which comprises about 75% of the weight of living organisms. Their energy is sufficiently low that once in contact with molecules, they do not cause bond breakage or rearrangements. To argue, however, as to why the visible spectrum is ideally suited to life on earth is to argue about precedence of chickens or eggs, inasmuch as visible light is the predominant form of EMR available to life now as well as when primordial life began (Wald, in Raven et al. 1976).

Insects are also physiologically sensitive to visible EMR. Insect spectral discrimination is achieved either with visual pigments that absorb specifically in a particular range of wavelengths or with one pigment mediated by filters (Prokopy and Owens 1983). Spectral-specific behavioral responses have been documented in a number of taxa but do not necessarily indicate that color vision is involved. True color vision, characterized by discrimination of hue independent of intensity, has been documented in far fewer taxa. The great majority of herbivorous insects tested, irrespective of the mechanism of perception, display visual sensitivity to wavelengths between 350 to 600 nm, peaking around 550 nm, the yellow region of the spectrum. Response to yellow color has been demonstrated in at least four orders of insect herbivores (Prokopy and Owens 1983). Prokopy and Owens speculate that yellow may be a "super-normal forage-type stimulus, emitting peak energy in the same bandwidth of the insect-visible spectrum as foliage emits peak energy but at a greater intensity." In the primarily herbivorous order Lepidoptera, the stemmata (simple eyes) of the larvae respond extensively to the green–blue-green region (Philogene and McNeil 1984), corresponding to most foliar hues.

Any change in the physical, nutritional, allelochemical, or populational status of a plant species can conceivably affect the insects associated with that plant. Although insects can and do respond both behaviorally and physiologically to light independent of their host plant (Philogene and McNeil 1984), it should be remembered that for herbivorous insects the host plant is a source of both food and shelter (Southwood 1973); thus effects of electromagnetic radiation on a plant will almost certainly have effects on its insect herbivores. Changes in growth form, photosynthetic efficiency and the distribution and abundance of secondary chemicals (allelochemicals) caused by EMR all stand to affect insect performance on a particular host. Some of these changes are general physiological stress responses; others, however, may be characteristic of exposure to particular wavelengths of EMR.

2. VISIBLE LIGHT (390–780 nm) AND NUTRITIONAL VALUE OF PLANTS

Light energy is, by convention, one section of the electromagnetic spectrum. What we perceive as visible light is only a narrow band, ranging from about 380 to 750 nm; ultraviolet light at the earth's surface extends from about 280 to 380 nm. The range of wavelengths that is ecologically important to plants is an even smaller subset of the electromagnetic spectrum. Wavelengths in the violet and red range (400 to 450 nm and 600 to 700 nm, respectively) are known as photosynthetically active radiation. This range corresponds to the absorption spectra of chlorophylls. Despite the ability of chlorophyll to absorb at wavelengths above 700 nm, photosynthetic efficiency is greatly reduced in this region of the spectrum. This reduced efficiency is due at least in part to the fact that chlorophyll is the only plant pigment that can absorb these wavelengths. Accessory pigments such as carotenoids absorb in the 400–550 nm region. Physical proximity of these pigments to chlorophyll, as well as overlap in absorption spectra, allow resonance transfer of energy with concomitant enhancement of photosynthetic efficiency (Krogmann, 1973). Blue light (380 to 500 nm) is important in regulating physiological processes such as phototropism and photocontrol of flowering, stomatal opening, seed germination and shifting of circadian rhythms (Schmidt 1984).

Inasmuch as photosynthetic efficiency determines the rate of accumulation of photosynthate, it also affects the suitability of plant tissue as food for insects. Shading, such as by a vegetational canopy, may reduce photosynthesis to such an extent that carbohydrates may affect the physical integrity of a plant in such a way as to leave it vulnerable to insect attack. Resistance of bread wheats to the wheat stem sawfly *Cephus cinctus* Norton, for example, is correlated with "stem solidness"; stem solidness in turn appeared to be correlated with the amount of seasonal sunshine. Holmes (1984) shaded resistant wheat with cotton sheeting and observed a significant increase in the percentage of infested stems among the shaded plants. The cotton sheeting screened out about 45% of the radiation from 380 to 650 nm and up to 49% in the higher wavelengths. When plants were shaded by red and yellow filters, which excluded up to 95% of the radiation at 575 nm and 60% at 600 nm, infestation rates were even higher (and stem solidness accordingly lower), suggesting that reduced photosynthetic efficiency is a factor in resistance. Decreased resistance was also observed under low light conditions for *Ostrinia nubilalis* (Hübner) (European corn borer) on field corn *Zea mays* L. (Manuwoto and Scriber 1985), although the exact causal mechanism was not determined.

Photosynthetic rate can affect the suitability of plants as food for insects in ways other than simply decreasing available carbohydrate. There is some evidence that the production of allelochemicals, plant chemicals not known to play any role in primary physiological processes but which mediate interactions between plants and other organisms, may be limited by available energy. Although the evidence is not overwhelming, it is suggestive. For example, Croteau et al. 1972 demonstrated that administering sucrose (a product of photosynthesis) to peppermint along with the biosynthetic precursor mevalonic acid significantly enhances its incorporation into mono- and sesquiterpenes. Exposure to elevated labeled CO_2 in the light produces similar results. Terpene synthesis in peppermint thus appears to be limited by available photosynthetic energy. Since both mono- and sesquiterpenes are antifeedants or toxins to a wide variety of herbivorous insects (Koul 1982), quantitative variation in these allelochemicals due to photosynthetic limitations may greatly affect resistance. In another example (Larsson et al. 1986), clones of the willow *Salix dasyclados* Wimm. were grown under three regimes: low light with access to nutrients (to limit carbon supply from photosynthesis), high light with restricted nutrients (to limit nitrogen), and high light with access to nutrients. Leaves from trees under low light conditions (and thus with limited carbon supply) produced significantly lower amounts of phenolic compounds and suffered approximately five times more feeding damage than leaves from other treatments when exposed to the willow beetle *Galerucella lineola* F. in bioassays.

3. SOLAR ULTRAVIOLET LIGHT (286–390 nm)

Light radiation outside the photosynthetically active wavelengths can nevertheless affect the relationship between plants and insect herbivores. Ultraviolet light, for example, is energy rich; the shorter the wavelength, the higher the energy content. UV light alone can be a stress factor for plants; it has been associated with impairment of cytoplasmic streaming, tumor induction, and genetic mutation (although many of these studies used artificial sources of UV light that emit primarily in the far-UV range, from 200 to 280 nm, not much of which ever makes it to the earth's surface) (Caldwell 1971). There is also some evidence that increased UV can inhibit or interfere with photosynthesis. At high altitudes, where UV intensities are greater, photosynthesis can be reduced up to 25%, presumably through inhibition of photophosphorylation and the Hill reaction, although photodestruction of chlorophyll may also be involved (Klein 1978).

3.1 Quantitative Effects of Ultraviolet Light on Plant Chemistry

Plants are not entirely without defenses against UV. These defenses take three forms (Caldwell 1971). First and foremost is UV absorption by pigments, waxes, and other secondary chemicals. Photoreactivation, or repair of damage caused by short wavelength radiation after subsequent exposure to longer wavelength radiation by means of light-activated enzymes, appears to occur as well. Also suspected to occur in plants is so-called "dark repair," in which lesions, primarily in DNA or RNA, are excised by specific enzymes that are active independent of light. Of these three mechanisms, accumulation of UV-absorbing pigments in the presence of UV light and other wavelengths is known to occur in a number of plant taxa and is presumed to result in protection from damaging wavelengths (Table 2). Photoinduction involves a number of biosynthetic pathways; several key biosynthetic enzymes are photoinduced by UV light. Chief among these is phenylalanine ammonia-lyase, a catalyst for the biosynthesis of a number of chemical classes, in-

Table 2. Plant Compounds Induced or Increased by Light

Plant Compound	Light Source	Plant Source
Alkaloids	Red and IR	Tobacco
Alkaloids	Visible	Lupines
Alkaloids	UV	Various Solanaceae
Anthocyanins	Visible	Many plants
Betacyanins	Red	Centrospermae
Cannabinoids	UV	Marijuana
Carotenoids	Blue	Many plants
Cardenolides	Visible	*Digitalis lanata*
DIMBOA	Visible	*Zea mays* L.
Flavonoids	UV	Many plants
Furanocoumarins	UV	Wild parsnip
Isoflavonoids	UV	Soybeans
Tannins	"Sunlight"	Oak
Terpenes	"Sunlight"	*Hymenaea courbaril* L.

From Berenbaum 1987.

cluding phenylpropanoids, coumarins, flavonoids, acetophenones, and lignans (Fig. 2). At least two other enzymes of phenylpropanoid metabolism and six of the 13 enzymes of flavone and flavonol glycoside pathways are photoinduced (Heller et al. 1979). Moreover some biosynthetic pathways are actually catalyzed by light; at least two steps in the biosynthesis of coumarin, a known toxin for herbivorous insects (Mansour 1982), are light-dependent isomerization reactions (Towers 1984). As a result of both photoinduction and photocatalysis, a wide variety of compounds increase in concentration in the presence of sunlight. Many of these compounds in turn also affect feeding behavior (Koul 1982, Bernays and Chapman 1977). Decreased production of glycoalkaloids in foliage under shaded conditions, for example, is thought to lead to reduced resistance in potato species (Tingey and Singh 1980).

Light may also play a role in determining the suitability of plant *parts* as food. Outer tissue layers of fruits or vegetables, for example, are almost invariably higher in flavonoids than are inner tissue layers. The total concentration of quercitin, a growth inhibitor for noctuids (Elliger et al. 1980), in outer epidermis of onion is two orders of magnitude higher than inner epidermis, and the outer leaves of lettuce and endive are one to two orders of magnitude greater in quercitin content than are inner leaves (Hermann 1976). The absence, as opposed to attenuation, of particular wavelengths, particularly UV light, may also greatly affect allelochemical content of plants. Quercitin concentrations of lettuce are 40 times greater in open air than in greenhouses (Hermann 1976), whose glass filters out UV-A radiation. Perhaps the greatly decreased flavonoid content of greenhouse-grown vegetables may render them particularly vulnerable to aphids, whiteflies, and

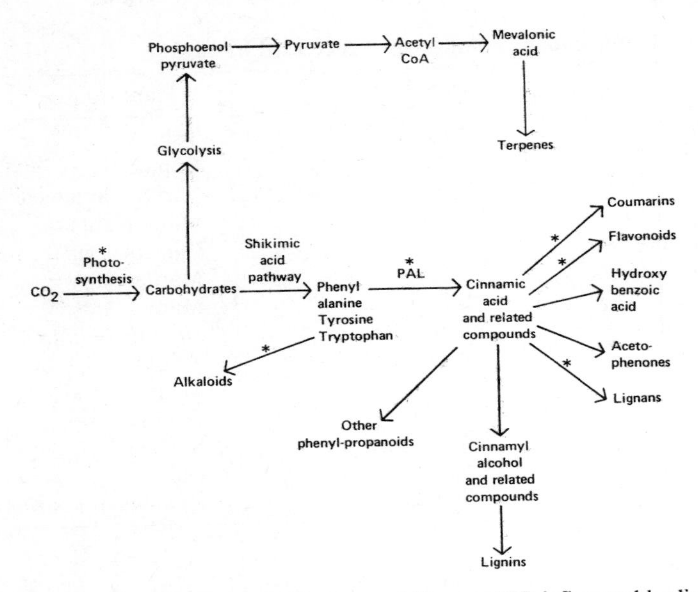

Figure 2. Biosynthetic pathways (Bonner and Varner 1965) influenced by light (*).

other greenhouse pests. UV light can affect the qualitative composition as well as the quantity of allelochemicals. The wild parsnip, *Pastinaca sativa* L., produces at least six different furanocoumarins, chemicals with known behavioral and toxicological effects on herbivorous insects. When wild parsnips are grown outdoors under a filter excluding UV, the absolute amounts as well as the relative concentrations of the five principal furanocoumarins change (Zangerl and Berenbaum 1987). Since the five furanocoumarins vary in toxicity (Berenbaum 1985), a plant growing in an environment low in UV light (e.g., under a vegetative canopy that transmits less than 10% of light in the UV region—Robberecht and Caldwell 1978) may be rendered more susceptible to attack.

Several studies have been done in which exposure to light was actually shown to affect plant suitability to insects. Soybeans, when exposed to far UV (254 nm), accumulate a variety of isoflavonoids, including glyceollins, which are known to be antimicrobial and are thought to possess other forms of biological activity (Hart et al. 1983). When UV-irradiated discs were presented to Mexican bean beetles along with nonirradiated discs, both larvae and adults showed a distinct preference for nonirradiated discs, as evidenced by significant differences in the number of feeding ridges on paired discs. This nonpreference for irradiated discs is consistent with the idea that UV-induced production of isoflavonoids reduces the palatability of soybean foliage to insects. However, in another study continuous illumination for eight weeks (wavelengths not specified) increased the susceptibility of a resistant soybean variety, *Glycine max* (L.) Merrill, concomitant with a decrease in a phenylpropanoid component (Khan et al. 1986). In yet another study other illumination-inducible resistance factors in soybean have been described for the bean fly, *Ophiomyia centrosematis* (de Meijere) (Chiang and Norris 1984).

Exposure to ultraviolet radiation may bring about physiological and chemical changes in plant tissue that renders it not necessarily more toxic to insects but rather less nutritionally suitable. Epicuticular waxes, for example, increase in quantity, often substantially, in plants exposed to UV light; irradiation by UV-B increased total wax by approximately 25% in at least three plant species (Steinmuller and Tevini 1985). Inasmuch as waxes are largely indigestible to herbivorous insects, an increase in the wax content of foliage may reduce its digestibility on a per weight basis.

It is clear that UV radiation affects changes in plant chemistry and that these changes alter the quality of plant tissue as food for insects. What is not so clear is whether insects are a driving force for these chemical changes, or if they are for once innocent bystanders of a sort. In coping with an abiotic stress factor, plants may inadvertently gain some protection against any biotic stress factors that may occur at the same time. The importance of insect herbivory as a selective force in modifying plant responses to light is a difficult, if not impossible, issue to resolve. Insects may affect plant responses to UV light as an environmental stress in the same way it has been suggested that they affect plant response to nutritional stress. Janzen (1974) suggested that, in conditions of limiting nutrition, retaining photosynthate and photosynthetic tissue is at a premium and plants therefore are likely to invest more heavily in defensive substances. If intense UV light is an indicator of poor environmental quality (indeed, intense sunlight may be associated with water deficits and other environmental unpleasantness), induction of UV-absorbing chemicals with antiherbivore properties may be a double benefit.

3.2 Qualitative Effects of Ultraviolet Light on Plant Chemistry

Although some effects of light may only indirectly mediate plant–insect interactions, at least one phenomenon is directly involved in determining allelochemical toxicity; this is the phenomenon of photactivation. Many plant chemicals can absorb photons of light energy at particular wavelengths. This energy can transform a molecule from its relatively nonreactive ground state to a highly reactive excited state in which electrons, via photon energy, have moved to higher energy orbitals. Excited states are generally short-lived as well as reactive; their absorbed energy can be released in three ways. A molecule can return to ground state by emitting a photon of light energy (i.e., it can fluoresce), it can return to ground state by emitting heat, or it can return to ground state by transferring its energy to another molecule (Giese 1976). If this energy is absorbed by a molecule critical to the physiological function of an organism such as an insect, toxicity can ensue. The

Table 3. Plant-Derived Phototoxins with Insecticidal Properties

Class	Source
Acetylenes	Compositae
Benzopyrans and furans	Compositae
Beta-carboline alkaloids	Rutaceae, Simaroubaceae
Extended quinones	Guttiferae, Polygonaceae
Furanocoumarins	Leguminosae, Moraceae, Rutaceae, Umbelliferae, Compositae, Solanaceae, Thymeliaceae
Furanochromones	Rutaceae, Umbelliferae
Furoquinoline alkaloids	Rutaceae
Benzylisoquinoline alkaloids	Berberidaceae, Rutaceae, Rubiaceae
Thiophenes	Compositae

From Berenbaum 1987.

furanocoumarins, for example, produced primarily by plants in the Rutaceae and Umbelliferae (Murray et al. 1982), form excited states that react preferentially with pyrimidine bases in DNA. Linear furanocoumarins, which can form crosslinks between DNA strands if sufficient UV energy is present, are thus generally toxic to most organisms (Berenbaum 1978 and references therein).

Photoactivation takes on another form when a molecule in an excited state interacts with ground state oxygen to form excited oxygen molecules such as singlet oxygen (Larson 1986). Singlet oxygen itself is highly reactive and can damage proteins, membrane lipids, and DNA. Plant products that transfer their excited state energy to oxygen in the presence of light include thiophenes found throughout the family Compositae (Downum and Rodriguez 1986) and extended quinones of the Guttiferae and Polygonaceae (Knox and Dodge 1985).

At least nine biosynthetically distinct classes of plant chemicals, produced by at least a dozen different plant families, are known to be phototoxic to insects (Table 3). In the case of furanocoumarins, toxicity to *Heliothis zea* (Boddie), the corn earworm, is directly proportional to the amount of UV light present (Berenbaum and Zangerl 1987). This increased efficacy in the presence of increased UV intensity, which is likely to hold true for other classes of phototoxins as well, may account for the observation (Downum and Rodriguez 1986, Downum 1986) that phototoxic plants seem particularly common in high light environments such as tropical forests and dry deserts. It may also account for variation in the local distribution of plant species. For example, Berenbaum (1981) found that umbelliferous plants containing phototoxic furanocoumarins are far more frequently found in high light environments than are confamilials that lack furanocoumarins altogether.

Phototoxicity, though, is a global phenomenon and can be found in habitats ranging from tropical forests to railroad rights of way and city parking lots. Phototoxic genera are found in the Umbelliferae, Compositae, and Leguminosae, three of the largest plant families. Phototoxicity does not necessarily occur in related plant families, however, nor is it attributable to biosynthetically related plant chemicals. Furanocoumarins, for example, are derived from the shikimic acid pathway, whereas polyacetylenes and thiophenes are acetate derived. Why so many plants have converged on a common mode of defense—use of solar energy to potentiate endogenous allelochemicals—may be a question of energetic efficiency. By using exogenous energy sources to activate chemicals, plants presumably conserve metabolic energy that can be allocated to other functions. Phototoxicity, then, may be part of a strategic defense package designed to obtain maximum chemical reactivity for minimum energy investment.

3.3 Human Influences on Plant Responses to Ultraviolet Light

With anthropogenic inputs into the earth's atmosphere in the form of nitric oxide emissions and chlorofluorocarbons increasing each year, it is important to identify potential problems associated with ozone depletion concomitant with these inputs. Chief among these problems is the increased amount of biologically damaging shortwave ultraviolet light (UV-B) that will reach the earth's surface due to the decreased ability of stratospheric ozone to screen out these wavelengths. By the end of the century, the depletion of ozone is estimated to be between 7 and 16% (NASA 1979, NRC 1979); a 5% decrease in ozone over the midcentral United States is estimated to bring about a 26% increase in UV at 297.5 nm (NAS 1973).

The adverse effects of UV-B radiation on plants have already been described (*see Sec. 3*). Although UV-B wavelengths are not used directly for photosynthesis, they can affect the process indirectly by altering enzyme efficiency; in addition plant allelochemicals change in response to UV exposure. One major consequence of UV-induced changes in photosynthetic efficiency and secondary chemistry in plant tissue is that these changes may affect plant resistance to damage by herbivorous insects. Predicting exactly what effects increased UV radiation will have on insect–plant interactions is a tricky business. Changes in photosynthesis can alter the nutritional compositon of plant tissue to affect insect growth and development, whereas changes in plant secondary chemistry directly affect the behavior and physiology of insects as they feed. Yield losses due to decreased photosynthetic efficiency may be more than offset, then, by increased resistance to insect pests. Although many studies have investigated the consequences of elevated UV on plant physiology, few if any have investigated the effects of these changes in physiology on plant-associated insects. In view of the fact that agricultural insect pests destroy almost 20% of crop biomass produced in the world today, examining the impact of increasing UV on interactions between crop plants and their insect associates is of critical importance to allow for the development of management practices for the future.

4. GAMMA RAYS (0.0001–0.14 nm)

Humans have intervened in the natural order of things and, courtesy of a variety of electronic or electrical devices, are greatly altering the distribution of EMR in the environment at both ends of the spectrum. Gamma rays result from nuclear testing; microwaves, radar, television, and radio waves emanate from electronic devices. Responses of plants to these forms of EMR, hitherto present in the environment in only very attentuated form, are varied and unpredictable.

Gamma rays are produced by radioactive nuclei. Although natural sources of radiation do exist, the chief source of gamma radiation on earth is human experimentation with or the use of nuclear energy. Plants vary widely in their sensitivity to gamma radiation. Some, like conifers (Grosch and Hopwood 1979), have sensitivities on the order of one kilorad (KR), comparable to that of mammals, whereas other species, notably bryophytes and other "lower" plants, can tolerate dosages in excess of 50 KR. Resistance in plants appears to be inversely correlated with chromosome size. Species with large chromosomes and with a large interphase volume appear to be highly sensitive. Physiognomy is another correlate with resistance to gamma radiation; stature is generally inversely correlated with resistance. Thallophytes, then, are more resistant than herbs, which in turn are more resistant than are shrubs and trees. Finally, actively dividing tissue (e.g., meristem) is, not surprisingly, sensitive to the disruptive effects of gamma radiation on the replication.

Ecologically, then, there are definite patterns of sensitivity among plants to ionizing radiation (Daniel and Platt 1968). Early successional species, with low stature, predominate in irradiated communities. Following exposure to ionizing radiation, forest communities revert to an earlier successional stage (Woodwell 1967, Woodwell and Oosting 1965). The same pertains in tall grass communities; following radiation, tall forbs are replaced by low annuals. Plants reproducing sexually are generally replaced in the community (due to seed sterility) by those species capable of vegetative propagation; by the same token, plants with underground stolons or rhizomes are able to recover faster from exposure. In general, gamma radiation decreases life span and the ability to withstand stresses from other sources, such as insect herbivory (Woodwell 1962).

Plant responses to ionizing radiation affect insect performance due to the changes brought about in plant physiology, morphology, and ecology. Phenological changes induced by ionizing radiation can interfere with host location by specialized herbivores. In one study of a field adjacent to an air-shielded nuclear reactor, about 50% of the plant species showed phenological rearrangement, with early flowering and seed production. Annuals exposed to 10 KR or more completed their life cycles up to an entire month faster. Leaves of hardwood trees and herbaceous plants fall one to three weeks earlier, relative to unexposed plants, after exposure to gamma radiation

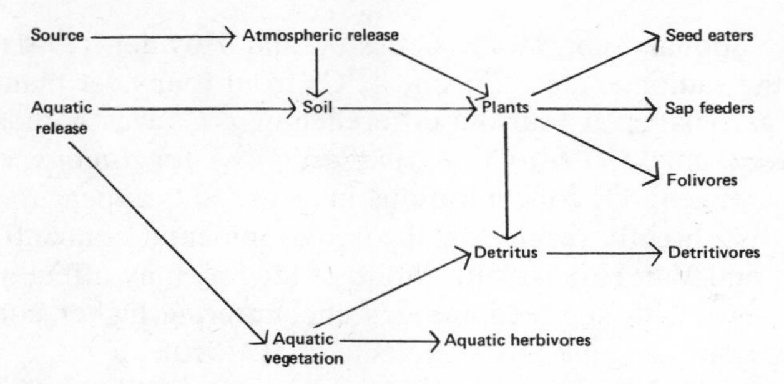

Figure 3. Radionuclide transfer pathways affecting herbivores in an ecosystem. (Modified from Kitchings et al. 1976)

(Grosch and Hopwood 1979), a time interval that may prevent leaf-mining insects from completing development before abscission. After leaf abscission, survivorship of leaf-mining insects drops dramatically (Faeth et al. 1981).

The efficacy of visual orientation to host plants by insects can be altered by color changes induced by ionizing radiation. Daniel and Platt (1968) showed that grasses and some forbs displayed increased red pigmentation, and yellowing of foliage occurred in at least three species of forbs. Since yellow is a general visual orientation cue for aphids and other homopterans (Prokopy and Owens 1983), plants exposed to radiation may be discovered more readily by insects and colonized more rapidly. Changes in plant morphology may also influence visual orientation by insects (Prokopy and Owens 1983). Developmental abnormalities, including leaf "dwarfing, thickening, altered shape or texture, puckering, abnormal curvature, marginal curling, distorted venation, fusions and mosaic-like color change" (Grosch and Hopwood 1979), can affect internal feeders such as leaf miners restricted to particular types of plant tissues.

Because ionizing radiation can change plant community composition, it can also influence insect community composition. Those species associated with late successional plants will be replaced as their hosts are replaced by early successional species. As a result the herbivore community may tend to exhibit more generalized feeding habits (Brown 1984). Insect density and diversity will likely be affected by the decrease in plant primary productivity resulting from exposure to radiation, inasmuch as primary production determines the amount of energy available to higher trophic levels.

Apart from ionizing radiation itself, radioactive fission products are processed differentially by plant species; concentration factors for algae and higher plants range from 10 to 3000 for the element strontium and from 50 to 25,000 for cesium, for example (Reichle et al. 1970). Insect exposure to radionuclides thus depends on the identity of its host plants and its host specificity (Fig. 3). In a study of insect–vegetation relationships in White Oaks Lake, Tennessee, a holding basin for low level radioactive wastes from

Oak Ridge National Laboratory, Crossley and Howden (1961) compared transfer of the radionuclides ^{90}Sr and ^{137}Cs from four host plant genera to their insect associates. A fourfold difference in ^{137}Cs was noted among the four genera examined (2.6 $\mu Ci \times 10^{-4}/g$ dry wt for *Juncus* vs. 10.4 for *Polygonum*). In general, concentrations in arthropods appear to be indeterminate, that is, directly proportional to environmental concentrations (see also Reichle and Van Hook 1970). Mode of feeding may affect assimilation of radionuclides, with sap-feeding bugs encountering higher concentration than folivores on the same hosts (Reichle et al. 1970).

Although the concentration of radionuclides in insect herbivores is readily documented, the importance of the presence of radionuclides in the bodies of insect herbivores is more difficult to assess. In most arthropods lacking calcified exoskeletons, biological turnover and excretion are rapid and efficient, with the result that the biological half-lives for virtually all major elements is from 10 to 100 days and is under 10 days for such elements as Rb, Sr, Y, Na, P, K, Ca, and Co (Reichle et al. 1970). Insects and related arthropods may be more resistant to ionizing radiation than are most vertebrates since their external skeletal system does not incorporate radionuclides as does vertebrate bone; persistence of radionuclides is in part a function of binding by tissue proteins or metallic cations (Grosch and Hopwood 1979).

5. ELECTRIC FIELDS

Natural variation in the intensity of electric fields varies on average from 120 to 150 V/m (DC atmospheric field) in fair weather to 10,000 V/m during electrical storms. The DC geomagnetic field is relatively stable at 0.5 G. At a frequency of 50 to 60 Hz, however, intensity not only of the atmospheric field, but also the geomagnetic field drops to 10^{-4} V/m and 10^{-8} G. In contrast, electric fields of human origin range from 10 to 250 V/m and 0.001 to 25 G in fields of 60 Hz. Near a hair dryer the magnetic field was measured at more than 10 G (Sheppard and Eisenbud 1977). Reports on the effects of naturally occurring variation in electric and magnetic fields on plants are very hard to find (short of the obvious effects, e.g., being struck by lightning). Several anecdotal reports, however, suggest that severe storms, particularly dust storms in which the normal electrical balance of the atmosphere is disturbed, may cause browning and other forms of leaf damage (Diprose et al. 1984).

Considerably more work has been done on human-derived variation in electromagnetic radiation. Magnetic field strengths ranging from several hundred to several thousand Gauss do have effects on plant growth and mitotic index (references in Sheppard and Eisenbud 1977), and relatively strong DC electric fields affect growth rates and seed germination. To some extent electric fields may enhance germination by providing free electrons to react with and inactivate free radicals that would otherwise damage cell membranes (Sheppard and Eisenbud 1977 and references therein). Electric fields in excess of 100 to 130 kV/m are destructive, causing leaf browning and cell death, perhaps by causing accelerated enzyme activity. One common response to enhanced electrical activity is a darkening of green color (Diprose et al. 1984); again change in leaf color may affect host finding by visually orienting plant herbivores. Overall, currents less than 10^{-16} A do not affect plants; those between 10^{-15} and 10^{-9} may stimulate plant growth (perhaps due to generation of ions), and higher currents can kill plants.

This sensitivity to electric fields has been utilized in weed control, whereby electric currents are applied directly to plant tissue. Discharges of 30 to 50 kV of a millionth of a second duration sufficed to kill a variety of young weeds within four to six days (Diprose et al. 1984).

6. MICROWAVES

Studies of the effects of microwave radiation on plants have been been conducted almost exclusively by individuals interested in developing new methods for weed control; these studies have recently been reviewed by Diprose et al. (1984). As a result the major conclusion to be drawn from these studies is that plants can indeed be killed by microwave radiation; broad-leaved species, perhaps due to their larger absorptive surface area, are more susceptible than are grasses at lower energy levels. As to how microwaves kill plants, there is little consensus and few data; heat per se may not play a role. In one study (Davis as cited by Diprose et al. 1984) mortality of microwave-irradiated seeds depended on mass, volume, moisture content, and ether-soluble content; that ether-soluble content affects susceptibility suggests that plant secondary compounds may affect plant responses to microwave irradiation.

Without more definitive information on the sublethal physiological effects of microwave radiation on plants, it is difficult even to speculate on the effects of microwaves on insect–plant interactions, other than to say that radiation-induced mortality of plant species will leave host-specific insects without a source of food and will restrict the feeding options of more generalized herbivores. Needless to say, considerably more research in this area needs to be conducted, even if microwave radiation is not a cost-effective method of weed control at the present time, in order to assess its impact on the associated insect fauna of targeted weed species and its impact on nontarget plants and insects.

7. CONCLUSIONS

Drawing conclusions on the effects of electromagnetic radiation on the interaction between herbivorous insects and their host plants is a risky business, since data are sorely lacking. Studies that are conspicuously rare are those actually focused on the effects of natural variation in intensity and spectral distribution of radiation and those investigating the interactive effects of radiation with other environmental factors. Much of the present body of knowledge, such as it is, is speculative, and based on predictions of insect behavior in the face of physiological, ecological, and morphological changes in plants induced by electromagnetic radiation. Although these predictions may indeed be valid, the possibility exists that synergistic interactions occur. For example, the response of insects to a UV-induced change in plant chemistry may be influenced by the insect's response to changes in UV light intensity independent of its host plant. Such studies of interactions are indeed of value inasmuch as variations in electromagnetic radiation are as much a part of the insect's environment as is its host plant. For future studies ecologically relevant wavelengths should be incorporated into any experimental design purporting to test the effects of a plant or plant chemical on an insect herbivore.

In one case in particular, EMR may be a potent mediator of insect–plant coevolution. For those plants producing phototoxic defensive compounds, the efficacy of their defense depends not only on variation in levels of allelochemicals, which can be controlled to some extent by the plant, but also on variation in exogenous light levels, which are independent of plant control. Insects may over evolutionary time evolve resistance not to the phytochemical per se but rather to light as a potentiating factor. In one survey, the fauna of phototoxic plants in the family Umbelliferae contained a disproportionately high percentage of concealed feeders—leaf-rolling lepidopterans, leaf-mining dipterans, stem-boring coleopterans and lepidopterans, and the like (Berenbaum 1978). Concealed feeding, or feeding in such a manner as to reduce exposure to photoactivating wavelengths of light, is an effective defense against phototoxic plants (Berenbaum 1987, Sandberg 1987). Habitat characteristics are particularly important for phototoxic plants as well; growing in a shady spot may result in greatly reduced fitness due to increased intensity of insect herbivory.

Light and other forms of electromagnetic radiation have been largely ignored by ecologists interested in plant–insect interactions, yet there is increasingly more evidence to indicate that it is not a factor to be taken lightly (as it were); incorporating EMR as a standard parameter in future investigations—along with humidity, temperature, and other environmental factors known to affect insect–plant interactions—may well serve to add considerable insight and to clarify many observations that are otherwise inexplicable. Moreover, human inputs into soil, water, and atmosphere are radically changing the distribution and abundance of electromagnetic radiation in the earth's environment. The effects of some forms of input, such as electricity and gamma radiation, on plant–insect interactions have been studied to some extent; the effects of other forms of inputs, such as microwave, x-ray, radio, radar, and television, are essentially unknown. Since humans interact on a global scale with plants and compete with the insects that eat them, it is of vital importance to understand the impacts of anthropogenic alteration of EMR on earth on plant–insect interactions and to modify our behavior, either from the point of view of plant propagation and pest management or from the point of view of reducing EMR release, accordingly, to retain our competitive edge.

ACKNOWLEDGMENTS

I thank E. Heininger, R. Larson, J. Neal, J. Nitao, S. Sandberg, and A. Zangerl of the University of Illinois at Urbana-Champaign for valuable discussion and comments on the manuscript. This work has been supported by National Science Foundation grant BSR 835-1407, and a John Simon Guggenheim Fellowship.

REFERENCES

Berenbaum, M. 1978. Toxicity of a furanocoumarin to armyworms: a case of biosynthetic escape from insect herbivory. *Science* 201: 532–534.

Berenbaum, M. 1981. Furanocoumarin distribution and insect herbivory in the Umbelliferae: plant chemistry and community structure. *Ecology* 62: 1254–1266.

Berenbaum, M. 1985. Brementown revisited: interactions among allelochemicals in plants. *Rec. Adv. Phytochem.* 19: 139–169.

Berenbaum, M. 1987. Charge of the light brigade; insect adaptations to phototoxins. *In* Heitz, J. R., and K. R. Downum, (eds.), *Light Activated Pesticides*. *ACS Symp. Ser.* 339, pp. 206–216.

Berenbaum, M., and A. Zangerl. 1987. Defense and detente in the coevolutionary arms race: synergisms, syntheses, and sundry other sins. *In* Spencer, K. (ed.), *Chemical coevolution*. Pergamon, New York (in press).

Bernays, E. A., and R. F. Chapman. 1977. Deterrent chemicals as a basis of oligophagy in *Locusta migratoria* (L.). *Ecol. Entomol.* 2: 1–18.

Bonner, J., and J. E. Varner. 1985. *Plant Biochemistry*. Academic Press, New York.

Brown, V. K. 1984. Secondary succession: insect–plant relationships. *Bioscience* 34: 710–716.

Caldwell, M. M. 1971. Solar UV radiation and the growth and development of higher plants. *In* Giese, A. C. (ed.), *Photophysiology*. Academic Press, New York, pp. 131–177.

Chiang, H. S., and D. M. Norris. 1984. "Purple stem," a new indicator of soybean stem resistance to beanflies (Diptera: Agromyzidae). *J. Econ. Entomol.* 77: 121–125.

Crossley, D. A., and H. F. Howden. 1961. Insect–vegetation relationships in an area contaminated by radioactive wastes. *Ecology* 42: 302–317.

Croteau, R., A. Burbott, and W. D. Loomis. 1972. Apparent energy deficiency in mono- and sesqui-terpene biosynthesis in peppermint. *Phytochemistry* 11: 2937–2948.

Daniel, C, and R. Platt. 1968. Direct and indirect effects of short term ionizing radiation on old-field succession. *Ecol. Monog*. 38: 1–29.

Diprose, M., F. Benson, and A. Willis. 1984. The effect of externally applied electrostatic fields, microwave radiation, and electric currents on plants and other organisms, with special reference to weed control. *Bot. Rev.* 50: 171–223.

Downum, K. 1986. Photoactivated biocides from higher plants. *In* Green, M., and P. Hedin, (eds.), *Natural Resistance of Plants to Pests. ACS Symp. Ser.* 296, pp. 197–205.

Downum, K., and E. Rodriguez. 1986. Toxicological action and ecological importance of plant photosensitizers. *J. Chem. Ecol.* 12: 823–834.

Elliger, C. A., B. C. Chan, and A. Waiss. 1980. Flavonoids as larval growth inhibitors *Naturwissenschaften* 67: 358–360.

Faeth, S., E. F. Connor, and D. Simberloff. 1981. Early leaf abscission: a neglected source of mortality for folivores. *Am. Nat.* 117: 409–415.

Giese, A. C. 1976. *Living with Our Sun's Ultraviolet Rays.* Plenum, New York.

Grosch, D., and L. Hopwood. 1979. *Biological Effects of Radiation.* Academic Press, New York.

Hart, S., M. Kogan, and J. Paxton. 1983. Effect of soybean phytoalexins on the herbivorous insects Mexican bean beetle and soybean looper. *J. Chem. Ecol.* 9: 657–672.

Heller, W., B. Egin-Buhler, S. Gardiner, H.-H. Knobloch, U. Matern, J. Ebel, and K. Hahlbrock. 1979. Enzymes of general phenylpropanoid metabolism and of flavonoid glycoside biosynthesis in parsley. *Plant Physiol.* 64: 371–373.

Hermann, K. 1976. Flavonols and flavones in food plants: a review. *J. Fd. Technol.* 11: 433–448.

Holmes, N. 1984. The effect of light on the resistance of hard red spring wheats to the wheat stem sawfly, *Cephus cinctus* (Hymenoptera: Cephidae). *Can Entomol.* 116: 677–684.

Janzen, D. H. 1974. Tropical blackwater rivers, animals and mast fruiting by the Dipterocarpaceae. *Biotropica* 66: 69–103.

Khan, Z. R., D. M. Norris, H. S. Chiang, N. Weiss, and S. A. Oosterwyk. 1986. Light-induced susceptibility in soybean to cabbage looper, *Trichoplusia ni* (Lepidoptera: Noctuidae). *Environ. Entomol.* 15: 803–808.

Kitchings, T., D. DiGregorio, and P. Van Voris. 1976. A review of the ecological parameters of radionuclide turnover in vertebrate food chains. *In, Ecol. Soc. Am. Spec. Publ.* 1, Cushing, C. E. (ed.), Dowden, Hutchinson and Ross, Stroudsburg, PA, pp. 304–313.

Klein, R. 1978. Plants and near-ultraviolet radiation. *Bot. Rev.* 44: 1–27.

Knox, J. P., and A. Dodge. 1985. Singlet oxygen and plants. *Phytochemistry* 24: 889–896.

Koul, O. 1982. Insect feeding deterrents in plants. *Ind. Rev. Life Sci.* 2: 97–125.

Krogmann, D. W. 1973. *The Biochemistry of Green Plants.* Prentice-Hall, Englewood Cliffs, NJ.

Larson, R. A. 1986. Insect defenses against phototoxic plant chemicals. *J. Chem. Ecol.* 12: 859–870.

Larsson, S., A. Wiren, L. Lundgren, and T. Ericsson. 1986. Effects of light and nutrient stress on leaf phenolic chemistry in *Salix dasyclados* and susceptibility to *Galerucella lineola* (Coleoptera). *Oikos* 47: 205–210.

Mansour, M. H. 1982. The chronic effects of some allelochemics on the larval development and adult reproductivity of the cotton leafworm, *Spodoptera littoralis* Boisd. *Zeitschr. Pflanzenkrank. und-Schutz* 89: 224–229.

Manuwoto, S., and J. M. Scriber. 1985. Neonate larval survival of European corn borers, *Ostrinia nubilalis*, on high and low DIMBOA genotypes of maize: effects of light intensity and degree of insect inbreeding. *Ag. Ecosyst. Environ.* 14: 221–236.

Murray, R. H., S. Brown, and J. Mendez. 1982. *The Natural Coumarins.* Wiley, London.

National Academy of Sciences (NAS). 1973. *Biological Impacts of Increased Intensities of Solar Ultraviolet Radiation.* Washington, DC.

National Aeronautics and Space Administration (NASA). 1979. The stratosphere: present and future. *NASA Ref. Publ.* 1049.

National Research Council (NRC). 1979. *Stratospheric Ozone Depletion by Halocarbons: Chemistry and Transport*. National Academy of Sciences, Washington, DC.

Philogene, B., and J. McNeil. 1984. The influence of light on the non-diapause related aspects of development and reproduction in insects. *Photochem. Photobiol.* 40: 753–761.

Prokopy, R., and E. Owens. 1983. Visual detection of plants by herbivorous insects. *Annu. Rev. Entomol.* 28: 337–364.

Raven, P., R. Evert, and H. Curtis. 1976. *Biology of Plants*. Worth, New York.

Reichle, D., P. Dunaway, and D. Nelson. 1970. Turnover and concentrations of radionuclides in food chains. *Nucl. Saf.* 11: 43–55.

Reichle, D., and R. I. van Hook. 1970. Radionuclide dynamics in insect food chains. *Manitoba Entomol.* 4: 22–32.

Richardson, J. 1977. *Dimensions of Ecology*. Williams and Wilkins, Baltimore.

Ricklefs, R. E. 1973. *Ecology*. Chiron, Portland, OR.

Robberecht, R., and M. Caldwell. 1978. Leaf epidermal transmittance of ultraviolet radiation and its implications for plant sensistivity to ultraviolet-radiation induced injury. *Oecologia* 32: 277–287.

Sandberg, S. 1987. Chemical ecology of *Hypericum perforatum* (Guttiferae) and its leaf tying tortricid (Lepidoptera) herbivores. M. S. thesis. University of Illinois at Urbana-Champaign.

Schmidt, W. 1984. Bluelight physiology. *Bioscience* 34: 698–704.

Sheppard, A., and M. Eisenbud. 1977. *Biological Effects of Electric and Magnetic Fields of Extremely Low Frequency*. NYU Press, New York.

Southwood, T. R. E. 1973. The insect/plant relationship—an evolutionary perspective. *Symp. R. Entomol. Soc. Lond.* 6: 3–30.

Steinmuller, D., and M. Tevini. 1985. Action of ultraviolet radiation (UV-B) upon cuticular waxes in some crop plants. *Planta* 164: 557–564.

Tingey, W. M., and S. R. Singh. 1980. Environmental factors influencing the magnitude and expression of resistance. *In* Maxwell, F., and P. Jennings (eds.), *Breeding Plants Resistant to Insects*. Wiley, New York, pp. 86–113.

Towers, G. H. N. 1984. Interactions of light with phytochemicals in some natural and novel systems. *Can. J. Bot.* 62: 2900–2911.

Woodwell, G. 1962. Effects of ionizing radiation on terrestrial ecocystems. *Science* 138: 572–577.

Woodwell, G. 1967. Radiation and the patterns of nature. *Science* 156: 461–470.

Woodwell, G., and J. Oosting. 1965. Effects of chronic gamma irradiation on the development of old field plant communities. *Rad. Biol.* 5: 205–222.

Zangerl, A., and M. Berenbaum. 1987. Furanocoumarins in wild parsnips: effects of photosynthetically active radiation, ultraviolet light and soil nutrients. *Ecology* 68: 516–520.

6

PLANT STRESS FROM ARTHROPODS: INSECTICIDE AND ACARICIDE EFFECTS ON INSECT, MITE, AND HOST PLANT BIOLOGY

Thomas J. Riley

Department of Entomology
Louisiana State University Agricultural Center
Baton Rouge, Louisiana

1. INTRODUCTION

The application of insecticides and acaricides is a widely practiced means of achieving quick, effective, and relatively inexpensive control of insect or mite pests of plants. Over time, there have been instances of rapid pest population increases following insecticide/acaricide application; populations that have led to increased damage to host plants and necessitated retreatment in order to bring about the desired level of insect/mite control. In most cases this "resurgence" of the pest species has been attributed to the destruction of natural enemies that would have normally kept the pest population below damage threshold levels. However, in several instances the use of an insecticide or acaricide has been implicated in having a direct effect on the pest itself, resulting in a change in its biology that affects its population dynamics and results in its resurgence. As a result plants are abruptly relieved of pressure from the pest arthropod after pesticide application but within a short time are subjected to an intensified attack due to the rapid return and buildup of the pest population.

Certain insecticides can also exert a dramatic positive effect on plant growth. In the absence of insects, mites or other pests, some have been shown to stimulate growth and subsequently increase the yields of crop plants. Investigations into the causes of these phenomena have shown that some insecticides and their metabolites can inhibit or stimulate cell division and elongation, resulting in an increase or decrease in growth and dry matter in the plant. Insecticides have also been shown to inhibit the degradation of growth-promoting enzymes. The plant's maintenance of these enzymes has thus been implicated as a cause for increased plant growth.

Precise information on the effects of insecticide and acaricide mediated changes in host plants on insects and mites that utilize these plants for food is scarce, and the exact causes of the observed changes in the arthropods and the plants are not entirely understood. However, it appears that in the few well-documented cases of insecticide- or acaricide-induced resurgence, either the pest is biologically predisposed to inflict greater than normal injury to the host plant, the host plant is affected and becomes more favorable or attractive to the pest, or there is a combination of these two phenomena.

The major goal of this chapter is to provide an overview of recent research specifically addressing the changes in the arthropod pest's biology that occur following exposure to sublethal amounts of insecticide or acaricide, and how these changes alone, or in combination with changes in the host plant, can contribute to resurgence of the pest and continued or increased stress to the host plant.

2. EFFECT OF INSECTICIDES ON INSECT BIOLOGY AND PEST RESURGENCE

The only well-documented example of insecticide-induced resurgence is that of the brown planthopper, *Nilaparvata lugens* (Stål), in rice. Significant increases in brown planthopper populations following insecticide application have become a major problem in rice insect management in tropical Asia (Heinrichs and Mochida 1984). Investigations into the causes of this phenomenon indicate that in addition to the reduction of natural enemies by the insecticides, other factors favored the resurgence of this pest (Reissig et al. 1982a). Direct stimulation of the insect by certain insecticides has been determined as a very important factor contributing to the rapid resurgence of the brown planthopper, following the use of certain insecticides initially applied to bring about its control (Heinrichs and Mochida 1984).

Reissig et al. (1982b) evaluated 39 insectides for their resurgence-inducing qualities when used for brown planthopper control in rice. Comparisons of the increases in brown planthopper populations that occurred after insecticide application indicated that 16 materials were found to contribute to resurgence (Table 1). Some insecticide treatments resulted in plants with 100% hopperburn, due to excessive planthopper feeding, whereas the untreated controls suffered only minor or no damage. The feeding rate of the brown planthopper was investigated by Chelliah and Heinrichs (1980) and found to increase on rice treated with deltamethrin, methyl parathion, and diazinon (Fig. 1). They also observed that the deltamethrin and methyl parathion treatments resulted in improved plant growth and that a greater number of brown planthoppers were attracted to these plants.

The reproductive capacity was another aspect of brown planthopper biology found to be positively affected when insects came into contact with certain insecticides (Chelliah and Heinrichs 1980, Chelliah et al. 1980). Adults reared from fifth instar nymphs that were able to survive the application of deltamethrin, diazinon, and methyl parathion produced more eggs than adults reared from nymphs not exposed to insecticides. The most dramatic increase in reproductive rate was found in the deltamethrin treatments where more nymphs hatched, the nymphal stadia were shortened and adult longevity was greater than for planthoppers on the untreated control plants (Table 2).

In field tests Heinrichs et al. (1982a) investigated brown planthopper oviposition on rice sprayed with insecticides. The number of brown planthoppers was greater in plots treated with deltamethrin and the number of eggs reached 340 per hill of deltamethrin treated rice as compared to 10 per hill in the untreated plants 80 days after transplanting (Fig. 2). The ratio of eggs in the treated versus control plots at 78 days after transplanting was 34:1, whereas the ratio of adults was 5:1. This agrees with the results of Chelliah and Heinrichs (1980) and indicates that the adults on the deltamethrin-treated plants were reproductively stimulated and deposited more eggs per female than did those on the untreated plants. The numbers of brown planthoppers in field tests was also found to increase, in relation to the rate of deltamethrin and methyl parathion used (Heinrichs et al. 1982b).

Carbofuran, a systemic carbamate insecticide, was also found to positively affect the number of brown planthoppers in rice (Heinrichs et al. 1982a). A comparison of three application methods indicated that foliar sprays resulted in greater planthopper populations followed by root zone application and a broadcast application of granules to the paddy water. The

Table 1. Insecticides That Caused Brown Planthopper (BPH), *N. lugens*, Resurgence[a] in Field Tests at IRRI, Los Baños, Philippines

Test No.	Treatment	Formulation[b]		BPH in Insecticide Treatment/BPH in Check	Hopperburn (%)[c]
1	Azinophos ethyl	40%	EC	4.6	0a
	Quinalphos	25%	EC	5.3	0a
	Penthoate	50%	EC	6.0	0a
	Methomyl	19.8%	EC	9.3	0a
	Check				0a
2	Diazinon	5%	G	33.2	100a
	Isazophos	5%	G	36.7	100a
	Carbofuran	3%	G	35.2	80a
	Check				0b
3	Tetrachlorvinphos	75%	WP	14.5	12d
	Methyl parathion	50%	EC	32.6	65b
	Monocrotophos	16.8%	EC	2.2	2e
	Pyridaphenthion	75%	WP	14.5	29c
	Cyanophenphos	40%	EC	71.5	91a
	Check				0e
4	Triazophos	40%	EC	5.5	78b
	Deltamethrin	31%	EC	5.5	100a
	Check				10c
5	Penncap-M	25%	EC	2.3	91b
	Fenvalerate	38%	EC	2.8	99a
	Check				13c

From Heinrichs and Mochida 1984.

[a] Insecticide treatments that had significantly higher BPH populations than the untreated control at the last sampling date as based on Duncan's multiple range test at the 5% level.

[b] EC and WP formulations applied as a spray at 0.75 kg AI/ha and G formulations broadcast at 1.0 kg AI/ha.

[c] Means within a column, under each test, that are followed by a common letter are not significantly different at the 5% level by Duncan's multiple range test.

variation in planthopper populations among the carbofuran treatments may have been due to differences in the levels of sublethal doses of the insecticides reaching the insects directly or systemically through the plant.

A change in the chemistry of rice plants treated with carbofuran may have also contributed to the increase in brown planthoppers in the carbofuran treatments (Venugopal and Litsinger 1983). Increases in the populations of the rice blue leafhopper, *Zygina maculifrons* (Motch.), have been attributed to the low calcium and carbohydrate levels and increased nitrogen concentrations in plants treated with systemic insecticides (Mani and Jayaraj 1976). They offered as an explanation the possibility that the reduction of calcium

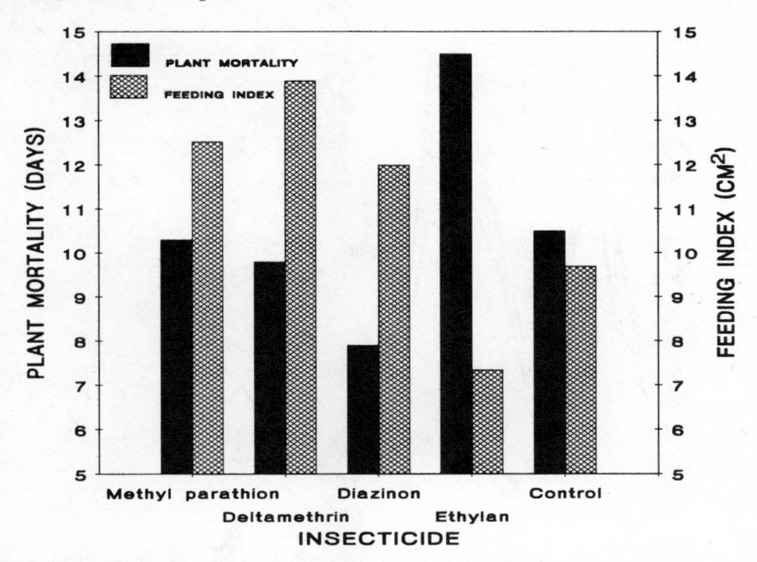

Figure 1. Effect of foliar spray of insecticides on feeding rate and damage by *N. lugens*. The feeding index is based on area of ninhydrin positive spots produced by honeydew on filter paper by five 5th-instar *N. lugens* nymphs feeding for 24 hours. Plant mortality, which indicates the number of days required for the plants to be killed, is the result of feeding damage caused by 20 nymphs placed on each plant at 15 days after the third spray application. (From Chelliah and Heinrichs 1980, reprinted with permission from *Environ. Entomol.* © 1980, Entomological Society of America)

Table 2. Influence of Foliar-Sprayed Insecticides on the Ovipositional Rate, Nymphal Duration, and Adult Longevity of *N. lugens*

Treatment	Reproductive Rate (no. of nymphs hatched)	Nymphal Duration (days)	Adult Longevity (days)
Methyl parathion	292ab	14.1ab	11.1ab
Deltamethrin	321a	13.6a	12.7a
Diazinon	275b	—	9.9b
Ethylan	125d	—	10.3b
Check	166c	14.7b	10.2b

From Chelliah and Heinrichs 1980.

Note: Means within a column followed by a common letter are not significantly different at the 5% level by the Duncan's multiple range test.

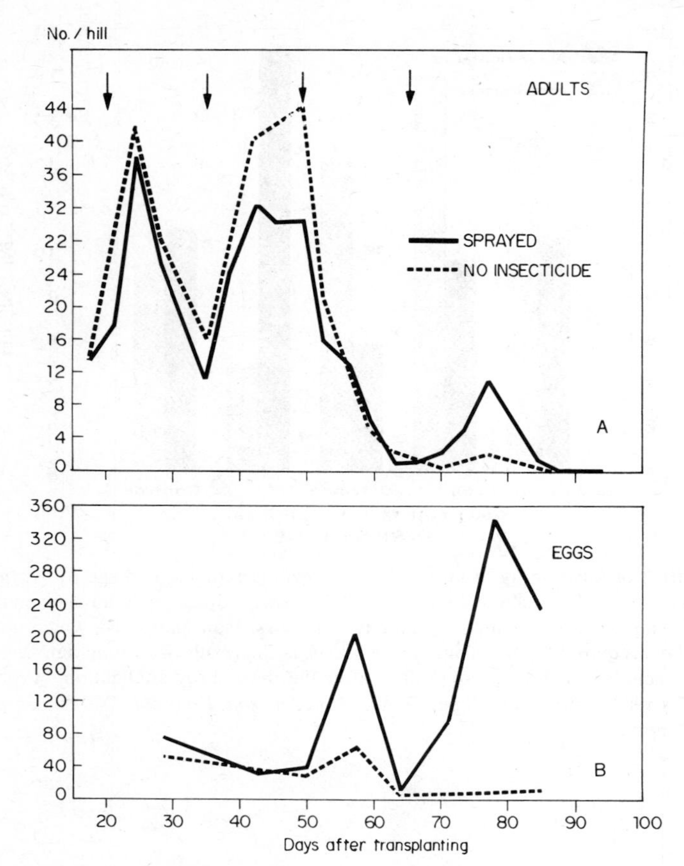

Figure 2. Populations of *N. lugens* adults (*A*) and eggs (*B*) in field plots of rice receiving no insecticide or applications of deltamethrin (0.025 kg AI/ha) at 20, 35, 50, and 56 DT (arrowed). Adults were collected by sampling 40 hills per plot with a D-vac suction machine. The egg population was determined by removing 10 hills per plot and counting the eggs in the laboratory. Variety IR22, wet season, 1978. (From Heinrichs et al. 1982a, reprinted with permission from *Environ. Entomol.* © 1982, Entomological Society of America)

and carbohydrates and the increase of nitrogen may have resulted in a weakening of the cell wall, allowing easier stylet insertion by the probing insects. A similar response by the brown planthopper could be partly responsible for the observed increases in its population following carbofuran treatment of rice.

Insecticides are an important factor in the relationship between the brown planthopper and rice because they can have a positive stimulatory effect on both organisms. The brown planthopper feeds at the base of the rice plant, where in certain situations it may be exposed only to sublethal levels of insecticides applied for its control. Thus the combination of feeding site preference of the pest, its capacity for positive reproductive stimulation by insecticides, and the attraction of the insect to rice exhibiting the enhanced growth and lush foliage that results from the application of certain insecticides have all apparently contributed to the resurgence of the brown planthopper in insecticide-treated rice.

Conversely, Javid and All (1984) reported a considerable reduction in the number of eggs produced per female fall armyworm, *Spodoptera frugiperda* (J. E. Smith), treated with sublethal doses of methomyl as 12-day-old larvae. In addition egg viability was also found to be less than that of adults from untreated larvae. However, in adults that survived larval treatment by doses greater than LD_{50}, egg viability was similar to the untreated controls although egg production was decreased. Thus, even though the number of eggs per female decreased, the fact that, at certain dosages, egg viability was similar to that of untreated insects is of some note. This response, if present under field conditions, could eventually become an important factor in the development of insecticide resistance in this species and result in a change in its capacity as a population to induce plant stress.

Fecundity in the pink bollworm, *Pectinophora gossypiella* (Saunders), was also found to be affected by low concentrations of otherwise lethal insecticides Bariola (1984). Treated moths were found to lay fewer eggs, and in some cases the eggs exhibited reduced hatching. Mating was also shown to be reduced by treating moths with sublethal amounts of insecticide. These responses are additional benefits realized when moths are not killed outright by an insecticide application because of their location within the cotton plant canopy. However, should the target insect species respond positively in some way to the sublethal amounts of insecticide, then resurgence such as occurs in the brown planthopper could result.

Changes in the nitrogen content of sugarbeet foliage has been shown to affect indirectly populations of beet armyworms, *Spodoptera exigua* (Hübner), through an increase in fecundity resulting from increased larval cannibalism (Al-Zubaidi and Capinera 1983). Cannibalism was shown to develop as compensation for an inadequate larval diet. In laboratory experiments fertilization rates that resulted in plant foliage with less than normal nitrogen levels were shown to cause an increase in the percent cannibalism among larvae. Should an insecticide cause a similar decrease in the nitrogen content of sugarbeet foliage, a similar affect on adult fecundity may occur and result in a change in the rate of armyworm population growth.

3. INSECTICIDE- AND ACARICIDE-ENHANCED STRESS FROM MITES

The potential for mite populations to rapidly exploit a favorable change in their environment is notorious. Such population explosions result in increased stress to host plants and are often attributed to a decrease in the predator population due to insecticide or acaricide application. However, the growth of mite populations can also result from direct exposure to acaricides and insecticides.

The controversy surrounding early reports of mite stimulation from insecticide and acaricide applications is reported by Bartlett (1968) and van de Vrie et al. (1972). Their reviews point out that the early research did not adequately explain whether chemical control efforts improved the nutritional status of the plants in favor of the mites or caused an increase in mite reproduction due to some direct stimulatory effect.

Dittrich et al. (1974) investigated chemical stimulation of mites by DDT and carbaryl in the context of the insecticide hormoligosis concept of Luckey (1968). This hypothesis states that "subharmful quantities of any stressing agent will be stimulatory to an organism by providing increased sensitivity to respond to changes in its environment and increased efficiency to develop new or better systems to fit a suboptimum environment." The research of Dittrich et al. (1974) did indeed show that low levels of insecticide residues, and not a change in plant physiology, were stimulatory to *Tetranychus urticae* Koch as measured by an increase in egg production and the proportion of females in the population. Similar stimulation of egg production and a population increase were reported by McCoy (1977) with the citrus rust mite, *Phyllocoptruta oleivora* (Ashmead), in citrus fruit treated with methidathion.

Recent research into pesticide-induced reproductive stimulation in the two-spotted spider mite in peanut and cotton has shown that the species is subject to resurgence attributed to a greater reproductive output. Boykin and Campbell (1982) using peanut, a very favorable host for the two-spotted spider mite, found that peanut leaves treated with carbaryl and carbaryl + mancozeb consistently resulted in slight increases in mite reproduction compared to mites on untreated leaves. They surmised that these same chemicals probably also reduced other population-limiting factors and that this combined with the slight increase in reproductive potential contributed to the observed increases in mite populations in treated peanut fields. They advised that chemical mite control programs should be designed around materials that are known to cause a reduction in mite reproduction.

Maggi and Leigh (1983) arrived at similar conclusions when studying the effects of methyl parathion treated cotton on the two-spotted spider mite. Their research showed that there was a small increase in oviposition in mites from methyl-parathion-treated cotton compared to those from untreated plants. They speculated that an increase, even though slight, when superimposed on other factors affecting mite populations, might result in substantial population increases over an entire growing season. They stated this would be especially detrimental in the cases where an increase in reproduction occurred early in the growing season, allowing the growth of mite populations to exceed the regulating effects of predators and other natural control agents. Based on their research, they suggest that methyl parathion not be used for early season pest control, except when the alternatives are unacceptable for use in a cotton management program.

The resurgence of the citrus red mite, *Panonychus citri* (McGregor) was investigated by Jones and Parrella (1984). They showed that in this species, as in the two-spotted spider mite, resurgence of the pest occurs in the absence of predators and that it was due to the positive stimulation of reproduction after exposure of the mites to sublethal levels of the insecticides malathion and permethrin. Their research again demonstrated that insecticide-induced increases in reproduction can be an important factor influencing the buildup of destructive levels of mite pests.

Two-spotted spider mites have also been shown to undergo substantial population increases on soybean plants treated with carbofuran (Mellors et al. 1984). In experiments using carbofuran applied in the soil at planting at the rates of 0.168, 1.68, and 16.8 kg AI/ha, treated plants were taller and had more nodes than those with no insecticide. At the 1.68 and 16.8 kg AI/ha rates, the plants were toxic to the Mexican bean beetle and also had reduced spider mite populations compared to the untreated plants. The 0.168 kg AI/ha rate of carbofuran was ineffective at controlling the Mexican bean beetle, and it resulted in mite populations that were about three times greater than those on the untreated plants.

In temperature- and water-stressed plants, apparently less carbofuran is taken up by the plants. Plants treated with carbofuran at 1.68 kg AI/ha and subjected to water stress and high temperatures were significantly less toxic to the Mexican bean beetle than were plants treated with same rate of carbofuran and provided with adequate water (Mellors et al. 1984).

Water stress and high temperatures have been shown to favor the buildup of damaging mite populations (Hollingsworth and Berry 1982, Tanigoshi et al. 1975). In the case of soil-applied carbofuran in soybean, high temperatures and less than adequate moisture lead to reduced levels of the insecticide in the plants. As the growing season progresses, levels of carbofuran in the plant may decline to the level that stimulates the two-spotted spider mite and leads to population growth. In addition the low moisture and high temperature conditions may also favor the development of the mite population. In fields treated at planting with carbofuran and subjected to drought stress later in the growing season, the combination of insecticide and environmentally induced mite stimulation may render the plants very susceptible to mite injury.

It is apparent that mite pests can be positively influenced by the application of insecticides to their respective host plants. The direct stimulation of reproduction through contact with treated plant parts and the stimulatory effect of the systemic insecticide carbofuran, whether acting directly on the mites or acting through enhanced plant growth and nutrition, are factors to be considered when management decisions are made in crops susceptible to mite attack.

4. INSECTICIDE AND ACARICIDE EFFECTS ON ARTHROPOD HOST PLANTS

Insecticides and acaricides are routinely applied to plants in order to suppress the detrimental activities of insects or mites. Since the widespread use of synthetic chemcial pesticides has become commonplace, several have been shown to promote positive changes in plant growth and physiology that could not be attributed to the control of insects and other pests. These altered plants are often larger, faster growing, and in some cases yield more than those grown in an equivalent favorable environment, free of insect or mite pressure. Insecticide- or acaricide-stimulated plants may contribute to insect and mite resurgence by being more attractive to the pest, or by providing more nutritionally favorable food. Evidence to support this premise is presently circumstantial and speculative; however, data showing that insecticides and acaricides can stimulate and change the growth and chemistry of plants do exist. This subject will now be discussed in the context of how it may contribute to insect/mite stress.

4.1 Effects on Plant Growth and Yield

Chlorinated hydrocarbon pesticides applied to plants have been observed to produce growth and yield increases not attributable to the control of insects, nematodes, or other pest organisms. Chapman and Allen (1948) found that low concentrations of DDT stimulated plant growth in several vegetable crops. They remarked that all parts of the plants were affected, resulting in greater plant height, and that the effect of DDT was similar to certain plant hormones.

Cotton treated with toxophene-DDT at weekly intervals, beginning with first squares and first flowers, produced an increase in yield and boll production but had no effect on the number of leaves and the dry weight of the plants (Brown et al. 1962).

Dieldrin caused increased fruit set in tomatoes that was not accountable to insect control. Reed (1964) and Hagley (1965) reported that a low rate of aldrin apparently stimulated growth in tomato seedlings.

In contrast to the growth stimulation described earlier, organochlorines can also cause reductions in seed germination plant growth and yield. Several vegetable plants were stunted, deformed, and showed necrosis and chlorosis when high rates of DDT were applied (Chapman and Allen 1948). However, this injury disappeared and the growth stimulation occurred as the rate of DDT was decreased. Rice seeds treated with gamma-BHC exhibited poor growth in the field, and yields were halved compared to untreated controls (Rolston et al. 1960). Aldrin and DDT at 16 kg AI/ha inhibited root growth and reduced the size of cauliflower, Chinese cabbage, and tomato plants, yet at 1.6 kg AI/ha no effect was observed, though aldrin "appeared to stimulate growth in tomato seedlings" (Hagley 1965).

The research of Brown et al. (1962) demonstrated that the number of leaves and the dry weight of cotton plants significantly increased following treatment with methyl-parathion. Hacskaylo and Scales (1959) observed that cotton plants sprayed with azinphos-methyl produced more flowers than untreated plants. Red potatoes treated with azinphos-methyl set more tubers and produced a yield significantly greater than untreated plants which was beyond that attributable to insect and disease control (Schulz 1961). Pless et al. (1971) reported more rapid growth and higher yields in burley tobacco grown in high phosphate soil treated with disulfoton, as compared to plants grown in similar soil with no insecticide. Sorghum has been shown to exhibit a response to the insecticide disulfoton that differs depending on the rate of chemical used (Thompson and Harvey 1980). A rate of 4.48 kg AI/ha applied at planting with the seed reduced the plant stand by 32%. Conversely, rates of 0.28, 0.56, 1.12, and 2.24 kg AI/ha resulted in yield increases, and these increases were not the result of insect, mite, or, disease control.

The carbamate insecticide-nematicide carbofuran as well as a few others have been shown to elicit a physiological response in several agriculturally important plants. These plants, when grown in the presence of the pesticides, exhibited enhanced growth and greater yields than plants grown in the absence of the pesticides. Apple (1971) reported yield increases of 4.0, 1.9, and 6.9% in corn planted in soil treated with carbofuran at 1.10, 2.25, and 4.50 kg AI/ha. Daynard et al. (1975) also reported a yield increase in corn treated with carbofuran that could not be attributed to control of the northern corn rootworm. Carbofuran was responsible for more rapid growth and a greater yield in burley tobacco grown in high phosphate soil when compared to plants grown in the same but untreated soil (Pless et al. 1971). Soybean plants treated with carbofuran and methomyl were taller and produced larger seeds than untreated plants (Wheeler and Bass 1971) and aldicarb was found to increase the dry weight of the greengram *Vigna radiata* (L.) (Yien et al. 1979).

Corn seedlings grown in controlled environmental chambers in carbofuran-treated soil emerged from the soil sooner, increased in height more rapidly, and had greater dry weights 16 days after emergence than did seedlings grown in the absence of carbofuran (Duveiller et al. 1981). Similar results were obtained by Venugopal and Litsinger (1983, 1984) when studying the effects of carbofuran on the growth of rice *Oryza sativa* L. Using hydroponic techniques and sterile soil in the complete absence of pest organisms, their research showed that carbofuran produced significantly greater root growth, greater plant height, an increase in the total number and number of productive tillers, more rapid maturation, and increased yield. The two major metabolites of carbofuran were also shown to produce the same positive increases in growth and yield in rice (Venugopal and Litsinger 1984).

However, in birdsfoot trefoil, *Lotus corniculatus* L., Belanger et al. (1985) found that the shoot length of carbofuran-treated plants was not increased and that the number of plants was reduced compared to the untreated control. In a pair of experiments the number of nodules was not affected, but the dry weight of individual nodules, when compared to the untreated controls, was greater in one experiment and less in another. These differences in the carbofuran-treated plants were attributed to the effect of the chemical on the physiology of the plant and not to its nematicidal activity.

The formamidine insecticide-acaricide chlordimeform has been observed to affect growth and increase yield when applied for early season insect control in cotton (Lincoln and Dean 1976, Phillips et al. 1977). This has led to its being used as a ''yield enhancer'' by some cotton producers. Changes in yield, ranging from yield increases to yield decreases, have been reported to result from early season chlordimeform treatment (Weaver and Bhardwaj 1985). However, there is disagreement in the literature as to whether the observed yield increases are due to early season insect control, a physiological response in the plant, or a combined effect. Experimental evidence indicates that physiological changes do take place in cotton plants treated with chlordimeform, but how these changes reduce or contribute to plant stress from insects and mites, and the effects on yield of these interactions remains to be determined.

Measurements taken from cotton plants three days after chlordimeform treatment have shown an increase in stomatal conductance of water vapor, transpiration rate, and apparent photosynthesis (Bauer and Cothren 1986). They did not, however, find any differences in leaf area, dry weight, plant height, square number, and square dry weight. Weaver and Bhardwaj (1985) have shown that in chlordimeform-treated plants, petioles tended to be smaller and leaf blade weight higher, and live square number was higher but there were more aborted squares than in untreated plants. Also, flowering rate was slower for the first two weeks but exceeded that of the untreated plants during the third week of flowering, under adequate moisture; however, under drought stress, blooming rate of the treated was below that of the untreated plants. Benedict et al. (1986) found that clordimeform increased early square and boll counts, total bolls and lint harvested, and resulted in an earlier harvest, but boll diameter and the amount of lint per boll was not affected.

The recent interest in chlordimeform as a yield-enhancing chemical presents an ideal opportunity for the development of research programs that address the area of insecticide, plant, and insect interactions. Research to date has not conclusively determined if insect control, plant stimulation, their combined effects, or an effect of the stimulated plant on the insect is responsible for the greater yields often associated with chlordimeform-treated cotton. Evidence strongly suggests that a physiological effect on the plant does in fact occur. This should be exploited to determine how these chemically altered plants may in turn affect the insects and mites that utilize them for food.

In general, plants that are growing more rapidly and producing more biomass due to insecticide-acaricide stimulation could favor insects and mites by providing an increased flow of nutritional material in the xylem and phloem. In cases of stimulation by systemic insecticides, the increased sap flow associated with the increase in plant growth could provide abundant food for the insect, and low levels of insecticide could directly stimulate the insect. Both of these factors could lead to rapid resurgence of the pest species.

An additional contributing factor to plant stress as a consequence of insecticide-acaricide application might be an increased or protracted period of attractiveness of crops, due to more prolific or prolonged flowering induced by chemicals applied for pest control purposes. Insects, particularly Lepidoptera, attracted to the flowers for nectar and then ovipositing, might be more readily attracted to these crops and, once there, remain to feed and oviposit for a much longer period, resulting in a greater than usual number of developing larvae in the field. If this is coupled with a reduced predator population due to the previous application of an insecticide, the pest population could reach damaging proportions very quickly.

4.2 Effects on Plant Growth-Promoting Enzymes

The carbamate insecticide carbofuran is a well-known example of a plant growth stimulating insecticide. Research into the mechanism of action responsible for the observed growth-stimulating effect of carbofuran has shown that it and its metabolites promote plant growth through the inhibition of the enzymic degradation of indole-3-acetic acid (IAA), a plant hormone necessary for growth (Lee 1976, 1977; Lee and Chapman 1977; Thompson et al. 1981). Increased plant growth results from the continued presence of IAA at the critical level required for growth promotion.

4.3 Effects on Plant Cells

The response of plant cells exposed to insecticides is variable and dependent on the chemical involved. Organochlorine insecticides have been shown to cause reduced cell division and various mitotic effects (Scholes 1955a, b) including polyploidy (Scholes 1953). Carbofuran and aldicarb increased the mitotic index, defined as the percentage of cells dividing, in cells from the root tips of the onion, *Allium cepa* L. (Thompson et al. 1981), and in sugarcane root cells (Miah and Akhtar 1985).

The introduction of polyploidy into commercial plant production as a result of exposure to pesticides could eventually introduce changes in the quality of host plants that enhance or reduce the damaging effects of insect and arthropod pests. The routine prophylactic application of chemical pesticides in plant-breeding nurseries should be carefully planned in order to avoid those materials known to induce genetic changes in the plants.

The introduction of any form of pesticide-induced, heritable genetic change into plants grown for genetically uniform seed or plant production is a situation that should be carefully evaluated, not only in relation to a change in the susceptibility of the plant to insects, mites, and other pests but also for changes in other important plant characteristics.

4.4 Effects on the Nutritional Quality of Host Plants

Nutritional value of the host plant may be changed by the application of an insecticide. Hacskaylo (1957), using the organophosphate insecticide, phorate, on cotton plants, reported increases in carbohydrates and phosphorus, and a decrease in nitrogen. Rao and Rao (1983) reported a nutritional change in pest free rice. They reported an increase in amino-N and a reduction in the total and reducing sugars in the plants, 15 days after application of several insecticides.

Also in rice, Venugopal and Litsinger (1983) reported higher amounts of N, K, Zn, and Fe in plants treated with carbofuran. Levels of P, Mg, and Ca remained the same as in untreated plants, but the amount of Cu and Mn decreased. Nitrogen fixation was greater in soil where plants were grown after treatment with carbofuran and other pesticides (Belanger et al. 1985, Rao et al. 1984). Using birdsfoot trefoil, Belanger et al. (1985) demonstrated that carbofuran increased nitrogen fixation, although nodule number was not affected.

Cotton treated with chlordimeform showed changes in the chemical composition of the leaf petioles (Weaver and Bhardwaj 1985). Concentrations of Ca, P, Mg, and K tended to be greater two weeks after chlordimeform treatment, especially in plants under drought stress. When adequate water was available, the differences between treated and untreated plants was less.

Changes in the physiology of arthropod host plants that affect their nutritional value may play an important role in the development of destructive pest populations. Plants that are more nutritious could attract more individuals and favor a rapid population explosion. Accessiblity of the food may also be important. Plants altered by insecticides may also favor the development of large insect populations if the plant is changed so as to allow a particular pest easier access to its most nutritious parts, such as suggested by Mani and Jayaraj (1976). Mobile arthropods that assess food value and quality by frequently "sampling" potential food sources may congregate on plants or in entire fields of plants, that are nutritionally or physically more suited for them.

5. SUMMARY AND PROSPECTUS

Many cases of insecticide- and acaricide-induced pest resurgence can be accounted for by the destruction of nontarget natural enemies by a pesticide. However, examples exist in which the post treatment increase in a pest species could not be explained by the elimination of natural control agents.

In these cases the pesticide itself has been shown to be responsible for changes in both the arthropod pest and the plant. The result is a pest that is biologically predisposed to inflict greater than normal injury to its host plant, a host plant that is more favorable to the pest, or a combination of these two phenomena.

The stimulating effect that certain insecticides and acaricides have on a select number of mite and insect species may become an important factor in arthropod pest management programs if more species are shown to elicit a similar response. The use of an insecticide or acaricide that induces pest resurgence through reproductive stimulation, or another form of population growth-inducing stimulus, is a situation that needs to be avoided. Presently, the cases of rice brown planthopper and two-spotted spider mite resurgence, induced through reproductive stimulation, serve as models of what can happen.

In the future the increasing costs of pest control will lead to the refinement of arthropod pest management programs, illuminating the need for more precise techniques of monitoring target and nontarget species. This will subsequently allow detection of chemically induced pest resurgence and bring more research effort to bear on the problem.

The registration process for insecticides and acaricides could also reach a point where resurgence-inducing qualities are evaluated with as much emphasis as efficacy against target species, host plant phytotoxicity, persistence of residues, and so on.

Changes in pest behavior that cause exposure to sublethal, resurgence-inducing doses of insecticides also warrant more research. This is true especially if undertaken in coordination with plant breeders, plant morphologists, and agricultural engineers in order to ensure that research effort is directed toward the delivery of a lethal amount of insecticide/acaricide to the target pest.

The reaction of plants to the application of insecticides and acaricides and the effects of the plant response on its insect and mite fauna remain virtually uninvestigated. Research to date has primarily addressed the effects of the chemicals on desirable plant characteristics. Future research should take this work one step further.

Enhanced plant growth from insecticide and acaricide stimulation could influence the ability of the plant to withstand insect feeding. It could also provide pests with a more lush, vigorous, nutritious, and attractive plant. The latter case could result in greater feeding, rapid population growth, and subsequently more damage from pest species. In addition any detrimental effects due to pesticides might also render the plant more susceptible to pest damage and other biotic and abiotic stresses, or in the case of plant vigor reduction, result in a decrease of food quality and subsequently retard development of large pest populations.

Through an understanding of how plant–insecticide interactions influence arthropod communities dependent on affected host plants, and the development of the methodology to quantify these effects, a more detailed picture will emerge of how host plant quality affects the status of pests and nontarget and beneficial arthropods. This along with research on the direct stimulation of pest species by insecticides and acaricides will allow a more precise understanding of the role of pesticides in the agroecosystem, greatly assisting in the planning, implementation, and evaluation of areawide pest management programs of the future.

REFERENCES

Al-Zubaidi, F. S., and J. L. Capinera. 1983. Application of different nitrogen levels to the host plant and cannibalistic behavior of beet armyworm, *Spodoptera exigua* (Hübner) (Lepidoptera: Noctuidae). *Environ. Entomol.* 12: 1687–1689.

Apple, J. W. 1971. Response of corn to granular insecticides applied to the row at planting. *J. Econ. Entomol.* 64: 1208–1211.

Bariola, L. A. 1984. Pink bollworms (Lepidoptera: Gelechiidae): effects of low concentrations of selected insecticides on mating and fecundity in the laboratory. *J. Econ. Entomol.* 77: 1278–1282.

Bartlett, B. R. 1968. Outbreaks of two-spotted spider mites and cotton aphids following pesticide treatment. I. Pest stimulation vs. natural enemy destruction as the cause of outbreaks. *J. Econ. Entomol.* 61: 297–303.

Bauer, P. J., and J. T. Cothren. 1986. Chlordimeform yield enhancement effect on cotton. *In Proc. Beltwide Cotton Prod. Res. Conf.* National Cotton Council, Memphis, pp. 520–521.

Belanger, G., J. E. Winch, and J. L. Townsend. 1985. Carbofuran effects on establishment of legumes in relation to plant growth, nitrogen fixation and soil nematodes. *Can. J. Plant Sci.* 65: 423–433.

Benedict, J. H., M. H. Walmsley, J. C. Segers, and M. F. Treacy. 1986. Yield enhancement and insect suppression with chlordimeform (Fundal) on dryland cotton. *J. Econ. Entomol.* 79: 238–242.

Boykin, L. S., and M. V. Campbell. 1982. Rate of population increase of the twospotted spider mite (Acari: Tetranychidae) on peanut leaves treated with pesticides. *J. Econ. Entomol.* 75: 966–971.

Brown, L. C., G. W. Cathey, and C. Lincoln. 1962. Growth and development of cotton as affected by toxaphene-DDT, methyl parathion; and calcium arsenate. *J. Econ. Entomol.* 55: 298–301.

Chapman, R. K., and T. C. Allen. 1948. Stimulation and suppression of some vegetable plants by DDT. *J. Econ. Entomol.* 41: 616–623.

Chelliah, S., and E. A. Heinrichs. 1980. Factors affecting insecticide-induced resurgence of the brown planthopper, *Nilaparvata lugens* on rice. *Environ. Entomol.* 9: 773–777.

Chelliah, S., L. T. Fabellar, and E. A. Heinrichs. 1980. Effect of sublethal doses of three insecticides on the reproduction rate of the brown planthopper, *Nilaparvata lugens* on rice. *Environ. Entomol.* 9: 778–780.

Daynard, T. B., C. R. Ellis, B. Bolwyn, and R. L. Misner. 1975. Effects of carbofuran on grain yield of corn. *Can. J. Plant Sci.* 55: 637–639.

Dittrich, V., P. Streibert, and P. A. Bathe. 1974. An old case reopened: mite stimulation by insecticide residues. *Environ. Entomol.* 3: 534–540.

Duveiller, E., M. Schroder, Ch. Verstraete, and J.-F. Ledent. 1981. Effet physiologique d'insecticides sur la croissance juvenile du mais. *Med. Fac. Landbouww.-Rijksuniv. Gent* 46: 535–542.

Hacskaylo, J. 1957. Growth and fruiting properties and carbohydrate, nitrogen, and phosphorus levels of cotton plants as influenced by Thimet. *J. Econ. Entomol.* 50: 280–284.

Hacskaylo, J., and A. L. Scales. 1959. Some effects of Guthion alone and in combination with DDT and of a dieldrin-DDT mixture on growth and fruiting of the cotton plant. *J. Econ. Entomol.* 52: 396–398.

Hagley, E. A. C. 1965. Effect of insecticides on the growth of vegetable seedlings. *J. Econ. Entomol.* 58: 777–778.

Heinrichs, E. A., and O. Mochida. 1984. From secondary to major pest status: the case of insecticide-induced rice brown planthopper, *Nilaparvata lugens*, resurgence. *Prot. Ecol.* 7: 201–218.

Heinrichs, E. A., G. B. Aquino, S. Chelliah, S. L. Valencia, and W. H. Reissig. 1982a. Resurgence of *Nilaparvata* lugens (Stål) populations as influenced by method and timing of insecticide applications in lowland rice. *Environ. Entomol.* 11: 78–84.

Heinrichs, E. A., W. H. Reissig, S. Valencia, and S. Chelliah. 1982b. Rates and effects of resurgence-inducing insecticides on populations of *Nilaparvata lugens* (Homoptera: Delphacidae) and its predators. *Environ. Entomol.* 11: 1269–1273.

Hollingsworth, C. S., and R. E. Berry. 1982. Twospotted spider mite (Acari: Tetranychidae) in peppermint: population dynamics and influence of cultural practices. *Environ. Entomol.* 11: 1280–1284.

Javid, A. M., and J. N. All. 1984. Effects of methomyl on weight and development of fall armyworm (Lepidoptera: Noctuidae). *Ann. Entomol. Soc. Am.* 77: 193–196.

Jones, V. P., and M. P. Parrella. 1984. The sublethal effects of selected insecticides on life table parameters of *Panonychus citri* (Acari: Tetranychidae). *Can. Entomol.* 116: 1033–1040.

Lee, T. T. 1976. Insecticide–plant interaction: carbofuran effect on indole-3-acetic acid metabolism and plant growth. *Life Sci.* 18: 205–210.

Lee, T. T. 1977. Promotion of plant growth and inhibition of enzymic degradation of indole-3-acetic acid by metabolites of carbofuran, a carbamate insecticide. *Can. J. Bot.* 55: 574–579.

Lee, T. T., and R. A. Chapman. 1977. Inhibition of enzymic oxidation of indole-3-acetic acid by metabolites of the insecticide carbofuran. *Phytochemistry* 16: 35–39.

Lincoln, C., and G. Dean. 1976. Yield and blooming of cotton as affected by insecticides. *Ark. Farm Res.* 25: 5.

Luckey, T. D. 1968. Insecticide hormoligosis. *J. Econ. Entomol.* 61: 7–12.

Maggi, V. L., and T. F. Leigh. 1983. Fecundity response of the two-spotted spider mite to cotton treated with methyl parathion or phosphoric acid. *J. Econ. Entomol.* 76: 20–25.

Mani, M., and S. Jayaraj. 1976. Biochemical investigation on the resurgence of rice blue leaf hopper *Zygina maculifrons* (Motch.). *Ind. J. Exp. Biol.* 14: 636–637.

McCoy, C. W. 1977. Resurgence of citrus rust mite populations following application of methidathion. *J. Econ. Entomol.* 70: 748–752.

Mellors, W. K., A. Allegro, and A. N. Hsu. 1984. Effects of carbofuran and water stress on growth of soybean plants and twospotted spider mite (Acari: Tetranychidae) populations under greenhouse conditions. *Environ. Entomol.* 13: 561–567.

Miah, M. A. H., and S. Akhtar. 1985. Effect of carbofuran on mitotic index of sugarcane roots. *Ind. J. Agric. Sci.* 55: 55–56.

Phillips, J. R., G. A. Herzog, and W. F. Nicholson. 1977. Effect of chlordimeform on fruiting characteristics and yield of cotton. *Ark. Farm Res.* 26: 4.

Pless, C. D., E. T. Cherry, and H. Morgan, Jr. 1971. Growth and yield of burley tobacco as affected by two systemic insecticides. *J. Econ. Entomol.* 64: 172–175.

Rao, P. R. M., and P. S. P. Rao. 1983. Effect of insecticides on growth, nutritional status and yield of rice plant. *Ind. J. Agric. Sci.* 53: 277–279.

Rao, J. L. N., J. S. Prasad, and V. R. Rao. 1984. Nitrogen fixation C_2H_2 reduction in the rice rhizosphere soil as influenced by pesticides and fertilizer nitrogen. *Curr. Sci.* 53: 1155–1156.

Reed, J. P. 1964. Tomato yield response owing to early applications of dieldrin. *J. Econ. Entomol.* 57: 292–294.

Reissig, W. H., E. A. Heinrichs, and S. L. Valencia. 1982a. Insecticide-induced resurgence of the brown planthopper, *Nilaparvata lugens*, on rice varieties with different levels of resistance. *Environ. Entomol.* 11: 165–168.

Reissig, W. H., E. A. Heinrichs, and S. L. Valencia. 1982b. Effects of insecticides on *Nilaparvata lugens* and its predators: spiders, *Microvelia atrolineata*, and *Cyrtorhinus lividipennis*. *Environ. Entomol.* 11: 193–199.

Rolston, L. H., P. Rouse, and V. Hall. 1960. Effect of insecticidal seed treatments on rice. *J. Kansas Entomol. Soc.* 33: 119–122.

Scholes, M. E. 1953. The effect of hexachlorocyclohexane on mitosis in roots of the onion (*Allium cepa*) and strawberry (*Fragaria vesca*). *J. Hortic. Sci.* 28: 49–68.

Scholes, M. E. 1955a. The effects of chlordane and toxaphene on mitosis in roots of onion (*Allium cepa* L.). *J. Hortic. Sci.* 30: 12–24.

Scholes, M. E. 1955b. The effect of aldrin, dieldrin, isodrin, endrin, and DDT on mitosis in roots of the onion (*Allium cepa* L.). *J. Hortic. Sci.* 30: 181–187.

Schulz, J. T. 1961. A physiological response of potato to foliar applications of the insecticide, guthion. *J. Econ. Entomol.* 54: 839–840.

Tanigoshi, L. K., S. C. Hoyt, R. W. Browne, and J. A. Logan. 1975. Influence of temperature on population increase of *Tetranychus mcdanieli* (Acarina: Tetranychidae). *Ann. Entomol. Soc. Am.* 68: 972–978.

Thompson, A. R., M. S. Taylor, G. H. Edmonds, and S. Lord. 1981. Laboratory experiments on effects of aldicarb and carbofuran on the growth of *Arabidopsis thaliana* and on mitosis in root tips of *Allium cepa*. *Med. Fac. Landbouww. Rijksuniv. Gent* 46: 543–552.

Thompson, C. A., and T. L. Harvey. 1980. Direct effect of the systemic insecticide disulfoton of yield of grain sorghum. *Prot. Ecol.* 2: 21–25.

van de Vrie, M., J. A. McMurtry, and C. B. Huffaker. 1972. Ecology of tetranychid mites and their natural enemies: a review. III. Biology ecology, and pest status, and host-plant relations of tetranychids. *Hilgardia* 41: 343–432.

Venugopal, M. S., and J. A. Litsinger. 1983. Carbofuran—a direct growth stimulant of rice. *In Proc. Rice Pest Management Seminar*. Tamil Nadu Agric. Univ., Coimbatore, India, pp. 120–135.

Venugopal, M. S., and J. A. Litsinger. 1984. Effect of carbofuran on rice growth. *Prot. Ecol.* 7: 313–317.

Weaver, J. B., and H. L. Bhardwaj. 1985. Growth regulating effect of chlordimeform in cotton. *In Proc. Beltwide Cotton Prod. Res. Conf.* National Cotton Council, Memphis, pp. 288–293.

Wheeler, B. A., and M. H. Bass. 1971. Effects of certain systemic insecticides on growth and yield of soybeans. *J. Econ. Entomol.* 64: 1219–1221.

Yein, B. R., H. Singh, and K. P. Goswami. 1979. Combined influence of pesticides and fertilizers on the nodulation and development of greengram. *Ind. J. Agric. Sci.* 49: 961–966.

7

THE EFFECTS OF PLANT GROWTH REGULATORS AND HERBICIDES ON HOST PLANT QUALITY TO INSECTS

Bruce C. Campbell

Plant Protection Research Unit
USDA-ARS
Albany, California

1. INTRODUCTION

The discovery that certain herbicides affect insect populations originated with field observations in the 1940s. Interest in herbicidal effects on insects increased as these observations showed that populations of certain species of pest insects actually increased after application of herbicides. This initiated research that attempted to determine whether herbicides directly affected insects or indirectly through biochemical changes in the host plant induced by the herbicides. More recently, interest in the chemical aspects of plant–insect interactions has fostered the idea that endogenous plant hormones or synthetic growth-regulating chemicals can be used to manipulate insect attacks on plants. The intent of this contemporary view on the use of plant growth regulators is to make the plants or crops resistant to insect attack by either physically altering plant structure (flowering, fruit set, etc.) or affecting biosynthetic pathways in secondary plant metabolism to manipulate concentrations of natural products that are feeding deterrents, attractants, or toxicants.

As with the herbicides, the original observations of insect populations being affected by plant growth regulator (PGR) applications to plants were fortuitous. In both cases, of herbicides and PGRs, knowledge of the chemical bases of their effect on plants has only recently provided some insight into explaining their indirect effect on insect behavior, physiology, metabolism, and population growth.

This chapter will review the literature on research pertaining to both direct and indirect effects of herbicides and PGRs on insects. The indirect effects are defined as changes occurring in the host plant that alter their quality as either a source of food or refuge to the insect. The direct effects are based on reports where herbicides, PGRs, or plant growth hormones have been shown to impact directly insect physiology, reproduction, and growth, for example, through ingestion from artificial diets, injection, or by topical application. It is not the intention of this chapter to review the numerous papers pertaining to natural products from insects (secretions, saliva, etc.) which have plant growth regulatory or hormonelike activity. For literature on this subject, see Detling and Dyer (1981). This chapter also will not cover the literature on the biochemical aspects of gall formation in plants induced by insects or microorganisms inoculated into plants by insects. For reviews of this subject, the reader is directed to Paclt (1980) and Elzen (1983). Instead, this chapter will be a review of the literature where herbicides, PGRs, or plant hormones have been deliberately applied to plants or insects and as a result have affected insect behavior, growth, or reproduction.

Research on the potential use of PGRs or herbicides as a strategy in insect pest management to facilitate the control of insects is still in progress, though there are already a few reviews on the effects of herbicides and PGRs on insects. One of the earliest reviews by Nickell (1982) covers some of the early research on the effects of 2,4-D and chlormequat chloride on insects. Recent reviews by Hare (1983), Kogan and Paxton (1983), and Campbell et al. (1984) have encompassed concepts pertaining to PGR- or hormone-induced changes in plant chemistry that affect insect population dynamics or induced host plant resistance, respectively.

This chapter is divided into three sections. Section 2 will focus on the direct and indirect effects of herbicides on insects. Section 3 deals with PGRs and hormones and their effects on insect–plant interactions. Section 4 discusses the potential advantages of incorporating PGRs into integrated pest management programs.

2. HERBICIDE EFFECTS ON INSECTS

It is well established that plant growth and development is, in part, controlled by endogenous plant hormones. However, the use of synthetic plant growth regulating chemicals has become an ever increasing aspect of horticultural and agricultural practices. Herbicides, as designated in this section, are chemicals which, in low concentration, are toxic to plants. It is well known, however, that herbicides applied at certain concentrations can produce specific, nonlethal changes in plants, and therefore can be considered under such circumstances as PGRs. In fact, herbicides were the first of these groups of synthetic chemicals to be investigated for their effect on insects. The initial study occurred in Louisiana sugarcane fields in the mid-1940s. It was observed that application of the then new phenoxy herbicide, 2,4-D, to control weeds actually resulted in greater infestation of sugarcane by the sugarcane borer, *Diatraea saccharalis* (F.). This increased infestation by the sugarcane borer was believed to have been a result of the reduction in the population of a trichogrammatid parasitoid, *Trichogramma minutum* Riley, due to the toxic effects of 2,4-D (Ingram et al. 1947). This study thus initiated research on the effects of directly applying herbicides to insects.

2.1. Direct Effects of Herbicides on Insects

Studies on the direct effects of herbicides on insects are summarized in Table 1. Most of the studies report that certain herbicides are toxic to insects. Most commonly mentioned was the fact that 2,4-D was toxic to predatory or parasitic insects, resulting in increased populations of pestiferous insects in treated plots. The first report of this side effect of herbicidal application to control weeds was the report by Ingram et al. (1947) dealing with the sugarcane borer and the reduction of its trichogrammatid parasitoid. This early report caused concern that 2,4-D was generally toxic to Hymenoptera. In the early 1950s it was observed that honeybee populations were lower in fields treated with 2,4-D as a result of it being toxic to bees (Leppick 1951, Johansen 1959). Concern over the toxicity of 2,4-D to bees was still being voiced in the mid-1960s (King 1964). Toxicity of 2,4-D was also observed to occur in early instar larvae of the wheat-stem sawfly, *Cephus cinctus* Norton, but the adults were unaffected (Gall and Dogger 1967).

In addition to the early concern of 2,4-D toxicity to Hymenoptera were reports in the late 1950s of increased aphid populations in 2,4-D-treated fields

Table 1. Summary of Studies of Direct Effects of Herbicides against Insects

Chemical Name	Common and/or Trade Name	Insect	Effect	Reference
3-Amino-1,2,4-triazole	Amitrole	*Heliothis zea*	Antifeedant	Dimetry and Mansour 1975
2,4-Dichlorophenoxyacetic acid	2,4-D	*Diatraea saccharalis*	Increased population of *D. saccharalis* due to toxic effect of 2,4-D to parasitoid	Ingram et al. 1947
		Rhopalosiphum padi, R. maidis, Macrosiphum avenae	Increased population due to toxicity to a predatory ladybird beetle	Adams and Drew 1965
		Cephus cinctus	Toxic to larvae, adults unaffected	Gall and Dogger 1967
		Apis mellifera	Toxic to bees	Leppick 1951, Johansen 1959, King 1964
		Aphids	Increase in aphids due to predator mortality	Simpson 1961
		Coccinellidae	Toxic to larvae	Adams 1960
		Agrotis ypsilon	When injected into larvae resulted in lower body weight but no effect on food consumption	El-Ibrashy and Mansour 1970
		Various soil insects	Slight mortality	Fox 1964
2,4,5-Trichlorophenoxyacetic acid	2,4,5-T	*Onychiurus quadriocellatus*	Direct application on springtails reduced reproduction and growth	Eijsackers 1978

1,1-Dimethyl-4,4-bipyridinium dichloride	Paraquat	Soil mesofauna	No effect on collembolans, mites or ants	Guerra et al. 1982
2-Chloro-4-(ethylamino)-6-(isopropylamino)-5-triazine	Atrazine	*Musca domestica*	Chemosterilant to male houseflies, causes reduced egg hatch and larval growth	Bořkovec et al. 1967
		Collembola and oribatid mites	No effect	Fox 1964, Sabatini et al. 1980
Trichloroacetic acid	TCA	Soil mesofauna (wireworms, Collembola, mites, etc.)	Variable effects	Fox 1964
N-(Phosphonomethyl)glycine	Glyphosate, Roundup	*H. zea*	No effect on larvae fed 150 ppm in artificial diet	Campbell et al. 1984
N-1-Naphthyl-phtalamic acid	Naptalam	*H. zea*	Antifeedant	Dimetry and Mansour 1975
4,6-Dinitro-2-*sec*-butylphenol	Dinoseb	*Geocoris punctipes, Coleomegilla maculata, Orius insidiosus*	Toxic in residual film or topical application	Stam et al. 1978
		H. zea	No effect at 25 ppm in artificial diets	Campbell et al. 1984
Monosodium methanearsonic acid	MSMA	*G. punctipes, C. maculata, O. insidiosus, Scymnus louisiana, Eretmocerus haldemani*	Not toxic	Stam et al. 1978
2-Chloro-2,6-diethyl-*N*-(methoxymethyl)acetanilide	Alachlor	*Trichogramma cacoeciae, Chrysopa carnea, Epistrophe balteata*	Directly toxic to *Trichogramma* and syrphid larvae but no effect on the green lacewing	Tanke and Franz 1978
2-Chloro-3-(4-chlorophenyl)methyl propionate	Chlorofenprop-methyl, Bidisin	*E. balteata*	Direct application is ovicidal	Tanke and Franz 1978

Table 1. *(continued)*

Chemical Name	Common and/or Trade Name	Insect	Effect	Reference
Methyl-*m*-hydroxy carbonilate, *m*-methyl carbanilate	Phenmedipham	*E. balteata*	Repellent to adult females decreasing oviposition on treated plants	Tanke and Franz 1978
?	UVON-Kombi	Various species of Carabidae	Reduction when applied to cultivated potato fields	Mueller 1972a,b
?	ELBANIL	Various species of Carabidae	Reduction when applied to cultivated potato fields	Mueller 1972a,b

(Maxwell and Harwood 1958). A study that soon followed this 1958 report showed that the direct spraying of 2,4-D (at herbicidal concentrations) on coccinellid larvae caused significant mortality (Adams 1960). Increases in aphid populations were later attributed to predator mortality as based on field observations (Simpson 1961). In 1965 Adams and Drew confirmed that an increase in grain aphids in plots of barley treated with 2,4-D was due to a reduction in the population of a predatory ladybird beetle, *Hippodamia tridecempunctata tibialis* (Say).

After the initial observation by Ingram et al. (1947) on 2,4-D toxicity to sugarcane borer, Fox (1948) reported that increased damage to wheat by the wireworm, *Ctenicera aeripennis destructor* (Brown), was associated with early season treatment of wheat fields with 2,4-D. This report was soon followed by a study in 1949 on the effects of herbicides on grasshopper populations of Canadian grasslands (Putnam 1949). A study by Fox (1964) showed that TCA and atrazine treatment of Canadian grasslands did not result in any apparent effects to mites and wireworms, whereas 2,4-D treatment caused some slight mortality. A later study by Sabatini et al. (1980) showed that atrazine had no toxic effect on either Collembola or oribatid mites inhabiting the soil of maize fields. Also use of paraquat in the Amazon had no effect on the soil invertebrate mesofauna (Guerra et al. 1982).

The remaining reports of direct herbicidal effects on insects are miscellaneous studies where herbicides were either fed to or injected into insects. Bořkovec et al. (1967) reported that atrazine was a chemosterilant to male houseflies. El-Ibrashy and Mansour (1970) observed only minor physiological effects of 2,4-D injected into larvae of the greasy cutworm, *Agrotis ypsilon* (Rott.), but saw no effects when it was fed to the larvae. Application of the related herbicides 2,4,5-T, on the springtail, *Onychiurus quadriocellatus* Gisin., adversely affected reproduction, feeding, and molting (Eijsackers 1978). Lastly, Dimetry and Mansour (1975) reported that amitrole and naptalam were feeding deterrents to *Heliothis zea* (Boddie), and in a different study it was observed that glyphosate fed to *H. zea* larvae at 150 ppm had no effect on behavior or growth (Campbell et al. 1984).

In the late 1960s use of herbicides for weed management expanded greatly. This expansion resulted from the additional use of herbicides for pre-emergence and postemergence weed control in order to facilitate either no-till or conservation tillage. This new usage of herbicides was accompanied by concerns of their potential environmental impact on natural biological control fauna. Most of these studies are considered in the following section, but some are mentioned here because of their direct toxicity to entomophagous insects. Widespread usage of the pre-emergence herbicides in cultivated potato fields in East Germany in the late 1960s resulted in a significant reduction of the carabid beetle population (Mueller 1972a,b). Large-scale use of the herbicides MSMA and dinoseb to control Johnson grass, *Sorghum halepense* (L.), in Louisiana cotton fields coincided with outbreaks of the bandedwing whitefly, *Trialeurodes abutilonea* (Haldeman). The toxicity of these herbicides to various predatory insects in cotton fields was examined by residual film tests and topical applications. Although MSMA was only marginally toxic, dinoseb was toxic to the predatory hemipterans *Geocoris punctipes* (Say), *Orius insidiosus* (Say), and a ladybird beetle, *Coleomegilla maculata* (DeGeer), at subinsecticidal (0.2–1.3 μg/insect) levels (Stam et al. 1978). A number of pre-emergence herbicides were found to affect the viability of various beneficial insects, by being directly toxic, repellent, or ovicidal (Tanke and Franz 1978). Alachlor was directly toxic to the parasitoid *Trichogramma cacoeciae* March and larvae of the syrphid *Epistrophe balteata* DEG, but not the green lacewing *Chrysopa carnea* Steph. Chlorfenpropmethyl was an ovicide to syrphid eggs and application of phenmedipham repelled oviposition by adult female syrphids on the treated foliage.

2.2 Effects of Herbicides on Host Plant Quality to Insects

Other than in the few studies reported here, where herbicides were directly applied to insects, reports of herbicidal toxicity to insects may be in error. Because of the obvious modifications herbicides induce in normal plant metabolism, a vast majority of herbicidal effects on insect populations are due to changes in the quality of the host plant as a food source. Studies that have either suggested or shown that differences in insect performance associated with herbicidal treatment are due to biochemical changes in the host plant are outlined in Table 2. However, despite the wide usage of herbicides, including some which have been in use for over 40 years, the detailed biochemical mode of action of many herbicides on plant metabolism is either unknown or unclear. Many herbicides are known to interfere with photosynthesis (e.g., 2,4-D) (van Assche and Carles 1982), but their effects on specific biosynthetic pathways of plants are largely unknown. In some instances the mode of action has been definitely determined (e.g., glyphosate), and in such cases speculations as to the chemical changes in the host plant that affect its herbivorous insect populations, or even the predator and parasitoid guild associated with the herbivore, are possible. Furthermore herbicidal treatment can also have an impact on the plant as a source of refuge or as a oviposition site for herbivorous insects by changing the physical structure of the host plant.

The herbicide 2,4-D has received the most attention where it has caused physical or chemical changes in the host plant, and thereby changes in the insect populations feeding on the treated plant. Initially, in 1949, Putnam (1949) noticed that the nymphal population of *Melanopus mexicanus mexicanus* Sauss. in Alberta increased in weed plots treated with 2,4-D. He conjectured that the herbicides caused an initial improvement in food quality of the plants to the grasshoppers. It was not until almost a decade later, with observations of high aphid populations in 2,4-D-treated fields, that the potential improvement of the nutritional quality of herbicide-treated plants was further investigated. Maxwell and Harwood (1958) first reported that treat-

Table 2. Studies on Changes in Insect Populations or Performance Resulting from Herbicidal-Induced Changes in the Host Plant

Chemical Name	Common/ Trade Name	Insect	Host Plant	Effect	Reference
3-Amino-1,2,4-triazole	Amitrole	*Acyrthosiphon pisum*	*Vicia faba*	No effect	Robinson 1961
3,6-Dicloro-*o*-anisic acid	Dicamba, Banvel D, Disugran	*Macrosiphum avenae, Schizaphis graminum, Rhopalosiphum padi*	*Hordeum vulgare*	0.22% applications increased aphid populations 18–38%	Hintz and Schulz 1969
4-Chloro-2-butynyl-*N*-3-chlorophenyl carbamate	Barban	*Macrosiphum avenae, Schizaphis graminum, Rhopalosiphum padi*	*Hordeum vulgare*	0.44% application increased aphid populations 7–34%	Hintz and Schulz 1969
2,4-Dichlorophenoxyacetic acid	2,4-D	*A. pisum*	*V. faba*	Significant increase	Maxwell and Harwood 1958
		A. pisum	*V. faba*	Increase due to 2,4-D induction of higher amino acid levels	Maxwell and Harwood 1960
		A. pisum	*V. faba*	No effect	Robinson 1959, 1960
		Oscinella frit	*Avena sativa*	No effect	Mellanby et al. 1959
		Melanopus m. mexicanus	Various weeds	Number of nymphs was higher in treated plots	Putnam 1949
		Circulifer tenellus	Snapbeans	Seed and foliage treatment had no effect on leafhopper population or incidence of curly top	Peary 1959
		Chilo suppressalis	*Oryza sativa*	Treatment of rice caused increased population of rice stem borer	Ishii and Hirano 1963
		R. maidis *Ostrinia nubilalis*	*Zea mays*	Increased population of aphids and corn borers associated with elevated levels of protein in treated corn	Oka and Pimentel 1976
		Pectinophora gossypiella	*Gossypium hirsutum*	Use of 2,4-D in combination with other PGRs terminated late-season vegetative growth and boll formation removing adequate food for larvae to achieve pupation. Reduced pink bollworm 87–96%	Kittock et al. 1973

Table 2. (*continued*)

Chemical Name	Common/ Trade Name	Insect	Host Plant	Effect	Reference
		Chrysomphalus ficus, *Lepidosaphes becki*	*Citrus sinensis*	In combination with malathion reduced leaf and fruit drop and improved control of pests	El-Sayed et al. 1982
4-(2,4-Dichlorophenoxyl)butyric acid	2,4-DB	*Cerotoma trifurcata*, *Heliothis zea*	*Glycine max*	Herbicidal effect on soybean plants made host plant unsuitable for bean leaf beetle and corn earworm	Agnello et al. 1986b
2-Methyl-4-chloro-phenoxyacetic acid	MCPA	*M. avenae*, *S. graminum*, *R. padi*	*Hordeum vulgare*	0.16% solution to barley increased aphid populations 2–34%	Hintz and Schulz 1969
		A. pisum	*V. faba*	No effect	Robinson 1959, 1960
		Spiders	*A. sativa*	Decrease in population	Raatikainen and Huhta 1968
		M. avenae	*Triticum sativa*	No effect	Rautapää 1972
		Macrosteles fascifrons	*O. sativa*	Caused movement of aster leafhopper from killed weeds onto rice seedlings, lowering grain yield	Way et al. 1984
N-(Phosphonomethyl)glycine	Glyphosate, Roundup	*S. graminum*	*Sorghum bicolor*	No effect at reducing resistance of sorghum to greenbugs	Dreyer et al. 1984
		H. zea	*Lycopersicon esculentum*	Reduced larval growth by 73% when applied at 150 ppm to tomatoes—not correlated with change in nutrients or allelochemicals	Campbell et al. 1984
4-cyano-2,6-diidophenol	Ioxynil	*M. avenae*	*T. sativa*	No effect	Rautapää 1972
2-(2-Methyl-4-chlorophenoxy)propionic acid	Mecoprop	*M. avenae*	*T. sativa*	No effect	Rautapää 1972
N-2,4-dimethyl-5-(trifluoromethyl)-sulfonylaminophenylacetamide	Mefluidide, Embark	*Cerotoma trifurcata*, *H. zea*	*Glycine max*	Bean leaf beetles and *H. zea* larvae higher in treated soybeans	Agnello et al. 1986b
		Epilachna varivestis	*G. max*, *Phaseolus lunatus*	Treated soybean plants are distasteful one day after application, then are normal one week later	Agnello et al. 1986a

?	Fluazifopbutyl	*C. trifurcata, H. zea*	*G. max*	Bean leaf beetle and *H. zea* larvae higher in treated fields	Agnello et al. 1986b
		E. varivestis	*G. max, P. lunatus*	Treated soybean leaves are rejected up to one week post-treatment. Both lima bean and soybean treated plants are preferred one to three weeks post-treatment. Pupal weight increased	Agnello et al. 1986a
?	Sethoxydim	*C. trifurcata*	*G. max*	No effect on pupal weight.	Agnello et al. 1986a
		H. zea	*P. lunatus*	Treated leaves of both plants preferred one to two weeks post-treatment	Agnello et al. 1986b
Sodium 5-[2-chloro-4-(trifluoromethyl)-phenoxyl]-2-nitrobenzoate	Acifluorfen, Blazer	*Geocoris punctipes*	*G. max*	Increased egg laying and viability of egg hatch from one day to three weeks post-treatment	Farlow and Pitre 1983
3-Isopropyl-1*H*-2,1,3-benzothiadiazin-4-(3*H*)-one-2,2-dioxide	Bentazon, Basagran	*Geocoris punctipes*	*G. max*	Increased egg laying and viability of egg hatch from one day to three weeks post-treatment	Farlow and Pitre 1983
4,6-Dinitro-2-*sec*-butyl phenol	Dinoseb	*M. avenae*	*T. sativa*	Reduced population growth. No effect on colonizing	Rautapää 1972
		H. zea	*Z. mays*	Reduced growth of larvae	Campbell et al. 1984, Hatley et al. 1977
			L. esculentum	No effect	Campbell et al. 1984
		Various predacious and parasitic arthropods	*G. hirsutum*	No effect	Stam et al. 1978
Monosodium methanearsonic acid	MSMA	*Trialeurodes abutilonia, Lygus lineolaris* and various predacious and parasitic insects	*G. hirsutum*	Increased whitefly (prey) and lygus bug populations by herbicide may have induced resultant increase in parasite and predator populations	Stam et al. 1978
3-(3,4-Di-chlorophenyl)-1,1-dimethylurea	Diuron	Various phytophagus, predacious and parasitic insects	*G. hirsutum*	No impact on field populations in cotton	Stam et al. 1978
2-Chloro-2,6-diethyl-*N*-(methoxymethyl)acetanilide	Alachlor	*Trichogramma cacoeciae*	Cultivated field plot	Acquisition of herbicide after systemic translocation in plant reduced degree of parasitization	Tanke and Franz 1978

ment of broad beans with 2,4-D resulted in increased populations of pea aphids. They (1960) later suggested that the increased pea aphid reproduction observed on 2,4-D-treated plants resulted from higher free amino acid levels, but they did not provide any evidence for this claim. Another report that appeared at this time stated that snap beans treated with 2,4-D had similar beet leafhopper populations and rates of infection of curly top virus, transmitted by this insect, as those observed for untreated plants (Peary 1959). In the same year Mellanby et al. (1959) also observed that there were no differences in frit fly infestations of oats between 2,4-D-treated and untreated plots. Robinson (1959, 1960) reported that broad beans treated with one of the phenoxy herbicides 2,4-D, 2,4,5-T, or MCPA resulted in no change in pea aphid populations. Moreover Robinson (1961) reported that there were no changes in pea aphid populations on broad beans treated with other herbicides that disrupt photosynthesis.

A study by Hintz and Schulz (1969), 10 years later, revealed that MCPA applied to barley in a 0.16% solution resulted in significant increases in English grain aphids, greenbugs, and oat bird cherry aphids by 2.3, 33.7, and 29.1%, respectively. They suggested that the lack of an effect of 2,4-D and MCPA on pea aphid populations as reported by Robinson (1959, 1960) was due to a low concentration of the herbicide application. Rautapää (1972) found that MCPA applied to wheat, either alone or with some experimental PGRs, had no effect on English grain aphids. However, elevated levels of protein in 2,4-D-treated corn were associated with increased populations of corn leaf aphids (Oka and Pimentel 1976). Hence, the effect of phenoxy herbicides on host plant quality to sap-feeding insects is variable. It appears in the literature that this may partly be a result of different times and rates of application and different plant species studied. An example of the untimely use of an herbicide causing severe problems was in California rice fields. Application of MCPA 52 to 58 days after rice seeding killed broadleaf weeds and sedges but resulted in the movement of aster leafhopper, *Macrosteles fascifrons* Stål, from these weeds onto the rice seedlings. This in turn caused a decrease in number of panicles per plant and a significant reduction in grain yield (Way et al. 1984). The phenoxy herbicides are effective at low dosages on broadleaf plants, but their effects on monocotyledons, at similar dosages, are more subtle. In fact, they often promote an auxinlike response in grasses, namely, root elongation or stimulation of shoot growth (Nickell 1982).

In addition to increases in sap-feeding insects, phenoxy-herbicide-treated plants are reported to cause increases in insects that bore into the meristematic tissues of graminaceous crops. These reports include increased rice stem borer populations in 2,4-D-treated rice (Ishii and Hirano 1963) and increased European corn borer infestation of corn (Oka and Pimentel 1976), the latter associated with higher levels of protein induced by 2,4-D. Hence the cases where increased insect populations occurred on 2,4-D, or other phenoxy-herbicide-treated plants, involved insects whose feeding strategies enabled them to exploit the chemical changes induced by the herbicides. The sap-feeding insects tapped into the phloem, thereby acquiring translocated nutrients, and the boring insects fed on plant tissues which responded in an auxinlike manner to these herbicides.

Phenoxy herbicides have also been used to physically change the host plant so as to diminish populations of pest insects. For example, when the diethylamine salt of 2,4-D is applied to late season cotton, the number of green, postharvest bolls is reduced by 90% (Kittock et al. 1973). This treatment resulted in an 87 to 96% reduction in the number of diapausing pink bollworm, and in a significant reduction of fruit yield. The 2,4-D herbicide has also been used to alter physically the host plant in order to improve exposure of pest insects to insecticides. El-Sayed et al. (1982) used 2,4-D to retard leaf and fruit drop on navel orange trees. This treatment improved the pesticidal effectiveness of malathion on two diaspids, *Chrysomphalus ficus* Ashmead and *Lepidosaphes becki* (New.), which are citrus pests. Lastly, for some undetermined reason, oat fields treated with MCPA had reduced spider populations (Raatikainen and Huhta 1968). This observation is somewhat anomalous as an increase in the prey population might be expected due to its toxicity to predators (Adams and Drew 1965, Simpson 1961).

There are a few studies showing that changes in aphid populations are due to changes in host plant chemistry induced by herbicides. Treatment of barley with 3,6-dichloro-*o*-anisic acid (Dicamba, Banvel D, or Disugran) resulted in significant increases in the grain aphids, the English grain aphid, the greenbug, and the oat bird cherry aphid (Hintz and Schulz 1969). These increases are probably associated with higher levels of sucrose that are generally induced by this herbicide when applied to monocotyledons (Thomas 1982). The increased sucrose levels stimulated increased sap-ingestion by these phloem-feeding aphids. Similarly, application of Barban to oats had the same effect (i.e., increased the populations of grain aphids) (Hintz and Schulz 1969). A somewhat curious noneffect on greenbug populations occurred when glyphosate was applied to sorghum (Dreyer et al. 1984). The herbicidal activity of glyphosate is due to its inhibition of the enzyme 5-enolpyruvylshikimate-3-phosphate synthase (Hoagland and Duke 1982). The inhibition of this enzyme disrupts synthesis of phenolics in plants. A number of phenolics, including *p*-hydroxybenzaldehyde, its acid, and dhurrin, a cyanogenic glycoside analogue, were previously isolated from sorghum and found to be feeding deterrents to greenbugs (Dreyer et al. 1981). It was expected that glyphosate-treated sorghum would have reduced levels of these feeding deterrents, and, as a result, greenbug populations would have increased. However, it was later shown that greenbugs are capable of avoiding these compounds (Campbell et al. 1982) by probing around intracellular vacuoles where these compounds are compartmentalized (Kojima et al. 1979). Despite the lack of effect of glyphosate treatment on aphid populations, the intent of the study was to use herbicides and/or PGRs to manipulate plant chemistry in order to study the chemical basis of insect–plant interactions.

In another study, when glyphosate was applied to tomato plants at 150 ppm, the growth of larvae of *H. zea* reared on those plants was reduced by 73% (Campbell et al. 1984). However, this reduction in larval growth was not correlated with any changes in simple carbohydrates (i.e., fructose, glucose, sucrose, and *myo*-inositol) or in secondary compounds known to be allelochemically active against *H. zea* (i.e., tomatine, rutin, and chlorogenic acid). Glyphosate had no effect on *H. zea* larval growth when incorporated into artificial diets (see Table 1); thus the reduction in larval growth on tomato must be associated with other unknown glyphosate-induced changes in plant chemistry.

Herbicides that are known to disrupt photosynthesis have had variable indirect effects on insects exposed to treated plants. The population growth of the English grain aphid was reduced on wheat treated with dinoseb (Rautapää 1972), a herbicide generally used to control annual weeds or as a yield enhancer for certain grain crops. The chemical basis for this response to the treated wheat was not determined, but the rate of colonization by alate aphids was not affected by dinoseb treatment. The specific mode of herbicidal action of dinoseb is not fully known. It does inhibit electron transport in both chloroplasts and mitochondria, thereby disrupting photophosphorylation and ATP generation (Moreland et al. 1982). The response of the aphids to treated wheat may therefore have been a result of reduced photosynthesis, thus reducing translocation of nutrients in the phloem. Similarly, growth of *H. zea* larvae reared on dinoseb-treated corn was reduced (Campbell et al. 1984, Hatley et al. 1977). But larvae reared on tomato plants treated with dinoseb had the same rates of growth as larvae reared on untreated plants (Campbell et al. 1984). Two other herbicides, ioxynil and mecoprop, which have similar herbicidal effects as dinoseb (i.e., disruption of photosynthesis) when applied to wheat, had no effect on English grain aphids (Rautapää 1972).

The variable responses of insects to the herbicide-treated plants reported in the aforementioned studies highlight the unpredictability of a generalized scenario in herbicide–plant–insect interactions. This variability is due to differences in herbicidal dose–response effects between plants, particularly in regard to how plant metabolism responds to the herbicides—as affected by variety, age, temperature, time of application, for example—and how changes in plant chemistry induced by the herbicides affect the feeding behavior or physiology of the various species of insects studied. A probable explanation for the relatively large number of studies in which there were herbicide-associated changes in plant–aphid interactions is that chemical changes induced in plant metabolism by herbicides are reflected in either the constituents translocated in the phloem-sap, the chief source of food for aphids, or changes in plant biopolymers through which aphids must probe (Campbell et al. 1986). Moreover aphid feeding and probing behavior are highly sensitive to even slight changes in phloem-sap chemistry (Pollard 1973).

The most recent studies, reporting effects of herbicides on plant–insect interactions, investigated three postemergence herbicides applied to soybeans (Agnello et al. 1986a, b). In these studies mefluidide, fluazifopbutyl, and sethoxydim were used to control Johnson grass in soybean fields over a three-year period. In plots treated with fluazifopbutyl and mefluidide, the bean leaf beetle population increased. However, densities of larvae of *H. zea* were significantly lower in sethoxydim-treated fields (Agnello et al. 1986b). The related study showed, further, that herbicide-treated soybeans were an improved food source to the Mexican bean beetle (MBB) (Agnello et al. 1986a). The feeding behavior of Mexican bean beetles was initially deterred (24 hours after application) by soybean foliage treated with mefluidide, sethoxydim, or fluazifopbutyl. However, two to three weeks after application, MBB adults preferred to feed on the herbicide-treated plants, with a concomitant increase in egg mass size on the treated plants. MBB pupal weights were significantly reduced and higher in cohorts reared on fluazifopbutyl-treated soybeans and lima beans, respectively. Mefluidide is known to be a growth retardant to legumes (Wu and Santelmann 1977), but the specific biosynthetic changes induced by these three herbicides on legumes have not been determined. It is known, however, that in sugarcane mefluidide-treatment increases sucrose content (Nickell 1982).

There are many studies showing that changes in either the chemistry or physical structure of the host plant affect both herbivorous insects, and indirectly the parasites and predators of these insects (Price et al. 1980). A final consideration of the impact of herbicides on plant–insect interactions is the potential role of herbicides in the interaction of parasites and predators with herbivorous insects. Herbicides not only have a direct effect on parasitic and predatory insects (see section 2.1), but by chemically inducing changes in the host plant, herbicides potentially affect the quality of herbivorous insects as hosts or prey for third trophic level insects.

An interesting study revealed that reduced populations of the parasitoid, *Trichogramma cacoeciae* March., resulted from acquisition of the herbicide, Alachlor, from the host insect. The host insect had ingested the herbicide after it was translocated throughout the plant (Tanke and Franz 1978). Alachlor is directly toxic to this parasitoid (see Table 1). Stam et al. (1978) found that treatment of cotton fields with MSMA, an herbicide used to control Johnson grass, resulted in higher populations of a predator, the convergent ladybird beetle, *Hippodamia convergens* Guérin-Méneville, and a parasitoid *Eretmocerus haldemani* Howard. These population increases were a result of increased populations of a prey insect, the banded wing whitefly, *Trialeurodes abutilonia* (Haldeman). However, MSMA treatment was also associated with increased populations of a second cotton pest, the lygus bug, *Lygus lineolaris* (Beauvois). Hence the benefits of increased predator and parasitoid populations may have been negated by the increased populations of both pest species. In this same study treatment of cotton fields with two other herbicides, dinoseb and diuron, had no effect on the phytophagous, predacious, or parasitic insects.

Application of two postemergence herbicides to cotton increased the population size of the big-eyed bug, *Geocoris punctipes* (Say), which can be a pest on cotton but is also predacious on other insects (Farlow and Pitre 1983). In this study both acifluorfen and bentazon treatments of cotton resulted in increased oviposition and improved egg viability in both greenhouse and field studies. The chemical basis for these results was not determined. Acifluorfen is activated in the presence of light by plant carotenoids, causing an oxidative chain reaction in membrane lipids including that of the chloroplasts (Orr and Hess 1982). The disruption of the integrity of cell and organelle membranes is associated with the release of ethylene, a natural plant growth regulator that initiates premature senescence in plants. This subject will be discussed further in the following section. In any case such a plant response is generally associated with activation of cell wall proteases and carbohydrases (Zeroni and Hall 1980), resulting in the breakdown of structural proteins and cell wall polysaccharides. The improved performance of *G. punctipes* may be due to the increased content of these breakdown products and the greater availability of nutrients in the herbicide-treated plants. Because *G. punctipes* is a facultative predator, this study is a rare example of how a predator can directly benefit from chemical changes induced in the plant by an herbicide.

3. EFFECTS OF PLANT GROWTH REGULATORS AND HORMONES ON INSECTS

The use of PGRs has progressively increased in both agriculture and horticulture, especially over the last decade. Whereas herbicides have always had a fundamental agricultural role for controlling weeds, the widespread use of PGRs has been relatively limited. Plant growth regulators (including plant hormones) are generally defined as either natural or synthetic compounds that, when applied to or released within a plant at low concentrations, alter plant metabolism or biosynthesis. This can result in a change in growth, structure, aging, resistance to pests, fertilization, fruiting, and so forth, without causing lethal effects (Nickell 1982). For obvious reasons the commercial use of PGRs is to increase yield, to improve quality or harvestability. Under certain circumstances some herbicides may be used as PGRs, depending on application rate, time, and plant species treated. The advantage of the commercial PGRs (also referred to as plant bioregulators) is that they can produce their effects, generally, by a single application using small amounts of active ingredient.

The plant hormones are essentially PGRs, except by definition they are naturally occurring compounds. Plant hormones are commonly divided into the auxins (e.g., indole-3-acetic acid), gibberellins, cytokinins, ethylene, inhibitors, and other miscellaneous hormones (e.g., triacontanol, the brassinolides) whose role in plant growth and physiology is more limited or highly specific. Interestingly, there have been a number of studies that suggest that certain plant hormones have a direct effect on the reproductive physiology of insects. Moreover the number of studies on the effect of PGRs on plant–insect interactions is much more extensive than that for the herbicides. The more specific biochemical or physiological changes produced by PGRs in plants simplifies any proposed chemical basis for explaining changes in insect populations that feed on plants treated with PGRs. The remainder of this chapter will be devoted to a review of these studies.

3.1 Direct Effects of Plant Growth Regulators and Hormones on Insects

Studies reporting the direct effects of plant growth regulators or plant hormones on insects or insect tissues are outlined in Table 3. One of the earliest interests in PGRs for insect–plant interactions was their potential role in explaining population fluctuations in insects. The role of natural plant hormones is still being investigated as to their effect on insect fecundity, physiological development, and initiation of diapause. It was initially reported by Ellis et al. (1965), of the Anti-locust Research Center in England, that diets deficient in gibberellic acid delayed sexual maturation in the desert locust, *Schistocerca gregaria* (Forskäl). When locusts fed on senescent plants with low levels of gibberellic acid, their rate of growth was slower, and they prematurely entered diapause. Similar observations were made for the cotton stainer, *Dysdercus cardinallis* Gerth. Abscisic acid (ABA), on the other hand, had no effect on either insect, even though it has some structural similarity to the insect juvenile hormone. Nation and Robinson (1966) also reported that gibberellic acid was a dietary requisite for normal larval and pupal development of honeybees.

It was later reported that gibberellins were a necessary requisite for locust maturation (Carlisle et al. 1969). When chlormequat chloride (CCC) was added to the diet, locust maturation was delayed. When applied to plants, CCC has activity that opposes the physiological effects of gibberellic acid. Based on these observations, it was proposed that levels of gibberellic acid in desert plants modulate the maturation, and therefore outbreaks, of the desert locust. A later study working with a grasshopper, *Aulocara elliotti* (Thomas), found an opposite correlation between gibberellic acid content and sexual maturity (Visscher-Neumann 1980). Higher levels of gibberellic acid and ABA lowered the fecundity of female grasshoppers and reduced the viability of eggs. However, in this study these plant hormones were incorporated into a grass diet at levels 100 to 1000 times higher than natural levels. Thus these hormonal levels may have induced significant biochemical changes in the grass diet, which in turn may have caused the reproductive physiological alterations in the grasshoppers.

Table 3. Studies That Investigated Direct Effects of Plant Growth Regulators (PGR) and Hormones on Insects

PGR	Arthropod/Insect	Effect	Reference
Gibberellic acid	*Schistocerca gregaria*	Diet deficient in GA_3 delayed maturation and caused adult diapause	Ellis et al. 1965
	Schistocerca gregaria	Necessary constituent in artificial diet for achieving maturation	Carlisle et al. 1969
	Apis mellifera	Necessary constituent in artificial diet for achieving maturation	Nation and Robinson 1966
	Aulocara elliotti	Retarded fecundity and viability of eggs when added to diet	Visscher-Neumann 1980
	Spodoptera frugiperda	Does not induce detoxicative enzyme activity	Yu 1983
	Bombyx mori	Addition to artificial diet increased larval weight	Kamada and Shipei 1984
	Heliothis zea	No effect on growth of larvae when incorporated into artificial diets	Campbell et al. 1984
	Drosophila hydei	When injected, caused chromosomal abnormalities	Alonso 1971
Abscisic acid	*S. gregaria*	No effect on growth or development	Carlisle et al. 1969
	Dysdercus cardinallis	No effect on growth or development	Carlisle et al. 1969
	Aulocara elliotti	When added to diet reduced fecundity and egg viability	Visscher-Neumann 1980
	Choristoneura fumiferana	No effect on growth at 10 ppm in diet	Eidt and Little 1970
	Tenebrio molitor	Slight juvenilizing effect when injected	Eidt and Little 1970
	Aphis fabae	Stimulated fecundity and rate of growth when added to leaf discs	Scheurer 1976
Ethylene	*Melanoplus sanguinipes*	Exposure of nymphs for 6 h retards growth whereas 12–24 h exposure accelerates growth. Reduced longevity of female adults. No effect on males	Chrominski 1982

	Anthonomus grandis	Acts as an attractant	Hedin et al. 1976
Indole-3-acetic acid	*Bombyx mori*	0.04 ppb in artificial diet increases larval weight; when sprayed on mulberry leaves, increases cocoon size 10%	Kamada and Shipei 1984
	Spodoptera frugiperda	Does not induce detoxicative enzymes	Yu 1983
	Lygus disponsi	High concentrations in artificial diet inhibit feeding	Hori 1980
6-Benzylaminopurine	*H. zea*	No effect on larval growth when incorporated at 500 ppm in artificial diets	Campbell et al. 1984
Carbaryl	Insects in general	Is an insecticide, but used for thinning apples	Hori 1980
Chlormequat chloride	*Agrotis ypsilon*	Growth inhibitor to larvae when ingested from diet or injected	El-Ibrashy and Mansour 1970
	Schistocerca gregaria, *D. cardinallis*	Has juvenilizing effect when added to artificial diets; teratogenic	Carlisle et al. 1969
	H. zea	No direct effect at 1000 ppm in artificial diets	Campbell et al. 1984
	Schizaphis graminum	No effect at 1.0% in artificial diet	Dreyer et al. 1984
Chlorphosphonium chloride	*Prodenia litura*	Antifeedant	Tahori et al. 1965b
Daminozide	*A. ypsilon*	Juvenilizing effect when injected into larvae	El-Ibrashy and Mansour 1970
Maleic hydrazide	*Acyrthosiphon pisum*	Reduced fecundity and caused mortality in nymphs	Bhalla and Robinson 1968
	H. zea	No effect at 500 ppm in artificial diet	Campbell et al. 1984
	Drosophila spp.	Does not cause chromosomal breaks	Fahmy et al. 1982
Mepiquat chloride	*S. graminum*	No effect at 1.0% in artificial diet	Dreyer et al. 1984
Oxamyl	Insects in general	An insecticide used for thinning apples	Byers 1978
Triacontanol	Aphids	Natural constituent of exocuticular wax	Campbell and Nes 1983

An early effort to use plant hormones to control an insect pest suggested that ABA be used to delay budbreak in spruce (Little and Eidt 1968), and thereby disrupt the life cycle of the spruce budworm, *Choristoneura fumiferana* (Clemens) (Eidt and Little 1968). ABA had no direct effect on spruce budworms when incorporated into an artificial diet at 10 ppm (Eidt and Little 1970). However, ABA had a slight juvenilizing effect on the yellow mealworm, *Tenebrio molitor* L., when 1 to 3μg were directly injected. Although ABA, applied at 500 ppm, delayed budding in balsam fir, no further information was provided as to the practicality of using ABA to control spruce budworms. In another study ABA ingested by the pea aphid from artificial diets stimulated fecundity and aphid growth (Scheurer 1976). As to whether increases in ABA concentrations in the host plant are signals to aphids of impending host plant senescence has not been reported.

Despite the significant changes in plant physiology induced by the plant hormone ethylene, few studies have investigated its effect on insects. Ethylene acts as an attractant to the boll weevil, *Anthonomus grandis* Boheman (Hedin et al. 1976). However, ethylene is ubiquitous throughout plants and probably does not play any role in attracting host-specific insects to a particular species of plant or crop. In a another study (Chrominski et al. 1982) six-hour exposures of the lesser migratory grasshopper, *Melanoplus sanguinipes* (Fabricius) nymphs, to ethylene retarded growth. Conversely, exposures of 12 to 24 hour increased the rate of growth of grasshopper nymphs but reduced longevity of adult females.

A number of miscellaneous studies have examined the effects of gibberellic acid and indole-3-acetic acid (IAA) on insect growth. Addition of either gibberellic acid or IAA at 0.04 ppb to the artificial diet of the silkworm, *Bombyx mori* (L.), increased larval weight. Spraying these hormones on mulberry leaves increased the eventual size of cocoons (Kamada and Shipei 1984). Cytokinins had no effect on growth of silkworm moths. Addition of gibberellic acid to artificial diets had no effect on larval growth of *H. zea* (Campbell et al. 1984). Neither of these plant hormones induced the detoxicative enzymes, glutathione-*S*-transferase or microsomal oxidases, in larvae of the fall armyworm, *Spodoptera frugiperda* (J. E. Smith) (Yu 1983). Relatively high concentrations of IAA in artificial diets deterred feeding by the bug, *Lygus disponsi* Linnavuori (Hori 1980). Metabolites of IAA in the excreta of *L. disponsi* fed on artificial diets incorporated with IAA matched the metabolites of IAA in the excreta of bugs that fed on rape flower buds. Alonso (1971) reported that injection of gibberellic acid into larvae of *Drosophila* resulted in abnormalities of salivary gland chromosomes.

The results of the preceding studies on the direct effects of plant hormones on insects present a somewhat equivocal picture of the role of plant hormones in insect–plant interactions. Similarly, a number of synthetic PGRs have been tested to determine their direct effects on insects. For the most part the direct effects of these PGRs have been studied in order to differentiate direct effects from effects that result from biochemical changes in host plants induced by the PGRs. Two examples of extreme effects of PGRs on insects are the insecticides carbaryl and oxamyl (Byers 1978). In addition to their true insecticidal activity, both of these compounds are abscission inducers. They have been used in commercial orchards to thin apples in order to improve the individual size and quality of the harvested apples.

A number of PGRs have been tested for their effects on larval growth of *H. zea*. A cytokinin, 6-benzylamino purine, a dwarfing agent, chlormequat chloride (CCC), and an antiauxin, maleic hydrazide, had no effect on *H. zea* growth and development when incorporated into artificial diets at a level as high as 1000 ppm (Campbell et al. 1984) (far in excess of concentrations used commercially). However, CCC inhibited the growth of the greasy cutworm, *Agrotis ypsilon* (Rott.), when ingested from artificial diets or injected (El-Ibrashy and Mansour 1970). Also CCC had no feeding deterrent activity against the greenbug, *Schizaphis graminum* (Rondani), at as high a concentration as 10,000 ppm in an artificial diet (Dreyer et al. 1984). As will be discussed in the next section, CCC is one of the few PGRs whose activity in plants induces resistance against aphids. Mepiquat chloride (PIX®), a PGR having a similar gibberellin-antagonistic effect as CCC, was also not a feeding deterrent to greenbugs (Dreyer et al. 1984).

The plant growth inhibitor, daminozide, was not an antifeedant but had some juvenilizing effect when injected into larvae of *A. ypsilon* (El-Ibrashy and Mansour 1970). Another plant growth inhibitor, chlorphosphonium chloride, was a feeding deterrent to the cotton leafworm, *Prodenia litura* F., when sprayed on cotton or waxbean foliage (Tahori et al. 1965b, c). Although maleic hydrazide had no effect on *H. zea* (Campbell et al. 1984), it was moderately toxic to pea aphids and reduced adult fecundity (Bhalla and Robinson 1968). The mode of antiauxin activity of maleic hydrazide is believed to be due to its structural similarity to uracil, which may be integral to disruption of transcription and/or translation of genetic information. Although maleic hydrazide causes chromosomal breakages in onion cells, it has no effect on the chromosomes of *Drosophila,* apparently because it is metabolized into inactive breakdown products (Fahmy et al. 1982).

Lastly, the long-chained fatty alcohol, triacontanol, was reported to be a growth-enhancing bioregulator for wheat (Ries et al. 1977). However, this reported activity is inconsistent, and other investigators have had difficulty in duplicating the originally reported effects of triacontanol. As to whether triacontanol is a true natural PGR is still in question. However, it is noteworthy that triacontanol is a normal constituent of insect exocuticular waxes and is also found in beeswax. It is normally synthesized by these insects (Campbell and Nes 1983).

3.2 Effects of Plant Growth Regulators on Host Plant Quality to Insects

The potential of using commercially available PGRs to control insects was a concept originally proposed by van Emden (1964). His proposal was chiefly based on observations that Brussels sprouts treated with chlormequat chloride (CCC) were less suitable host plants to cabbage aphids, *Brevicoryne brassicae* (L.), and green peach aphids, *Myzus persicae* (Sulzer), than untreated plants. Furthermore the commercial qualities of the treated plants were improved by the CCC treatment. This early proposal stimulated efforts to develop other PGRs that trigger plant resistance against insects and plant pathogens. To date, however, CCC has been the most extensively studied PGR for inducing host plant resistance to insects. There are, however, a number of other PGRs which have been studied to determine their effect on biochemical interactions between plants and insects. These studies are listed in Table 4. Many studies have attempted to use PGRs to deliberately alter the host plant quality to insects. However, in some cases, observations of changes in the quality of host plants to insects induced by PGRs, were originally fortuitous. For the most part the use of PGRs to alter host plant–insect interactions can be divided into two basic strategies, namely, (1) altering the physical structure or growth of the plant to eliminate resources (food, oviposition sites, refuge, etc.) and (2) inducing biochemical changes in the plant that reduce its nutritional value, increase allelochemicals, or change chemical cues for host plant recognition by the insect.

3.2.1 Use of PGRs to Alter the Host Plant Physically

One of the first attempts to use PGRs in controlling pest insects occurred in the 1960s. Tamaki and Weeks (1968) used the defoliant *S,S,S*-tributylphosphorotrithioate to defoliate peach trees in order to control green peach aphids. This treatment reduced the overall number of aphids by removing the food resource of the aphids. The treatment also increased predation by syrphid larvae due to the higher density of aphids on remaining, unfallen leaves. To reduce the number of overwintering pink bollworms *Pectinophora gossypiella* (Saunders), Adkisson (1962) used the defoliants orthoarsenic acid and *S,S,S*-tributylphosphorotrithioate to desiccate late season cotton bolls. Both of these PGRs were already in use as a preharvest treatment. These treatments removed leaves from the cotton plant to facilitate machine harvesting of lint. Since this original study, a number of further attempts have been made to use PGRs to control pink bollworm. In all cases the strategy was to terminate cotton growth in order to starve the larvae before they entered diapause. Kittock et al. (1975, 1980) used *S,S,S*-tributylphosphorotrithioate, 2,4-D, and 3,4-dichloroisothiazole-5-carboxylic acid to remove late season bolls from cotton for the control of pink bollworm. The plant growth suppressant chlormequat chloride (CCC) was also used to prevent fruiting in late season cotton for the same purpose (Kittock et al. 1973, Bariola et al. 1976). The use of the more severe growth suppressants (e.g., 2,4-D) and defoliants (e.g., ethephon) reduced lint yields (Kittock et al. 1973). Mixtures of PGRs such as CCC and 3,4-dichloroisothiazole-5-carboxylic acid gave the best results (Bariola et al. 1976). The use of these PGRs was

Table 4. Studies in Which Application of Plant Growth Regulators (PGR) Affected Insect–Host Plant Interactions or Altered Insect Development and Population Growth

PGR	Insect	Plant	Effect	Reference
Abscisic acid	*Aulocara elliotti*	Grass	Reduced fecundity and egg viability	Visscher-Neumann 1980
	Choristoneura fumiferana	Balsam fir	500 ppm sprayed on balsam fir delayed budding and spruce budworm infestation	Eidt and Little 1970
Ethylene	Insects in general	Plants in general	Damage by herbivorous insects to plants causes release of ethylene	Morgan et al. 1983, Abeles 1973
Gibberellic acid	*Ceratitus capitata* *Anastrepha suspensa*	Citrus fruit	50 ppm applied to fruit sustains resistance to fruit flies	Greany, unpublished
	Tetranychus urticae	Snapbean plants	15 ppm applied to plants reduced population size	Eichmeier and Guyer 1960
	T. urticae, Panonychus ulmi	Beans, apple trees, cotton	Treated snapbeans and apple trees reduced mite populations; treated cotton gave variable results	Rodriguez and Campbell 1961
	Schistocerca gregaria	Desert plants	Senescent vegetation, low in gibberelllic acid, induced adult diapause	Ellis et al. 1965
	A. elliotti	Grass	Reduced fecundity and egg viability	Visscher-Neumann 1980
	Aphis fabae	*Vicia faba*	Reduced fecundity and population growth rate	Honeyborne 1969
	Acyrthosiphon pisum	*V. faba*	No effect	Robinson 1960
	Liriomyza trifolii	Chrysanthemum	Treated leaves had a higher density of leaf miners	Knodel-Montz et al. 1983

Table 4. (*continued*)

PGR	Insect	Plant	Effect	Reference
Indole-3-acetic acid	*Bombyx mori*	Mulberry leaves	ng quantities sprayed on leaves increased cocoon size 10%	Kamada and Shipei 1984
	Lygus disponsi	Rape flower bud	Excreta had auxinlike activity and contained metabolites of IAA	Hori 1980
1-Amino-2-nitrocyclo-pentane-1-carboxylic acid	*Brevicoryne brassicae* *Myzus persicae*	Brussels sprouts	500–1000 ppm foliar spray reduced growth rate of aphids; not related to any change in soluble N_2	van Emden 1969b
L-Amino-oxy-phenyl-propionic acid	*Heliothis zea*	*Gossypium hirsutum*	10^{-3} *M* applied to cotton reduced condensed tannins, but no effect on growth of *H. zea* larvae	Campbell et al. 1984
o-Arsenic acid	*Pectinophora gossypiella*	*G. hirsutum*	Reduced pink bollworm population in diapause by defoliation of late season cotton	Adkisson 1962
6-Benzyladenine	*Amphorophora agathonica*	Red Raspberry	Treatment of excised leaflets with 10 ppb delayed senescence and loss of resistance; no effect on susceptible varieties	Brodel and Schaefers 1980
	Aphis rubicola	Red Raspberry	No effect on susceptible, intact plants	Brodel and Schaefers 1980
Dimethyl-β-bromo-ethylene sulfonium bromide	*A. pisum*	*Pisum sativum*	Treatment of pea seeds resulted in reduced aphid infestations in budding pea plants	Protopopova and Dolgopolova 1982
Chlorflurenol	*C. fumifera*	Balsam fir	Used to delay budbreak to control spruce budworm; equivocal results	Eidt and Little 1970

Chlormequat chloride	*Agrotis ypsilon*	Castor bean	Reduced larval growth and pupal weight	El-Ibrashy and Mansour 1970
	Prodenia litura	*G. hirsutum*	Reduced larval growth	Tahori et al. 1965b
	Cecidophyopsis ribis	Black currant	Reduced infestation and increased number of buds	Smith and Corke 1966, Smith 1969
	Aphis varians	Black currant	Reduced infestation and increased number of buds	Smith 1969
	P. gossypiella	*G. hirsutum*	Terminated late-season fruiting reducing food for overwintering diapause	Kittock et al. 1973, Bariola et al. 1976
	H. zea *Heliothis virescens*	*G. hirsutum*	Terminated late-season fruiting reducing food for overwintering diapause	Thomas et al. 1979
	H. zea	*Lycopersicon esculentum*	Reduced larval growth by 37–63% when applied at 250 and 1000 ppm. No effect on allelochemical or nutritional compounds	Campbell et al. 1984
	T. urticae	Chrysanthemum	In some cases populations were reduced	Worthing 1969
	Myzus persicae	Chrysanthemum	In some cases populations were reduced	Worthing 1969
	Brevicoryne brassicae, *M. persicae*	Brussels sprouts	Production of alatae reduced and nymphal mortality higher on treated plants	van Emden 1964a
	Brevicoryne brassicae, *M. persicae*	Brussels sprouts	Production of alatae reduced and nymphal mortality higher on treated plants: reports reduction in amino acids is cause of effect	van Emden and Wearing 1965
	Brevicoryne brassicae, *M. persicae*	Brussels sprouts	1% soil drench or foliar spray reduced growth rate 47 and 52%, respectively	van Emden 1969a

Table 4. (*continued*)

PGR	Insect	Plant	Effect	Reference
	Megoura viciae	*V. faba*	Foliar spray or soil drench had no effect on aphids	Adolphi 1967
	Aphis nerii	Oleander	Reduced population of oleander aphid	Tahori et al. 1965a,c
	B. brassicae *A. fabae*	Brussels sprouts *V. faba*	Reduced population of aphids on treated plants: increase in soluble N_2 and sugar in treated plants	Honeyborne 1969
	A. pisum	*V. faba*	Rate of reproduction reduced; levels of soluble N_2 reduced	Yule et al. 1966
	A. pisum	*Pisum sativum*	Treatment of pea seeds with CCC resulted in reduced infestation of pea plants during budding	Protopopova and Dolgopolova 1982
	A. pisum	*Medicago sativa*	Reduction of aphids on treated plants associated with changes caused in the neutral sugar composition of pectic substances	Campbell et al. unpublished
	Schizaphis graminum	*Sorghum bicolor*	0.1% treatment of seedlings made mature plants resistant to greenbugs; methoxy content of pectin increased in treated plants	Dreyer et al. 1984
	Hyperomyzus lactucae	Black currant	Treated plants had reduced populations of aphids	Singer and Smith 1976
	Grain aphids	*Triticum sativa*	1% application on spring wheat had no effect on colonization by aphids and on aphid growth	Rautapää 1972, Latteur 1976

Chlorphosphonium chloride	*Prodenia litura*	*Glycine max*, Wax beans	Deterred feeding of larvae	Tahori et al. 1965b,c
	Aphis nerii	Oleander	Reduced infestation	Tahori et al. 1965a,c
	M. persicae, T. urticae	Chrysanthemum	Reduced infestations by aphids and mites	Worthing 1969
Cycloheximide	*H. zea*	*L. esculentum*	No effect on larval growth and development	Campbell et al. 1984
Daminozide	*Cecidophyopsis ribis, A. varians*	Black currant	Reduced populations of mites and aphids on treated plants	Smith and Corke 1966
	Aphis pomi	Apple	Reduced infestation	Hall 1972
	Psylla pyricola	Pear	2000 ppm, 30–50 days post-bloom reduced terminal shoot growth; lowered psyllid damage due to feeding and honeydew; 35–57% decline in psyllid population after 3 yr	Westigard et al. 1980
	A. fabae, B. brassicae	Brussels sprouts	At 5000 ppm increased infestation of *B. brassicae*, but caused a decline in infestation by *A. fabae*	Honeyborne 1969
	T. urticae, M. persicae	Chrysanthemum	Application had no effect on numbers of mites or aphids	Worthing 1969
	Liriomyza trifolii	Chrysanthemum	Reduced number of eggs oviposited by a leaf miner	Knodel-Montz et al. 1983
	A. nerii	Oleander	Reduced population size of oleander aphid	Tahori et al. 1965a,c
S,S,S-Tributylphosphorotrithioate	*P. gossypiella*	*G. hirsutum*	Used to defoliate late season cotton to reduce availability of food for diapausing pink bollworm	Adkisson 1962
	Myzus persicae	Peach trees	Defoliation of trees reduced overall number of aphids and increased efficiency of predatory syrphids	Tamaki and Weeks 1968

Table 4. (*continued*)

PGR	Insect	Plant	Effect	Reference
Potassium 3,4-dichloroisothiazole-5-carboxylic acid	*P. gossypiella*	*G. hirsutum*	Reduced number of immature bolls as food for pink bollworm to enter diapause	Kittock et al. 1975, 1980
α-(4-Chlorophenyl)-α-(1-methylethyl)-5-pyrimidine methanol	*Neohetina eichhorniae*	*Eichhornia crassipes*	Use of retardant on water hyacinth improved performance of weed control by weevils	Center et al. 1982
Ethephon	*H. zea*	*Lycopersicon esculentum*	No effect on growth of larvae fed on treated foliage	Campbell et al. 1984
	P. gossypiella	*G. hirsutum*	Reduced green bolls to deprive pink bollworm of food to enter diapause; too severe effect on cotton	Kittock et al. 1973
	Lymantria dispar	Oak forest	Used as defoliant to deprive gypsy moth larvae of food. Reduced larval population by 50% and number of fall egg masses by 85%; severe mortality of treated trees	Sterrett et al. 1984
	Apis mellifera	*Cucurbita maxima*	Application to 2 + 4 leaf stage of pumpkin improved cross-pollination by bees and increased production of hybrid seeds	Hume and Lovell 1981
Ethylene bisnitrourathane	*A. fabae*	*V. faba*	Fecundity and size of aphids reduced and pigmentation of aphids affected; soluble N_2 reduced	Honeyborne 1969
Maleic hydrazide	*H. zea*	*L. esculentum*	Reduced growth of larvae when applied to foliage	Campbell et al. 1984

	A. pisum	*V. faba*	Significant mortality in nymphs and adults at 2000 ppm	Robinson 1959, 1960
	Choristoneura fumiferana	Balsam fir	Used to delay budbreak to control spruce budworm	Eidt and Little 1970
	A. pisum	*V. faba*	Reduced adult fecundity, larval mortality increased	Yule et al. 1966
Mepiquat chloride	*H. virescens*	*G. hirsutum*	Improved growth of budworms related to decrease in allelochemicals and increase in nutrients of treated cotton	Hedin et al. 1984
	H. zea	*G. hirsutum*	Bollworm reduction associated with increase in terpenoids and condensed tannin in treated plants	Zummo et al. 1984
	S. graminum	*S. bicolor*	Susceptible varieties made resistant to greenbugs by treating 2-leaf stage seedlings	Dreyer et al. 1984
1-Naphthyl acetic acid	*H. zea*	*L. esculentum*	No effect on *H. zea* larvae	Campbell et al. 1984

later (late 1970s) found to be effective in controlling the size of the overwintering population of other cotton pests such as the cotton bollworm, *H. zea,* the tobacco budworm, *H. virescens,* and the boll weevil, *A. grandis* (Thomas et al. 1979).

A second attempt to use PGRs for controlling insects emphasized the use of natural PGRs. In a series of studies performed by Eidt and Little (Eidt and Little 1968, Little and Eidt 1968), abscisic acid was used to delay budbreak in balsam fir. By delaying budbreak, they hoped to produce asynchrony between emergence and ovipositioning of spruce budworm and availability of food resources (Eidt and Little 1968, 1970). Although some success was achieved with this strategy under controlled, laboratory conditions, there were no further reports of success under natural conditions. A more recent effort at controlling another forest pest, the gypsy moth, *Lymantria dispar* (L.) used the defoliant, ethephon, in an oak forest (Sterrett et al. 1984). Ethephon and endothal [7-oxabicyclo(2,2,1)heptane-2,3,-dicarboxylic acid] were used to defoliate 60 to 80% of the trees in the treated area. Although this defoliation resulted in a 50 and 85% reduction in the larval population and numbers of egg masses, respectively, oak tree mortality was twice as high in the treated plots.

The treatment of pear trees with daminozide was used to control pear psylla, *Psylla pyricola* Foerster (Westigard et al. 1980). By retarding shoot growth in Bartlett pears with one application of daminozide at 2000 ppm, pear psylla populations were reduced by 35% and fruit damage reduced by 57% over a three-year period. During the first year posttreatment, fruit bud formation and yield actually increased. This is a good example of where the judicious use of a PGR resulted in enhancing plant production and, on a long-term basis, reduced a pest population.

3.2.2 Use of PGRs to Influence the Chemical Basis of Insect–Plant Interactions

Both natural and commercial PGRs have, reportedly, been used to alter the normal metabolism of a host plant, and thus the population dynamics of respective herbivorous insects. Researchers in this area have generally speculated that PGRs affect the chemical basis of insect–plant interactions in two ways. First, PGRs can alter the nutritional quality of the host plant by reducing the amount of available protein or amino acids. Second, PGRs can trigger the biosynthesis of allelochemicals in the plant, which will be both toxic and deter feeding by herbivorous insects. However, one of the most thoroughly studied effects of a PGR on the chemical basis of insect–plant interactions (that of chlormequat chloride) shows that the chemical basis is neither nutritional nor allelochemical but results from altering chemical constituents in the host plant that may serve as feeding cues (Campbell et al. 1986). Unfortunately, most reports of effects of PGRs on insect–plant interactions do not explore the chemistry of these effects and are only observations of insect performance on the treated plants.

Only a few studies have attempted to analyze changes in host plant chemistry induced by natural PGRs. A recent controversy in the chemical ecology of insect–plant interactions concerns the reports of so-called "talking trees." Briefly, refoliated trees that had been defoliated by herbivorous insects, or trees adjacent to the defoliated trees, have reportedly been shown to contain higher amounts of allelochemicals (condensed tannins, total phenolics, etc.) (Schultz and Baldwin 1982, Rhoades 1983). These higher levels of allelochemicals were associated with lower infestations by the next generation of insects. Hence it was proposed that trees being defoliated by insects signal adjacent trees to increase biosynthesis of feeding deterrents. However, this explanation is debatable (Greig-Smith 1986). In any case, the "signal" emitted by the defoliated tree may be ethylene (although within-plant elicitors may include cell wall fragmentary oligosaccharides; see Ryan et al. 1986, references therein). There is evidence that insect damage to plants causes the synthesis and release of ethylene (Abeles 1973, Morgan et al. 1983). Ethylene induction of biosynthesis of plant defensive chemicals, however, is well known in the field of plant pathology (Yang and Pratt 1978). Generally, it has been shown that ethylene can induce synthesis of lignin to contain infections by plant pathogens (Vance et al. 1980). In fact, ethylene can stimulate activity of phenylalanine ammonia lyase (PAL), an enzyme that is integral to the conversion of aromatic amino acids to precursors for phenolic and flavonoid biosynthesis (Rhodes and Wooltorton 1978). Alternatively, ethylene treatment made cell wall polysaccharides of oak leaves resistant to degradation by fungal cell-wall-degrading enzymes (Gaballe and Galston 1979). This observation is related to the induction of resistance to aphids in plants by the ethylenelike PGRs, ethylene bisnitrourathane, chlormequat chloride, and mepiquat chloride, which is discussed later in this section.

Gibberellic acid is another plant hormone that has been shown to affect the chemical basis of insect–plant interactions. There are numerous reports where application of gibberellic acid to plants resulted in lowering infestations of mites (Eichmeier and Guyer 1960, Rodriguez and Campbell 1961), aphids (Honeyborne 1969) (although in one case a low application rate resulted in no effect on aphids; Robinson 1960) and a grasshopper (Visscher-Neumann 1980). Alternatively, Ellis et al. (1965) reported that senescent foliage low in gibberellic acid was associated with slower growth in the desert locust. Application of 50 ppm of gibberellic acid to *Citrus* fruit prior to colorbreak sustains resistance against Caribbean and Mediterranean fruit flies (Greany and Rössler 1985). Larvae of these fruit flies are sensitive to the essential oils in the fruit peel (Greany et al. 1985). Application of gibberellic acid to *Citrus* inhibits the rate at which the peel softens. Hence, these larvae are subjected to deleterious peels oils (Coggins et al. 1969) for longer periods of time, reducing their survival and emergence through the peel. However, application of a gibberellic acid promoter to chrysanthemum resulted in increased oviposition and a higher density of leaf miners (Knodel-Montz et al. 1983).

Two other plant hormones that affect insects when applied to plants are abscisic acid and indole-3-acetic acid (IAA). Abscisic acid, applied at very high levels to a grass diet, reduced fecundity and egg viability in a grasshopper (Visscher-Neumann 1980). Application of 10% IAA to mulberry leaves improved the growth of silkworms (Kamada and Shipei 1984). In neither case was it determined whether these effects resulted directly from the plant hormone or from biochemical changes induced by the hormone in the plant. In another study, it was shown that excreta from lygus bugs feeding on rape flowers had auxinlike activity and contained metabolites of IAA (Hori 1980). The role that this might play in the interaction between the insect and its host plant was not discussed.

Of the synthetic PGRs, chlormequat chloride (CCC) has received the most attention with regard to its effect on reducing aphid infestations on cereal grasses. Likewise, PGRs having similar modes of action (i.e., gibberellin-antagonists) as CCC have also received moderate attention as to their effect on insect–plant interactions. There are a handful of other synthetic PGRs that have been shown to affect insect–plant interactions. The chemical basis of these effects have rarely been defined.

Perhaps the most widely used plant bioregulator is CCC. It is commonly used in Europe as an antilodging agent for wheat and barley. Almost 90% of the wheat in Germany is treated with CCC. In most of the investigations where the effects of CCC on insect–plant interactions were studied, CCC treatment resulted in reducing the respective herbivorous insect population. With regard to larvae of Lepidoptera, treatment of castor bean with CCC reduced larval growth and pupal weight of the greasy cutworm (El-Ibrashy and Mansour 1970). Similar results occurred in studies by Tahori et al. (1965b) with larvae of the cotton leafworm fed on CCC-treated cotton. Treatment of tomatoes with 250 ppm and 1000 ppm CCC reduced growth of larvae of *H. zea* by 37 and 63%, respectively (Campbell et al. 1984). However, CCC treatment had no effect on levels of nutrients (i.e., fructose, glucose, sucrose, or *myo*-inositol) and allelochemicals (α-tomatine, rutin, or chlorogenic acid). In a different study it was reported that CCC treatment of tomatoes promoted higher levels of rutin (Jablonowski 1972), a growth inhibitor of *H. zea* (Elliger et al. 1980). Treating plants with CCC has also reduced mite infestations on chrysanthemum (Worthing 1969) and black currant (Smith and Corke 1966, Smith 1969). However, most of the reports of CCC treatment of plants affecting insects have centered on aphids.

Initial reports that CCC reduced aphid infestations were made in the 1960s. Van Emden (1964, 1969a) reported that treating Brussel sprouts with CCC reduced infestations of cabbage aphids, and of green peach aphids. The reduction in aphids was related to lower levels of free amino acids in CCC-treated plants (van Emden and Wearing 1965). Tahori et al. (1965a, c) next reported that CCC treatment of oleander reduced populations of oleander aphids, *Aphis nerii* (Boyer). Other reports of aphid reduction by CCC treatment of the host plant include *Aphis varians* Patch on black currant

(Smith 1969), *M. persicae* on chrysanthemum, *Hyperomyzus lactucae* (L.) on black currant (Singer and Smith 1976), and pea aphids, *A. pisum,* on peas (Protopopova and Dolgopolova 1982). However, there are studies reporting that CCC treatment of the host plant had no effect on aphid populations. These studies include black bean aphid, *Megoura viciae* Buckton, on broadbean (Adolphi 1967) and various grain aphids on wheat (Latteur 1976, Rautapää 1972).

The chemical basis for the effects of CCC on aphid plant interactions was originally thought to be a result of lowering the nutritive value of the host plant. This was thought to occur by reduction of free amino acids (van Emden 1969b) or total soluble nitrogen in treated plants (Yule et al. 1966). Nevertheless, CCC treatment of wheat resulted in reduced soluble nitrogen levels (Linser et al. 1965), but aphid numbers were not reduced (Rautapää 1972, Latteur 1976). Conversely, CCC treatment of Brussel sprouts resulted in higher soluble nitrogen and sugar levels with reduced infestations of *B. brassicae* (Honeyborne 1969). Khan and Faust (1966) reported that treatment of barley coleoptiles with CCC resulted in higher soluble nitrogen levels in grown barley plants. Other researchers have shown that CCC treatment increases levels of free amino acids (Castro and Gutierrez 1979, Gasser and Thorburn 1972) and macronutrients (Castro 1978) in plants. Hence it is debatable whether CCC lowers the nutritive value of host plants to aphids.

The chemical basis for the effect of CCC on aphid–plant interactions may be due to changes induced by CCC on plant intercellular pectin. Certain structural aspects of plant intercellular pectins play a significant role in the basis of aphid–plant interactions (Dreyer and Campbell 1984, Campbell and Dreyer 1985). In particular, the ability of greenbugs, *Schizaphis graminum* (Rondani), to ingest phloem-sap from their host plant was found to be associated with the rate that pectinases from the aphid could break down host plant pectin. In particular, pectin in greenbug-resistant varieties of sorghum had a high methyl ester content that reduced the ability of greenbug pectinases to depolymerize this pectin (Dreyer and Campbell 1984). It was recently shown that treatment of seedlings of sorghum varieties, normally susceptible to greenbugs, with CCC resulted in resistance of grown plants to greenbugs (Dreyer et al. 1984). The chemical basis of this induced resistance from the CCC treatment was shown to be a result of increased methyl ester content of the pectin from these sorghum plants.

It has been shown that CCC also alters the pectin chemistry of other plants. Wheat treated with CCC had an 80% increase of pectin (Blaim and Preszlakowska 1967) and altered the chemistry of pectic substances (Blaim et al. 1970, Blaim and Szynal 1968). Alfalfa treated with CCC had pectin with a neutral sugar composition entirely different from that of untreated plants (B. C. Campbell, K. C. Jones, D. L. Dreyer, unpubl.), and CCC treatment of tomatoes also resulted in elevated levels of pectin (Blaim et al. 1968). However, CCC apparently does not affect hemicellulose, cellulose, and protein content of cell walls in wheat (Preszlakowska 1974). Structural changes in pectin also affect aphid feeding behavior. Breakdown products from the action of aphid polysaccharases on cell wall polysaccharides act as probing cues for host plant recognition (Campbell et al. 1986). Thus PGR-induced changes in the structure of these polysaccharides will also influence gustatory responses of aphids.

The chemical basis for the reduction of aphids on plants treated with other gibberellin-antagonistic PGRs is probably due to similar chemical changes in pectin as noted for CCC. These other PGRs include dimethyl-β-bromoethylenesulfonium bromide, which reduced pea aphids on peas (Protopopova and Dolgopolova 1982); daminozide, which reduced aphids and mites on black currant (Smith 1969), apple aphid on apples (Hall 1972), bean and cabbage aphids on Brussels sprouts (Honeyborne 1969), green peach aphid (Worthing 1969), and leaf miners (Knodel-Montz et al. 1983) on chrysanthemum and oleander aphid on oleander (Tahori et al. 1965a, c); mepiquat chloride, which reduced greenbugs on sorghum (Dreyer et al. 1984); and ethylene bisnitrourathane, which reduced bean aphids on broad bean (Honeyborne 1969).

Mepiquat chloride has been commonly used on cotton to stunt vegetative growth and increase plant density for greater yields per acre. Application of mepiquat chloride is generally associated with reduced cotton bollworm populations (Zummo et al. 1984). Reduction in bollworm growth has been associated with increased content of condensed tannins and terpenoids in cotton treated with mepiquat chloride (Zummo et al. (1984). Conversely, in cotton that was drought-stressed, mepiquat chloride improved growth of another cotton pest, the tobacco budworm, *H. virescens* (Hedin et al. 1984). This improved growth was associated with increased nutritive value of the treated cotton. Protein, lipid, and mineral content in treated cotton increased by 14, 42, and 26%, respectively. Allelochemical content declined with flavonoids and tannins decreasing by 31 and 38%, respectively.

Another experimental PGR, L-amino-oxy-phenylproprionic acid (L-AOPP), was used to influence *H. zea* on cotton (Campbell et al. 1984). L-AOPP is an inhibitor of PAL-activity (Hanson 1981), and its application to cotton at $10^{-3}M$ in 0.1 M sucrose reduced condensed tannin in cotton plants. At lower doses (10^{-6} M), it actually promoted PAL activity, with concomitant increases in condensed tannins. Levels of the flavonoid, gossypetrin were unaffected by either treatment. However, in neither case was growth of *H. zea* larvae affected when they were fed treated plants. Similarly, ethephon, 1-naphthyl acetic acid and cycloheximide had no effect on larvae of *H. zea* which fed on treated plants (Campbell et al. 1984).

In general, the remaining studies of the effects of PGR treatment on insect–plant interactions cited in Table 4 did not provide a chemical basis for any observed effects. However, it can be noted that plants treated with growth-retarding PGRs usually showed reduced growth of herbivorous insects, especially aphids. Maleic hydrazide treatment of host plants reduced population growth of the pea aphid, *Acyrthosiphon pisum* (Robinson 1959,

1960; Yule et al. 1966) and larval growth of *H. zea* (Campbell et al. 1984). Chlorphosphonium chloride treatment of respective host plants reduced infestations by the oleander aphid (Tahori et al. 1965a, 1965c), green peach aphid, two-spotted spider mite (Worthing 1969), and deterred feeding by the cotton leafworm (Tahori et al. 1965b). Treatment of Brussel sprouts with the amino acid antagonist, 1-amino-2-nitrocyclopentane-1-carboxylic acid, severely affected the plants and reduced growth of the aphids, *B. brassicae* and *M. persicae* (van Emden 1969b). However, use of the growth retardant EL-509, α-(4-chlorophenyl)-α-(1-methylethyl)-5-pyrimidine methanol on water hyacinth, *Eichhornia crassipes* (Mart.) Solms, improved the control of this weed by the water hyacinth weevil, *Neohetina eichhorniae* Warner (Center et al. 1982). Use of the cytokinin 6-benzyladenine (BA), retarded senescence in excised leaflets so that resistance against an aphid vector of mosaic virus, *Amphorophora agathonica* Hottes, was maintained in red raspberry (Brodel and Schaefers 1980). However, application of BA to susceptible varieties of red raspberry had no effect on infestations by another aphid, *Aphis rubicola* Oestlund, (Brodel 1979). Lastly, Hume and Lovell (1981) reported that treatment of pumpkin seedlings with ethephon improved cross-pollination by bees, resulting in greater seed production.

4. THE OUTLOOK FOR USE OF PGRS IN INSECT PEST MANAGEMENT

Current methods for improving host plant resistance toward insects rely on the time-consuming process of plant breeding. Plant breeding for resistance has been especially effective against aphid pests. But all too often, new biotypes of aphids arise that overcome the resistance mechanisms. The exploitation of plant biotechnology in recent years may eventually lead to the direct introduction of resistant genes into crops. However, to this day, the specific biochemical bases of resistance in crop plants toward insects and pathogens is still generally unknown. Thus insect control, for the most part, is still dependent on the use of pesticides.

The use of selective PGRs to alter plant chemistry and induce resistance to insect attack has great applied potential. The concept of using PGRs for this purpose is, however, not new. It is somewhat perplexing that this strategy in pest control has been overlooked. Such a strategy would serve as a means of controlling pestiferous insects, without causing direct toxicity to beneficial insects. Furthermore PGRs could be used to induce resistance in crop plants that are already agronomically suitable in terms of productivity, harvestability and taste without having to rely on the time-consuming process of breeding for insect resistance. Hence it is probable that PGRs can be used to "kill two birds with one stone." Namely, they could be used to manipulate crop growth to improve fruit set, flowering, taste, and harvesting. They could simultaneously make crops resistant to insect attack or modulate the life cycle of the pest population so as to increase its exposure to deleterious climatic or biotic agents. The question must be asked then: Why is there not more widespread commercial use of PGRs for insect control? Certainly, one reason is that PGRs have been unpredictable in inducing crop resistance to insects. But probably the chief reason for the lack of developing PGRs for insect control lies in the traditional screening techniques in the agricultural chemistry industry. Most blind screenings in industry of new, synthetic compounds use independent tests for insecticidal, herbicidal, fungicidal or plant growth regulatory activity. At the industrial level it may be impractical to perform cross-screenings of PGR induction of insect resistance. Perhaps more attention should be given to the possibility of cross-screenings in light of the very high costs for eventual registration (>$30 million) (Jung 1985) of agricultural chemicals of limited or specialized usage. In addition to their commercial value, however, PGRs can be very useful in the research of chemical bases of insect–plant interactions. PGRs whose mode of action modulates a particular biosynthetic pathway in a host plant can be used to change levels of either nutritional or allelochemical constituents. Such changes may then be used to determine the significance of one or a group of plant compounds in the chemistry of insect–plant associations. Such knowledge may prove to be eventually useful for improving host plant resistance, using plant breeding or genetic engineering techniques.

REFERENCES

Abeles, F. B. 1973. *Ethylene in Plant Biology*. Academic Press, New York.

Adams, J. B. 1960. Effects of spraying 2,4-D amine on coccinellid larvae. *Can. J. Zool.* 38: 285–288.

Adams, J. B., and M. E. Drew. 1965. Grain aphids in New Brunswick. III. Aphid populations in herbicide treated oat fields. *Can. J. Zool.* 43: 789–794.

Adkisson, P. L. 1962. Timing of defoliants and desiccants to reduce populations of the pink bollworm in diapause. *J. Econ. Entomol.* 55: 949–951.

Adolphi, H. 1967. Untersuchungen über Nebenwirkungen von Chlorcholinchlorid (CCC). *Z. Pflanzenk. Pflanzenpath. Pflanzenschutz* 74: 684–688.

Agnello, A. M., J. R. Bradley, Jr., and J. W. Van Duyn. 1986a. Plant-mediated effects of postemergence herbicides on *Epilachna varivestis* (Coleoptera: Coccinellidae). *Environ. Entomol.* 15: 216–220.

Agnello, A. M., J. W. Van Duyn, and J. R. Bradley, Jr. 1986b. Influence of postemergence herbicides on populations of bean leaf beetle, *Cerotoma trifurcata* (Coleoptera: Chrysomelidae), and corn earworm, *Heliothis zea* (Lepidoptera: Noctuidae), in soybean. *J. Econ. Entomol.* 79: 261–265.

Alonso, C. 1971. The effects of gibberellic acid upon developmental processes in *Drosophila hydei*. *Entomol. Exp. Appl.* 14: 73–82.

Bariola, L. A., D. L. Kittock, H. F. Arle, P. V. Vail, and T. J. Henneberry. 1976. Controlling pink bollworms: effects of chemical termination of cotton fruiting on populations of diapausing larvae. *J. Econ. Entomol.* 69: 633–636.

Bhalla, O. P., and A. G. Robinson. 1968. Effects of chemosterilants and growth regulators on the pea aphid fed an artificial diet. *J. Econ. Entomol.* 61: 552–555.

Blaim, K., and M. Przeszlakowska. 1967. Influence of CCC on the content of pectic substance in wheat stalks. *Bull. Acad. Polon. Sci., Sér. Sci. Biol.* 15: 445–448.

Blaim, K., and J. Szynal. 1968. Influence of (2-chloroethyl) trimethyl ammonium chloride (CCC) and calcium ions on wheat seedling growth. *Bull. Acad. Polon. Sci., Sér. Sci. Biol.* 16: 735–738.

Blaim, K., M. Przeszlakowska, and Z. Zebrowski. 1968. Influence of some plant growth regulators on the content and distribution of pectic substances in the stems and leaves of tomato plants. *Ann. Univ. M. Curie-Sklodowska, Sec. AA* 23: 293–299.

Blaim, K., M. Przeszlakowska, and J. Szynal. 1970. Studies on the mechanism of action of (2-chloroethyl) trimethyl ammonium chloride (CCC) on the wheat growth and on its resistance to lodging. *Zesz. Nauk. Univ. M. Kopernika, Sér. biol.* 13: 211–216.

Bořkovec, A. B., G. C. LaBrecque, and A. B. DeMilo. 1967. *S*-Triazine herbicides as chemosterilants of house flies. *J. Econ. Entomol.* 60: 893–894.

Brodel, C. F. 1979. Biological Studies of *Aphis rubicola* Oestlund (Aphididae) on red raspberry. Ph.D. dissertation. Cornell University, Ithaca, NY.

Brodel, C. F., and G. A. Schaefers. 1980. Use of excised leaflets of red raspberry to screen for potential nonpreference resistance to *Amphorophora agathonica*. *HortScience* 15: 513–514.

Byers, R. E. 1978. Chemical thinning of spur "Golden Delicious" and "Starkrimson Delicious" with sevin and vydate. *HortScience* 13: 59–61.

Campbell, B. C., and D. L. Dreyer. 1985. Host-plant resistance of sorghum: differential hydrolysis of sorghum pectic substances by polysaccharases of greenbug biotypes (*Schizaphis graminum*, Homoptera: Aphididae) *Arch. Insect. Biochem. Physiol.* 2: 203–215.

Campbell, B. C., and W. D. Nes. 1983. A reappraisal of sterol biosynthesis and metabolism in aphids. *J. Insect Physiol.* 29: 149–156.

Campbell, B. C., D. L. McLean, M. G. Kinsey, K. C. Jones, and D. L. Dreyer. 1982. Probing behavior of the greenbug (*Schizaphis graminum*, biotype C) on resistant and susceptible varieties of sorghum. *Entomol. Exp. Appl.* 31: 140–146.

Campbell, B. C., B. G. Chan, L. L. Creasy, D. L. Dreyer, L. B. Rabin, and A. C. Waiss, Jr. 1984. Bioregulation of host plant resistance to insects. *In* Ory, R. L. and F. R. Rittig, (eds.), *Bioregulators: Chemistry and Uses. Am. Chem. Soc. Symp. Ser.* 257, Washington, DC, pp. 193–203.

Campbell, B. C., K. C. Jones, and D. L. Dreyer. 1986. Discriminative behavioral responses by aphids to various plant matrix polysaccharides. *Entomol. Exp. Appl.* 41: 17–24.

Carlisle, D. B., P. E. Ellis, and D. J. Osborne. 1969. Effects of plant growth regulators on locusts and cotton stainer bugs. *J. Sci. Fd Agric.* 20: 391–393.

Castro, P. R. C. 1978. Effect of growth regulators on macronutrient levels in tomato (*Lycopersicon esculentum* Mill.). *An. Esc. Super. Agric., "Luiz de Queiroz," Univ. São Paulo* 35: 1–17.

Castro, P. R. C., and L. E. Gutierrez. 1979. Effects of growth regulators on the amino acid and phenolic compound contents in cotton (*Gossypium hirsutum* L. cv. "IAC-17"). *An. Esc. Super. Agric., "Luiz de Queiroz," Univ. São Paulo* 36: 89–97.

Center, T. D., K. K. Steward, and M. C. Bruner. 1982. Control of water hyacinth (*Eichhornia crassipes*) with *Neochetina eichhorniae* (Coleoptera: Curculionidae) and growth retardant. *Weed Sci.* 30: 453–457.

Chrominski, A., S. Visscher-Neumann, and R. Jurenka. 1982. Exposure to ethylene changes nymphal growth rate and female longevity in the grasshopper *Melanoplus sanguinipes*. *Naturwissenschaften* 69: 45–46.

Coggins, C. W., Jr., R. W. Scora, L. N. Lewis, and J. C. F. Knapp. 1969. Gibberellin delayed senescence and essential oil changes in the navel orange rind. *J. Agric. Fd Chem.* 17: 807–809.

Detling, J. K., and M. I. Dyer. 1981. Evidence for potential plant growth regulators in grasshoppers. *Ecology* 62: 485–488.

Dimetry, N. Z., and M. H. Mansour. 1975. Laboratory evaluation of 2 plant growth regulators as anti-feedants against the American bollworm *Heliothis armigera* (Hbn.) *Z. Pflanzenkr. Pflanzenschutz* 82: 561–565.

Dreyer, D. L., and B. C. Campbell. 1984. Association of the degree of methylation of intercellular pectin with plant resistance to aphids and induction of aphid biotypes. *Experientia* 40: 224–226.

Dreyer, D. L., J. C. Reese, and K. C. Jones. 1981. Aphid feeding deterrents in sorghum: bioassay, isolation, and characterization. *J. Chem. Ecol.* 7: 273–284.

Dreyer, D. L., B. C. Campbell, and K. C. Jones. 1984. Effect of bioregulator-treated sorghum on greenbug fecundity and feeding behavior: implications for host-plant resistance. *Phytochemistry* 23: 1593–1596.

Eichmeier, J., and G. J. Guyer. 1960. An evaluation of the rate of reproduction of the two spotted spider mite reared on gibberellin-treated bean plants. *J. Econ. Entomol.* 53: 661–664.

Eidt, D. C., and C. H. A. Little. 1968. Insect control by artificially prolonging plant dormancy—a new approach. *Can. Entomol.* 100: 1278–1279.

Eidt, D. C., and C. H. A. Little. 1970. Insect control through induced host-insect asynchrony: a progress report. *J. Econ. Entomol.* 63: 1966–1968.

Eijsackers, J. H. 1978. Side effects of the herbicide 2,4,5-T affecting reproduction, food consumption, and moulting of the springtail *Onychiurus quadriocellatus* Gisin. *Z. Angew. Entomol.* 85: 341–360.

El-Ibrashy, M. T., and M. H. Mansour. 1970. Hormonal action of certain biologically active compounds in *Agrotis ypsilon* larvae. *Experientia* 26: 1095–1096.

Elliger, C. A., B. C. Chan, and A. C. Waiss, Jr. 1980. Flavonoids as larval growth inhibitors. *Naturwissenschaften* 67:358–359.

Ellis, P. E., D. B. Carlisle, and D. J. Osborne. 1965. Desert locusts: sexual maturation delayed by feeding on senescent vegetation. *Science* 149: 546–547.

El-Sayed, M. M., M. I. Abdel-Megeed, A. A. Selim, E. M. K. Hussein, and A. El-Sisi. 1982. Field evaluation of compatible pesticides and plant growth regulators and local formulated emulsifiable concentrates. *Ain Shams. Univ. Fac. Agric. Res. Bull.* 0(1709): 1–10.

Elzen, G. W. 1983. Cytokinins and insect galls. *Comp. Biochem. Physiol.* A. 76A: 17–19.

Fahmy, A. I., M. M. Nawar, A. M. Hablas, and M. A. Sherif. 1982. Causes behind the differential effectiveness of maleic hydrazide in onion and *Drosophila. Ain Shams. Univ. Fac. Univ. Res. Bull.* 0(2041): 1–15.

Farlow, R. A., and H. N. Pitre. 1983. Bioactivity of the post-emergence herbicides acifluorfen and bentazon on *Geocoris punctipes* (Say) (Hemiptera: Lygaeidae). *J. Econ. Entomol.* 76: 200–203.

Fox, C. J. S. 1964. The effects of five herbicides on the numbers of certain invertebrate animals in grassland soil. *Can. J. Plant Sci.* 44: 405–409.

Fox, W. B. 1948. 2,4-D as a factor in increasing wireworm damage of wheat. *Sci. Agric.* 28: 423–424.

Gaballe, G. T., and A. W. Galston. 1979. Wound induced resistance to cellulose: Ethylene as effector. *Plant Physiol.* 63 (Suppl.): 369.

Gall, A., and J. R. Dogger. 1967. Effect of 2,4-D on wheat stem sawfly. *J. Econ. Entomol.* 60: 75–77.

Gasser, J. K. R., and M. A. P. Thorburn. 1972. Growth, composition, and nutrient uptake of spring wheat. Effects of fertilizer-nitrogen, irrigation and CCC (2-chloroethyltrimethylammonium chloride) on dry matter and nitrogen, phosphorus, potassium, calcium, magnesium, and sodium. *J. Agric. Sci.* 78 (Pt. 3): 393–404.

Greany, P. D., and Y. Rössler. 1985. Enhancement of citrus resistance to tephritid fruit flies. Final Report Bard Project No. US-375-081. Bet Dagan, Israel.

Greany, P. D., P. E. Shaw. P. L. Davis, and T. T. Hatton. 1985. Senescence-related susceptibility of Marsh grapefruit to laboratory infestation by *Anastrepha suspensa* (Diptera: Tephritidae). *Fla. Entomol.* 68: 144–150.

Greig-Smith, P. 1986. The trees bite back. *New Scientist* 1506: 33–35.

Guerra, R. T., C. R. Bueno, and H. O. Schubart. 1982. Preliminary evaluation on the effects of application of the herbicides paraquat and conventional plowing on the mesofauna of the soil in the region of Manaus Amazonas Brazil. *Acta Amazonica* 12: 7–14.

Hall, F. R. 1972. Influence of alar on populations of European red mite and apple aphid on apples. *J. Econ. Entomol.* 65: 1751–1753.

Hanson, K. R. 1981. Phenylalanine ammonia-lyase: a model for the cooperativity kinetics induced by D- and L-phenylalanine. *Arch. Biochem. Biophys.* 211: 564–574.

Hare, F. D. 1983. Manipulation of host suitability for herbivore pest management. *In* Denno, R. F., and M. S. McClure (eds.), *Variable Plants and Herbivores in Natural and Managed Systems*. Academic Press, New York, pp. 665–680.

Hatley, D. E., L. G. Hermann, and A. J. Ohlrogge. 1977. The response of corn (*Zea mays* L.) to foliar applications of 4,6-dinitro-*o*-*sec*-butylphenol (DNBP). *Proc. Plant Growth Regul. Work. Group* 4: 156–163.

Hedin, P. A., A. C. Thompson, and R. C. Gueldner. 1976. Cotton plant and insect constituents that control boll weevil behavior and development. *In* Wallace, J. W., and R. K. Mansell (eds.), *Biochemical Interactions between Plants and Insects. Recent Adv. Phytochem.*, Vol. 10. Plenum, New York, pp. 271–350.

Hedin, P. A., J. N. Jenkins, J. C. McCarty, Jr., J. E. Mulrooney, W. L. Parrott, A. Borazjani, C. H. Graves, Jr., and T. H. Filer. 1984. Effects of 1,1-dimethylpiperidinium chloride on the pests and allelochemicals of cotton and pecan. *In* Ory, R. L., and F. R. Rittig (eds.), *Bioregulators: Chemistry and Uses. Am. Chem. Soc. Symp. Ser.* 257. Washington, DC, pp. 171–191.

Hintz, S. D., and J. T. Schulz. 1969. The effect of selected herbicides on cereal aphids under greenhouse conditions. *Proc. N. Cent. Branch Entomol. Soc. Am.* 24: 114–117.

Hoagland, R. E., and S. O. Duke. 1982. Biochemical effects of glyphosate [*N*-(Phosphonomethyl) glycine]. *In* Moreland, D. E., J. B. St. John, and F. D. Hess (eds.), *Biochemical Responses Induced by Herbicides. Am. Chem. Soc. Symp. Ser.* 181, Washington, DC, pp. 175–205.

Honeyborne, C. H. B. 1969. Performance of *Aphis fabae* and *Brevicoryne brassicae* on plants treated with growth regulators. *J. Sci. Fd. Agric.* 20: 388–389.

Hori, K. 1980. Metabolism of ingested auxins in the bug *Lygus disponsi:* indole compounds appearing in excreta of bugs fed with host plants and the effect of IAA on feeding. *Appl. Entomol. Zool.* 15: 123–128.

Hume, R. J., and P. H. Lovell. 1981. Reduction of the cost involved in hybrid seed production of pumpkins (*Cucurbita maxima*). *New Zeal. J. Exp. Agric.* 9: 209–210.

Ingram, J. W., E. K. Bynum, and L. J. Charpentier. 1947. Effect of 2,4-D on sugarcane borer. *J. Econ. Entomol.* 40: 745–746.

Ishii, S., and C. Hirano. 1963. Growth responses of larvae of the rice stem borer to rice plants treated with 2,4-D. *Entomol. Exp. Appl.* 6: 257–262.

Jablonowski, W. 1972. Effect of chlorocholine chloride (CCC) on choline and rutin content in tomato fruits. *Bromatal. Chem. Toksykol.* 5: 27–32.

Johansen, C. 1959. Bee poisoning, a hazard of applying agricultural chemicals. *Wash. Agric. Exp. Sta. Circ.* 356.

Jung, J. 1985. Plant bioregulators: overview, use, and development. *In* P. A. Hedin (ed.), *Bioregulators for Pest Control. Am. Chem. Soc. Symp. Ser.* 276. Washington, DC, pp. 95–107.

Kamada, M., and I. Shipei. 1984. Growth promoting effect of plant hormones on silkworms. *Nippon Nogeikagaku Kaishi* 58: 779–784.

Khan, A. A., and M. A. Faust. 1966. Effect of cycocel and its analogues on growth and soluble protein content of young barley seedlings. *Nature* 211: 1215–1216.

King, C. C. 1964. Effects of herbicides on honeybees. *Glean. Bee Cult.* 92: 230–233.

Kittock, D. L., J. R. Mauney, H. F. Arle, and L. A. Bariola. 1973. Termination of late season cotton fruiting with growth regulators as an insect control technique. *J. Environ. Quality* 2: 405–408.

Kittock, D. L., L. A. Bariola, H. F. Arle, and P. V. Vail. 1975. Evaluation of chemical termination of cotton fruiting for pink bollworm control by estimating number of immature bolls at harvest. *Cotton Growing Rev.* 52: 224–227.

Kittock, D. L., H. F. Arle, T. J. Henneberry, L. A. Bariola, and V. T. Walhood. 1980. Timing late-season fruiting termination of cotton with potassium 3,4-dichloroisothiazole-5-carboxylate. *Crop Sci.* 20: 330–333.

Knodel-Montz, J. J., S. L. Poe, and R. E. Lyons. 1983. Effects of plant growth hormones on oviposition site selection of *Liriomyza trifolii* (Burgess) (Diptera: Agromyzidae). *Va. J. Sci.* 34: 102.

Kogan, M., and J. Paxton. 1983. Natural inducers of plant resistance to insects. *In* Hedin, P. A. (ed.), *Plant Resistance to Insects. Am. Chem. Soc. Symp. Ser.* 208. Washington, DC, pp. 153–172.

Kojima, M., J. E. Poulton, S. S. Theyer, and E. E. Conn. 1979. Tissue distributions of dhurrin and of enzymes involved in its metabolism in leaves of *Sorghum bicolor. Plant Physiol.* 63: 1022–1028.

Latteur, G. 1976. *Les Pucerons des Cereales: Biologie, Nuisance, Ennemis.* Memoire 3, Centre de Recherches Agronomiques de l'Etat Gemblous.

Leppick, E. E. 1951. New insecticides are disastrous to bees. *Am. Bee J.* 91: 462–463.

Linser, H., K. H. Neumann, and H. El Damaty. 1965. Preliminary investigation of the action of (2-chloroethyl)-trimethyl ammonium chloride on the composition of the soluble *N*-fraction and protein fraction of young wheat plants. *Nature* 206: 893–895.

Little, C. H. A., and D. C. Eidt. 1968. Effect of abscisic acid on budbreak and transpiration in woody species. *Nature* 220: 498–499.

Maxwell, R. C., and R. F. Harwood. 1958. Increased reproduction of aphids on plants affected by the herbicide 2,4-dichlorophenoxyacetic acid. *Bull. Entomol. Soc. Am.* 4: 100.

Maxwell, R. C., and R. F. Harwood. 1960. Increased reproduction of pea aphids on broad beans treated with 2,4-D, *Ann. Entomol. Soc. Am.* 53: 199–205.

Mellanby, K., R. A. French, and J. Riches. 1959. Herbicide spray and frit fly attack on oats. *Entomol. Exp. Appl.* 2: 319–320.

Moreland, D. E., S. C. Huber, and W. P. Novitzky. 1982. Interaction of herbicides with cellular and liposome membranes. *In* Moreland, D. E., J. B. St. John, and F. D. Hess (eds.) *Biochemical Responses Induced by Herbicides. Am. Chem. Soc. Symp. Ser.* 181. Washington, DC, pp. 79–96.

Morgan, P. W., R. D. Powell, M. P. Grisham, and W. L. Sterling. 1983. Induction of stress ethylene synthesis in cotton by fleahoppers. *Plant Physiol.* 72 (suppl. 1): 38.

Mueller, G. 1972a. Faunistic-ecologic research on the Coleoptera of the cultivated areas near the coast at Greifswald: the effect of the herbicides UVON-Kombi (II) and ELBANIL (III) on the epigeic fauna of cultivated area. *Pedobiologia* 12: 169–211.

Mueller, G. 1972b. Changes in the beetle fauna of the upper surface of the soil of cultivated fields after application of herbicides (Coleoptera). *Folia Entomol. Hung.* 25: 297–305.

Nation, J. I., and F. A. Robinson. 1966. Gibberellic acid; effects of feeding on an artificial diet for honey bees. *Science* 152: 1765–1766.

Nickell, L. G. 1982. *Plant Growth Regulators*. Springer-Verlag, New York.

Oka, I. N., and D. Pimentel. 1976. Herbicide (2,4-D) increases insect and pathogen pests on corn. *Science* 193: 239–240.

Orr, G. L., and F. D. Hess. 1982. Proposed site(s) of action of new diphenyl ether herbicides. *In* Moreland, D. E., J. B. St. John, and F. D. Hess (eds.), *Biochemical Responses Induced by Herbicides. Am. Chem. Soc. Symp. Ser.* 181. Washington, DC, pp. 131–152.

Paclt, J. 1980. Further theses on the general biological interpretation of plant galls. *Beitr. Biol. Pflanz.* 55: 101–118.

Peary, W. E. 1959. Laboratory tests for control of beet leafhoppers on snap beans grown for seed. *J. Econ. Entomol.* 52: 700–703.

Pollard, D. G. 1973. Plant penetration by feeding aphids (Hemiptera, Aphidoidea): a review. *Bull. Entomol. Res.* 62: 631–714.

Price, P. W., C. E. Bouton, P. Gross, B. A. McPheron, J. N. Thompson, and A. E. Weis. 1980. Interactions among three trophic levels: influence of plants on interactions between insect herbivores and natural enemies. *Annu. Rev. Ecol. Syst.* 11: 41–65.

Protopopova, E. G., and L. N. Dolgopolova. 1982. Growth regulators for pea aphid control. *Khimya V Selskom Khozyaistve* 20: 54–56.

Przeszlakowska, M. 1974. Changes in the cell walls of wheat culms treated with 2-chloroethyltrimethyl ammonium chloride (CCC). *Acta Agrobotanica* 27: 19–28.

Putnam, L. G. 1949. The survival of grasshopper nymphs on vegetation treated with 2,4-D. *Sci. Agric.* 29: 396–399.

Raatikainen, M., and V. Huhta. 1968. On the spider fauna of Finnish oat fields. *Ann. Zool. Fenn.* 5: 254–261.

Rautapää, J. 1972. Effect of herbicides and chlormequat chloride on host-plant selection and population growth of *Macrosiphum avenae* (F.) (Hom., Aphididae). *Ann. Agric. Fenn.* 11: 135–140.

Rhoades, D. F. 1983. Herbivore population dynamics and plant chemistry. *In* Denno, R. F., and M. S. McClure (eds.), *Variable Plants and Herbivores in Natural and Managed Systems*. Academic Press, New York, pp. 3–55.

Rhodes, J. M., and L. S. C. Wooltorton. 1978. The biochemistry of phenolic compounds in wounded plant storage tissues. *In* Kahl, G. (ed.), *Biochemistry of Wounded Plant Tissues*. de Gruyter, New York, pp. 243–286.

Ries, S. K., V. Wert, C. C. Sweeley, and R. A. Leavitt. 1977. Triacontanol: a new naturally occurring plant growth regulator. *Science* 195: 1339–1341.

Robinson, A. G. 1959. Note on fecundity of the pea aphid, *Acyrthosiphon pisum* (Harris), caged on plants of broad bean, *Vicia faba* L., treated with various plant growth regulators. *Can. Entomol.* 91: 527–528.

Robinson, A. G. 1960. Effect of maleic hydrazide and other plant growth regulators on the pea aphid, *Acyrthosiphon pisum* (Harris), caged on broad beans, *Vicia faba* L. *Can. Entomol.* 92: 494–499.

Robinson, A. G. 1961. Effects of amitrole, zytron, and other herbicides or plant growth regulators on the pea aphid, *Acyrthosiphon pisum* (Harris), caged on broad beans, *Vicia faba* L. *Can. J. Plant Sci.* 41: 413–417.

Rodriguez, J. G., and J. M. Campbell. 1961. Effects of gibberellin on nutrition of the mites, *Tetranychus telarius* and *Panonychus ulmi*. *J. Econ. Entomol.* 54: 984–987.

Ryan, C. A., P. D. Bishop, J. S. Graham, R. M. Broadway, and S. S. Duffey. 1986. Plant and fungal cell wall fragments activate expression of proteinase inhibitor genes for plant defense. *J. Chem. Ecol.* 12: 1025–1036.

Sabatini, M. A., A. Pederzoli, B. Fratello, and R. Bertolani. 1980. Microarthropod communities in soil treated with atrazine. *Boll. Zool.* 46: 333–342.

Scheurer, S. 1976. The influence of phytohormones and growth regulating substances on insect development processes. *In* Jermy, T. (ed.), *The Host-Plant in Relation to Insect Behaviour and Reproduction*. Plenum, New York, pp. 255–259.

Schultz, J. C., and I. T. Baldwin, 1982. Oak leaf quality declines in response to defoliation by gypsy moth larvae. *Science* 217: 149–151.

Simpson, G. W. 1961. Improper use of herbicides may increase aphid damage. *Maine Farm Res.* 9: 13–14.

Singer, M. C., and B. D. Smith. 1976. Use of the plant growth regulator, chlormequat chloride, to control the aphid, *Hyperomyzus lactucae,* on black currants, *Ann. Appl. Biol.* 82: 407–414.

Smith, B. D. 1969. Spectra of activity of plant growth retardants against various parasites of one host species. *J. Sci. Fd Agric.* 20: 398–400.

Smith, B. D., and A. T. K. Corke. 1966. Effect of (2-chloroethyl)-trimethylammonium chloride on the eriophyid gall mite *Cecidophyopsis ribis* Nal. and three fungus diseases (*Pseudopeziz ribis, Sphaerotheca mors-uvae, Botrytis cinerea*) of black currant. *Nature* 212: 643–644.

Stam, P. A., D. F. Clower, J. B. Graves, and P. E. Schilling. 1978. Effects of certain herbicides on some insects and spiders found in Louisiana cotton fields. *J. Econ. Entomol.* 71: 477–480.

Sterrett, J. P., K. Boyd, R. E. Webb, and R. H. Hodgson. 1984. Chemical defoliation of an oak forest: an attempt to control gypsy moth. *J. Ga. Entomol. Soc.* 19: 304–310.

Tahori, A. S., A. H. Halevy, and G. Zeidler. 1965a. Effect of some plant growth retardants on the oleander aphid, *Aphis nerii* (Boyer). *J. Sci. Fd Agric.* 16: 568–569.

Tahori, A. S., G. Zeidler, and A. H. Halevy. 1965b. Effect of some plant growth retardants on the feeding of the cotton leafworm. *J. Sci. Fd Agric.* 16: 570–572.

Tahori, A. S., G. Zeidler, and A. H. Halevy. 1965c. Phosphon (2,4-dichlorobenzyltributyl phosphonium chloride) as insect antifeeding compound. *Naturwissenschaften* 52: 191–192.

Tamaki, G., and R. E. Weeks. 1968. Use of chemical defoliants on peach trees in an integrated program to suppress populations of green peach aphids. *J. Econ. Entomol.* 61: 431–435.

Tanke, W., and J. M. Franz. 1978. Side-effects of herbicides on beneficial insects. *Entomophaga* 23: 275–280.

Thomas, R. O., T. C. Cleveland, and G. W. Cathey. 1979. Chemical plant growth suppressants for reducing late-season cotton bollworm-budworm feeding sites. *Crop Sci.* 19: 861–863.

Thomas, T. H. 1982. *Plant Growth Regulator Potential and Practice*. BCPC, London.

van Assche, C. J., and P. M. Carles. 1982. Photosystem II inhibiting chemicals: molecular interaction between inhibitors and a common target. *In* Moreland, D. E., J. B. St. John, and F. D. Hess (eds.), *Biochemical Responses Induced by Herbicides. Am. Chem. Soc. Symp. Ser.* 181. Washington, DC, pp. 1–21.

van Emden, H. F. 1964. Effect of (2-chloroethyl)trimethylammonium chloride on the rate of increase of the cabbage aphid (*Brevicoryne brassicae* (L.). *Nature* 201: 946–948.

van Emden, H. F. 1969a. Plant resistance to *Myzus persicae* induced by a plant regulator and measured by aphid relative growth rate. *Entomol. Exp. Appl.* 12: 125–131.

van Emden, H. F. 1969b. Plant resistance to aphids induced by chemicals. *J. Sci. Fd Agric.* 20: 385–387.

van Emden, H. F., and C. H. Wearing. 1965. The role of the aphid host plant in delaying economic damage levels in crops. *Proc. Assoc. Appl. Biol.* 56: 323–324.

Vance, C. P., T. K. Kirk, R. T. Sherwood. 1980. Lignification as a mechanism of disease resistance. *Annu. Rev. Phytopathol.* 18: 259–288.

Visscher-Neumann, S. 1980. Regulation of grasshopper fecundity, longevity and egg viability. *Experientia* 36: 130–131.

Way, M. O., A. A. Grigarick, and S. E. Mahr. 1984. The aster leafhopper (Homoptera: Cicadellidae) in California rice: herbicide treatment affects population density and induced infestations reduce grain yield. *J. Econ. Entomol.* 77: 936–942.

Westigard, P. H., P. B. Lombard, R. B. Allen, and J. G. Strang. 1980. Pear Psylla: population suppression through host plant modification using daminozide. *Environ. Entomol.* 9: 275–277.

Worthing, C. R. 1969. Use of growth retardants on chrysanthemums: effect on pest populations. *J. Sci. Fd Agric.* 20: 394–397.

Wu, C.-H., and P. W. Santelmann. 1977. Influence of six plant growth regulators on Spanish peanuts. *Agron. J.* 69: 521–522.

Yang, S. F., and H. K. Pratt. 1978. The physiology of ethylene in wounded plant tissues. *In* Kahl, G. (ed.), *Biochemistry of Wounded Plant Tissues.* de Gruyter, New York, pp. 595–622.

Yu, S. J. 1983. Induction of detoxifying enzymes by allelochemicals and host plants in the fall armyworm (*Spodoptera frugiperda*). *Pestic. Biochem. Physiol.* 19: 330–336.

Yule, W. N., E. W. Parups, and I. Hoffman. 1966. Toxicology of plant-translocated maleic hydrazide: lack of effects on insect reproduction. *J. Agric. Fd Chem.* 14: 407–409.

Zeroni, M., and M. A. Hall. 1980. Molecular effects of hormone treatment on tissue. *In* MacMillan, J. (ed.), *Hormonal Regulation of Development I: Molecular Aspects of Plant Hormones.* Springer-Verlag, New York, pp. 511–586.

Zummo, G. R., J. H. Benedict, and J. C. Segers. 1984. Effect of a plant growth regulator mepiquat-chloride on host plant resistance in cotton to bollworms (Lepidoptera: Noctuidae). *J. Econ. Entomol.* 77: 922–924.

8

INSECT POPULATIONS ON HOST PLANTS SUBJECTED TO AIR POLLUTION

Patrick R. Hughes

Boyce Thompson Institute for Plant Research
Cornell University
Ithaca, New York

1. INTRODUCTION

1.1 General Remarks

Historically, most studies concerning the impact of air pollution on plant–insect relations have been observational, relating some measure of insect presence, such as damage or numbers, to some measure of pollutant presence, such as foliar concentration, damage, or distance from a point source. Typically, these early reports described outbreaks of forest insects in the vicinity of industrial facilities. Unfortunately, concurrent information was not published concerning the frequency of such outbreaks in unaffected areas or how frequently outbreaks were not found in areas similarly impacted by pollution. Also the composition of the pollutant was complex and generally not well delineated. Therefore, though these studies provided the suggestion that airborne pollutants could alter the success of herbivorous insects and were important in stimulating later research, the evidence was, at best, correlative with no foundation in controlled experimentation.

Virtually no controlled studies of pollutant-related alterations of plant–insect relations can be found in the literature prior to 1980, except for those concerned with the direct effects of various chemicals on the insects. However, research in this area has undergone two significant shifts in the last decade. First, more emphasis has been placed on effects in agricultural systems and nonforest habitats, and, second, efforts have turned from observational to experimental studies with the objective of demonstrating cause–effect relationships and elucidating underlying mechanisms. These studies identified more positively the potential economic and ecologic importance of the pollutant–plant–insect interaction and were the forerunners of a now widespread and rapidly growing interest in the subject.

1.2 Purpose and Scope

This chapter is not intended to be a catalog of the available literature on this subject, but rather it is hoped to serve as an update and synthesis of what is currently known, with some interpretation of the collective meaning of these results. Additionally, the discussion will attempt to identify the present directions of research in the area, the problems investigators might need to consider in undertaking such research, and the directions the research might take in the immediate future. If successful, this approach will bring the interested reader up to the present state of understanding, identify weak points in the present knowledge, and help to generate new ideas.

The following discussion will be limited to airborne pollutants and will be concerned primarily with the indirect effects of these compounds on herbivorous insects through their interaction with the biotic environment. Thus the great amount of literature devoted to effects of waterborne pollutants on plants and aquatic ecosystems will be excluded. Furthermore the chapter will touch only lightly on the vast amount of information concerning pesticides as pollutants. The effects of airborne pesticides and water pollutants on insects are largely direct through toxicity, and the reader is referred to the literature for more information on these subjects.

1.3 Approach

The ensuing discussion will begin by defining the term ''air pollutant'' and will continue with a summary of past correlative and experimental evidence showing that these air pollutants can, in fact, affect insect populations. Possible mechanisms by which this occurs will then be outlined, and one mechanism that has been studied in detail will be discussed. A general synthesis of what is known on the subject will then be attempted, and some consideration given to placing this work in perspective with the larger subject of plant stress. Some of the specialized methodology required for studies of pollutant impact on plant–insect relations and the need for attention to quality assurance in these studies will also be described. Finally, some thoughts will be presented concerning more immediate research needs.

2. AIR POLLUTANTS

2.1 Definition

Although most people feel that they understand what the term ''air pollution'' means, a concise, biologically meaningful definition is somewhat elusive. The Engineers' Joint Council defined air pollution as follows:

> Air pollution means the presence in the outdoor atmosphere of one or more contaminants, such as dust, fumes, gas, mist, odor, smoke, or vapor in quan-

Table 1. Composition of Nonpolluted, Dry Air near Sea Level

Component	Concentration (ppm)	Component	Concentration (ppm)
Nitrogen	780,900	Hydrogen	0.5
Oxygen	209,400	NO	0.33
Argon	9,300	CO	0.1
CO_2	332	Xenon	0.08
Neon	18	Ozone	0.02
Helium	5.2	Ammonia	0.01
Methane	1.5	NO_2	0.001
Krypton	1	SO_2	0.0002

Reprinted with permission from "Cleaning Our Environment—A Chemical Perspective." Copyright 1978 American Chemical Society.

> tities, of characteristics, and of duration, such as to be injurious to human, plant, or animal life or to property, or which unreasonably interferes with the comfortable enjoyment of life and property (Perkins 1974).

This definition is quite broad, but it constitutes the core of the typical legislative definition. In the Clean Air Act, as amended through July 1981, Title III Section 302 defines air pollutant in the following manner:

> The term air pollutant means any air pollution agent or combination of such agents including any physical, chemical, biological, radioactive (including source material, special nuclear material, and by-product material) substance or matter which is emitted into or otherwise enters the ambient air.

In practice, a distinction is frequently made between contaminants of anthropogenic origin and those of natural origin, inasmuch as the former are generally controllable whereas the latter are not. Of course, from the biological perspective, the origin of a particular pollutant is of little importance. Hence, for the purpose of this chapter, an air pollutant will be considered as any material, whether of anthropogenic or natural origin, present in the air in such abnormally high concentration that it causes alteration of the biochemical or physiological state of plants or animals with which it comes into contact. This definition therefore includes both substances undetectable in clean air, such as peroxyacetyl nitrate, and those found in clean air but at much lower concentrations than in the polluted air, such as oxides of sulfur and nitrogen, ozone, and carbon monoxide (Table 1).

2.2 Common Pollutants

A thorough discussion of the properties and occurrence of the many air pollutants that have been studied is outside the scope of this chapter; the

Table 2. Partial List of Naturally Occurring and Anthropogenic Air Pollutants

Sulfur Dioxide	Miscellaneous Phytotoxic Pollutants
Fluorides	Unsaturated hydrocarbons
Hydrogen fluoride	Agricultural chemicals
Silicon tetrafluoride	Hydrogen chloride
Calcium fluoride	Ammonia
Cryolite	Hydrogen sulfide
Photochemical Oxidants	Particulates
Ozone	Heavy metals
Oxides of nitrogen	Dust
Peroxyacetyl nitrate (PAN)	Volcanic ash
Ambient oxidant complex	Acidic precipitation

reader is referred to texts such as that by Perkins (1974) for a detailed treatment of the general subject and to the National Academy of Sciences publications (1971, 1977, 1978) for in depth consideration of several major contaminants. However, some introduction is required to provide the less specialized reader with a general background and framework in which to better understand the ensuing discussion. Hence a brief description of some of the more important or common pollutants, major sources of these, and some pertinent properties is included here.

Air contaminants can be classified as primary pollutants, which are emitted from combustion and numerous industrial processes, or secondary pollutants, which are products of chemical reactions in the atmosphere between primary pollutants and hydrocarbons in the presence of sunshine. Primary pollutants most often affect vegetation in localized areas because they are produced by point sources and their concentration away from the source is reduced by dilution and dispersion. Secondary pollutants, however, are more important regionally because they are being produced throughout the air mass.

A partial list of both naturally occurring and anthropogenic air pollutants is presented in Table 2, and major sources of some of the most common air pollutants are listed in Table 3. Pollutants considered to be most important in terms of phytotoxicity include ozone, sulfur dioxide (SO_2), peroxyacetyl nitrate (PAN), nitrogen oxides (NO_x), agricultural chemicals, fluoride (F), and ethylene (Heck et al. 1973). In addition, numerous other elements such as metals and compounds such as hydrocarbons can be important alone or in combinations.

2.2.1 Sulfur Dioxide

Worldwide, SO_2 has been one of the most important of the major air pollutants, being produced in large quantity by combustion of coal and oil as well as by many industrial processes. It is also emitted in large quantities

Table 3. Major Sources of Some Common Air Pollutants

Pollutant	Sources[a]
Sulfur dioxide	Fuel combustion in stationary sources
	Industrial processes
	Transportation
	Solid waste disposal
Fluoride	Manufacture of steel
	Manufacture of brick and tiles
	Manufacture of aluminum
	Combustion of coal
	Manufacture of phosphate fertilizers
	Manufacture of elemental phosphorous
	Manufacture of glass and frit
	Welding operations
Oxides of nitrogen	Transportation
	Fuel combustion in stationary sources
	Solid waste disposal
	Agricultural burning
	Industrial processes
	Biological (bacterial) action

[a] Compiled from Applied Science Associates 1976; National Academy of Sciences 1971, 1977.

from natural sources, such as active volcanoes and fumaroles and vents. Effects of gaseous SO_2 on plant biochemistry, physiology, growth, and productivity are well documented (Applied Science Associates 1976, Amundson 1983, Koziol and Whatley 1984).

In addition to its importance as a gaseous pollutant, SO_2 is important in the formation of acidic precipitation (see Section 2.2.4 on acidic precipitation).

2.2.2 Fluoride

Fluoride, which comes from such varied sources as volcanoes, dust from fluoride-containing soils, combustion of coal and many industrial processes, is one of the most phytotoxic of the common pollutants and has the additional important property of accumulating in plants (Weinstein 1977). The forms of fluoride that affect plants most are the airborne compounds, such as hydrogen fluoride and silicon tetra fluoride, and the water soluble compounds, such as sodium fluoride and aluminum fluoride (National Academy of Sciences 1971).

Fluoride taken up by the plant moves in the transpiration stream and ultimately accumulates in highest concentrations in the apical or marginal portions of leaves. This fluoride is generally immobile, likely being bound as an insoluble complex or salt. Effects of fluoride appear to result primarily from its activity as an inhibitor of enzymes, and susceptibility to injury varies considerably with plant species and conditions (Table 4).

2.2.3 Oxides of Nitrogen and Photochemical Oxidants

The secondary pollutants, also called photochemical oxidants, include oxides of nitrogen, ozone, peroxyacyl nitrates, and an unidentified complex of oxidants (Applied Science Associates 1976, Heck et al. 1970). Of the six oxides of nitrogen, only two—nitric oxide (NO) and nitrogen dioxide (NO_2)—are important as air pollutants. Injury to plants attributed to oxides of nitrogen is thought to be caused primarily by NO_2, but NO is important in reactions leading to the production of PAN and increased concentrations of O_3.

Oxides of nitrogen are produced both anthropogenically and biologically (Table 3). Bacterial action is estimated to produce approximately 10 times more NO than the combined NO_x emissions related to human activities (Applied Science Associates 1976).

O_3 is considered to cause more plant damage in the United States than any other pollutant. Although O_3 is formed naturally by the action of sunlight on oxygen in the upper atmosphere and by electrical discharges during thunderstorms, these sources are not considered to be important to plant injury. Damaging concentrations occur primarily as a result of a series of reactions in the air (Figure 1). Catalyzed by sunshine, NO_2 forms NO and atomic oxygen. The atomic oxygen combines with oxygen in air to form O_3. With no other reactants, the O_3 would recombine with NO to form NO_2 and O_2, and no increase would occur. However, in the presence of hydrocarbons, NO reacts to produce PANs, thus removing NO from the cycle and allowing for the buildup of damaging concentrations of O_3 and PANs.

2.2.4 Acidic Precipitation

Acidic precipitation, which consists of both aqueous (liquid or frozen) acidic deposition and dry deposition of acid particles and acid-forming gases (Hileman 1981), has been a growing concern worldwide in recent years. Acidic precipitation is defined as atmospheric deposition with a hydrogen ion concentration greater than 2.5 μeq/L (i.e., pH <5.6) (Evans 1984, Hileman 1981). Precipitation in the northeastern United States generally ranges from pH 3.0 to 6.0, with only a fraction of 1% of all rainfalls having a pH below 3 or above 5.5 (Evans 1984).

Acidic precipitation is caused primarily by oxides of sulfur, most of which originate from ore smelting and burning of coal and petroleum products, and oxides of nitrogen, approximately 40% of which come from transportation.

Table 4. Relative Sensitivity of Various Plants to Fluoride

Sensitive	Intermediate	Tolerant
	Crops, Garden Plants, and Weeds	
Barley (young)	Alfalfa	Alfalfa
Chickweed	Barley (mature)	Asparagus
Corn, sweet	Bean, snap	Bean, snap
Grass, crab	Blueberry	Burdock
Grass, Johnson	Cane, sugar	Cabbage
Lamb's quarters	Carrot	Cauliflower
Milo maize	Chickweed	Celery
Oats (young)	Clover, crimson	Cotton
Pigweed	Clover, sweet	Cucumber
Potato, sweet	Clover, yellow	Dock
Smartweed	Corn, field	Eggplant
Sorghum	Goldenrod	Nightshade
St. John's wort	Grass, crab	Oats (mature)
Wheat	Grass, Johnson	Plantain
	Lamb's quarters	Potato
	Nettle-leaf goosefoot	Purslane
	Oats (young and mature)	Rye (mature)
	Pea	Soybean
	Pepper, bell	Spinach
	Potato	Squash, summer
	Rye (young)	Sugarcane
	Sorghum	Tobacco
	Soybean	Tomato
	Spinach	Wheat (mature)
	Strawberry	
	Sunflower	
	Tomato	
	Wheat (young and mature)	
	Ornamental Shrubs and Flowers	
Azalea	Aster	Bridalwreath
Gladiolus	Azalea	Camellia
Iris	Cherry, flowering	Chrysanthemum
Jerusalem cherry	Cherry, Jerusalem	Firethorn
Tulip	Dahlia	Mock-orange
	Geranium	Petunia
	Lilac	Privet
	Narcissus	Pyracantha
	Peony	Virginia creeper
	Plum, flowering	
	Rhododendron	
	Rose	
	Sweet William	
	Violet	

Table 4. (*continued*)

Sensitive	Intermediate	Tolerant
	Trees	
Apricot, Chinese	Apple	Apple
Apricot, Royal	Apricot, Moorpark	Arborvitae
Blueberry	Apricot, Tilton	Ash, European Mountain
Boxelder	Ash, European Mountain	Ash, Modesto
Douglas fir	Ash, green	Birch, cutleaf
Grape, European	Aspen, Quaking	Birch, white
Grape, Oregon	Cherry, Bing	Cherry, flowering
Larch, western	Cherry, Choke	Coffee
Peach (fruit)	Cherry, Royal Ann	Currant
Pine, eastern white (young needles)	Cherry, sweet	Dogwood
Pine, Lodgepole (young needles)	Fir, Grand	Elderberry
Pine, Mugo (young needles)	Grape, Concord	Elm, American
Pine, Ponderosa (young needles)	Grapefruit (fruit)	Elm, Chinese
Pine, Scotch (young needles)	Lemon	Juniper
Plum, Bradshaw	Linden, European	Laurel, Mountain
Prune, Italian	Linden, little-leaf	Linden, American
Spruce, blue	Maple, hedge	Locust, black
Serviceberry	Maple, Norway	Mock-orange
	Maple, silver	Oak
	Mulberry, red	Olive, Russian
	Orange	Pear
	Peach (foliage)	Pine, eastern white (mature needles)
	Plum, flowering	Pine, Lodgepole (mature needles)
	Poplar, Carolina	Pine, Mugo (mature needles)
	Poplar, Lombardy	Pine, Ponderosa (mature needles)
	Raspberry, red	Pine, Scotch (mature needles)
	Serviceberry	Planetree, London
	Spruce, white (young needles)	Poplar, balsam
	Sumac, Staghorn	Raspberry, red
	Sumac, smooth	Sweetgum
	Tangerine	Sycamore
	Walnut, black	Tree-of-Heaven
	Walnut, English	Willow
	Yew, spreading Japanese	

Compiled from Applied Science Associates 1976, National Academy of Sciences 1971, Treshow and Pack 1970.

Note: Plants may be listed under more than one level of sensitivity in the various references.

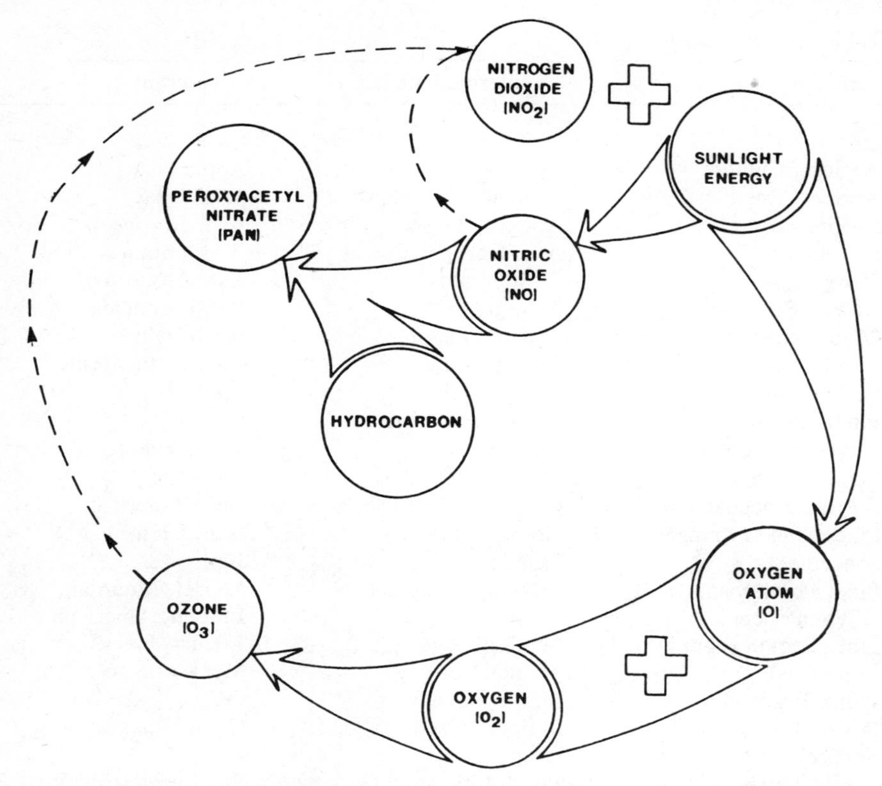

Figure 1. Atmospheric NO_2 photolytic cycle with hydrocarbons. (From Applied Science Associates, Inc., 1976)

If not buffered by soil constituents, such as carbonate rock, the precipitation acidifies the soil and water, changing soil properties and causing leaching of cationic nutrients and increased mobility of toxic cations. Other possible mechanisms by which acidic deposition can affect plants include alteration of foliar integrity, root growth, soil microbial activity, resistance to pests, seed germination, establishment of seedlings, and interactions with other abiotic stresses (Anonymous 1986). A diagrammatic representation of major mechanisms linking emissions, atmospheric deposition, and environmental responses is presented in Figure 2.

Although effects of acidic deposition on some segments of ecosystems appear to be indisputable, effects on vegetation still need clarification. Amathor (1984) reviewed available experimental evidence and concluded that plants are probably not affected directly by ambient acid rain. Most likely, acidity affects plants indirectly by alteration of soil properties and activity of soil microbes, and such alterations occur primarily in soils with poor buffering capacities (Amathor 1984, Evans 1984). There is some evidence

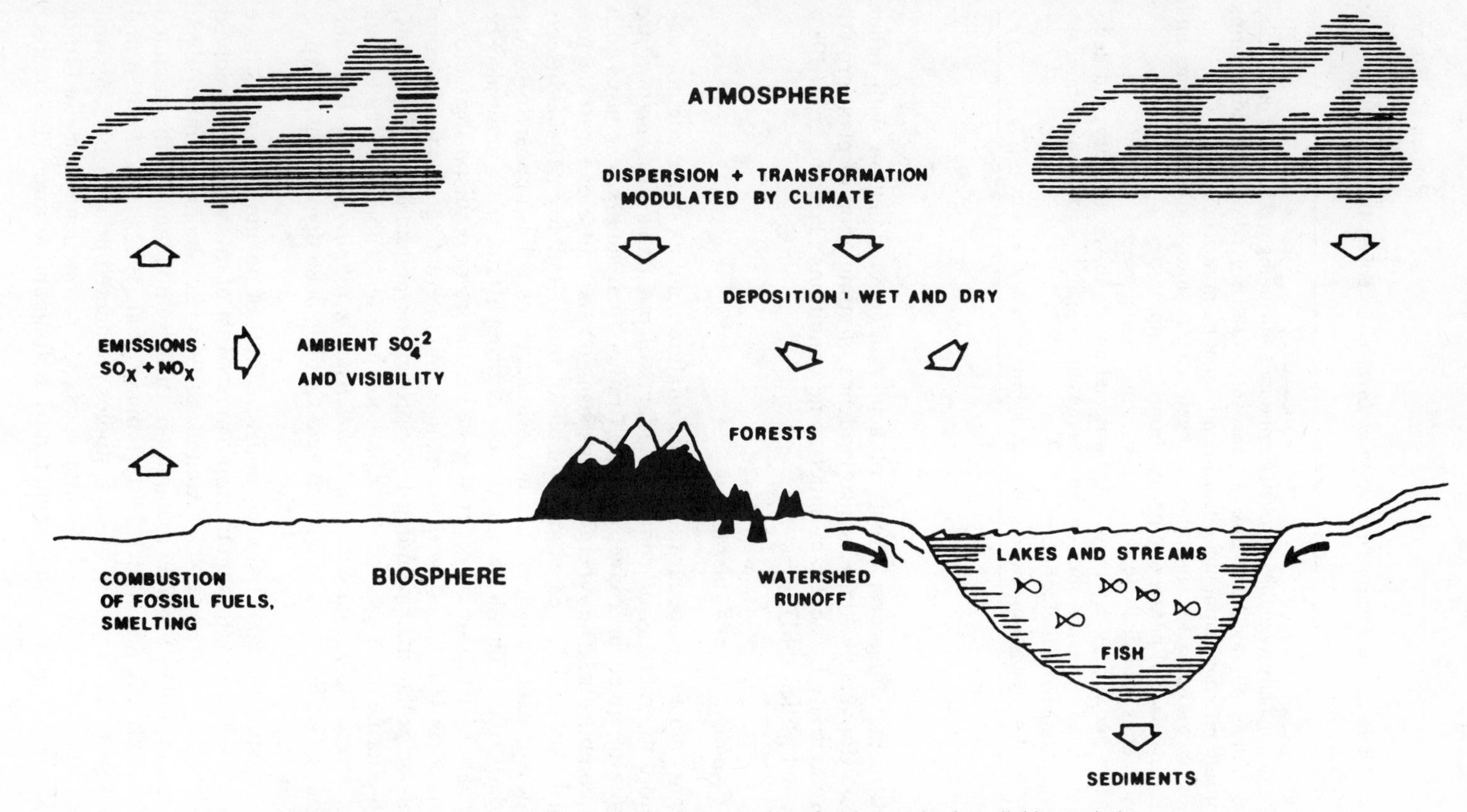

Figure 2. Acid deposition: diagrammatic representation of major mechanisms linking emissions, atmospheric deposition, and environmental responses. (Reprinted from "Acid Deposition: Long-Term Trends," 1986, with permission of the National Academy Press, Washington, DC.)

Table 5. Air Pollution Agents Thought to be Involved in Forest Decline in North America

1. Ozone
2. Total biologically available nitrogen compounds, including wet and dry deposition of all biologically available gaseous, aerosol, and dissolved or suspended forms of nitrate nitrogen, ammonia nitrogen and ammonium nitrogen
3. Other phytotoxic gases, including nitrogen oxides, sulfur dioxide, fluorine, and both peroxyacetyl nitrate and peroxyproprionyl nitrate
4. Toxic metals, especially lead, cadmium, zinc, and copper
5. Nutrient and acidity-determining cations and anions in wet and dry acid deposition, including potassium, sodium, magnesium, calcium, hydrogen, nitrate, sulfate, phosphate, and chloride
6. Growth-altering organic chemicals such as ethylene and aniline

From Hinrichsen 1986.

of detrimental long-term effects of acid rain in conjunction with other stresses. Conversely, there is also evidence of stimulation of plant growth and productivity by acid rain under some conditions (Troiano et al. 1983, Irving and Miller 1981).

2.2.5 Combinations and Interactions

Pollutants most often occur in complex mixtures, not as single compounds. Therefore interactions or combined effects of pollutants in a mixture are generally involved in a given situation rather than simple responses to a single component. The emerging consensus on the cause of forest decline in North America serves as an example of this (Hinrichsen 1986). Insects and known plant pathogens are not primary causes but appear to play a secondary role. Climatic factors, such as drought and frost, may also be involved but, again, are considered to be secondary or predisposing factors. A primary cause is felt to be the combined and interacting effects of a complex of air pollutants, including O_3, nitrogenous pollutants, toxic metals, growth-altering organics, and other phytotoxic gases (Table 5). Greater than additive response of yellow poplar (*Liriodendron tulipifera* L.) to the combination of O_3, SO_2, and simulated acid rain has been demonstrated experimentally (Chappelka et al. 1985).

Synergistic, antagonistic, and additive effects of pollutant mixtures have also received some attention in crop plants. Most of this work has involved mixtures of O_3, SO_2, and NO_2 (Amundson 1983, Lefohn and Ormrod 1984, Irving and Miller 1984) and combinations of these pollutants with simulated acidic rain (Irving and Miller 1981, Troiano et al. 1982, 1983, Lefohn and Ormrod 1984). A summary of such studies conducted prior to 1982, as well as an excellent discussion of research needs and recommendations for future work, are given in the report prepared by Lefohn and Ormrod (1984). Rec-

Table 6. Relative Sensitivity of Some Plants to SO_2

Sensitive		Tolerant
	Crops, Garden Plants, and Weeds	
Alfalfa	Mustard, wild	
Barley	Nightshade	
Beans	Oats	
Beets	Okra	
Bindweed	Peas	
Bluegrass, annual	Pepper	
Broccoli	Pepper bush, Brazilian	
Brussels sprouts	Plantain	
Buckwheat	Potato, sweet	
Careless weed	Pumpkin	
Carrot	Radish	
Chard, Swiss	Ragweed	
Chickweed	Rhubarb	
Clover, red	Rye	
Clover, yellow	Safflower	
Cotton	Soybean	
Cucumber	Spinach	
Dock, curly	Squash	
Endive	Sunflower	
Fern, bracken	Thistle, Sow	
Fleabane	Tomato	
Kohlrabi	Turnips	
Lettuce, cultivated	Velvet-weed	
Lettuce, prickly	Wheat	
Mallow		
Mustard, cultivated		
	Ornamental Shrubs and Flowers	
Aster	Sweet pea	Lilac
Bachelor Button	Sweet William	
Cosmos	Tulip	
Four O'Clock	Verbena	
Gladioli	Violet	
Morning Glory	Zinnia	
	Trees	
Alder	Larch	Cedar
Apple	Mulberry	Citrus
Aspen, large-toothed	Pear	Linden
Aspen, trembling	Pine, Austrian	Maple
Birch, white	Pine, Jack	
Catalpa	Pine, Ponderosa	
Douglas fir	Pine, red	
Elm, American	Pine, white	
Hazel	Willow	

Compiled from Applied Science Associates 1976, Barrett and Benedict 1970.

Table 7. Relative Sensitivity of Some Plants to NO_2

Sensitive	Intermediate	Tolerant
	Crops, Garden Plants, and Weeds	
Alfalfa	Bean, bush	Asparagus
Barley	Bean, field	Bean, bush
Carrot	Celery	Cabbage
Celery	Cheeseweed	Carrot
Clover, red	Chickweed	Grass, Kentucky Blue
Clover, spring	Corn, Sweet	Kohlrabi
Leek	Dandelion	Lamb's-quarters
Lettuce	Grass, Annual Blue	Neetle-leaf goosefoot
Mustard	Potato	Onion
Oats	Rye	Pigweed
Parsley	Tomato	
Peas	Wheat	
Pinto bean		
Rhubarb		
Sunflower		
Tobacco		
Vetch, spring		
	Ornamental Shrubs and Flowers	
Azalea	Bottle Brush	Carissa
Bougainvillaea	Brittlewood	Croton
Brittlewood	Dahlia	Dahlia
Hibiscus, Chinese	Fuchsia	Daisy
Oleander	Gardenia	Gladiolus
Pyracantha	Ixora	Heath
Rose	Jasmine, Cape	Ixora
Snapdragon	Ligustrum	Juniper, Shore
Sweet pea	Petunia	Lily of the Valley
Tuberous begonia	Pittosporum, Japanese	Plantain Lily
	Rhododendron, Catawba	Rose
	Trees	
Apple	Fir, white	Beech
Birch, European white	Fir, Nikko	Elder
Larch, European	Grapefruit	Elm, Scotch
Larch, Japanese	Linden, little-leaf	Ginkgo
Pear	Maple, Norway	Hornbean, European
	Maple, Japanese	Locust, black
	Orange	Oak, English
	Spruce, white	Pine, Austrian
	Spruce, Colorado Blue	Yew, English
	Tangelo	

Compiled from Applied Science Associates 1976, National Academy of Sciences 1977, Taylor and MacLean 1970.

Note: Plants may be listed under more than one level of sensitivity in the various references.

Table 8. Relative Sensitivity of Some Plants to O_3

Sensitive	Intermediate	Tolerant
	Crops, Garden Plants, and Weeds	
Alfalfa	Cabbage	Beet
Barley	Carrot	Cotton
Bean	Corn, field	Descurainia
Buckwheat	Cowpea	Jerusalem cherry
Citrus	Cucumber	Lamb's-quarters
Clover, red	Endive	Lettuce
Corn, sweet	Hypericum	Mint
Grape	Parsley	Piggy-back plant
Grass, bent	Parsnip	Rice
Grass, Brome	Pea	Strawberry
Grass, crab	Peanut	
Grass, orchard	Pepper	
Muskmelon	Sorghum	
Oat	Stevia	
Onion	Timothy	
Potato	Turnip	
Radish		
Rye		
Safflower		
Smartweed		
Soybean		
Spinach		
Tobacco		
Tomato		
Wheat		
	Ornamental Shrubs and Flowers	
Azalea, campfire	Begonia	Arborvitae
Azalea, Hino	Carnation	Azalea, Chinese
Azalea, Korean	Chrysanthemum	Euonymus, dwarf winged
Azalea, snow	Forsythia, Lynwood Gold	Firethorne, Laland's
Bridalwreath	Privet, common	Fuchsia
Browallia	Redbud, eastern	Geranium
Coleus	Rhododendron, Catawbiense Album	Gladiolus
Cotoneaster, rock	Rhododendron, Nova Zembia	Gloxinia
Cotoneaster, spreading	Rhododendron, Roseum Elegans	Holly, American (Male)
Lilac, Chinese		Holly, American (Female)
Lilac, common		
Petunia		
Privet, Londense		

Table 8. *(continued)*

Sensitive	Intermediate	Tolerant
	Ornamental Shrubs and Flowers—(continued)	
Snowberry	Silverberry	Holly, Hetz's Japanese
	Viburnum, Linden	Impatiens
	Viburnum, tea	Ivy, English
		Juniper, western
		Marigold
		Pachysandra
		Pagoda, Japanese
		Periwinkle
		Pieris, Japanese
		Privet, Amur North
		Rhododendron, Carolina
		Snapdragon
		Spirea
		Viburnum, Koreanspice
		Virginia creeper
		Yew, dense
		Yew, Hatfield's Pyramidal
		Zinnia
	Trees	
Ash, green	Alder	Apricot
Ash, white	Apple, crab	Avocado
Aspen, quaking	Apricot, Chinese	Beech, European
Cherry, Bing	Boxelder	Birch, European white
Grape, Concord	Catalpa	Box, Japanese
Locust, Honey	Cedar, incense	Dogwood, gray
Maple, silver	Cherry, Lambert	Dogwood, white
Mountain Ash, European	Elm, Chinese	Fir, balsam
Oak, Gambel	Fir, white	Fir, Douglas
Oak, white	Fir, big-cone Douglas	Fir, white
Pine, Austrian	Grape, Thompson Seedless	Gum, black
Pine, Coulter	Gum, sweet	Hemlock, eastern
Pine, eastern white	Honeysuckle, blue-leaf	Lemon
Pine, Jack	Larch, European	Laurel, Mountain
Pine, Jeffrey	Larch, Japanese	Linden, American
Pine, Monterey	Mock-orange, sweet	Linden, little-leaf
Pine, Ponderosa	Oak, black	Locust, black
Pine, Virginia	Oak, pin	Maple, Norway
Poplar, hybrid	Oak, scarlet	Maple, sugar
Poplar, Tulip	Pine, Knobcone	Maple, red
		Mimosa

Table 8. ***(continued)***

Sensitive	Intermediate	Tolerant
	Trees—(continued)	
Sumac	Pine, Lodgepole	Oak, Burr
Sycamore, American	Pine, Pitch	Oak, English
Tree-of-Heaven	Pine, Scotch	Oak, northern red
Walnut, English	Pine, sugar	Oak, Shingle
	Pine, Torrey	Pachysandra
	Poinsettia	Pagoda, Japanese
	Walnut, English	Peach
	Willow, Weeping	Pear, Bartlett
		Pine, Digger
		Pine, Singleleaf Pinyon
		Pine, red
		Redwood
		Sequoia, giant
		Spruce, Black Hills
		Spruce, Colorado Blue
		Spruce, Norway
		Spruce, white
		Walnut, black

Compiled from Applied Science Associates 1976, Hill et al. 1970.

ommendations concerning biological effects include directing attention to realistic exposure regimes, modes of action, and sources of variation in plant response (genetically determined and environmentally mediated). Reinert (1984) also reviewed this topic and emphasized several needs, including the need to consider sequential as well as simultaneous exposures to pollutants, interactions of acid rain with gaseous pollutants, and the influence of plant nutrition on plant response. Clearly, interactions between pollutants in a mixture are very important to effects of the pollution on the plant, and although the subject is complex, it more realistically reflects the stress to which the plant is subjected.

2.2.6 Plant Sensitivities

The sensitivity of a given plant to a particular pollutant is important in that it determines whether or not exposure to the pollutant is likely to alter plant metabolism sufficiently to affect plant–insect relations. Sensitivities of many plant species to common pollutants, based primarily on formation of visible symptoms of injury, are given in Tables 4 and 6 through 9. The rankings are based on experimental and subjective observations by different investigators, and therefore may vary and can be considered only as approximate. Some plants occur in more than one column because different investigators

Table 9. Relative Sensitivity of Various Plants to PAN

Sensitive	Intermediate	Tolerant
	Crops and Weeds	
Bean	Alfalfa	Bean, lima
Celery	Barley	Broccoli
Chard, Swiss	Beet, sugar	Cabbage
Chickweed	Beet, table	Cauliflower
Clover	Carrot	Corn
Endive	Cheeseweed	Cotton
Grass, Annual Blue	Dock, sour	Cucumber
Ground-cherry	Lamb's-quarters	Onion
Jimson weed	Soybean	Radish
Lettuce	Spinach	Rhubarb
Mustard	Tobacco	Sorghum
Nettle, little-leaf	Wheat	Squash
Oat		Strawberry
Pepper		
Pigweed		
Tomato		
Wild oat		
Sensitive	**Tolerant**	
	Ornamentals and Trees[a]	
Aster	Apple	Lily
Dahlia	Arborvitae	Locust, honey
Fuchsia	Ash, green	Maple, Norway
Mimulus	Ash, white	Maple, silver
Mint	Azalea	Maple, sugar
Petunia	Basswood	Mountain ash, American
Primrose	Begonia	Narcissus
Ranunculus	Birch, European white	Oak, English
Sweet basil	Bromiliads	Oak, northern red
	Cactus	Oak, pin
	Calendula	Oak, white
	Camellia	Orchids
	Carnation	Periwinkle
	Chrysanthemum	Pine, Austrian
	Coleus	Pine, eastern white
	Cyclamen	Pine, Pitch
	Dogwood	Pine, red
	Fir, balsam	Pine, Scotch
	Fir, Douglas	Pine, Virginia
	Fir, white	Poplar, hybrid
	Gum, sweet	Poplar, Tulip
	Hemlock, eastern	Snapdragon
	Ivy	Spruce, Black Hills
	Larch, European	Spruce, blue
	Larch, Japanese	Spruce, Norway
	Lilac, common	Spruce, white

From Applied Science Associates 1976.

[a] None reported to be intermediate in sensitivity.

reported different susceptibilities. The sensitivity of a plant to pollutants depends on many factors, including the particular pollutant, the plant species and genetic line, the plant stage, the environmental conditions under which exposure occurs, and other pollutants to which it is also exposed (either sequentially or simultaneously). These factors must be kept in mind whenever considering how pollutants might affect plants and the plant–insect relation.

3. EVIDENCE OF POLLUTANT-INDUCED CHANGES IN INSECT POPULATION DENSITIES

3.1 Correlative

Virtually all early studies of pollutant effects on insect populations were observational and concerned outbreaks of forest insects in the vicinity of industrial facilities. Observations have been made of more than 50 insect and mite species in a variety of forest types, and several good reviews of this literature are available (Baltensweiler 1985, Sierpinski 1985, Alstad et al. 1982, Hay 1977, Heagle 1973). In general, negative correlations between pollutant and insect levels have been found in areas of very high pollutant concentration and positive correlations in areas of moderate or low concentration. Changes in the dominant insect species from primary pests that infest healthy trees to secondary pests and saprophytes that infest weak and dead trees have been observed as the health of trees in affected areas declined.

Many of these studies were prompted by "outbreaks" of one or more insect species near an industrial site. A good example concerns outbreaks of the saddleback looper (*Ectropus crepuscularia* [Schiff.]) and spruce budworm (*Choristoneura orea*) in the area of an alumina reduction smelter at Kitimat, British Columbia, which occurred in the 1960s (Reid, Collins and Associates 1976). The plume from the smelter was channeled primarily up and down a valley, traveling principally in a northerly direction from spring through early fall and in a southerly direction in the winter. A plot of the fluoride content of the trees in the area showed this same pattern (Figure 3), as did a plot of the area defoliated by these two insects (Figure 4). This led to the suggestion that there was a relationship between the pollution and the pest outbreaks, and the implication that fluoride, the most phytotoxic pollutant identified in the plume, was responsible.

Another example is a recent study in which the population density of pine bark bugs (*Aradus cinnamomeus* Panz.) on Scots pine (*Pinus sylvestris* L.) was studied along an air pollutant gradient in an industrialized area of Finland (Heliovaara and Vaisanen 1986). Population density was found to increase to a maximum at 1 to 2 km from the factories and then to decrease rapidly toward the source of pollution. From their results the authors presented a

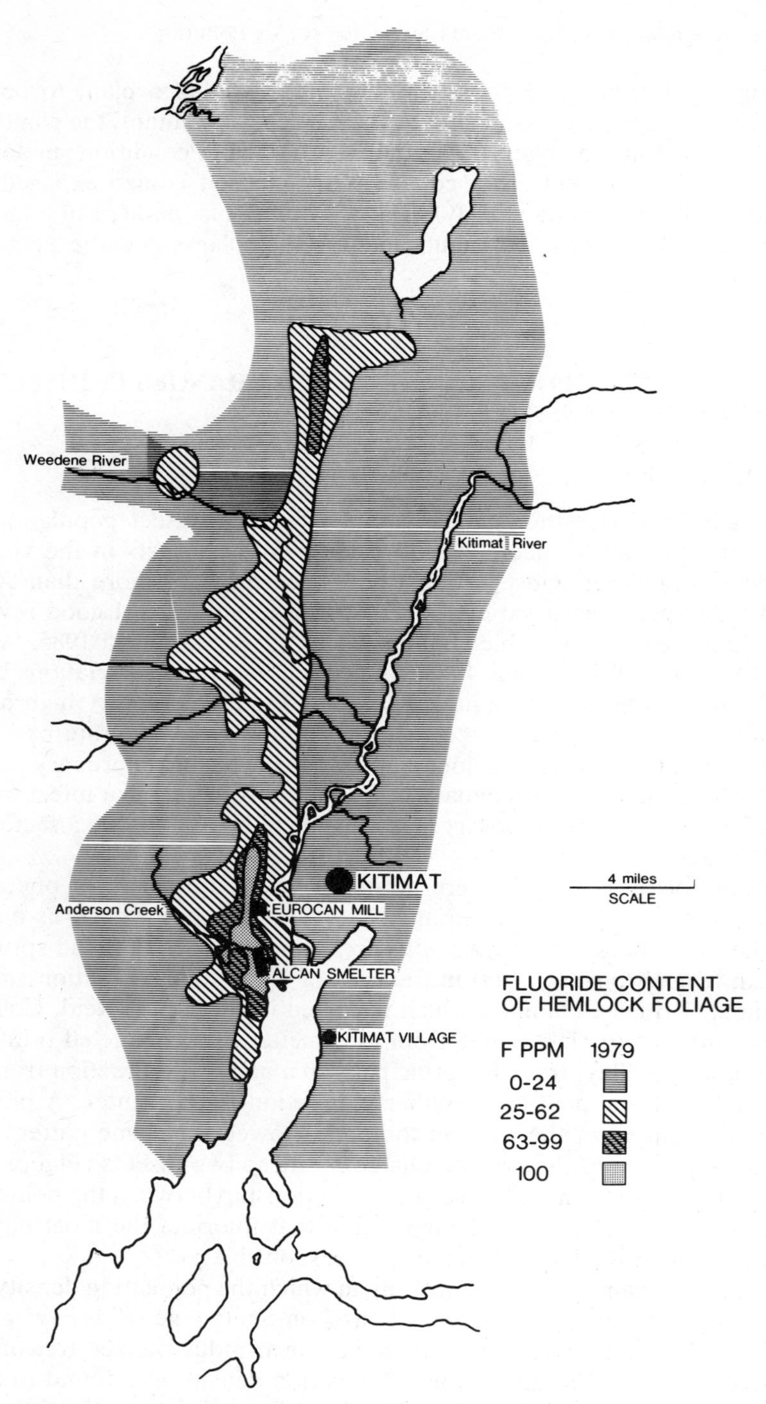

Figure 3. Fluoride content of hemlock foliage near an aluminum smelter at Kitimat, British Columbia. (Adapted from Reid, Collins and Associates, Ltd., 1976)

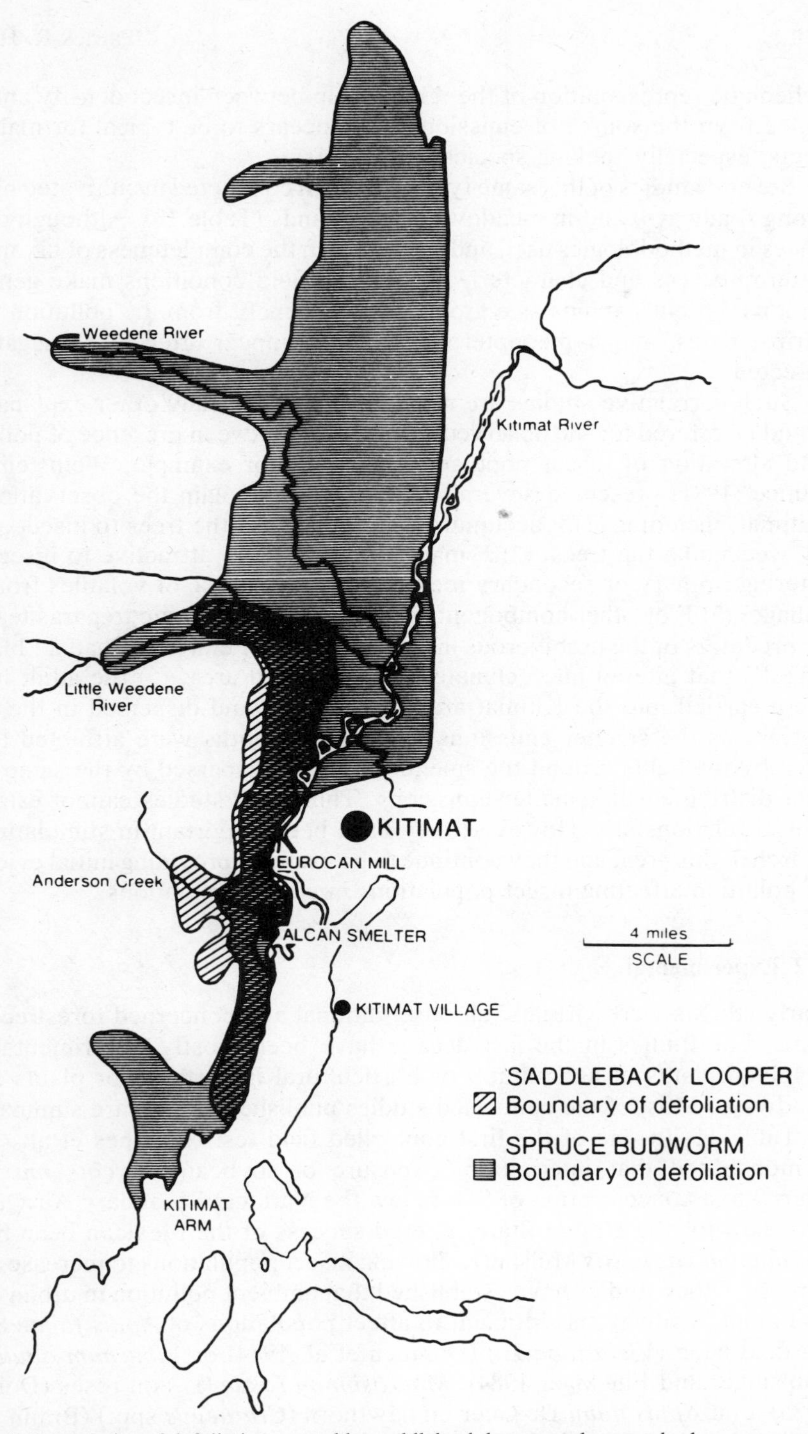

Figure 4. Boundaries of defoliation caused by saddleback looper and spruce budworm near an aluminum smelter at Kitimat, British Columbia. (Adapted from Reid, Collins and Associates, Ltd., 1976)

schematic representation of the relationship between insect density and distance from the source of emissions that appears to be typical for many insects, especially sucking species (Figure 5).

Several studies of this same type have been conducted in cultivated plants, along roadways, and in meadows or grasslands (Table 10). Although differences in methodologies used and variations in the completeness of taxonomic determinations and characterization of the field conditions make generalizations difficult, aphids as a group seem to benefit from the pollution while thrips, mites, and hymenopterous parasites appear often to be negatively affected.

Such correlative studies are not definitive, and many other explanations could be offered for the observed associations between presence of pollution and alteration of insect population density. For example, Weinstein and Bunce (1981) presented several hypotheses to explain the observations at Kitimat, including (1) F accumulation predisposed the trees to insect attack by weakening the trees, (2) F made the trees more attractive to insects by altering primary or secondary metabolism and release of volatiles from the foliage, (3) F or other components of the plume were toxic to parasites and/or predators of the herbivorous insects, (4) smelter emissions had a "blanket effect" that altered microclimate in the affected area, (5) the adult moths were carried into the Kitimat area by the wind and dispersed in the same pattern as the smelter emissions, and (6) the moths were attracted to the area by the lights around the smelter and then dispersed by the same wind that distributed the smelter emissions. Thus such studies cannot establish causal relationships. However, they have been important in stimulating research in this area, and they continue to be useful in providing initial evidence of pollution affecting insect populations in specific situations.

3.2 Experimental

Early studies were virtually all observational and concerned forest ecosystems, but studies in the last decade have been mostly experimental and involved plants of agricultural or horticultural importance or plants along roadways. Most of the controlled studies published to date are summarized in Table 11. In one of the first controlled field tests, Hughes et al. (1983) demonstrated that intermittent exposure of soybean (*Glycine max* [L.] Merr.) to a concentration of SO_2 below the National Secondary Air Quality Standard for the United States altered success of the Mexican bean beetle (*Epilachna varivestis* Mulsant), allowing insect populations to increase more rapidly. Other studies have established that ambient pollution in urban areas and along roadways is sufficient to affect populations of *Aphis fabae* Scop. on field bean (*Vicia faba* L.) (Dohmen et al. 1984) or *Viburnum opulus* L. (Bolsinger and Flückiger 1984), *Macrosiphon rosae* (L.) on roses (Dohmen 1985), and *Aphis pomi* De Geer on hawthorn (*Crataegus* spp.) (Braun et al. 1981; Braun and Flückiger 1984a, b). Although only a small number of insect

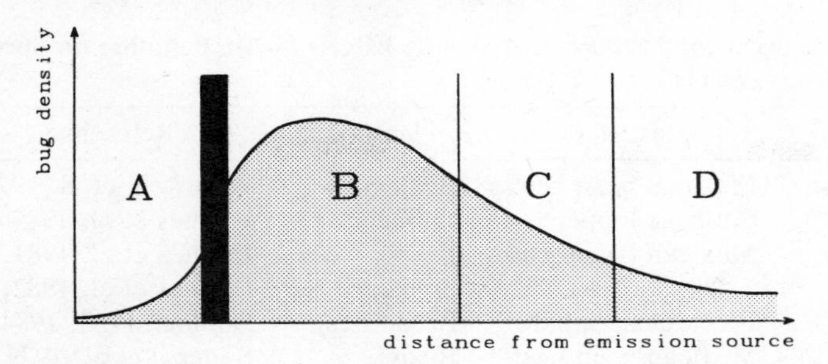

Figure 5. Schematized illustration of the occurrence of *Aradus cinnamomeus* along the gradient of pollutant concentration. A = lethal area with no or very few bugs, B = Area of serious injury by *Aradus*, C = area of high population density without serious injury, D = natural population density. (From Kari Heliovaara and Rauno Vaisanen 1986: Industrial air pollution and the pine bark bug, *Aradus cinnamomeus* Panz. [Het., Aradidae]. *J. Appl. Ent.* 101:476.)

Table 10. Studies Correlating Insect Populations with Presence of Air Pollution in Nonforest Habitats

Plant(s)	Insect(s)/Mites	Source of Pollution	Reference
Citrus	Mites and scales	Dust	Bartlett 1951
Apple	*Laspeyresia pomonella* (L.)	Dust	Callenbach 1939, 1940
Apple	*Aphis gossypii*, Glover, *Panonychus ulmi* (Koch)	Sulfur plant	Przybylski 1967
Apple, wheat, and grasses	General survey	Sulfur plant	Przybylski 1974
Hawthorn	*Aphis pomi* De Geer	Road traffic	Flückiger et al. 1978, Braun and Flückiger 1984
Apple, wheat	General Survey	Road traffic	Przybylski 1979
Hawthorn	*Sprageidus similis* (Fuess.)	Road traffic	Port and Thompson 1980
European beech	*Phalera bucephala* (L.)		Port and Hooton 1982
Hawthorn	*A. pomi, Dysaphis* sp., psyllids	Road traffic	Braun et al. 1981
Grassland	Thrips	ZAPS (SO_2)	Rice et al. 1979
Grassland	Acrididae	ZAPS (SO_2)	Leetham et al. 1979, McNary et al. 1981
Wheat	General survey	Volcanic ash	Klostermeyer et al. 1981

Table 11. Experimental Studies Concerning Effects of Air Pollution on Success of Herbivorous Insects

Pollutant(s)	Insect(s)	Plant(s)	Reference
HF	Mexican bean beetle	Bean	Weinstein et al. 1973
	Cabbage looper	Cabbage	Hughes et al. 1985b
SO_2	Mexican bean beetle	Beans	Hughes et al. 1981
		Soybean	Hughes et al. 1982, 1983
	Black bean aphid	Broad bean	Dohmen et al. 1984
O_3	Mexican bean beetle	Bean	Benepal et al. 1979
		Soybean	Endress and Post 1985
	Gypsy moth	White oak	Jeffords and Endress 1984
	Tomato pinworn	Tomato	Trumble et al. 1987
NO_2	Black bean aphid	Broad bean	Dohmen et al. 1984
Ambient air	Black bean aphid	Broad bean	Dohmen et al. 1984
	Rose aphid	Rose	Dohmen 1985
	Black bean aphid	*Viburnum*	Bolsinger and Flückiger 1984
	Green apple aphid	Hawthorn	Braun and Flückiger 1984, 1985
Acid rain	Comminutors (mites, springtails)	Humus	Hagvar and Abrahamsen 1977, Hagvar and Amundsen 1981, Abrahamsen et al. 1980
	Large copper butterfly	Great water dock	Bink 1986
SO_2, O_3, acid rain	Elm leaf beetle	Elm	Hall et al. 1988

species are represented in these studies, the results have demonstrated conclusively that ambient pollution is affecting insect populations in a variety of habitats and plant types.

Accompanying the recent concern about increase in atmospheric CO_2 has been interest in the consequence of elevated CO_2 to herbivores. Studies of cabbage looper (*Trichoplusia ni* [Huebner]) on lima bean (*Phaseolus lunatus* L.) and soybean looper (*Pseudoplusia includens* [Walker]) on soybean have shown a trend of increased consumption and decreased growth of the insects on plants grown in high CO_2 atmospheres (Lincoln et al. 1986, Osbrink et al. 1987). This appears to be caused by a reduction in leaf nutrient concentration, notably nitrogen, and reduced nitrogen utilization efficiency by the larvae (Lincoln et al. 1986).

4. MECHANISMS

An understanding of the mechanisms by which herbivore success is altered in the presence of air pollution is necessary both to understanding the re-

Table 12. Studies Concerning Direct Effects of Air Pollutants on Phytophagous Insects

Pollutant(s)	Insect/Mite	Reference
	Gases	
SO_2, NO, NO_2, CH_4, CO	*Oncopeltus fasciatus* (Dallas)	Feir 1978; Feir and Hale 1983a, b
CO	*Carausius morosus* Brunner	Baker and Wright 1977
SO_2	*Drosophila melanogaster* Meigen	Ginevan 1976, Ginevan and Lane 1978
HF	*D. melanogaster*	Gerdes et al. 1971
	Particulates	
Volcanic ash	*Laspeyresia pomonella* (L.)	Howell 1981
	Psyllla pyricola (Foerster)	Fye 1983
	Megachile rotundata (Fabricius)	Kish 1984
Fluoride salts	*Bombyx mori* (L.)	Fuji and Honda 1972, Imai and Sato 1974
Sodium fluoride	*Scotia segetum* Den. et Schiff.	Weismann and Svatarakova 1973, 1974, 1975, 1976
	Mamestra brassicae (L.)	
	Leptinotarsa decemlineata (Say)	
	Tetranychus urticae Koch	
Heavy metals	*Onychiurus armatus* (Tullb.)	Bengtsson et al. 1985
Sorptive dusts	*D. melanogaster*	Ebeling 1971
	Callosobruchus chinensis (L.)	

lationship and to identifying when the interaction might be important. Nevertheless, virtually no research has been directed toward identifying these mechanisms. Possible mechanisms have been outlined in several reports (Laurence et al. 1983, Hughes 1984, Hughes and Laurence 1984), and the following summary of these suggestions is included to help stimulate ideas and research.

4.1 Direct Effects

Pollutants could affect the abundance or distribution of plant-feeding insects *directly* through toxicity, stimulation of metabolism, or alteration of behavior. The toxicity of specific pollutants has been investigated in many studies, but relatively few phytophagous insects have been examined (Table 12). In general, these studies indicate that phytophagous insects are relatively insensitive to most gaseous pollutants and water-insoluble particulates, particularly at concentrations commonly occurring in the environment, but they might be affected more readily by ingestion of water-soluble forms. However, at very high concentrations most pollutants, both gases and particulates, exhibit toxicity to insects.

Stimulation of animal metabolism by low levels of toxins, termed hormesis or hormoligosis (Luckey 1959), is a well-known phenomenon. Although such stimulation of insect metabolism by pesticides has been demonstrated (Leigh and Wynholds 1980, Lowery and Sears 1986), it has not been studied extensively with air pollutants. Only one study was found concerning a phytophagous insect. In this study, growth of the large milkweed bug (*Oncopeltus fasciatus* [Dallas]) was stimulated by intermittent exposure to SO_2, CO, and NO; SO_2 also stimulated reproduction (Feir and Hale 1983a). In studies with nonphytophagous insects, ozone greatly stimulated oviposition by *Musca domestica* L., *Stomoxys calcitrans* (L.), and *Drosophila melanogaster* Meigen and resulted in significant increases in adult populations of these flies as compared to controls (Levy et al. 1972, Beard 1965). Similarly, egg production by *Tribolium confusum* Jacquelin duVal was stimulated significantly by exposure to sublethal doses of sodium fluoride (Johansson and Johansson 1972). More such examples can be expected among phytophagous insects as additional studies are conducted.

No data are available on alteration of insect behavior by air pollutants. Based on currently published data, one would expect repellency of insects by the pollutants to be an important factor only at very high concentrations of pollution or in rare cases. Although alteration of behavior by inhaled gases is known among insects (e.g., Ribbands 1950, Baker and Wright 1977, Ginevan et al. 1980), this has not been studied specifically in the context of pollutants and plant-feeding insects.

4.2 Indirect Effects

Pollutants could also affect insects *indirectly* by (1) affecting behavior/physiology of predators, parasites, or pathogens, (2) altering the microclimate or microhabitat, (3) inducing changes in the host plant chemistry/morphology, or (4) altering plant abundance or distribution.

4.2.1 Effects on Predators, Parasites, and Pathogens

Many reports have suggested that populations of predators and parasites of plant-feeding insects might be reduced by pollutants (Alstad et al. 1982), especially SO_2 (Hillmann 1972, Hillmann and Benton 1972, Przybylski 1968) and inert dusts (Bartlett 1951, Callenbach 1940). Dust-related mortality of the *Prospaltella* parasitoid is especially implicated in outbreaks of the black pineleaf scale, *Nuculaspis californica* (Coleman) (Edmunds 1973). However, Petters and Mettus (1982) found that females of the hymenopterous parasite, *Bracon hebetor* Say, were very insensitive to SO_2 and concluded that putative SO_2 effects on parasitic wasps in nature are probably not due to effects on fecundity or fertility. More recent studies suggest that, at least in some cases, dust- or drought-related changes in leaf temperature and host plant physiology may also be involved in pest population increases previously attributed to parasite mortality caused by dust.

As with parasites the evidence that pollution reduces the effectiveness of predators is indirect. Populations of ground beetles were reduced in the vicinity of a kraft paper mill, and a negative correlation was found between the rate of fallout of sulphate and population density of the beetles (Freitag and Hastings 1973, Freitag et al. 1973). Volcanic ash falling in pear orchards reduced populations of adult *Deraeocoris brevis* (Uhler) and *Anthocorus* spp., predators of pear psylla (Fye 1983).

The extent to which potentially toxic pollutants are concentrated within plant-feeding species and through trophic levels seems to be quite variable (Dewey 1973, Joosse and Buker 1979, Beyer and Moore 1980, Hopkin et al. 1985, Hughes et al. 1985b, Hughes et al. 1987). In general, the sequence of increasing concentrations appears to be herbivores–omnivores–predators–scavengers (Buse 1986, Kettle et al. 1983, Dewey 1973). However, to date no biological consequences have been attributed to this accumulation where it does occur.

Heavy emissions of pollution might also increase susceptibility of insects to predation, as evidenced by the phenomenon of industrial melanism. Industrial melanism, which has been described in at least 70 species of moths in Europe, United Kingdom, and North America (Kettlewell 1955, Owen 1961, Mikkola 1975) is the change from predominantly pale (i.e., typical) colored forms of the insect to darker forms containing heavy melanism and is associated with air pollution. This has been studied most extensively in the peppered moth, *Biston betularia* L., in the United Kingdom. The moths rest exposed on tree trunks and depend on cryptic coloration to avoid predation (Kettlewell 1955, 1956, 1958). Pollution, notably SO_2, reduces the occurrence of light-colored lichens on the tree bark, leaving the pale forms of the moth very apparent to predators. Selective predation of pale and melanic forms in polluted and nonpolluted areas, respectively, has been demonstrated experimentally (Kettlewell 1955, 1956, 1958; Bishop 1972).

A growing body of literature now demonstrates that both parasites and predators of polyphagous insects can be affected indirectly by the quality of the food plant upon which their hosts have fed. For example, wing morphology of the coccinellid *Adonia variegata* (Goeze) was affected when it fed on *Aphis nerii* Fonscolombe if the aphid was feeding on *Cionura erecta* (L.) but not if it was feeding on another asclepiad, *Cynanchum acutum*. L. (Pasteels 1978). In other studies growth of two pupal parasitoids of the gypsy moth (*Lymantria dispar* [L.]), one of which also parasitizes the fall webworm (*Hyphantria cunea* [Drury]), was found to vary depending on the diet of their hosts (Greenblatt and Barbosa 1980, 1981). Although the mechanism remains to be demonstrated definitively, data suggest that plant allelochemicals ingested by the phytophagous insects are responsible for some of these effects on the predators and parasites (Barbosa and Saunders 1985). Whatever the specific mechanism(s) might be, alteration of the plant by air pollution could alter the quality of the herbivorous insect as a host for predators and parasites.

Air pollutants might also affect viruses, bacteria, and fungi that are pathogenic to insects. These pathogens could be affected directly by toxicity or indirectly by pollutant-related changes in dew and precipitation. Although direct toxicity to entomopathogens has not been demonstrated to date, air pollutants have been found to be toxic to some pathogens of plants (Laurence 1981, Laurence et al. 1983). Since the pH values of foliage surfaces are important in inactivation of entomopathogenic viruses (Ali and Sikorowski 1986, and references therein), air pollutants might affect the pathogens by altering the pH of dew or precipitation on the leaves. Other studies have also implicated certain basic cations, especially magnesium, calcium, and potassium, as contributing to inactivation of baculoviruses (Andrews and Sikorowski 1973; Elleman and Entwistle 1982, 1985a,b); air pollution could affect concentrations of these elements in the dew and precipitation.

Indirect effects of pollution on success of insect pathogens through alteration of host plants is also possible. Kunimi and Aruga (1974) showed that larvae of the fall webworm, *H. cunea*, were less susceptible to infection by nuclear and cytoplasmic polyhedrosis viruses when grown on a diet containing mulberry leaf powder than when grown on a diet without the powder. More recently, Ramoska and Todd (1985) have demonstrated that susceptibility of adult chinch bugs, *Blissus leucopterus* (Say), to the fungus *Beauveria bassiana* (Bals.) Vuill. is strongly affected by the host plant, with infection and fungal growth being much lower when the insects feed on sorghum and corn than when they feed on wheat. Apparently, a plant-produced fungal inhibitor is providing some protection to the insects. Allelochemicals, such as terpenoids and phenolics, in the host plant are known to be antibiotic to entomopathogens, especially bacteria and fungi (Barbosa and Saunders 1985). Air pollutants can affect production of these compounds by plants (Hughes and Laurence 1984) and therefore may indirectly alter insect susceptibility to the pathogens through their effects on the plant.

4.2.2 Effects on Microclimate or Microhabitat

Air pollution can be expected to alter microclimate and microhabitat in any of several ways. For example, pollutant-induced changes in leaf morphology, such as trichome size and density (Mishra 1982, Gupta and Ghouse 1986), structure of the wax layer(s) (Shriner 1974, 1980), or stomatal density or pore size (Mishra 1982, Gupta and Ghouse 1986), can affect both microclimate and microhabitat (see Section 4.2.3). In addition, particulates such as dust can alter leaf characteristics, including reflectance, temperature, and stomatal conductance (Armbrust 1986, Manning 1971, Anda 1986, Flückiger et al. 1977), thus changing microclimate and microhabitat. Stomatal conductance can also be affected by physiological changes induced by pollutants (Mansfield 1976). Although any of these changes might be expected to affect some insects, virtually no experimental work has been directed toward demonstrating their involvement in this manner except in studies with tetranychid mites (Fleschner 1958). In these studies mite populations increased in the presence of dust and the increase was not due to any changes in biological control factors or to physical or chemical effects on mite reproduction or longevity.

4.2.3 Effects on Host Plants

Insects rely on plants for nutrition to support growth and reproduction, for behavioral cues needed in host location, feeding, and ovipositing, and for substrate and protection from the environment. Thus air pollution can affect insects indirectly by altering any of the properties of the host plant important to these processes. Such effects have been the focus of much of the recent work concerning pollutant–plant–insect interactions, and many possible mechanisms by which this might occur are outlined in Table 13. Most of these mechanisms have been reviewed in detail by Hughes and Laurence (1984) and Laurence et al. (1983). Research to date has focused primarily on demonstrating that pollutant-induced changes in the plant are affecting particular insects, with little attention being given to the specific mechanisms involved. However, a considerable amount of information is available on specific changes induced in plants by air pollutants as well as on the importance of many of these characteristics (chemical and physical) to particular insects. Consideration of this information together, though admittedly speculative, may be defensible in terms of the ideas and working hypotheses thus stimulated.

Host Vulnerability to Discovery. Pollutants can affect the availability of a plant to insects by altering the actual or apparent population density of the plant. In the simplest sense this could be done by the pollutant stunting or killing the plant. It might also occur by the pollution affecting the plant's ability to compete with other plants, thus affecting species diversity and community structure, or by alteration of the properties serving as behavioral cues for location, identification, and acceptance of the plant as a host by the insect. For example, injurious levels of pollutants can cause color changes in plants (Malhotra and Blauel 1980), and orientation of many insects, including aphids and butterflies, is known to be strongly influenced by color. Host location by many insects feeding on conifers is affected by terpenes released into the air by the trees (Wood 1982, Städler 1974), and emission of these volatiles by balsam fir increased following exposure to SO_2 (Renwick and Potter 1981). In earlier studies exposure of spruce trees to smoke or SO_2 decreased the volatile oil content of needles, indicating a

Table 13. Major Mechanisms by Which Pollutants Might Alter Plants as Hosts for Insects

A. *Host Vulnerability to Discovery*

1. Host density
2. Behavioral cues (chemical and physical)
   a. Color
   b. Surface morphology
   c. Surface chemicals
   d. Volatile chemicals
   e. Nonvolatile primary and secondary metabolites
3. Plant species diversity/community structure

B. *Host Nutritional Quality*

1. Plant nutrition
2. Primary metabolite levels
3. Secondary metabolite levels
4. Plant water balance
5. Metabolic activity (hormesis)
6. Plant hormones

C. *Host Defenses*

1. Constitutive
   a. Surface morphology
   b. Toughness
   c. Secondary metabolites
2. Induced
   a. Phytoalexins
   b. Translocated induced compounds

D. *Alteration of Gene Expression*

similar loss to the atmosphere (Dässler 1964, Cvrkal 1959). Ethane, which acts as a strong repellant to the beetle *Monochamus alternatus* Hope (Sumimoto et al. 1975), increased in red pine (*Pinus resinosa* Ait.) and paper birch (*Betula papyrifera* Marsh.) in response to injury by SO_2 (Kimmerer and Kozlowski 1982).

Apparency/availability of suitable host material can also be altered by pollutant-induced changes in plant species diversity and community composition or by changes in availability of a particular plant structure on which the insect depends. In a study near a thermal power plant, the number of genera and species of trees, shrubs, herbs, and grasses increased at increasing distances from the facility, with the trees being affected first, then the shrubs, and lastly the herbs and grasses (Pandey 1983). Falkengren-Grerup (1986) observed changes in distribution of some forest herb species that appeared to be attributable to increased nitrogen deposition and acidity from pollution. Sympatric divergence of sweet vernal grass (*Anthoxanthum odoratum* L.) in response to heavy metal contamination of the soil has been demonstrated (Antonovics 1971). Because habitat heterogeneity and complexity of plant architecture affect insect populations and number of species (Lawton and Schroder 1977, Price 1983), such changes would be expected to produce significant changes in insect populations and communities. In addition, pollutants can affect root-mycorrhizal associations (Stroo et al. 1988; van Noordwijk and Hairiah 1986; Reich et al. 1985, 1986; McCool and Menge 1983), which can significantly affect plant competition and, therefore, species composition of a community (Allen and Allen 1984, 1986).

In addition, plant population density, species diversity, and community composition might also be affected by pollutant effects on pollinators. Numerous reports have cited negative effects of various pollutants on pollinators. Honeybees are biological magnifiers of airborne contaminants, with large quantities accumulating in the tissues of adults (Bromenshenk and Carlson 1975, Bromenshenk 1979a, Pratt and Sikorski 1982). A wide range of pollutants has been reported to cause mortality of bees, including fluoride, lead, arsenic, and sulfur dioxide (Knowlton et al 1950, Maurizio and Staub 1956, Galob 1958, Hillmann 1972), and prolonged exposure of sweat bees to gaseous SO_2 affected flight behavior (Ginevan et al. 1980).

Exposure to pollutants frequently causes changes in the way in which plants partition incoming resources (see Lechowicz 1987). Although all plant parts can be affected, root growth is generally reduced to a greater extent than that of shoots. Leaf area frequently increases slightly, but flower and seed production may decrease. These changes might reduce resource availability to specialized insects, such as flower or seed feeders.

Host Nutritional Quality. Pollutants can alter the nutritional quality of the plant as a host for herbivorous insects in many ways. They can change the nutrition of the plant itself. Depending on the soil type and properties, prolonged acidic deposition can lower the soil pH (Ziegler 1973, Nyborg 1978), affecting microbial activity involved in nutrient cycling (Klein and Alexander 1986, Strojan 1978, Bromenshenk 1979b, Smith 1975, Hagvar and Amundsen 1981), elevating levels of elements such as iron, manganese, and aluminium to toxic concentrations (Rorison 1980), reducing availability of elements such as phosphorous, calcium, magnesium, and molybdenum (Weinstein and Alscher-Herman 1982, Rorison 1980), and detrimentally affecting mycorrhizal associations (Stroo et al. 1988; van Noordwijk and Hairiah 1986; Reich et al. 1985, 1986; McCool and Menge 1983). Also, microbes involved in biological nitrogen fixation can be sensitive to pollutants, with different effects depending on the conditions (Nyborg 1978, Shriner and Johnston 1981, Rao and Rao 1986). On the other hand, pollutants such as oxides of sulfur or nitrogen, as well as some metals, can serve a beneficial role as needed nutrients for plants under certain conditions (Thomas et al. 1943; Nyborg

1978; Cowling and Lockyer 1976; Faller 1971, 1972; Troiano and Leone 1977), especially in lower producing, nonagricultural systems deficient in these elements (Elkiey and Ormrod 1981, Bell and Clough 1973, Rowland 1986). In addition acidic precipitation can increase leaching of nutrients from foliage, including K, Ca, Mg, P, and N (Wood and Bormann 1974, Fairfax and Lepp 1975, Henderson et al. 1977, Evans et al. 1981).

Nutritional quality of the host can also be altered by pollutant-induced changes in levels of primary or secondary metabolites. Free amino acids and organic acids generally increase in plants exposed to moderate concentrations of many pollutants (Table 14). Reducing sugars also generally increase, but sucrose and foliar lipids tend to decrease. However, these effects can vary according to the concentration of the pollutants and the duration of exposure.

Many "secondary" plant metabolites affecting the acceptability or quality of plants as hosts for insects can also be affected by air pollutants. For example, flavonoids, which are known to inhibit insect growth (Elliger et al. 1980), increased in foliage of alfalfa exposed to O_3 (Jones and Pell 1981, Hurwitz et al. 1979). Phenolics increased in spruce exposed to SO_2 (Grill et al. 1975), but increased survival of *Sacchiphantes abietis* (L.) on spruce exposed to atmospheric pollutants was correlated with a diminished ability of the trees to produce a specific defensive reaction that appears to involve phenolics (Thalenhorst 1974). Phenolics are important to insects in diverse ways, including as allelochemics (Beck and Reese 1976, Duffey 1980, Hedin et al. 1977, Isman and Duffey 1982, Johnson et al. 1985), and nutrients (Bernays and Woodhead 1982a, b; Schopf et al. 1982).

Host quality can be changed by pollutant-related effects on plant water balance. Water balance can be affected by physical interference with stomata, such as occlusion by dust (Flückiger et al. 1977), or by biochemical and physiological changes that affect stomatal function (Hallgren 1978; Evans 1982a, b, 1984). Water balance can also be affected by pollutant-induced alteration of root-mycorrhizal associations. Acid rain and SO_2 can decrease mycorrhizal infection of roots (Reich et al. 1985, 1986; Stroo et al. 1988), which could significantly reduce uptake of water by plants (Allen and Allen 1986). Numerous reports have shown changes in insect reproduction on drought-stressed plants (e.g., Kennedy et al. 1958, Kennedy and Booth 1959, Wearing and van Emden 1967), and the importance of water as a nutrient has also been demonstrated (Scriber 1978, Reese 1978, Scriber and Slansky 1981).

Pollutants can alter suitability of the plant as a host for herbivorous insects by affecting pathogenic and epiphytic microbes. A vast literature describes the many effects of pollutants on the pathogens and microflora of leaves (Laurence 1981, Laurence et al. 1983, Shriner 1980, Heagle 1973, Manning 1975). Both beneficial and detrimental effects on plant-feeding insects have been attributed to the plant–microbe interaction (Maramorosch and Jensen 1963, Lewis 1979), and the possibility of microbial-induced resistance to

Table 14. Effects of Pollutants on Foliar Concentrations of Some Primary Plant Metabolites

Metabolite(s)	Pollutant	Plant(s)	Effect	Reference
Amino acids	HF	Bean, tomato	+	Weinstein 1961
		Soybean	+	Yang and Miller 1963a, b
	O_3	Bean, tobacco, Beet, corn, Barley, rye	+	Tomlinson and Rich 1967
	HF, SO_2	Clover, bean, Beet, rye, "Bulbous" plants	+	Arndt 1970
	SO_2	Bean	+	Godzik and Linskens 1974
	Exhaust (auto)	Hawthorn	+	Flückiger et al. 1978, Braun & Flückiger 1985
	SO_2, NO_2	Spruce	+	Zedler et al. 1986
Amines	SO_2	Pea	+	Priebe et al. 1978; Jager et al. 1985
Foliar lipids	SO_2	Jack pine, Lodgepole pine	−	Malhotra and Khan 1978
Organic acids	HF	Tomato, bean	+	Weinstein 1961
		Soybean	+	Yang and Miller 1963a, b
Free sugars	HF	Tomato, bean	−	Weinstein 1961
	HF	Soybean	+	Yang and Miller 1963b
	SO_2	Bean	+	Koziol and Jordan 1978
Sucrose	HF	Soybean	−	Yang and Miller 1963a
	SO_2	Pea	−	Kostir et al. 1970
Reducing sugars	HF	Soybean	+	Yang and Miller 1963a
	SO_2	Pea	+	Kostir et al. 1970
	SO_2	Spruce	+	Grill et al. 1975
	SO_2	Pine	−	L'Hirondelle and Addison 1985
	O_3	Potato	+	Pell et al. 1980
	Exhaust	Hawthorn	+	Flückiger et al. 1978
Total nonstructural CHO	O_3	Clover	−	Blum et al. 1982

insects is receiving some attention (McIntyre et al. 1980, Benedict and Bird 1981). The mechanisms are not understood, but, clearly, interactions of pollutants with these microbes can significantly affect the plant–insect relation.

At certain doses pollutants can alter host plants by stimulating growth. This enhancement of metabolism by low doses of toxicant, or hormesis (see earlier discussion), has been demonstrated to occur in a wide variety of plants in response to O_3, SO_2, and fluoride (Bennett et al. 1974). Success of insects on such faster growing plants might be enhanced (Mattson 1980). Pollutants might also affect plant growth by affecting biosynthesis and translocation of plant hormones or acting as hormones themselves. There is some evidence that SO_2 interacts with plant hormones (Hallgren 1978), and sulfite has been shown to enhance oxidation of indole-3-acetic acid *in vitro* (Meudt 1971, Yeh et al. 1971, Yang and Saleh 1973). Production of ethylene, a very potent hormone affecting many processes in plants, is increased in many species by O_3 as well as SO_2 (Hughes and Laurence 1984). In addition, ethylene is emitted directly into the atmosphere as a pollutant from a variety of sources (Sawada and Totsuka 1986). Also, some secondary compounds affected by pollution appear to function as endogenous regulators in plants (Levin 1971, Seigler and Price 1976, Smith and Banks 1986).

Host Defenses. Pollutants can also affect plant defenses against herbivores. As described in the previous paragraphs, pollutants can affect surface morphology, leaf toughness, and secondary metabolites that serve as constitutive defenses against insects. In addition, by their effects on plant metabolism, they can affect the ability of the plant to respond to damage by herbivory, namely, the induced defenses such as phytoalexins and translocated compounds that deter herbivory.

Alteration of Gene Expression. Although not investigated extensively to date, pollutants might also affect plants as hosts for insects by alteration of gene expression. A number of stresses, including lack of oxygen, high temperature, salinity, UV light, and drought, are known to induce expression of specific genes (Sachs and Ho 1986). Among the pollutants, heavy metals (particularly cadmium and copper) have been shown to induce de novo synthesis of a metal-binding protein or proteins (Wagner 1984; Grill et al. 1985a, b; Sachs and Ho 1986).

4.3 Specific Mechanisms

Although many studies have shown that pollutant-induced changes in host plants can alter success of certain insects, virtually nothing is known about specific mechanisms involved. The most completely studied system is that of the effects of SO_2 on success of Mexican bean beetle (MBB) on soybean. Both laboratory and field studies have shown that growth, rate of development, and fecundity of the insect are increased on plants exposed to SO_2 (Hughes et al. 1982, 1983). In addition, adults prefer to feed on foliage that has been exposed to the pollutant (Hughes et al. 1982). Using pupal weight as a biological indicator of pollutant-induced change in the plant, subsequent studies showed a monotonic relationship between ambient concentration of the pollutant to which the plant was exposed and effect on the insect (Hughes et al. 1985a). Weight gain of the insects on the treated plants (relative to that

Table 15. Growth, Rate of Development, and Survivorship of Mexican Bean Beetle on SO_2-Fumigated, GSH-Augmented, or Control Soybean Leaves

	Pupal Weight (mg)		Days to Adult		Survivorship (%)	
Treatment	SO_2	GSH	SO_2	GSH	SO_2	GSH
Control	27.2	25.8	22.8	37.7	97	96
Treated	31.7	36.9	21.8	35.7	100	98
P	0.01	0.01	0.05	0.01	ns	ns

Note: n = 25–35 insects per group; data were analyzed by Student's *t*-test. Ambient temperature was 27°C in SO_2 test and 24°C in GSH test (Hughes and Chiment, unpubl.).

of the control insects) increased with increasing pollutant to a maximum on plants exposed to approximately 780 $\mu g\ SO_2/m^3$; as pollutant concentration continued to be increased above 780 $\mu g/m^3$, the weight difference decreased, with insects fed on plants fumigated with about 2000 $\mu g\ SO_2/m^3$ being no different in weight from controls. When reared on plants that were (1) freshly fumigated, (2) fumigated and allowed a period in which to recover, or (3) nonfumigated (controls), weight gain by the MBB was maximum on the freshly fumigated plants, minimum on the control plants, and intermediate on the plants given a recovery period following fumigation. Thus, whatever change occurred in the plant upon fumigation, the plant recovered to some degree when the pollutant was removed, but recovery was slow relative to response time.

In investigations of the mechanism by which these effects are produced, foliar concentration of reduced glutathione (GSH) was found to change with fumigation in the same manner as insect growth, being monotonic in relation to ambient SO_2 with a maximum at approximately 790 $\mu g/m^3$ and returning to control levels at a concentration of about 2000 $\mu g/m^3$ (Chiment et al. 1986). When foliar GSH was artificially increased twofold (i.e., to approximately the same level as in plants fumigated with 790 $\mu g\ SO_2/m^3$) by allowing excised leaves to imbibe a solution of GSH through their petioles, all of the changes in insect growth and reproduction observed on fumigated plants were also produced (Tables 15 and 16). In addition adults strongly preferred to feed on the GSH-treated leaves when given a choice, but GSH did not stimulate feeding when offered alone or in combination with sucrose in filter-disc feeding assays (Hughes and Chiment, unpublished). Thus, if the change in foliar GSH is not causally related to altered success of MBB, they are at least very strongly correlated.

Change in foliar GSH is a common response to oxidative pollutants as well as other stresses, such as drought, salinity, and cold stress (Shevyakova and Loshadkina 1965, Levitt 1972, Gamble and Burke 1984, Hughes and Johnson, unpubl.). GSH was elevated in SO_2-affected foliage of spruce (*Picea abies* [L.]), pine (*Pinus sylvestris* L.), larch (*Larix decidua* Mill), and birch (*Betula pendula* Roth) (Grill et al. 1979, 1980, 1982), and a mixture of

Table 16. Mean Fecundity and Longevity of Mexican Bean Beetle on SO_2-Fumigated, GSH-Augmented, or Control Soybean Leaves

	Preoviposition Period (days)		Eggs/Female (lifetime)		Longevity (days)	
Treatment	SO_2	GSH	SO_2	GSH	SO_2	GSH
Control	10.4	20.2	75	126	24	60
Treated	9.5	17.4	135	239	25	69
P	ns	0.05	0.05	0.01	ns	ns

Note: n = 5–10 insects per group in SO_2 test and 10–20 per group in GSH test; data were analyzed by Student's *t*-test. Ambient temperature was 27°C in SO_2 test and 24°C in GSH test (Hughes and Chiment, unpubl.).

SO_2 and O_3 caused a greater increase of GSH in spruce needles than SO_2 alone (Smidt 1984). In crop plants Guri (1983) found a decrease in GSH in bean (*Phaseolus vulgaris* L.) following exposure to O_3. Hughes and Chiment (unpubl.) found apparent initial decreases followed by marked increases in O_3-fumigated bean, radish, and soybean, and increases up to 700% in corn, spinach, cabbage, and cotton fumigated with SO_2.

Although increased GSH concentration may often reflect stress within the plant, it is not a general predictor of impact of the stress on insects. Foliar GSH increased in *P. vulgaris* fumigated with SO_2 (Hughes and Chiment unpublished), but success of the MBB was little affected on these plants (Hughes et al. 1981). Growth, rate of development, and survivorship of cabbage loopers (*Trichoplusia ni*) were unaffected when larvae were reared on soybean leaves artificially augmented with GSH (Table 17).

5. METHODOLOGY

Research directed toward identifying and elucidating interactions between pollutants, plants, and insects requires an interdisciplinary approach most

Table 17. Growth and Development of Cabbage Looper on GSH-Augmented or Control Soybean Leaves

	Pupal Weight (mg)		Development Time (days)	
	Males	Females	Males	Females
Control	279	268	21.2	20.2
GSH-augmented	272	269	20.9	20.1
P	ns	ns	ns	ns

Note: n = 10–15 insects per group; data were analyzed by Student's *t*-test (Hughes and Chiment, unpubl.).

easily accomplished by collaboration between scientists with expertise in air pollution, plant sciences, and entomology. This collaboration is not always possible, and even when such arrangements are realizable, there is a great advantage to the entomologists of having an understanding of the types of methodology available to implement the plant treatments. Toward this end the following brief discussion of methods is included. This is intended only as a point of beginning, and the reader is referred to a two-volume series prepared by Hogsett et al. (1987a, b) for a comprehensive description and evaluation of exposure systems, monitoring equipment, and microenvironmental sampling.

5.1 Criteria for Characterizing Environmental and Biological Parameters of a System

Although most entomologists are careful to adequately describe the biological attributes of the insects used in studies (age, stage, previous host history, rearing conditions, etc.), they are frequently less diligent in describing the environmental conditions under which tests are conducted or the plant characteristics or cultural methods used to grow the plants. Studies of pollutant–plant–insect relationships must include proper characterization of four categories of attributes: (1) features of pollutant exposure, (2) edaphic, climatic, and atmospheric environment (both during exposure periods and during nonexposure periods), (3) biological features of the plant, and (4) biological features of the insects being used. Careful delineation of these characteristics is essential to identifying interactions affecting response variables, to relating results of different studies, and to duplicating experimental conditions.

Criteria of importance in evaluating the performance of exposure systems (Table 18) were described in detail by Hogsett et al. (1987a); this list also serves as an excellent starting point for identifying features one might consider when conducting studies of pollutant–plant–insect interactions. Table 19 lists guidelines recommended by a committee of experts for measuring and reporting environmental parameters in studies using plant growth chambers (Krizek 1982). These standards were proposed to enhance uniformity of research conducted in controlled environments and to facilitate comparison of results obtained by different laboratories. Their adoption by researchers investigating pollutant–plant–insect relationships would serve these functions within the field as well as enable results to be related to the plant physiology literature.

5.2 Exposure Systems

Exposure systems have been developed to fulfill a wide variety of needs involving investigation of either dry or wet deposition. Those involving dry deposition are adapted to study effects of either gaseous pollutants or particulates, whereas those used in studies of wet deposition are modified to

Table 18. Potentially Relevant Criteria to Use in Evaluating Exposure System Performance

A. Chemical Properties of the Pollutant in the Atmosphere
   1. Dry deposition
      a. Spatial features
         (1) Concentration distribution (uniform versus nonuniform)
            (a) Vertical plane (e.g., profiles)
            (b) Horizontal plane (e.g., patches)
      b. Temporal features
         (1) Exposure profiles
            (a) Square wave
            (b) Time series
         (2) Daily exposure regimes
            (a) Diurnal
            (b) Noctural
         (3) Seasonal exposure regimes
         (4) Annual exposure regimes
            (a) Growing season
            (b) Winter season
         (5) Episodes
            (a) Episode duration
            (b) Respite duration
            (c) Stochasticity/periodicity of events
   2. Wet deposition
      a. Solution chemistry (predroplet formation)
         (1) Cations
         (2) Anions
         (3) Organics
         (4) Reactive chemical species
         (5) Acidity/pH
         (6) Conductivity
      b. Incident hydrometeor chemistry [same chemical species as (a) above]
         (1) Spatial features
         (2) Temporal features
            (a) Exposure profiles
            (b) Daily exposure regimes
            (c) Seasonal exposure regimes
            (d) Annual exposure regimes
            (e) Episodes
      c. Incident hydrometeor physics
         (1) Spatial features
            (a) Horizontal-plane distribution
            (b) Vertical-plane distribution
         (2) Temporal features
            (a) Rate
            (b) Duration
            (c) Frequency
            (d) Time of day/season
            (e) Interevent features

Table 18. (*continued*)

- d. Incident hydrometeor physical features
  - (1) Droplet size statistics
  - (2) Terminal velocity
  - (3) Chemical-specific deposition rates
    - (a) Spatial distribution
    - (b) Temporal distribution
      - (1) Event
      - (2) Seasonal
      - (3) Annual
- 3. Combination of pollutants (dry/dry, wet/wet, wet/dry)
  - a. Characteristics of co-occurrence
  - b. Characteristics of independence
  - c. Relationship between wet deposition event and antecedent dry deposition period

B. Physical and Chemical Features of Environment

- 1. Atmosphere
  - a. Radiation
    - (1) Spectral quality, including the ultraviolet component (PAR)
    - (2) Quantity
  - b. Temperature/moisture
    - (1) Air temperature
    - (2) Leaf temperature
    - (3) Leaf-to-air differentials (vapor pressure deficit)
    - (4) Dew and frost formation
    - (5) Thermoperiod
  - c. Turbulence and mixing
    - (1) Canopy aerodynamic resistance to turbulent transfer
    - (2) Leaf boundary layer resistance to H_2O
    - (3) Wind speed
  - d. Turnover time of air reservoir
  - e. Ambient hydrometeor deposition
  - f. Trace gases
    - (1) Carbon dioxide
    - (2) Ethylene
    - (3) Water vapor
- 2. Soil
  - a. Chemistry
    - (1) Cation exchange capacity
    - (2) Partial pressure of oxygen
    - (3) Soil solution chemistry
    - (4) Soil nutrient analysis
  - b. Physics
    - (1) Temperature
    - (2) Water potential
    - (3) Porosity/solution infiltration

Table 18. ***(continued)***

C. Biological Features of the Environment
   1. Vegetation
      a. Structure
         (1) Leaf area/leaf area index
         (2) Canopy architecture
         (3) Leaf developmental stage/phenology
         (4) Canopy height relative to chamber
         (5) Canopy girth relative to chamber
         (6) Belowground root architecture
         (7) Mycorrhizae
      b. Function
         (1) Net photosynthesis
         (2) Transpiration
         (3) Leaf water potential
         (4) Leaf conductance to H_2O
         (5) Growth rates
         (6) Biomass partitioning above and below ground
         (7) Chlorophyll content
   2. Deposition of pollutants
      a. Wet deposition
         (1) Foliar interception of hydrometeor
            (a) Horizontal-plane variation
            (b) Vertical-plane variation
         (2) Foliar retention of hydrometeor
         (3) Chemical processing of intercepted hydrometeor
            (a) Through-fall chemistry
            (b) Stem-flow chemistry
         (4) Deposition to soil surface
         (5) Chemical processing of hydrometeor chemicals in soil/litter
      b. Dry deposition rates and amounts
         (1) Deposition to individual leaves
            (a) Leaf surface
               (1) Dry surface
               (2) Wet surface
            (b) Leaf interior
         (2) Deposition to canopy
         (3) Deposition to chamber walls/surfaces
            (a) Wet
            (b) Dry
         (4) Deposition to soil/litter surface

D. Hardware
   1. Dry deposition
      a. Chamber air handling
         (1) Air exchange rate—turnover time
         (2) Air movement—recirculation, single-pass
         (3) Equilibration time
         (4) Direction of air flow

Table 18. (*continued*)

(5) Air pressure
(6) Air leakage
(7) Air filtration
b. Chamber characteristics
(1) Pollutant sorption
(2) Size
(3) Volume-to-leaf area ratio
c. Pollutant concentration
d. Environment
(1) Monitoring
(2) Control
e. Data acquisition
2. Wet deposition
a. Hydrometeor generation
(1) Makeup water conditioning
(2) Generation of droplets
(3) Distribution/dispensing
(4) Monitoring/control
b. Environment
(1) Monitoring
(2) Control
c. Data acquisition

From Hogsett et al. 1987a.
Note: Criteria will vary depending on deposition mode and experiment design.

investigate simulated rainfall, simulated mist/fog/cloud water, or other aerosol simulations; a few systems have been developed to study combinations of dry and wet depositions (see Hogsett et al. 1987a, b).

Many systems have been designed for investigating the effects of gaseous exposures on plants. Systems for use indoors range from simple units placed in a greenhouse or controlled-environment chamber to modified or specially built controlled-environment chambers (see Hogsett et al. 1987a, b). Rogers et al. (1977) designed a system of chambers utilizing the concept of a continuous stirred tank reactor to permit appraisal of uptake of gaseous pollutants by plants and provide very uniform environmental conditions between experimental and control groups. This system was expanded by Heck et al. (1978) to four and nine units, and a similar system of 10 units with high intensity lights and temperature/humidity control has been constructed at the Boyce Thompson Institute for Plant Research (see Hughes et al. 1985a). Recently, Mueller and Garsed (1984) described a microprocessor-based system for controlling fluctuating concentrations of SO_2 over long periods in chamber studies; this system can be used for other gases and has also been expanded for use with mixtures of gases.

Outdoor systems are more varied and generally fall into one of three

Table 19. Proposed Guidelines for Measuring and Reporting Environment for Plant Studies

Parameter	Typically Used Unit	Measurements: Where to Take	Measurements: When to Take	Measurements: What to Report
		Radiation: PAR (Photosynthetically active radiation)		
1. Photosynthetic photon flux density (PPFD) 400–700 nm with cosine correction. *or*	μmol/s/m^2 [a] *or* μE/s/m^2	At top of plant canopy. Obtain average over plant-growing area.	At start and finish of each study and biweekly if studies extend beyond 14 days.	Average over containers at start of study. Decrease or fluctuation from average over course of study. Wavebands measured.
2. Photosynthetic irradiance 400–700 nm with cosine correction.	W/m^2			
Total irradiance with cosine correction. Indicate bandwidth.	W/m^2	At top of plant canopy.	At start of each study.	Average over containers. Wavebands measured.
		Radiation: Spectral Distribution		
1. Spectral photon flux density $\lambda_1 - \lambda_2$ nm in <20 nm bandwidths with cosine correction. *or*	μmol/s/m^2/nm ($\lambda_1 - \lambda_2$ nm) (quanta) *or*	At top of plant in center of growing area.	At start of each study as a minimum.	Spectral distribution of radiation with integral ($\lambda_1 - \lambda_2$) at start of study. Source of radiation and instrument/sensor.
2. Spectral irradiance (spectral energy flux density) $\lambda_1 - \lambda_2$ in <20 nm bandwidths with cosine correction.	W/m^2/nm ($\lambda_1 - \lambda_2$ nm)			

Illuminance 380–780 nm with cosine correction.[b]	klx	At top of plant canopy.	At start of each study.	Average over containers. Wavebands measured.
		Temperature		
Air Shielded and aspirated (≧3 m/s) device.	°C	At top of plant canopy. Obtain average over plant growing area.	Hourly over the period of the study. (Continuous measurement advisable.)	Average of hourly average values for the light and dark periods of the study with range of variation over the growing area.
Soil or liquid	°C	In center of container.	Hourly during the first 24 h of the study. Start immediately after watering (monitoring over the course of the study advisable).	Average of hourly average values for the light and dark periods for the first day (or over entire period of the study). Location of measurement.
		Atmospheric Moisture		
Shielded and aspirated (≧3 m/s) psychrometer, dew-point sensor or infrared analyzer.	% RH, dew-point temperature, or g/m^3	At top of plant canopy in center of plant growing area.	Once during each light and dark period, taken at least 1 h after light changes. Monitoring over the course of the study advisable.	Average of once daily readings for both light and dark periods with range of diurnal variation over the period of the study (or average of hourly values if taken).

Table 19. ***(continued)***

Parameter	Typically Used Unit	Measurements		
		Where to Take	When to Take	What to Report
Air Velocity				
	m/s	At top of plant canopy. Obtain maximum and minimum readings over plant growing area.	At start and end of studies. Take 10 successive readings at each location and average.	Average and range of readings over containers at start and end of the study.
Carbon Dioxide				
	$mmol/m^3$	At top of plant canopy.	Hourly over the period of the study.	Average of hourly average readings and range of daily average readings over the period of the study.
Watering				
	liter	N/A	At time of additions.	Frequency of watering. Amount of water added per day and/or range in soil moisture content between waterings.
Substrate				
	N/A	N/A	N/A	Type of soil and amendments. Components of soilless substrate. Container dimensions.

	Nutrition		
Solid media: mol/m^3 or mol/kg Liquid culture: mol or mol/L	N/A	At times of nutrient additions.	Nutrients added to solid media. Concentration of nutrients in liquid additions and solution culture. Amount and frequency of solution addition and renewal.
	pH		
pH units	In saturated media, extract from media, or solution of liquid culture.	Start and end of studies in solid media. Daily in liquid culture and before each pH adjustment.	Mode and range during study.
	Electrical Conductivity		
dS/m (decisiemens per meter)[c]	In saturated media, extract from media, or solution of liquid culture.	Start and end of studies in solid media. Daily in liquid culture.	Average and range during study.

From Krizek, 1982.

Note: Proposed by the North Central Region Committee (NCR-101) on Growth Chamber Use.

[a] The first is preferred because it follows the SI convention. However, since 1 Einstein = 1 mol of photons, the values are equivalent. It is inaccurate to report that "radiation values are *xx.x* $\mu mol/s/m^2$." This is wrong for the same reason that reporting mol/kg is wrong without associating that value and units with the element (i.e., K was 300 mol/kg). Thus "the PPFD was 320 $\mu mol/s/m^2$" is correct since it specifically associates a definition (i.e., photons within a certain waveband) with the value and units.

[b] Report with PAR reading *only* for historical comparison.

[c] 1 dS/m = 1 mmho/cm (millimho per centimeter).

categories: (1) plume exposures, (2) air exclusion systems, or (3) outdoor chambers. Plume exposures involve the injection of the pollutant into ambient air and use of local wind movement to disperse and carry the pollutant over the area being treated. The delivery devices are simple tubes with small holes, providing a system with minimum impact on the plant environment. Among the simplest of the plume exposure systems is the Zonal Air Pollution System (ZAPS), which consists of a network of 1-in. aluminum pipes set parallel to the ground at a height of 76 cm with 0.79 mm horizontal holes every 3 m for release of fumigant (Lee and Lewis 1978); compressors powered by a generator provided continuous air flow through the pipes. Many variations of this concept have been designed to increase efficiency, portability, and control (Hogsett et al. 1987a, b); recent innovations include control of release by a microcomputer that also controls monitoring (Greenwood et al. 1982), permitting release rate to be adjusted according to measured concentrations in the field, wind speed, and wind direction. McCleod et al. (1985) summarized open-air fumigation techniques and used computer simulation of gas dispersion to develop a system design that provides uniform exposure over an area of about 64 m^2.

Air exclusion systems are designed to shield plants from ambient pollution by enveloping them in filtered air. The systems employ blowers connected to large diameter, inflatable polyethylene tubes lying between rows of plants to pass charcoal-filtered air over the plants; the size, number, and orientation of holes in the tubing can be manipulated according to objectives of the study. These systems can be adapted to excludje ambient air, produce linear gradients of ambient air, or produce linear gradients of single pollutants or mixtures (e.g., Laurence et al. 1982, Thompson and Olszyk 1985).

As with the other types of systems, outdoor chambers occur in a wide variety of designs, ranging from very low chambers with open tops to larger chambers with semiopen tops to large, closed domes. The purpose of the chambers is to permit control over plant exposure to pollutants by excluding ambient air while blowing filtered air or filtered air with measured amounts of contaminants over the plants. A widely used design is the open-top chamber, the essential components of which have been listed by Hogsett et al. (1987a) as "(1) chamber covering permitting light penetration; (2) a high capacity blower to inject air; (3) charcoal and/or particulate filters to remove ambient pollutants; and (4) some structure for dispersing and mixing air entering the chamber, e.g., a plenum." Typical open-top chambers (3-m diameter) in field use in wheat are shown in Figure 6; the bottom half of the chamber is a double-walled plenum that distributes incoming air around the chamber. An octagonal closed-top chamber (2.1 m tall by 2.5 m across) comparable to open-top chambers in cost of construction has been developed recently (Musselman et al. 1986); these chambers permit much more precise control and determination of pollution dose as well as more uniform distribution of the pollutants within the chambers.

Selection of the exposure system for a particular study will be dependent

Figure 6. Open-top chambers used to assess effects of O_3 on yield of winter wheat (Ithaca, NY). A = filter unit and blower, B = clear vinyl covering on aluminum frame, C = double-walled plenum (lower portion of each chamber), and D = Teflon sampling tubes leading to air monitoring system. (Photo by Charles Harrington, Office of University Publications, Cornell University)

on many factors, including the specific goals of the study and the resources available. As a rule indoor chambers provide the best control of environmental conditions and pollutant exposure. However, the conditions differ from ambient field conditions, which can be very important to many studies. In general, the clear chambers developed for field use provide a realistic environment but still differ from ambient conditions by having slightly higher air temperatures, slightly lower morning and afternoon light intensities, different air-flow pattern over plant canopy, lower evaporative water loss, and interference with dew formation (e.g., Olszyk et al. 1980). Plume exposure systems provide minimal disturbance of biota and microclimate, while permitting treatment of large plots but have the disadvantages of not being able to operate at lower than ambient levels of pollution and being greatly dependent on wind (the direction, speed, and timing of which are not controllable); however, recent modifications of these systems, including computer-controlled release and feedback monitoring, will likely increase their utility. Air exclusion systems present the advantages of allowing graded exposures and the use of mixtures with either concurrent or countercurrent gradients, while not producing rain shadows or interfering with dew formation, not affecting access by natural biota, having virtually no effect on light intensity, relative humidity, or temperatures of air, soil, or leaves (sum-

Figure 7. Experimental site at the verge of a motorway near Basel, Switzerland, showing four closed-top perspex chambers used to study effects of roadside pollution on aphids. The two chambers in the foreground were provided with filtered air (passed through activated charcoal and Purafil) while the two in the background received ambient air. (Photo courtesy of M. Bolsinger)

mer or fall), and having essentially no system effects when the ducts are deflated. However, their use is currently somewhat limited to lower growing crops; they can alter temperatures of soil, air, and leaf during winter; introduction of the pollutant under the canopy is unnatural; they are vulnerable to wind; an extensive monitoring network is required within the plot; questions remain about the degree of air exclusion above the ducts (Laurence et al. 1982, Thompson and Olszyk 1985).

Although the preceding types of exposure systems were developed for studying effects of pollutants on plants, virtually every basic design has been employed in the relatively few investigations of pollutant–plant–insect relations. Indoor studies have employed slightly modified controlled environment chambers (e.g., Weinstein et al. 1961; Hughes et al. 1981, 1982, 1985a), CSTR chambers (Jeffords and Endress 1984, Endress and Post 1985, Hughes et al. 1985b), and small Plexiglass or glass chambers (Dohmen et al. 1984). In outdoor studies, McNary et al. (1981) used the Zonal Air Pollution System to treat 0.52-ha plots of prairie grassland in an investigation of the effects of SO_2 on grasshopper densities. A modification of the air exclusion system of Laurence et al. (1982) was used to investigate the effect of SO_2 on success of Mexican bean beetles on soybean under field conditions (Hughes et al. 1983). Other outdoor studies have used open-top chambers (Hall et al. 1988),

Table 20. Recommendations for Exposure Regimes to Be Used in Wet and Dry Deposition Exposure Studies

1. The exposure regimes should reflect the episodic (seasonal, diurnal) nature of pollutant occurrence.
2. The exposure regimes should be representative (frequency of occurrences) of the study area or the area where the study species is indigenous.
3. The exposure regimes should consider the daily physiological function of the plant, but the exposure should follow ambient air quality and not be arbitrarily terminated.
4. The pollutant regimes should yield a range of pollutant(s) concentrations above and below the current ambient concentration(s).
5. Exposure regimes should use realistic frequency and concentrations of pollutant co-occurrence.
6. Data for the development of realistic exposure regimes for combinations of dry and wet deposition are limited.
7. Data for the development of realistic exposure regimes for particles and aerosols are limited.
8. Additional monitoring data are needed to characterize the concentration distributions for the dry deposition of particles and aerosols.

From Hogsett et al. 1987a.

closed-top chambers (Trumble et al. 1987), or small Plexiglass or glass chambers (Bolsinger and Flückiger 1984; Braun and Flückiger 1984, 1985; Dohmen et al. 1984; Dohmen 1985) (Figure 7).

5.3 Exposure Regimes

Of particular importance to research in this area is the selection of exposure regimes. Too easily, conclusions can be made or strongly implied based on results obtained from regimes that are totally unrealistic or not properly understood. Hogsett et al. (1987a) provide an excellent discussion of this topic, which will not be reproduced here. The experts participating in the workshop concluded that (1) exposure regimes should not be standardized, but there should be guidelines for developing the regimes, (2) exposure regimes should be region/site specific, and (3) exposure regimes should be selected to test experimental hypotheses. Table 20 summarizes some of the group's recommendations for exposure regimes, and Table 21 lists the types of exposure regimes currently being used, their uses, and their limitations.

5.4 Quality Assurance (QA)

The U.S. Environmental Protection Agency has implemented a quality assurance/quality control (QA/QC) policy for extramural research projects that is worth considering, at least in part, in other environmentally related work.

Table 21. Types of Exposure Regimes Currently Used in Fumigation Studies with Gaseous Pollutants

	Regimes			
	"Artificial"	Ambient	"Modified" Ambient	Simulated Ambient
Definition	Exposures were not designed to represent the temporal variation in concentration.	Exposures represent the temporal variation in concentrations.	Exposures represent the current temporal variation in ambient concentrations.	Exposures represent the temporal variation in ambient concentrations.
Uses	When insufficient monitoring data to characterize ambient air quality. *or* Rapid-response monitoring equipment not available. *or* Limited time resolution in the monitoring data. Suitable for short-term studies and some screening studies.	Assess the impact of current air quality at a site. Suitable for long-term studies.	Suitable for long-term studies. Exposures are coupled to the environment. Can be used to develop exposure–response functions. Can be used to create a range of concentrations above and below ambient levels.	Suitable for both long- and short-term studies in the field and lab. Exposures can be replicated in time and space. Can be used to develop exposure–response functions. Can be used to create a range of concentrations. Can be used to test hypothesis concerning the characteristics of exposure.
Limitations	Not suitable for long-term studies. Difficult to extrapolate the data to ambient conditions.	Cannot be used to replicate the exposure in time or space. Cannot be used to develop exposure–response functions.	Not suitable for greenhouse and laboratory studies. Exposures cannot be replicated in time or space.	Exposures may not be coupled to the environment.

From Hogsett et al. 1987a.

Established to ensure that all data collected are of known and documented quality, this policy calls for the development of QA/QC plans detailing such aspects as sampling procedures; sample custody; analytical procedures; data recording and analysis; assessment of data precision, accuracy, and completeness; internal checks for quality control; calibration procedures, frequency, and records; preventive maintenance; and corrective actions. A line of responsibility is fixed, and records are kept in a manner suitable for auditing. Although outside audits are not likely to be sought by most researchers, development of such a plan would ensure that QA/QC problems were thought through before research was initiated and would ensure measurements of defendable precision and accuracy. Among other benefits this would allow duplication of experimental conditions and comparisons between results of different laboratories to be accomplished more reliably.

6. SYNTHESIS

6.1 Perspective

The available data demonstrate that air pollutants are affecting population densities of at least some insects and suggest that the interaction is not uncommon. Although the number of carefully studied examples is too few to permit much generalization, sucking insects, particularly aphids, seem to be easily affected, and so-called "secondary" pests (i.e., those for which the plant is a marginal host) also seem more likely to be affected.

Air pollution is just one of many abiotic stresses with which plants must cope, as apparent from the various chapters of this book. Therefore the question of how air pollutant-induced changes in plants affect plant-feeding insects should be considered as part of a larger question of how plants respond to their environment and the consequences of these responses to phytophagous insects. What might we expect these consequences to be? It becomes quite apparent that a single change in the environment can cause many changes in the plant, and the consequences to plant-feeding insects will depend on the particular plant species and insect species involved. Whatever species are involved, whether population density of the insect is positively or negatively impacted will depend on the net effect of all of these changes in the plant, not on the effect of just one change. This is illustrated in Figure 8. Each stress or change in the environment will cause an array of changes in a given plant species (represented by the circles), and these arrays can overlap to varying degrees, as depicted by the shaded areas. For example, foliar concentration of soluble amino acids might be increased by all three stresses, though only two of them cause increased leaf temperature and only one of them causes a change in phenolic concentration. An insect population might expand more rapidly in response to simultaneous increases in both nitrogen and temperature than to nitrogen increases alone, but it

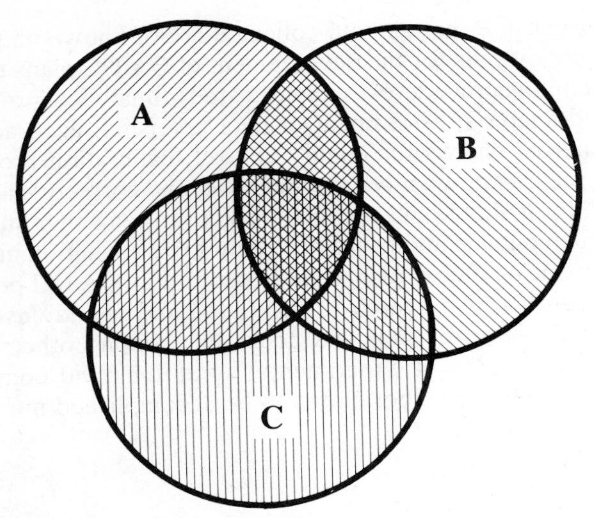

Figure 8. Diagrammatic representation of the physiological states or changes in physiological states of plants subjected to different environmental conditions.

might expand more slowly in the presence of increased phenolics even though nitrogen and leaf temperature are both elevated.

As a more practical illustration of this, growth, rate of development, and fecundity of the Mexican bean beetle on soybean were increased by exposure of the plants to SO_2 (Hughes et al. 1982, 1983, 1985a), and these effects were strongly correlated with increase in GSH (Chiment et al. 1986). However, these parameters were not affected when pinto bean (*Phaseolus vulgaris*) was used as the host rather than soybean (Hughes et al. 1981), even though GSH also increased in the fumigated pinto bean (Hughes and Chiment, unpublished). Similar examples can be cited with other stresses, such as drought stress of brussels sprouts which resulted in increased reproduction by the green peach aphid, *Myzus persicae*, and decreased reproduction by the cabbage aphid, *Brevicoryne brassicae* (Wearing and van Emden 1967).

Thus research in this area should be designed first to determine whether or not exposure of the plant affects growth, reproduction, or behavior of an insect, and if so, a systematic investigation of the mechanism(s) involved should be undertaken. Although current thought might dictate that the status of a certain class or classes of plant compounds be examined even in the absence of supporting evidence of their involvement, great care should be exercised not to imply more from the analyses than is warranted.

For the sake of stimulating thought, one might consider the hypothesis that the relation between plants and plant-feeding insects frequently is not the constant evolutionary "tug-of-war" so often stated, but rather the plants are evolving to cope with disease, competition, and abiotic environment—insofar as the host is concerned, the insects are evolving to cope with the changes in the plant. More simply stated, in general, plants might be affecting insect evolution far more than herbivorous insects are affecting plant evolution.

6.2 Workshop on the Effects of Air Pollutants on Plant–Pest Interactions

Approaching the subject from a very holistic perspective, a workshop was conducted to assess the state of knowledge and research directions in the area of pollutant–plant–pest interactions. The participants represented a wide array of specialties and each contributed their own particular expertise to an examination of what the problem might be, how this might impact at population and community levels, the questions that need to be addressed now, and the most promising approaches to future investigations. The following is largely extracted from the findings of this group (Bedford 1987) and provides a good summary of some current expert opinions in this subject area.

Four major themes emerged from the discussions. First, the role of air pollution in plant–pest dynamics is very poorly understood at this time. To what degree are current pest problems related to air pollution, under what conditions are air pollutants likely to be important, what are the mechanisms by which air pollutants are affecting plant–pest interactions, and what are the characteristics of pollutant exposures that are significant to the plant–pest relationship? Very little is known in answer to any of these questions.

A second major point upon which the participants agreed is that air pollutants are operating in the presence of many other "stresses" and at various levels of biological organization. Future research needs to consider the impact of multiple stresses and needs to consider impacts at population, community, and ecosystem levels.

A third theme emerging from the discussions is the need to define clearly the objectives of a study in order to develop the proper design and identify the appropriate variables to be measured. Two major objectives between which distinction must be made are (1) establishing cause and effect, and (2) identifying mechanisms. Cause-and-effect relationships can be investigated by looking for correlations between pollutant presence and insect presence, by conducting controlled tests (both laboratory and field) with and without addition of pollutant and by conducting controlled tests *in situ* in which ambient pollutants are included or removed. An understanding of mechanisms, which generally requires different experimental designs, is necessary in order to permit modeling, predictions, and insight into interactions with multiple stresses.

Related to the need to establish clear objectives for a study is the need to identify critical features of the system and the variables within each feature that must be characterized. Four classes of features that must be evaluated are (1) physical and chemical characteristics of the environment, (2) biological and chemical features of the plants and plant pests, (3) population level properties of the plants and plant pests, and (4) features of the community and ecosystem. Some important variables identified within each of these classes are listed in Table 22. Although the findings of the workshop apply to both herbivorous insects and plant pathogens, for simplicity the remainder of this discussion will refer only to insects.

An effort was made to develop a conceptual framework for viewing the effect of plant stress on the plant–insect interaction; such a framework could be used as the basis for formulating specific working hypotheses. The dominant view of those involved was that pollutants, alone or together with other stresses, would most likely affect insect populations indirectly through alteration of the host plant rather than directly. The system could be viewed as follows:

1. Environmental changes cause changes in allocation by plants, governed in part by altered gene expression. These changes include biochemical/physiological shifts (i.e., changes in chemical allocation) and morphological changes (i.e., changes in structural allocation).
2. The interaction can be spatial (i.e., one set of environmental conditions at one location, another at another location) or temporal (i.e., environmental conditions changing through time at one location).
3. The changes induced in plants by stress can be reversible (e.g., most biochemical/physiological shifts) or irreversible (e.g., morphological changes).
4. Driving the system would be the tendency of plants, in response to environmental conditions, to allocate their resources in such a manner as to increase acquisition of the resource(s) currently most limiting to plant growth and/or reproduction and to resist changes in the internal environment that would endanger cell function.

A set of three variables arose as the most promising candidates to measure when examining the relation of changes in insect populations to plant allocation patterns and biochemical status: (1) foliar carbon-to-nitrogen (C/N) ratio, (2) secondary defense compounds, and (3) the ratio of shoot biomass to root biomass (i.e., the shoot-to-root ratio).

Under this resource allocation/biochemical status approach, the general hypothesis was formulated: the biochemical status of plants determines their response to air pollutants and other stresses; the intensity of the stress, or stresses, causes adjustments in biochemical status that alter the distribution and abundance of pest populations. Two specific working hypotheses were derived from this general supposition: (1) the rate of damage, or intensity of stress, relative to the plant's potential rate of allocational response determines rate of recovery, and (2) plants allocate resources in a pattern that works to overcome that resource which most limits plant growth (i.e., growth or allocation will occur in a direction which allows the plant to deal with or compensate for a stress). The hypothesis relating rate of "recovery" of a

Table 22. Four Classes of Critical Features to Be Considered in Air Pollutant–Plant–Pest Interactions

A. Physical and Chemical Characteristics of the Environment
   1. Climate (temperature, humidity, wind, insolation)
   2. Site (soil, exposure)
   3. Geomorphology (landforms, parent material)
   4. Geochemistry (buffering capacity, hydrology, watershed)
   5. Light penetration (in gaps and layers)
   6. Pollutants (distribution in time and space)
B. Biological and Chemical Features of the Individual Organisms
   1. Quality of host plant
      a. Attractiveness (everything about plant that makes it attractive to insect or pathogen actively seeking host)
      b. Surface characteristics (especially of different organs of plants)
      c. Structural characteristics (toughness, architecture, way in which different parts of plant are assembled)
      d. Biochemical status (C, N, defense compounds)
      e. Plant history (stress, age, size)
   2. Vulnerability of host plant
      a. Physiological status (carbon, nitrogen)
      b. Resource availability (water, nutrients, light)
      c. Phenology (age determinant or indeterminant, juvenile, mature)
      d. Vigor (crown rates, sapwood area, root starch, storage reserves)
      e. Defense compounds (toxins, lignin, digestibilities)
   3. Insects and pathogens
      a. Dispersal and host-finding mechanisms
      b. Initial establishment processes
      c. Growth/utilization/biomass conversion
      d. Reproductive output and mechanisms (sexual, asexual)
      e. Detoxication/toxin production capacity (part of overall competitive abilities)
      f. Parasite/disease/hyperparasite load
C. Population Level Features
   1. Host
      a. Age and size distribution
      b. Spatial distribution and geographic extent
      c. Genetic frequency of sensitive individuals, structure of sensitivity, diversity of gene pools
      d. Infection status (history)
      e. Infestation
   2. Insect/pathogen
      a. Sex ratio as it affects number of females
      b. Age and size distribution
      c. Spatial distribution and dispersal
      d. Density (innoculum and survival characteristics)
      e. Proportion virulent
      f. Fecundity and survivorship

Table 22. (*continued*)

D. Community and Ecosystem Level Features
  1. Tight interspecific interactions
    a. Symbionts (e.g., mycorrhizae, N-fixers)
    b. Vectors
    c. Natural enemies of pest
  2. Diffuse interspecific interactions
    a. Species composition and distribution (especially plants) and interacting species
    b. Community successional status and age of community
    c. Microbial community dynamics (e.g., free-living N-fixers, decompsers)
    d. Competitive interactions (e.g., pest–pest)

Adapted from Bedford 1987.

plant to the frequency/intensity of the stress and the ability to adjust allocation of resources is illustrated diagrammatically in Figure 9.

Other major areas identified as requiring attention in studies of pollutant–plant–pest interactions included (1) epidemiological characteristics of the system (i.e., population level, community level, and system level parameters), (2) effects on natural enemies of phytophagous insects, and (3) effects on the structure, chemistry, and microflora of plant surfaces.

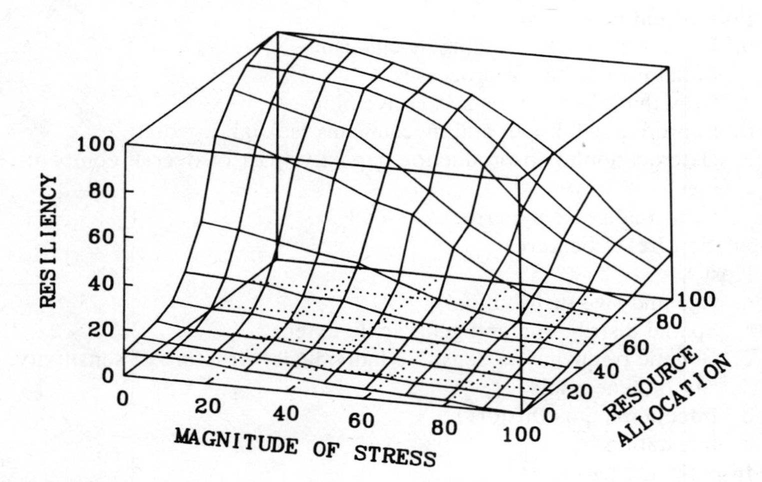

Figure 9. A conceptual representation of the hypothesis that the resilience of a plant (its ability to adjust to or recover from a ''stress'') is a function of the magnitude of the stress (as determined by its intensity, duration, and frequency) and the capacity of the plant to shift resource allocation (which is dependent on both plant and environment). The response surface would be expected to differ for different plant species and different environmental conditions. All scales are relative. Allocational parameters to be measured include (1) foliar carbon-to-nitrogen ratio (C/N), (2) secondary defense compounds (2° chemistry), and (3) ratio of shoot biomass to root biomass (shoot:root ratio). (From Bedford 1987)

Five approaches to research in this field identified by the workshop participants as potentially most productive at this time were:

1. Identification of model plant–pest or plant–pathogen systems that are well understood and can be manipulated; stress these systems experimentally with specific pollutants.
2. Determination of how stress affects specific biochemical traits to which particular insects or pathogens respond (this approach is pest specific).
3. Determination of key aspects of plant biochemical status that indicate stress, as well as the level and nature of stress. Determine if easily measurable traits that reflect indicator levels or thresholds for other less easily measured traits exist. Determine whether thresholds exist in these traits that indicate a decline in recovery ability (this approaches the problem more from the plant end and looks for indicators of susceptibility).
4. Study specific plants that are of widespread concern (e.g., red spruce).
5. Identification of the systems at risk, based on susceptibility due to level of stress, plant allocational dynamics, population level parameters, and/or community interactions; use models to identify the relative importance of these factors in different systems (this approach is based on system vulnerability).

An emphasis is placed on the use of model systems, and suggested characteristics of the ideal experimental system are given in Table 23.

One would expect the preceding hypotheses to change or be fine-tuned and new concepts to emerge as more investigations are undertaken. However, the results of this workshop serve as an excellent common starting point and source of ideas for researchers in this area.

6.3 Conclusions and Suggestions for Additional Research

Clearly, there is sufficient evidence to be concerned about air pollutants affecting plant productivity significantly through alteration of plant–insect (as well as plant–disease) interactions. Most work has involved short-term treatments, single pollutants, and, frequently, high concentrations. Also much of the past experimental work has focused on individual species under controlled conditions. However, the problem is more complex and requires integration of many factors to understand the actual and potential impact on ecosystems, to identify high risk situations, or to make predictions. Our comprehension of the problem is seriously limited by a lack of understanding of effects of long-term exposure to low concentrations, effects of mixtures of pollutants, interactions with other stress factors, or population level and community level effects. To overcome these deficiencies, model experi-

Table 23. Suggested Characteristics of Ideal Experimental System for Studying Effects of Air Pollutants on Plant–Pest Interactions

- The system should be long-term in nature.
- Storage mechanisms should be involved in the pest–host–pollutant interaction.
- Plants in the long-term model system should have genetic diversity that influences demographic changes in the population.
- There should be intraspecific competition in the model system.
- The system should lend itself to study of resource allocation in the host.
- The host in the system should have a spectrum of well-described foliar and soil-borne pests and parasitoids.
- The host should have some symbiotic relationships with other organisms.
- The host should be insect pollinated to allow pollinator studies.
- The host plant should be economically and/or ecologically important.
- The plant–pest system should be amenable to controlled experimentation.
- The host should have more than one use, such that the different uses cause different ecological relationships to develop.
- There should be enough experience with the system that experimental design criteria are available.
- There should be diversity in the host for resistance and sensitivity to pollutants and pests.
- There should be documented responses in the system to changes in mineral nutrient status.
- The system should have well-known phenological characteristics.
- The model system should be amenable to both laboratory and field studies, short- and long-term studies, and mechanistic and correlative studies.

From Bedford 1987.

mental systems need to be defined and conceptual frameworks need to be developed within which hypotheses can be formulated and tested.

Thus, though much has been accomplished in this area of research, much remains to be done, and a few specific topics appear as particularly outstanding in immediate importance. These include a better understanding of (1) the biologically relevant characteristics of pollutant exposure, (2) the impact of combinations of pollutants rather than single components, (3) interactions with other stresses, (4) the prevalence of interaction with plant–insect relations, (5) the mechanisms involved, and (6) population level and community level effects. A multidisciplinary, collaborative approach in future research would greatly enhance progress in this area.

REFERENCES

Abrahamsen, G., J. Hovland, and S. Hagvar. 1980. Effects of artificial acid rain and liming on soil organisms and the decomposition of organic matter. *In* Hutchinson, T. C., and M. Havas (eds.), *Effects of Acid Precipitation on Terrestrial Ecosystems*. Plenum, New York, pp. 341–362.

Ali, S., and P. P. Sikorowski. 1986. Effects of cotton leaf surfaces on the nuclear polyhedrosis virus of *Heliothis zea* and *Heliothis virescens* (Lepidoptera: Noctuidae). *J. Econ. Entomol.* 79: 364–367.

Allen, E. B., and M. F. Allen. 1984. Competition between plants of different successional stages: mycorrhizae as regulators. *Can. J. Bot.* 62: 2625–2629.

Allen, E. B., and M. F. Allen. 1986. Water relations of xeric grasses in the field: interactions of mycorrhizas and competition. *New Phytol.* 104: 559–571.

Alstad, D. N., G. F. Edmunds, Jr., and L. H. Weinstein. 1982. Effects of air pollutants on insect populations. *Annu. Rev. Entomol.* 27: 369–384.

Amathor, J. S. 1984. Does acid rain directly influence plant growth? *Environ. Pollut.* 36: 1–6.

American Chemical Society. 1978. Cleaning Our Environment: A Chemical Perspective. A report by the Committee on Environmental Improvement, Washington, DC.

Amundson, R. G. 1983. Yield reduction of soybean due to exposure to sulfur dioxide and nitrogen dioxide in combination. *J. Environ. Qual.* 12: 454–459.

Anda, A. 1986. Effect of cement kiln dust on the radiation balance and yields of plants. *Environ. Pollut.* 40: 249–256.

Andrews, G. L., and P. P. Sikorowski. 1973. Effects of cotton leaf surfaces on the nuclear polyhedrosis virus of *Heliothis zea* and *Heliothis virescens* (Lepidoptera: Noctuidae). *J. Invertebr. Pathol.* 22: 290–291.

Anonymous. 1986. Summary and Synthesis. *Acid Deposition: Long-Term Trends*. National Academy Press, Washington, DC, pp. 1–47.

Antonovics, J. 1971. The effects of a heterogeneous environment on the genetics of natural populations. *Am. Scientist* 59: 593–599.

Applied Science Associates, Inc. 1976. *Diagnosing Vegetation Injury Caused by Air Pollution*. Environmental Protection Agency, Washington, DC.

Armbrust, D. V. 1986. Effects of particulates (dust) on cotton growth, photosynthesis, and respiration. *Agron. J.* 78: 1078–1081.

Arndt, U. 1970. Concentration changes of free amino acids in plants under the effect of hydrogen fluoride and sulfur dioxide. *Staub-Reinhalt. Luft.* 30: 28–32.

Baker, G. M., and E. A. Wright. 1977. Effects of carbon monoxide on insects. *Bull. Environ. Contam. Toxicol.* 17: 98–104.

Baltensweiler, W. 1985. "Waldsterben": forest pests and air pollution. *Z. Angew. Entomol.* 99: 77–85.

Barbosa, P., and J. A. Saunders. 1985. Plant allelochemicals: Linkages between herbivores and their natural enemies. *In* Cooper-Driver, G. A., T. Swain, and E. E. Conn (eds.), *Chemically Mediated Interactions between Plants and Other Organisms*. Plenum, New York, pp. 107–137.

Barrett, T. W., and H. M. Benedict. 1970. Sulfur Dioxide. *In* Jacobson, J. S., and A. C. Hill (eds.), *Recognition of Air Pollution Injury to Vegetation: A Pictorial Atlas*. Air Pollution Control Association. Pittsburgh, PA, pp. C1–C17.

Bartlett, B. R. 1951. The action of certain "inert" dust materials on parasitic Hymenoptera. *J. Econ. Entomol.* 44: 891–896.

Beard, R. L. 1965. Observations on house flies in high-ozone environments. *Ann. Entomol. Soc. Am.* 58: 404–405.

Beck, S. D., and J. C. Reese. 1976. Insect–plant interactions: Nutrition and metabolism. *In* Wallace, J. W., and R. L. Mansell (eds.), *Biochemical Interaction between Plants and Insects*. Plenum, New York, pp. 41–92.

Bedford, B. L. (ed.). 1987. Modification of plant–pest interactions by air pollutants. Proc. Intl. Workshop. *ERC Report No. 117*, Ecosystems Research Center, Ithaca, NY.

Bell, J. N. B., and W. S. Clough. 1973. Depression of yield in ryegrass exposed to sulphur dioxide. *Nature* 241: 47–49.

Benedict, J. H., and L. S. Bird. 1981. Relationship of microorganisms within the plant and resistance to insects and diseases. *Proc. Beltwide Cotton Prod. Res. Conf.*, Memphis, pp. 149–150.

Benepal, P. S., M. Rangappa, and J. A. Dunning. 1979. Interaction between ozone and Mexican bean beetle feeding damage on beans. *Annu. Rep. Bean Improvement Cooperative* 22: 50.

Bengtsson, G., T. Gunnarsson, and S. Rundgren. 1985. Influence of metals on reproduction, mortality and population growth in *Onychiurus armatus* (Collembola). *J. Appl. Ecol.* 22: 967–978.

Bennett, J. P., H. M. Resh, and V. C. Runeckles. 1974. Apparent stimulations of plant growth by air pollutants. *Can. J. Bot.* 52: 35–41.

Bernays, E. A., and S. Woodhead. 1982a. Plant phenols utilized as nutrients by a phytophagous insect. *Science* 216: 201–203.

Bernays, E. A., and S. Woodhead. 1982b. Incorporation of dietary phenols into the cuticle in the tree locust *Anacridium melanorhodon*. *J. Insect Physiol.* 28: 601–606.

Beyer, W. N., and J. Moore. 1980. Lead residues in eastern tent caterpillars (*Malacosoma americanum*) and their host plant (*Prunus serotina*) close to a major highway. *Environ. Entomol.* 9: 10–12.

Bink, F. A. 1986. Acid stress in *Rumex hydrolapathum* (Polygonaceae) and its influence on the phytophage *Lycaena dispar* (Lepidoptera: Lycaenidae). *Oecologia* (Berlin) 70: 447–451.

Bishop, J. A. 1972. An experimental study of the cline of industrial melanism in *Biston betularia* (L.) (Lepidoptera) between urban Liverpool and rural North Wales. *J. Anim. Ecol.* 41: 209–243.

Blum, U., G. R. Smith, and R. C. Fites. 1982. Effects of multiple O_3 exposures on carbohydrate and mineral contents of ladino clover. *Environ. Expl. Bot.* 22: 143–154.

Bolsinger, M., and W. Flückiger. 1984. Effect of air pollution at a motorway on the infestation of *Viburnum opulus* by *Aphis fabae*. *Eur. J. For. Pathol.* 14: 256–260.

Braun, S., and W. Flückiger. 1984. Increased population of the aphid *Aphis pomi* at a motorway. I. Field evaluation. *Environ. Pollut.* (Ser. A) 33: 107–120.

Braun, S., and W. Flückiger. 1985. Increased population of the aphid *Aphis pomi* at a motorway. III. The effect of exhaust gases. *Environ. Pollut.* (Ser. A) 39: 183–192.

Braun, S., W. Flückiger, and J. J. Oertli. 1981. Einfluss einer Autobahn auf den Befall von Weissdorn. *Mitt. dtsch. Ges. Allg. Angew. Entomol.* 3: 138–139.

Bromenshenk, J. J. 1979a. Honeybees and other insects as indicators of pollution impact from the Colstrip power plants. *In* Preston, E. M. and T. L. Gullett (eds.), *The Bioenvironmental Impact of a Coal-Fired Power Plant. 4th Int. Rep.* EPA-600/3:79-044, Colstrip, MT, pp. 215–239.

Bromenshenk, J. J. 1979b. Responses of ground-dwelling insects to sulfur dioxide. *In* Preston, E. M. and T. L. Gullett (eds.), *Bioenvironmental Impact of a Coal-Fired Power Plant. 4th Int. Rep.* EPA-600/3:79-044, Colstrip, MT, pp. 673–703.

Bromenshenk, J. J., and C. E. Carlson. 1975. Impact on insect pollinators. *In* Smith, W. H., and L. S. Dochinger (eds.), *Air Pollution and Metropolitan Woody Vegetation*. Pinchot Inst. Consortium for Environmental Forestry Studies. USDA Forest Service, Delaware, OH, pp. 26–28.

Buse, A. 1986. Fluoride accumulation in invertebrates near an aluminum reduction plant in Wales. *Environ. Pollut.* 41: 199–217.

Callenbach, J. A. 1939. Road dust as a detriment to codling moth control. *Wisconsin Hortic.* 28: 131–132.

Callenbach, J. A. 1940. Influence of road dust upon codling moth control. *J. Econ. Entomol.* 33: 803–807.

Chappelka III, A. H., B. I. Chevone, and T. E. Burk. 1985. Growth response of yellow-poplar (*Liriodendron tulipifera* L.) seedlings to ozone, sulfur dioxide, and simulated acidic precipitation, alone and in combination. *Environ. Expl. Bot.* 25: 233–244.

Chiment, J. J., R. Alscher, and P. R. Hughes. 1986. Glutathione as an indicator of SO_2-induced stress in soybean. *Environ. Expl. Bot.* 26: 147–152.

Cowling, D. W., and D. R. Lockyer. 1976. Growth of perennial ryegrass (*Lolium perenne* L.) exposed to a low concentration of sulphur dioxide. *J. Expl. Bot.* 27: 411–417.

Cvrkal, H. 1959. Biochemicka diagnoza smrku v kourovych oblastech. *Berichte aus der tscheshoslowakischen Akademie der Landwirtschaftswissenschaften, Sektion Forstwesen*, 5: 1033–1048.

Dässler, H. G. 1964. The effect of SO_2 on the terpene content of spruce needles. *Flora* 154: 376–382.

Dewey, J. E. 1973. Accumulation of fluorides by insects near an emission source in western Montana. *Environ. Entomol.* 2: 179–182.

Dohmen, G. P. 1985. Secondary effects of air pollution: enhanced aphid growth. *Environ. Pollut.* 39: 227–234.

Dohmen, G. P., S. McNeill, and J. N. B. Bell. 1984. Air pollution increases *Aphis fabae* pest potential. *Nature* 307: 52–53.

Duffey, S. S. 1980. Sequestration of plant natural products by insects. *Annu. Rev. Entomol.* 25: 447–477.

Ebeling, W. 1971. Sorptive dusts for pest control. *Annu. Rev. Entomol.* 16: 123–158.

Edmunds, Jr., G. F. 1973. Ecology of black pineleaf scale (Homoptera: Diaspididae). *Environ. Entomol.* 2: 765–777.

Elkiey, T., and D. P. Ormrod. 1981. Sulphur and nitrogen nutrition and misting effects on the response of bluegrass to ozone, sulphur dioxide or their mixture. *Water, Air, and Soil Pollut.* 16: 177–186.

Elleman, C. J., and P. F. Entwistle. 1982. A study of glands on cotton responsible for the high pH and cation concentration of the leaf surface. *Ann. Appl. Biol.* 100: 553–558.

Elleman, C. J., and P. F. Entwistle. 1985a. Inactivation of a nuclear polyhedrosis virus on cotton by the substances produced by the cotton leaf surface glands. *Ann. Appl. Biol.* 106: 83–92.

Elleman, C. J., and P. F. Entwistle. 1985b. The effect of magnesium ions on the solubility of polyhedral inclusion bodies and its possible role in the inactivation of the nuclear polyhedrosis virus of *Spodoptera littoralis* by the cotton leaf gland exudate. *Ann. Appl. Biol.* 106: 93–100.

Elliger, C. A., B. C. Chan, and A. C. Waiss, Jr. 1980. Flavonoids as larval growth inhibitors. *Naturwissenschaften* 67: 358–360.

Endress, A. G., and S. L. Post. 1985. Altered feeding preference of Mexican bean beetle *Epilachna varivestis* for ozonated soybean foliage. *Environ. Pollut.* 39: 9–16.

Evans, L. S. 1982a. Effects of acidity in precipitation on terrestrial vegetation. *Water, Air, and Soil Pollut.* 18: 395–403.

Evans, L. S. 1982b. Biological effects of acidity in precipitation on vegetation: A review. *Environ. Expl. Bot.* 22: 155–169.

Evans, L. S. 1984. Botanical aspects of acidic precipitation. *Bot. Rev.* 50: 449–490.

Evans, L. S., T. M. Curry, and K. F. Lewin. 1981. Responses of leaves of *Phaseolus vulgaris* L. to simulated acidic rain. *New Phytol.* 88: 403–420.

Fairfax, J. A. W., and N. W. Lepp. 1975. Effect of simulated ''acid rain'' on cation loss from leaves. *Nature* 255: 324–325.

Falkengren-Grerup, U. 1986. Soil acidification and vegetation changes in deciduous forest in southern Sweden. *Oecologia* (Berlin) 70: 339–347.

Faller, N. 1971. Effects of atmospheric SO_2 on plants. *Sulfur Inst. J.* 6: 5–7.

Faller, N. 1972. Schwefeldioxid, Schwefelwasserstoff, nitrose Gase und Ammoniak als ausschliessliche S- bzw. *N*-Quellen der hoheren Pflanze. *Zeitschrift für Plfanzenernahrung und Bodenkunde* 131: 120–130.

Feir, D. 1978. Effects of air pollutants on insect growth and reproduction. *Physiologist* 21: 6.

Feir, D., and R. Hale. 1983a. Responses of the large milkweed bug, *Oncopeltus fasciatus* (Hemiptera: Lygaeidae) to high levels of air pollutants. *Intl. J. Environ. Studies* 20: 269–273.

Feir, D., and R. Hale. 1983b. Growth and reproduction of an insect model in controlled mixtures of air pollutants. *Intl. J. Environ. Studies* 20: 223–228.

Fleschner, C. A. 1958. Field approach to population studies of tetranychid mites on citrus and avocado in California. *In Proc. 10th Intl. Cong. Entomol.*, pp. 669–674.

Flückiger, W., H. Fluckiger-Keller, J. J. Oertli, and R. Guggenheim. 1977. Verschmutzung von Blatt- und Nadeloberflachen im Nahbereich einer Autobahn und deren Einfluss auf den stomataren Diffusionswiderstand. *Eur. J. For. Pathol.* 7: 358–364.

Flückiger, W., J. J. Oertli, and W. Baltensweiler. 1978. Observations of an aphid infestation on hawthorn in the vicinity of a motorway. *Naturwissenschaften* 65: 654–655.

Freitag, R., and L. Hastings. 1973. Kraft mill fallout and ground beetle populations. *Atmospheric Environ.* 7: 587–588.

Freitag, R., L. Hastings, W. R. Mercer, and A. Smith. 1973. Ground beetle populations near a kraft mill. *Can. Entomol.* 105: 299–310.

Fuji, M., and S. Honda. 1972. The relative oral toxicity of some fluorine compounds for silkworm larvae. *J. Sericult. Sci. Japan* 41: 104–110.

Fye, R. E. 1983. Impact of volcanic ash on pear psylla (Homoptera: Psyllidae) and associated predators. *Environ. Entomol.* 12: 222–226.

Galob, R. 1958. Poisoning of bees by dust containing lead. *Slov. Ceb.* 60: 57–60.

Gamble, P. E., and J. J. Burke. 1984. Effect of water stress on chloroplast antioxidant system. *Plant Physiol.* 76: 615–621.

Gerdes, R. A., J. D. Smith, and H. G. Applegate. 1971. The effects of atmospheric hydrogen fluoride upon *Drosophila melanogaster*-I. *Atmospheric Environ.* 5: 113–116.

Ginevan, M. E. 1976. Potential effects of sulfur dioxide air pollution on insects: an experimental study of *Drosophila melanogaster*, Ph.D. dissertation. Dept. of Entomol., University of Kansas, Lawrence.

Ginevan, M. E., and D. D. Lane. 1978. Effects of sulfur dioxide in air on the fruit fly, *Drosophila melanogaster*. *Environ. Sci. and Technol.* 12: 828–831.

Ginevan, M. E., D. D. Lane, and L. Greenberg. 1980. Ambient air concentration of sulfur dioxide affects flight activity in bees. *Proc. Natl. Acad. Sci. USA* 77: 5631–5633.

Godzik, S., and H. F. Linskens. 1974. Concentration changes of free amino acids in primary bean leaves after continuous and interrupted SO_2 fumigation and recovery. *Environ. Pollut.* 7: 25–38.

Greenblatt, J. A., and P. Barbosa. 1980. Interpopulation quality in gypsy moths with implications for success of two pupal parasitoids: *Brachymeria intermedia* (Nees) and *Coccygomimus turionellae* (L.). *Ecol. Entomol.* 5: 31–38.

Greenblatt, J. A., and P. Barbosa. 1981. Effects of host's diet on two pupal parasitoids of the gypsy moth: *Brachymeria intermedia* (Nees) and *Coccygomimus turionellae* (L.). *J. Appl. Ecol.* 18: 1–10.

Greenwood, P., A. Greenhalgh, C. Baker, and M. Unsworth. 1982. A computer-controlled system for exposing field crops to gaseous air pollutants. *Atmospheric Environ.* 16: 2261–2266.

Grill, D., H. Esterbauer, and G. Beck. 1975. Untersuchungen an phenolischen substanzen und glucose in SO_2-geschadigten fichtennadeln. *Phytopathol. Z.* 82: 182–184.

Grill, D., H. Esterbauer, and U. Klosch. 1979. Effect of sulphur dioxide on glutathione in leaves of plants. *Environ. Pollut.* 19: 187–194.

Grill, D., H. Esterbauer, M. Scharner, and C. H. Felgitsch. 1980. Effect of sulfur-dioxide on protein-SH in needles of *Picea abies*. *Eur. J. For. Pathol.* 10: 263–267.

Grill, D., H. Esterbauer, and K. Hellig. 1982. Further studies on the effect of SO_2-pollution on the sulfhydril-system of plants. *Phytopathol. Z.* 104: 264–271.

Grill, E., E.-L. Winnacker, and M. H. Zenk. 1985a. Phytochelatins: the principal heavy-metal complexing peptides of higher plants. *Science* 230: 674–676.

Grill, E., M. H. Zenk, and E.-L. Winnacker. 1985b. Induction of heavy metal-sequestering phytochelatin by cadmium in cell cultures of *Rauvolfia serpentina*. *Naturwissenschaften* 72: 432–433.

Gupta, M. C., and A. K. M. Ghouse. 1986. The effects of coal-smoke pollutants on the leaf epidermal architecture in *Solanum melongena* L. variety pusa purple long. *Environ. Pollut.* 41: 315–321.

Guri, A. 1983. Variation in glutathione and ascorbic acid content among selected cultivars of *Phaseolus vulgaris* prior to and after exposure to ozone. *Can. J. Plant. Sci.* 63: 733–737.

Hagvar, S., and G. Abrahamsen. 1977. Effect of artificial acid rain on Enchytraeidae, Collembola and Acarina in coniferous forest soil, and on Enchytraeidae in sphagnum bog—preliminary results. *Ecol. Bull.* (Stockholm) 25: 568–570.

Hagvar, S., and T. Amundsen. 1981. Effects of liming and artificial acid rain on the mite (Acari) fauna in coniferous forest. *Oikos* 37: 7–20.

Hall, R. W., J. H. Barger, and A. M. Townsend. 1988. Effects of simulated acid rain, ozone and sulfur dioxide on suitability of elms for elm leaf beetle. *J. Arboric.* 14: 61–66.

Hallgren, J. E. 1978. Physiological and biochemical effects of sulfur dioxide on plants. *In* Nriagu, J. O. (ed.), *Sulfur in the Environment. Part II. Ecological Impacts*. Wiley, New York, pp. 163–209.

Hay, J. C. 1977. Bibliography on Arthropoda and air pollution. *Forest Service Gen. Tech. Rep.* NE-24: 1–16.

Heagle, A. S. 1973. Interactions between air pollutants and plant parasites. *Annu. Rev. Phytopathol.* 11: 365–388.

Heck, W. W., R. H. Daines, and I. J. Hindawi. 1970. Other phytotoxic pollutants. *In* Jacobson, J. S., and A. C. Hill. (eds.), *Recognition of Air Pollution Injury to Vegetation: A Pictorial Atlas*. Air Pollution Control Association, Pittsburgh, PA, pp. F1–F24.

Heck, W. W., O. C. Taylor, and H. Heggestad. 1973. Air pollution research needs: herbaceous and ornamental plants and agriculturally generated pollutants. *J. Air Pollut. Control Assoc.* 23: 257–266.

Heck, W. W., R. B. Philbeck, and J. A. Dunning. 1978. A continuous stirred tank reactor (CSTR) system for exposing plants to gaseous air contaminants. *Agric. Res. Serv. U.S. Dept. of Agric.* 1–8.

Hedin, P. A., J. N. Jenkins, and F. G. Maxwell. 1977. Behavioral and developmental factors affecting host plant resistance to insects. *In* Hedin, P. A. (ed.), *Host Plant Resistance to Pests. Am. Chem. Soc. Symp. Ser.* 62, Washington, DC, pp. 231–275.

Heliovaara, K., and R. Vaisanen. 1986. Industrial air pollution and the pine bark bug, *Aradus cinnamomeus* Panz. (Het., Aradidae). *J. Appl. Entomol.* 101: 469–478.

Henderson, G. S., W. F. Harris, D. E. Todd Jr, and T. Grizzard. 1977. Quantity and chemistry of throughfall as influenced by forest-type and season. *J. Ecol.* 65: 365–374.

Hileman, B. 1981. Acid precipitation. *Environ. Sci. and Technol.* 15: 1119–1124.

Hill, A. C., H. E. Heggestad, and S. N. Linzon. 1970. Ozone. *In* Jacobson, J. S., and A. C. Hill. (eds.), *Recognition of Air Pollution Injury to Vegetation: A Pictorial Atlas*. Air Pollution Control Association, Pittsburgh, PA, pp. B1–B22.

Hillmann, R. C. 1972. Biological effects of air pollution on insects, emphasizing the reactions of the honey bee (*Apis mellifera* L.) to sulfur dioxide. Ph.D. dissertation. Pennsylvania State University, PA.

Hillmann, R. C., and A. W. Benton. 1972. Biological effects of air pollution on insects, emphasizing the reactions of the honey bee (*Apis mellifera* L.). *Elisha Mitchell Sci. Soc.* 88: 195.

Hinrichsen, D. 1986. Multiple pollutants and forest decline. *Ambio* 15: 258–265.

Hogsett, W. E., D. P. Ormrod, D. Olszyk, G. E. Taylor Jr., and D. T. Tingey. (eds.) 1987a. *Air Pollution Exposure Systems and Experimental Protocols: A Review and Evaluation of Performance*, Vol. 1. EPA Publ. (EPA/600/3-87/037a).

Hogsett, W. E., D. P. Ormrod, D. Olszyk, G. E. Taylor Jr., and D. T. Tingey. (eds.) 1987b. *Air Pollution Exposure Systems and Experimental Protocols. Appendices: Descriptions of Facilities*, Vol. 2. EPA Publ. (EPA/600/3-87/037b).

Hopkin, S. P., M. H. Martin, and S. J. Moss. 1985. Heavy metals in isopods from the supralittoral zone on the southern shore of the Severn Estuary, UK. *Environ. Pollut.* (Ser. B) 9: 239–254.

Howell, J. F. 1981. Codling moth: the effects of volcanic ash from the eruption of Mt. St. Helens on egg, larval, and adult survival. *Melandria* 37: 50–55.

Hughes, P. R. 1984. Effects of air pollution on plant-feeding insects, *In Illinois Climate: Trends, Impacts, and Issues. Proc. 12th Annu. ENR Conf.* Urbana-Champaign, Illinois, Dept. of Energy and Natural Resources Document No. 84/05, pp. 104–116.

Hughes, P. R., and J. A. Laurence. 1984. Relationship of biochemical effects of air pollutants on plants to environmental problems: insect and microbial interactions. *In* Koziol, M. J., and F. R. Whatley, (eds.), *Gaseous Air Pollutants and Plant Metabolism*, Butterworths, London, pp. 361–377.

Hughes, P. R., J. E. Potter, and L. H. Weinstein. 1981. Effects of air pollutants on plant–insect interactions: reactions of the Mexican bean beetle to SO_2-fumigated pinto beans. *Environ. Entomol.* 10: 741–744.

Hughes, P. R., J. E. Potter, and L. H. Weinstein. 1982. Effects of air pollution on plant–insect interactions: increased susceptibility of greenhouse-grown soybeans to the Mexican bean beetle after plant exposure to SO_2. *Environ. Entomol.* 11: 173–176.

Hughes, P. R., A. I. Dickie, and M. A. Penton. 1983. Increased success of Mexican bean beetle on field-grown soybeans exposed to sulfur dioxide. *J. Environ. Qual.* 12: 565–568.

Hughes, P. R., J. J. Chiment, and A. I. Dickie. 1985a. Effect of pollutant dose on the response of Mexican bean beetle (Coleoptera: Coccinellidae) to SO_2-induced changes in soybean. *Environ. Entomol.* 14: 718–721.

Hughes, P. R., L. H. Weinstein, L. M. Johnson, and A. R. Braun. 1985b. Fluoride transfer in the environment: Accumulation and effects on cabbage looper *Trichoplusia ni* of fluoride from water soluble salts and HF-fumigated leaves. *Environ. Pollut.* 37: 175–192.

Hughes, P. R., L. H. Weinstein, S. H. Wettlaufer, J. J. Chiment, G. J. Doss, T. W. Culliney, W. H. Gutenmann, C. A. Bache, and D. J. Lisk. 1987. Effect of fertilization with municipal sludge on the glutathione, polyamine, and cadmium content of cole crops and associated loopers. *J. Food Agric. Chem.* 35: 50–54.

Hurwitz, B., E. J. Pell, and R. T. Sherwood. 1979. Status of coumestrol and 4′,7-dihydroxyflavone in alfalfa foliage exposed to ozone. *Phytopathology* 69: 810–813.

Imai, S., and S. Sato. 1974. On the black spots observed in the integument of silkworms poisoned by fluorine compounds. *Japan. Soc. Air Pollut.* 9: 401.

Irving, P. M., and J. E. Miller. 1981. Productivity of field-grown soybeans exposed to acid rain and sulfur dioxide alone and in combination. *J. Environ. Qual.* 10: 473–478.

Irving, P. M., and J. E. Miller. 1984. Synergistic effect on field-grown soybeans from combinations of sulfur dioxide and nitrogen dioxide. *Can. J. Bot.* 62: 840–846.

Isman, M. B., and S. S. Duffey. 1982. Phenolic compounds in foliage of commercial tomato cultivars as growth inhibitors to the fruitworm, *Heliothis zea*. *J. Am. Soc. Hortic. Sci.* 107: 167–170.

Jager, H.-J., J. Bender, and L. Grunhage. 1985. Metabolic responses of plants differing in SO_2 sensitivity towards SO_2 fumigation. *Environ. Pollut.* 39: 317–335.

Jeffords, M. R., and A. G. Endress. 1984. Possible role of ozone in tree defoliation by the gypsy moth (Lepidoptera: Lymantriidae). *Environ. Entomol.* 13: 1249–1252.

Johansson, T. S. K., and M. P. Johansson. 1972. Sublethal doses of sodium fluoride affecting fecundity of confused flour beetles. *J. Econ. Entomol.* 65: 356–357.

Johnson, N. D., S. A. Brain, and P. R. Ehrlich. 1985. The role of leaf resin in the interaction between *Eriodictyon californicum* (Hydrophyllaceae) and its herbivore, *Trirhabda diducta* (Chrysomelidae). *Oecologia* (Berlin) 66: 106–110.

Jones, J. V., and E. J. Pell. 1981. The influence of ozone on the presence of isoflavones in alfalfa foliage. *J. Air Pollut. Control Assoc.* 31: 885–886.

Joosse, E. N. G., and J. B. Buker. 1979. Uptake and excretion of lead by litter-dwelling Collembola. *Environ. Pollut.* 18: 235–240.

Kennedy, J. S., and C. O. Booth. 1959. Responses of *Aphis fabae* Scop. to water shortage in host plants in the field. *Entomol. Exp. Appl.* 2: 1–11.

Kennedy, J. S., K. P. Lamb, and C. O. Booth. 1958. Responses of *Aphis fabae* Scop. to water shortage in host plants in pots. *Entomol. Exp. Appl.* 1: 274–291.

Kettle, A., G. Port, and A. Davison. 1983. The impact of fluoride pollution on invertebrates. *Bull. Brit. Ecol. Soc.* 14: 110–111.

Kettlewell, H. B. D. 1955. Selection experiments on industrial melanism in the lepidoptera. *Heredity* 9: 323–342.

Kettlewell, H. B. D. 1956. Further selection experiments on industrial melanism in the lepidoptera. *Heredity* 10: 287–301.

Kettlewell, H. B. D. 1958. A survey of the frequencies of *Biston betularia* (L.) (Lep.) and its melanic forms in Great Britain. *Heredity* 12: 51–73.

Kimmerer, T. W., and T. T. Kozlowski. 1982. Ethylene, ethane, acetaldehyde, and ethanol production by plants under stress. *Plant Physiol.* 69: 840–847.

Kish, L. P. 1984. Effect of Mount St. Helens ash on alfalfa leafcutting bee longevity and reproduction. *J. Idaho Acad. Sci.* 20: 11–17.

Klein, T. M., and M. Alexander. 1986. Effect of the quantity and duration of application of simulated acid precipitation on nitrogen mineralization and nitrification in a forest soil. *Water, Air, and Soil Pollut.* 28: 309–318.

Klostermeyer, E. C., L. D. Corpus, and C. L. Campbell. 1981. Population changes in arthropods in wheat following volcanic ash fall-out. *Melanderia* 37: 45–49.

Knowlton, G. F., A. P. Sturtevant, and C. J. Sorenson. 1950. Adult honey bee losses in Utah as related to arsenic poisoning. *Utah Agr. Exp. Sta.* 340: 5–30.

Kostir, J., I. Machackova, V. Jiracek, and E. Buchar. 1970. Effect of sulfur dioxide on the content of free sugars and amino acids in pea seedlings. *Experientia* 26: 604–605.

Koziol, M. J., and C. F. Jordan. 1978. Changes in carbohydrate levels in red kidney bean (*Phaseolus vulgaris* L.) exposed to sulphur dioxide. *J. Expl. Bot.* 29: 1037–1043.

Koziol, M. J., and F. R. Whatley. 1984. *Gaseous Air Pollutants and Plant Metabolism*. Butterworths, London.

Krizek, D. T. 1982. Guidelines for measuring and reporting environmental conditions in controlled-environment studies. *Physiol. Plant*. 56: 231–235.

Kunimi, Y., and H. Aruga. 1974. Susceptibility to infection with nuclear- and cytoplasmic-polyhedrosis viruses of the fall webworm, *Hyphantria cunea* Drury, reared on several artificial diets. *Jap. J. Appl. Entomol. Zool.* 18: 1–4.

Laurence, J. A. 1981. Effects of air pollutants on plant–pathogen interactions. *J. Plant Dis. and Prot.* 87: 156–172.

Laurence, J. A., D. C. MacLean, R. H. Mandl, R. E. Schneider, and K. S. Hansen. 1982. Field tests of a linear gradient system for exposure of row crops to SO_2 and HF. *Water, Air, and Soil Pollut.* 17: 299–407.

Laurence, J. A., P. R. Hughes, L. H. Weinstein, G. T. Geballe, and W. H. Smith. 1983. Impact of air pollution on plant–pest interactions: Implications of current research and strategy for future studies. *ERC Report No. 20*. Ecosystems Research Center, Ithaca, NY.

Lawton, J. H., and D. Schroder. 1977. Effects of plant type, size of geographical range and taxonomic isolation on number of insect species associated with British plants. *Nature* 265: 137–140.

Lechowicz, M. J. 1987. Resource allocation by plants under air pollution stress. *Bot. Rev.* 53: 281–300.

Lee, J. J., and R. A. Lewis. 1978. Zonal air pollution system: design and performance. *In* Preston, E. M., and R. A. Lewis (eds.), *The Bioenvironmental Impact of a Coal-Fired Power Plant. 3d Int. Rep.* EPA-600/3-78-021. Colstrip, MT, pp. 322–344.

Leetham, J. W., T. J. McNary and J. L. Dodd. 1979. Effects of controlled levels of SO_2 on invertebrate consumers. *In* Preston, E. M. and T. L. Gullett (eds.), *The Bioenvironmental Impact of a Coal-Fired Power Plant. 4th Int. Rep.* EPA-600/3-79-044. Colstrip, MT, pp. 723–763.

Lefohn, A. S., and D. P. Ormrod. 1984. A review and assessment of the effects of pollutant mixtures on vegetation—research recommendations. EPA-600/3-84-037.

Leigh, T. F., and P. F. Wynholds. 1980. Organophosphorus insecticides stimulated egg laying of mites reared on the treated cotton plants. *California Agriculture* 34: 14–15.

Levin, D. A. 1971. Plant phenolics: an ecological perspective. *Am. Nat.* 105: 157–181.

Levitt, J. 1972. *Responses of Plants to Environmental Stresses*. Academic Press, New York.

Levy, R., Y. J. Chiu, and H. L. Cromroy. 1972. Effects of ozone on three species of Diptera. *Environ. Entomol.* 1: 608–611.

Lewis, A. C. 1979. Feeding preference for diseased and wilted sunflower in the grasshopper, *Melanoplus differentialis*. *Entomol. Exp. Appl.* 26: 202–207.

L'Hirondelle, S. J., and P. A. Addison. 1985. Effects of SO_2 on leaf conductance, xylem tension, fructose and sulphur levels of jack pine seedlings. *Environ. Pollut.* 39: 373–386.

Lincoln, D. E., D. Couvet, and N. Sionit. 1986. Response of an insect to host plants grown in carbon dioxide enriched atmospheres. *Oecologia* (Berlin) 69: 556–560.

Lowery, D. T., and M. K. Sears. 1986. Effect of exposure to the insecticide azinphosmethyl on reproduction of green peach aphid (Homoptera: Aphididae). *J. Econ. Entomol.* 79: 1534–1538.

Luckey, T. D. 1959. Antibiotics in nutrition. *In* Goldberg, H. S. (ed.), *Antibiotics: Their Chemistry and Non-Medical Uses*. Van Nostrand, Princeton, NJ, pp. 173–321.

Malhotra, S. S., and R. A. Blauel. 1980. Diagnosis of air pollutant and natural stress symptoms on forest vegetation in western Canada. *Information Report NOR-X-228*. Environ. Canada, Can. For Serv., North. For. Res. Cent., Edmonton, Alberta.

Malhotra, S. S., and A. A. Khan. 1978. Effects of sulphur dioxide fumigation on lipid biosynthesis in pine needles. *Phytochemistry* 17: 241–244.

Manning, W. J. 1971. Effects of limestone dust on leaf condition, foliar disease incidence, and leaf surface microflora of native plants. *Environ. Pollut.* 2: 69–76.

Manning, W. J. 1975. Interactions between air pollutants and fungal, bacterial and viral plant pathogens. *Environ. Pollut.* 9: 87–90.

Mansfield, T. A. 1976. *Effects of Air Pollutants on Plants*. Cambridge Univ. P., Cambridge.

Maramorosch, K., and D. D. Jensen. 1963. Harmful and beneficial effects of plant viruses in insects. *Annu. Rev. Microbiol.* 17: 495–530.

Mattson, Jr., W. J. 1980. Herbivory in relation to plant nitrogen content. *Annu. Rev. Ecol. Syst.* 11: 119–161.

Maurizio, A., and M. Staub. 1956. Bienenvergiftungen mit fluorhaltigen Industrieabgasen in der Schweiz. *Schweizerische Bienen-Zeitung* 79: 476–486.

McCool, P. M., and J. A. Menge. 1983. Influence of ozone on carbon partitioning in tomato: potential role of carbon flow in regulation of the mycorrhizal symbiosis under conditions of stress. *New Phytol.* 94: 241–247.

McIntyre, J. L., J. A. Dodds, and J. D. Hare. 1980. Induced resistance in plants may protect from insects and pathogens. *Frontiers of Plant Sci.* 33: 4–5.

McLeod, A. R., J. E. Fackrell, and K. Alexander. 1985. Open-air fumigation of field crops: criteria and design for a new experimental system. *Atmospheric Environ.* 19: 1639–1649.

McNary, T. J., D. G. Milchunas, J. W. Leetham, W. K. Lauenroth, and J. L. Dodd. 1981. Effect of controlled low levels of SO_2 on grasshopper densities on a northern mixed-grass prairie. *J. Econ. Entomol.* 74: 91–93.

Meudt, W. J. 1971. Interactions of sulfite and manganous ion with peroxidase oxidation products of indole-3-acetic acid. *Phytochemistry* 10: 2103–2109.

Mikkola, K. 1975. Frequencies of melanic forms of *Oligia* moths (Lepidoptera, Noctuidae) as a measure of atmospheric pollution in Finland. *Ann. Zool. Fennici* 12: 197–204.

Mishra, L. C. 1982. Effect of environmental pollution on the morphology and leaf epidermis of *Commelina bengalensis* Linn. *Environ. Pollut.* 28: 281–284.

Mueller, P. W., and S. G. Garsed. 1984. Microprocessor-controlled system for exposing plants to fluctuating concentrations of sulphur dioxide. *New Phytol.* 97: 165–173.

Musselman, R. C., P. M. McCool, R. J. Oshima, and R. R. Teso. 1986. Field chambers for assessing crop loss from air pollutants. *J. Environ. Qual.* 15: 153–157.

National Academy of Sciences (NAS). 1971. Fluorides. Committee on Biologic Effects of Atmospheric Pollutants, Washington, DC.

National Academy of Sciences (NAS). 1977. Nitrogen Oxides. Commitee on Medical and Biologic Effects of Environmental Pollutants, Washington, DC.

National Academy of Sciences (NAS). 1978. Sulfur Oxides. Committee on Sulfur Oxides, Washington, DC.

Nyborg, M. 1978. Sulfur pollution and soils. *In* Nriagu, J. O. (ed.), *Sulfur in the Environment. Part II. Ecological Impacts*. Wiley, New York, pp. 359–390.

Olszyk, D. M., T. W. Tibbitts, and W. M. Hertzberg. 1980. Environment in open-top field chambers utilized for air pollution studies. *J. Environ. Qual.* 9: 610–615.

Osbrink, W. L. A., J. T. Trumble, and R. E. Wagner. 1987. Host suitability of *Phaseolus lunata* for *Trichoplusia ni* (Lepidoptera: Noctuidae) in controlled carbon dioxide atmospheres. *Environ. Entomol.* 16: 639–644.

Owen, D. F. 1961. Industrial melanism in North American moths. *Am. Nat.* 95: 227–233.

Pandey, S. N. 1983. Impact of thermal power plant emissions on vegetation and soil. *Water, Air, and Soil Pollut.* 19: 87–100.

Pasteels, J. M. 1978. Apterous and brachypterous coccinellids at the end of the food chain, *Cionura erecta* (Asclepiadaceae)-*Aphis nerii*. *Entomol. Exp. Appl.* 24: 579–584.

Pell, E. J., W. C. Weissberger, and J. J. Speroni. 1980. Impact of ozone on quantity and quality of greenhouse-grown potato plants. *Environ. Sci. and Technol.* 14: 568–571.

Perkins, H. C. 1974. *Air Pollution.* McGraw-Hill, New York.

Petters, R. M., and R. V. Mettus. 1982. Reproductive performance of *Bracon hebetor* females following acute exposure to sulphur dioxide in air. *Environ. Pollut.* 27: 155–163.

Port, G. R., and C. Hooton. 1982. Some effects of pollution on roadside fauna. In Visser, J. H., and A. K. Minks (eds.), *Proc. 5th Intl. Symp. Insect–Plant Relationships*, Pudoc, Wageningen, pp. 449–450.

Port, G. R., and J. R. Thompson. 1980. Outbreaks of insect herbivores on plants along motorways in the United Kingdom. *J. Appl. Ecol.* 17: 649–656.

Pratt, C. R., R. S. Sikorski. 1982. Lead content of wildflowers and honey bees (*Apis mellifera*) along a roadway: possible contamination of a simple food chain. *Proc. Pennsylvania Acad. Sci.*, pp. 151–152.

Price, P. W. 1983. Hypotheses on organization and evolution in herbivorous insect communities. *In* Denno, R. F., and M. S. McClure (eds.), *Variable Plants and Herbivores in Natural and Managed Systems.* Academic Press, New York, pp. 559–596.

Priebe, A., H. Klein, and H.-J. Jager. 1978. Role of polyamines in SO_2-polluted pea plants. *J. Expl. Bot.* 29: 1045–1050.

Przybylski, Z. 1967. Effect of gases and vapors of SO_2, SO_3, and H_2SO_4 on fruit trees and certain harmful insects. *Postepy Nauk Roln.* 14: 111–118.

Przybylski, Z. 1968. Results of consecutive observations of effect of SO_2, SO_3, and H_2SO_4 gases and vapors on trees, shrubs, and entomofauna of orchards in the vicinity of sulfur mines and sulfur processing plant in Machow. *Postepy Nauk Roln.* 15: 131–138.

Przybylski, Z. 1974. Resultats D'observations relatives a l'influence des gaz et vapeurs de soufre sur les arbres fruitiers, arbustes et arthropodes aux alentours des usines et mines de soufre dans la region de Tarnobrzeg. *Environ. Pollut.* 6: 67–74.

Przybylski, Z. 1979. The effects of automobile exhaust gases on the arthropods of cultivated plants, meadows and orchards. *Environ. Pollut.* 19: 157–161.

Ramoska, W. A., and T. Todd. 1985. Variation in efficacy and viability of *Beauveria bassiana* in the chinch bug (Hemiptera: Lygaeidae) as a result of feeding activity on selected host plants. *Environ. Entomol.* 14: 146–148.

Rao, J. L. N., and V. R. Rao. 1986. Nitrogen fixation (C_2H_2 reduction) as influenced by sulphate in paddy soils. *J. Agric. Sci., Camb.* 106: 331–336.

Reese, J. C. 1978. Chronic effects of plant allelochemics on insect nutritional physiology. *Entomol. Exp. Appl.* 24: 425–431.

Reich, P. B., A. W. Schoettle, H. F. Stroo, J. Troiano, and R. G. Amundson. 1985. Effects of O_3, SO_2, and acidic rain on mycorrhizal infection in northern red oak seedlings. *Can. J. Bot.* 63: 2049–2055.

Reich, P. B., A. W. Schoettle, H. F. Stroo, and R. G. Amundson. 1986. Acid rain and ozone influence mycorrhizal infection in tree seedlings. *J. Air Pollut. Control Assoc.* 36: 724–726.

Reid, Collins and Associates Ltd. 1976. Fluoride emissions and forest growth. Report to Aluminum Company of Canada Ltd, Vancouver, BC.

Reinert, R. A. 1984. Plant reponse to air pollutant mixtures. *Annu. Rev. Phytopathol.* 22: 421–442.

Renwick, J. A. A., and J. Potter. 1981. Effects of sulfur dioxide on volatile terpene emission from balsam fir. *J. Air Pollut. Control Assoc.* 31: 65–66.

Ribbands, C. R. 1950. Changes in the behaviour of honey-bees following their recovery from anaesthesia. *J. Exptl. Biol.* 27: 302–310.

Rice, P. M., L. H. Pye, R. Boldi, J. O'Loughlin, P. C. Tourangeau, and C. C. Gordon. 1979. The effects of "low level SO_2" exposure on sulfur accumulation and various plant life responses of some major grassland species on the ZAPS sites. *In* Preston, E. M. and T. L. Gullett (eds.), *The Bioenvironmental Impact of a Coal-Fired Power Plant. 4th Int. Rep.* EPA-600/3: 79-044. Colstrip, MT, pp. 494–591.

Rogers, H. H., H. E. Jeffries, E. P. Stahel, W. W. Heck, L. A. Ripperton, and A. M. Witherspoon. 1977. Measuring air pollutant uptake by plants: a direct kinetic technique. *J. Air Pollut. Control Assoc.* 27: 1192–1197.

Rorison, I. H. 1980. The effects of soil acidity on nutrient availability and plant response. *In* Hutchinson, T. C., and M. Havas (eds.), *Effects of Acid Precipitation on Terrestrial Ecosystems*. Plenum, New York, pp. 283–303.

Rowland, A. J. 1986. Nitrogen uptake, assimilation and transport in barley in the presence of atmospheric nitrogen dioxide. *Plant and Soil* 91: 353–356.

Sachs, M. M., and T. H. D. Ho. 1986. Alteration of gene expression during environmental stress in plants. *Annu. Rev. Plant Physiol.* 37: 363–376.

Sawada, S., and T. Totsuka. 1986. Natural and anthropogenic sources and fate of atmospheric ethylene. *Atmospheric Environ.* 20: 821–832.

Schopf, R., C. Mignat, and P. Hedden. 1982. As to the food quality of spruce needles for forest damaging insects. *Z. Angew. Entomol.* 93: 244–257.

Scriber, J. M. 1978. The effects of larval feeding specialization and plant growth form on the consumption and utilization of plant biomass and nitrogen: an ecological consideration. *Entomol. Exp. Appl.* 24: 494–510.

Scriber, J. M. and F. Slansky, Jr. 1981. The nutritional ecology of immature insects. *Annu. Rev. Entomol.* 26: 183–211.

Seigler, D. S., and P. W. Price. 1976. Secondary compounds in plants: primary functions. *Am. Nat.* 110: 101–105.

Shevyakova, N. I. and A. P. Loshadkina. 1965. Variation of the sulfhydryl group content in plants under conditions of salinization. *Soviet Plant Physiol.* 12: 280–286.

Shriner, D. S. 1974. Effects of simulated rain acidified with sulfuric acid on host–parasite interactions. Ph.D. dissertation. North Carolina State University, Raleigh.

Shriner, D. S. 1980. Vegetation surfaces: a platform for pollutant/parasite interactions. *In* Toribara, T. Y., M. W. Miller, and P. E. Morrow (eds.), *Polluted Rain. Enviromental Science Research*, Vol. 17. Plenum, New York, pp. 259–272.

Shriner, D. S., and J. W. Johnston. 1981. Effects of simulated, acidified rain on nodulation of leguminous plants by *Rhizobium* spp. *Environ. Expl. Bot.* 21: 199–209.

Sierpinski, Z. 1985. Luftverunreinigungen und Forstschadlinge. *Z. Angew. Entomol.* 99: 1–6.

Smidt, V. S. 1984. Begasungsversuche mit SO_2 und Ozon an jungen Fichten. *Eur. J. For. Pathol.* 14: 241–248.

Smith, D. A., and S. W. Banks. 1986. Biosynthesis, elicitation and biological activity of isoflavonoid phytoalexins. *Phytochemistry* 25: 979–995.

Smith, W. W. 1975. Depressed litter decomposition. *In* Smith, W. H., and L. S. Dochinger (eds.), *Air Pollution and Metropolitan Woody Vegetation*. Pinchot Inst., Consortium for Environmental Forestry Studies. USDA Forest Service, Delaware, OH, pp. 23–24.

Städler, E. 1974. Host plant stimuli affecting oviposition behavior of the eastern spruce budworm. *Entomol. Exp. Appl.* 17: 176–188.

Strojan, C. L. 1978. The impact of zinc smelter emissions on forest litter arthropods. *Oikos* 31: 41–46.

Stroo, H. F., P. B. Reich, A. W. Schoettle, and R. G. Amundson. 1988. Effects of ozone and acid rain in white pine (*Pinus strobus* L.) seedlings grown in five soils. II. Mycorrhizal infection. *Can. J. Bot.* (in press).

Sumimoto, M., M. Shiraga, and T. Kondo. 1975. Ethane in pine needles preventing the feeding of the beetle, *Monochamus alternatus*. *J. Insect Physiol.* 21: 713–722.

Taylor, O. C., and D. C. MacLean. 1970. Nitrogen oxides and the peroxyacyl nitrates. *In* Jacobson, J. S., and A. C. Hill (eds.), *Recognition of Air Pollution Injury to Vegetation: A Pictorial Atlas*. Air Pollution Control Association, Pittsburgh, PA, pp. E1–E14.

Thalenhorst, W. 1974. Untersuchungen über den Einfluss fluorhaltiger Abgase auf die Disposition der Fichte für den Befall durch die Gallenlaus *Sacchiphantes abietis* (L.). *Z. Pflanzenkrankh. Pflanzenschutz* 81: 717–727.

Thomas, M. D., R. H. Hendricks, T. R. Collier and G. R. Hill. 1943. The utilization of sulphate and sulphur dioxide for the sulphur nutrition of alfalfa. *Plant Physiol.* 18: 345–371.

Thompson, C. R. and D. M. Olszyk. 1985. A field air-exclusion system for measuring the effects of air pollutants on crops. Electric Power Research Institute Report EPRI EA-4203 RP 1908-3.

Tomlinson, H., and S. Rich. 1967. Metabolic changes in free amino acids of bean leaves exposed to ozone. *Phytopathology* 57: 972–974.

Treshow, M., and M. R. Pack. 1970. Fluoride. *In* Jacobson, J. S., and A. C. Hill (eds.), *Recognition of Air Pollution Injury to Vegetation: A Pictorial Atlas*. Air Pollution Control Association, Pittsburgh, PA, pp. D1–D17.

Troiano, J. J., and I. A. Leone. 1977. Changes in growth rate and nitrogen content of tomato plants after exposure to NO_2. *Phytopathology* 67: 1130–1133.

Troiano, J., L. Colavito, L. Heller, and D. C. McCune. 1982. Viability, vigor, and maturity of seed harvested from two soybean cultivars exposed to simulated acidic rain and photochemical oxidants. *Agric. and Environ.* 7: 275–283.

Troiano, J., L. Colavito, L. Heller, D. C. McCune, and J. S. Jacobson. 1983. Effects of acidity of simulated rain and its joint action with ambient ozone on measures of biomass and yield in soybean. *Environ. Expl. Bot.* 23: 113–119.

Trumble, J. T., J. D. Hare, R. C. Musselman, and P. M. McCool. 1987. Ozone-induced changes in host-plant suitability: interactions of *Keiferia lycopersicella* and *Lycopersicon esculentum*. *J. Chem. Ecol.* 13: 203–218.

van Noordwijk, M., and K. Hairiah. 1986. Mycorrhizal infection in relation to soil pH and soil phosphorus content in a rain forest of northern Sumatra. *Plant and Soil* 96: 299–302.

Wagner, G. J. 1984. Characterization of a cadmium-binding complex of cabbage leaves. *Plant Physiol.* 76: 797–805.

Wearing, C. H., and H. F. van Emden. 1967. Studies on the relations of insect and host plant. *Nature* 213: 1051–1052.

Weinstein, L. H. 1961. Effects of atmospheric fluoride on metabolic constituents of tomato and bean leaves. *Contrib. Boyce Thompson Inst.* 21: 215–231.

Weinstein, L. H. 1977. Fluoride and Plant Life. *J. Occupational Med.* 19: 49–78.

Weinstein, L. H., and R. Alscher-Herman. 1982. Physiological effects of fluoride on higher plants. In Unsworth, M. H., and D. P. Ormrod (eds.), *Effects of Gaseous Air Pollution in Agriculture and Horticulture*. Butterworths, London, pp. 139–165.

Weinstein, L. H., and H. W. F. Bunce. 1981. Impact of emissions from an alumina reduction smelter on the forests at Kitimat, B. C.: A synoptic view. *Air Pollut. Control Assoc.*, 74th Annu. Meeting.

Weinstein, L. H., D. C. McCune, J. F. Mancini, and P. van Leuken. 1973. Effects of hydrogen fluoride fumigation of bean plants on the growth, development, and reproduction of the Mexican bean beetle. *Proc. 3rd Intl. Clean Air Cong.*, Düsseldorf, pp. A150–A153.

Weismann, L., and L. Svatarakova. 1973. Influence of sodium fluoride on behaviour of caterpillars *Scotia segetum* Den. et Schiff. *Biologia* (Bratislava) 28: 105–109.

Weismann, L., and L. Svatarakova. 1974. Toxicity of sodium fluoride on some species of harmful insects. *Biologia* (Bratislava) 29: 847–852.

Weismann, L., and L. Svatarakova. 1975. Paralyzujuci ucinok $CaCl_2$ na toxicke posobenie NaF pre larvy *Leptinotarsa decemlineata* Say. *Biologia* (Bratislava) 30: 841–845.

Weismann, L., and L. Svatarakova. 1976. Toxickost fluoridu sodneho pre roztocca *Tetranychus urticae* Koch. *Biologia* (Bratislava) 31: 125–132.

Wood, D. L. 1982. The role of pheromones, kairomones, and allomones in the host selection and colonization behavior of bark beetles. *Annu. Rev. Entomol.* 27: 411–446.

Wood, T., and F. H. Bormann. 1974. The effects of an artificial acid mist upon the growth of *Betula alleghaniensis* Britt. *Environ. Pollut.* 7: 259–268.

Yang, S. F., and G. W. Miller. 1963a. Biochemical studies on the effect of fluoride on higher plants. I. Metabolism of carbohydrates, organic acids and amino acids. *Biochem. J.* 88: 505–509.

Yang, S. F., and G. W. Miller. 1963b. Biochemical studies on the effect of fluoride on higher plants. III. The effect of fluoride on dark carbon dioxide fixation. *Biochem. J.* 88: 517–522.

Yang, S. F., and M. A. Saleh. 1973. Destruction of indole-3-acetic acid during the aerobic oxidation of sulfite. *Phytochemistry* 12: 1463–1466.

Yeh, R., D. Hemphill, Jr., and H. M. Sell. 1971. The effects of sodium bisulfite, manganous chloride and 2,4-dichlorophenol on peroxidase-catalyzed oxidation of indole-3-acetaldehyde. *Can. J. Biochem.* 49: 162–165.

Zedler, B., R. Plarre, and G. M. Rothe. 1986. Impact of atmospheric pollution on the protein and amino acid metabolism of spruce *Picea abies* trees. *Environ. Pollut.* 40: 193–212.

Ziegler, I. 1973. The effect of air-polluting gases on plant metabolism. *Environ. Qual. Saf.* 2: 182–208.

9

EFFECTS OF MECHANICAL DAMAGE TO PLANTS ON INSECT POPULATIONS

C. Michael Smith

Department of Entomology
Louisiana State University Agricultural Center
Baton Rouge, Louisiana

1. INTRODUCTION

Green plants exist in an environment containing many different types of stresses, of which herbivores generally comprise only a small fraction. Mechanically disrupting plant tissues—whether by shaking, rubbing, or wounding by insect feeding—can affect plant growth and the development of associated insect populations. Abiotic stresses such as atmospheric dust, lightning, and wind can adversely affect plant growth and indirectly affect insects feeding on these plants.

Research in the areas of entomology, plant pathology, and plant physiology has progressed independently for several years, but recent studies indicate similarities in the response of plants to pathogens, insects, and various abiotic stresses. The intent of this chapter is to evaluate plant responses to various types of mechanical damage and to determine the effect of these responses on populations of insects associated with damaged plants.

2. EFFECTS OF PLANT WOUNDING ON INSECT POPULATIONS

2.1 Increased Plant Resistance to Insect Attack

The phenomenon of increased or induced plant resistance to insects is currently under more intense study than in the past. Kogan and Paxton (1983) define induced resistance as the "quantitative or qualitative enhancement of a plant's defense mechanism against pests in response to extrinsic physical or chemical stimuli." Previous reviews (Rhoades 1979, Edwards and Wratten 1983) note that wound-induced responses in plants resulting from mechanical damage may be part of a general defensive reaction, since some wound-evolved phytochemicals are similar to those developed during pathogen infection (Crute et al. 1985). Edwards and Wratten (1983) classed the responses of plant tissues to wounding as (1) cellular chemical changes, (2) changes in cells adjacent to the damaged tissue, and (3) generalized changes apparent in a plant part or the entire plant. Similar classifications exist for plant responses to pathogen invasion (De Wit 1985). For purposes of this review, wounding will refer to plant tissue abrasion, cutting, or tearing resulting from mechanical damage or from insect feeding that causes similar injury symptoms.

Within plants, the wounding process is marked by rapid, large-scale changes in protein (Davies and Schuster 1981), lipid (Galliard 1978), and phenol (Rhodes and Wooltorton 1978) metabolism. Damage to some plants also causes formation of the wound hormone "traumatic" acid (English and Bonner 1937) which induces cellular proliferation in castor bean (Scott et al. 1961) and tumor formation in tomato (Treshow 1955).

Damage to food plants adversely affects insects feeding on the plants (Table 1). Mechanical abrasion of cotton cotyledons and feeding by the two-spotted spider mite, *Tetranychus urticae* Koch, induce reductions in the populations of *T. urticae* feeding on cotton (Fig. 1) (Karban and Carey 1984, Karban 1985). Damaging soybean leaves by perforation with a metal file and wooden dowel (Fig. 2) has similar effects on larvae of the soybean looper, *Pseudoplusia includens* (Walker). Larvae-fed damaged leaves suffer significant reductions in growth rate (Reynolds and Smith 1985), weight, and increased mortality (Smith 1985). Damage to leaves of squash, *Cucurbita mos-*

Table 1. Occurrence, Time to Induction, and Duration of Wound-Induced Resistance in Plants to Insects

Plant	Method of Induction[a]	Induced Resistance: Time to Induction	Induced Resistance: Duration	Arthropods Affected	References
Alder	I	27 days	44 days	*Malacosoma californicum pluviale*	Rhoades 1983
Apple	M	35 days	?	*Tetranychus urticae*	Ferree and Hall 1981
Beet	I	24 days	42 days	*Pegomya betae*	Rottger and Klingauf 1976
Birch	I	1 yr	1–3 yrs	*Lymantria dispar*	Wallner and Walton 1979
	M	2 days	2 yrs	*Oporinia autumnata*	Haukioja and Niemela 1977
	I	1 yr	2–3 yrs	*Rheumaptera hastata*	Werner 1979
	I	$\frac{1}{4}$ days	60 days	*Spodoptera littoralis*	Wratten et al. 1984
Cotton	I	2 days	28 days	*Heliothis* spp.	Guerra 1981
	I,M	33 days	?	*Tetranychus urticae*	Karban and Carey 1984, Karban 1985
Maple	M	3 days	?	Lepidoptera larvae	Baldwin and Schultz 1983
Oak	M	30–60 days	1 yr	*L. dispar*	Schultz and Baldwin 1982
	I	40 days	120 days	*Phyllonorycter* spp.	West 1985
Pine	M	7 days	200 days	*Dendroctonus frontalis*	Nebeker and Hodges 1983
	I	?	1 yr	*Neodipridon sertifer*	Thielges 1968
Poplar	M	1 day	?	Lepidoptera larvae	Baldwin and Schultz 1983
Soybean	M	1 day	30 days	*Pseudoplusia includens*	Reynolds and Smith 1985, Smith 1985
Tomato	M	2 days	7 days	*S. littoralis*	Edwards et al. 1985
Willow	I,M	4–7 days	90 days	*Plagiodera versicolora*	Raupp and Denno 1984
	I	35 days	?	*Hyphantria cunea*	Rhoades 1983
	I	11 days	>15 days	*M. c. pluviale*	Rhoades 1983

[a] I = insect induced; M = mechanically induced.

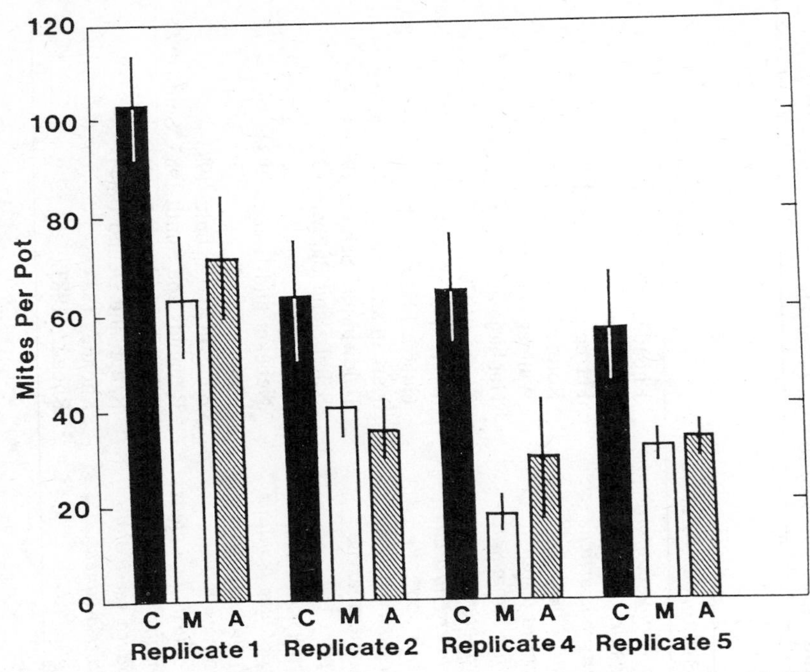

Figure 1. Induction of resistance in cotton to *T. urticae*. Mean numbers of mites of all stages per pot and standard errors. Treatments: C = controls; M = previously exposed to mites; A = mechanically abraded. (From Karban 1985. Reproduced with permission of Dr. W. Junk Publishers.)

chata, causes the movement of defensive substances to the damaged areas within only 40 minutes, and inhibits feeding by the coccinellid *Epilachna tredecimnotata* Latreille (Carroll and Hoffman 1980). In this interaction however, the insect has evolved the ability to carve out a circular "food patch" in the host plant leaf, which is cut off from the supply of plant defenses.

Phytochemicals produced by damaged plants may be detrimental to insects. Green and Ryan (1972) first discovered that mechanically wounding tomato leaves stimulates the release of a proteinase inhibitor, inducing factor (PIIF) into the vascular transport system of damaged plants. PIIF initiates the accumulation of proteinase (chymotrypsin) inhibitors (Ryan 1974) and decreases the production of protoplasts (Walker-Simmons et al. 1984). This primitive immune response occurs in most flowering crop plants (Walker-Simmons and Ryan 1977). Though PIIF itself has not been bioassayed, feeding of *Spodoptera littoralis* (Walker) larvae on damaged tomato leaves decreases by ninefold within eight hours after damage, and within 24 hours, leaves adjacent to initially damaged leaves promote similar adverse effects on larval feeding (Edwards et al. 1985). Broadway et al. (1986) demonstrated

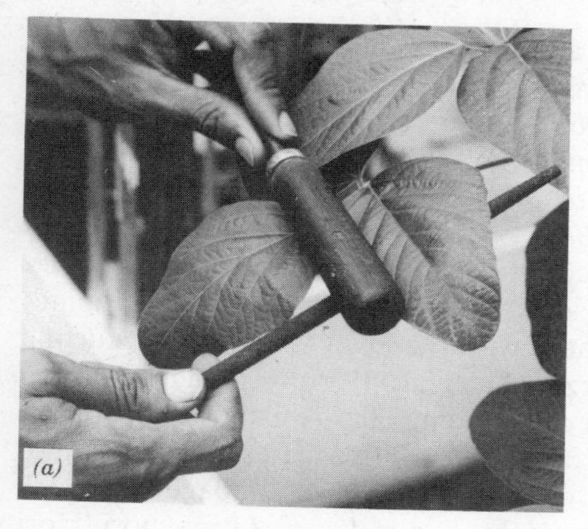

Figure 2. (*A*) Mechanical wounding of soybean leaves to induce resistance to *Pseudoplusia includens*. (*B*) Wounded leaves.

that feeding of the beet armyworm, *Spodoptera exigua* (Hubner), on tomato foliage caused a decline in the quality of foliage and a related reduction in the weight of larvae feeding on the foliage of insect-damaged plants. Tomato proteinase inhibitors, induced by herbivory, were strongly implicated as antiherbivore allelochemicals.

Phenols and associated compounds increase greatly in concentration around the damaged tissues of many plants when cellular lipids are broken down (Sucoff et al. 1967). Rhodes and Wooltorton (1978) note that wounding induces the oxidation of preformed endogenous phenols, which are further oxidized to produce toxic quinones, or the synthesis of either mono- or diphenols, which act as defensive compounds.

Miles (1968) suggests a chemically based interaction between plant feeding bugs, which inject polyphenoloxidases into host tissues, and the quinone-producing ability of wounded plant tissues. If polyphenoloxidase activity is high, plant quinones are oxidized, cell lysis occurs, and the plant becomes susceptible. However, if plant quinone content is high, a hypersensitive reaction occurs and insect feeding attempts are deterred. Such reactions occur in response to insect wounding of insect resistant cultivars of apple (Bramstedt 1938), grape (Henke 1963), spruce (Rohfritsch 1981, Tjia and Houston 1975), and *Solanum dulcamara* (Westphal et al. 1981).

Larvae of the beet fly, *Pegomya betae* Curtis, suffer increased mortality during development on sugar beet, *Beta vulgaris*, plants previously infested by beet flies. Mortality is positively correlated to polyphenoloxidase activity (Rottger and Klingauf 1976). Host selection by the bird cherry-oat aphid, *Rhopalosiphum padi* (L.), is associated with the level of phenols occurring in different wheat cultivars (Leszczynski et al. 1985). Aphid infestation causes an increase in the monophenol content of a resistant cultivar but increases polyphenol content in a susceptible cultivar (Leszczynski 1985). Phenolic levels also increase after damage to Chinese cabbage, *Brassica chinensis*, and sugar beet, *Beta vulgaris*, by *Lygus disponsi* Linavuori (Hori 1973, Hori and Atalay 1980), and following damage to cotton, *Gossypium hirsutum* L., by larvae of the cotton bollworm, *Heliothis zea* (Boddie) (Guerra 1981). Of 15 compounds detected in damaged cotton foliage, three have antibiotic effects on *H. zea* larvae (Guerra 1981).

Mechanical damage to other crop plants also stimulates enzymatic reactions that release phytochemicals detrimental to insects. Cyanide, released as hydrogen cyanide in damaged tissues of sorghum and cassava, is responsible for resistance to *Locusta* spp. (Bernays et al. 1977, Woodhead and Bernays 1977). DIMBOA (2,4-dihydroxy-7-methoxy-1-4-(2*H*) benzoxazin-3-one), an allelochemic in corn foliage conferring resistance to the European corn borer, *Ostrinia nubilalis* (Hubner) (Klun et al. 1967), is produced after tissue damage and release of DIMBOA from an inactive glycosidic form (Wahlroos and Virtanen 1959). Damage to roots of sweet potato by feeding of the sweet potato weevil, *Cylas formicarius* (F.), produces tissue necroses that give rise to the allelochemical ipomeamarone (Akazawa et al. 1960). Necrotic tissues promote the formation of furano-terpenoids (Uritani et al. 1975) and the coumarin derivatives scopoletin and umbelliferone. Although these compounds have phytoalexin properties, their insecticidal effects are unknown.

Several species of trees acquire induced resistance which adversely affects arthropod pests after defoliation or mechanical damage (Table 1). Resistance is induced within six to 48 hours and may last for from two months to several years (Baldwin and Schultz 1983, Benz 1974, Niemela et al. 1979, Wratten et al. 1984).

Scoring (cutting) of the outer bark of apple, *Malus domestica* Borkh., trees decreases populations of *T. urticae* feeding on foliage of scored trees (Ferree and Hall 1981). Development of the spear-marked black moth, *Rheumaptera hastata* (L.), decreases when larvae are fed foliage from paper birch trees, *Betula papyrifera* Marsh., which have been previously defoliated (Werner 1979). Larvae of the geometrid, *Oporinia autumnata* (Bkh.), fed damaged leaves of *B. pubescens* Ladeb. suffer delayed development (Haukioja and Niemala (1977) as do larvae of the gypsy moth, *Lymantria dispar* (L.), fed damaged leaves of *B. populifolia* Marsh (Wallner and Walton 1979). Similarly, previously damaged foliage of *B. pubescens* and *B. pendula* is much less palatable to *Spodoptera littoralis* (Walker) larvae than undamaged foliage (Wratten et al. 1984).

The same phenomenon exists for *L. dispar* larvae feeding on foliage from previously defoliated red oak, *Quercus rubra* L., (Schultz and Baldwin 1982) and black oak, *Q. velutina* Lam. (Wallner and Walton 1979). Survivorship and development of the lepidopteran leaf miner, *Phyllonorycter harrisella* (L.), is also reduced in proportion to the defoliation of the oak, *Q. robur*, on which it feeds (West 1985). Removal of the bark (bole scarring) from the trunk of the loblolly pine, *Pinus taeda* L., increases resistance to attack by the southern pine beetle, *Dendroctonus frontalis* Zimmerman, for several months (Nebeker and Hodges 1983). Larvae of the imported willow leaf beetle, *Plagiodera versicolora* (Laich.), fed damaged leaves of *Salix alba* L. and *S. babylonica* L. develop more slowly than those fed normal leaves (Raupp and Denno 1984). Similarly, fall webworms, *Hyphantria cunea* (Drury), fed damaged Sitka willow, *Salix sitchensis* Sanson, leaves develop more slowly than those fed normal leaves (Rhoades 1983).

As discussed previously, phenol concentration of tree foliage increases markedly after damage, and partially explains retarded growth in larvae fed damaged birch, maple, and poplar leaves (Baldwin and Schultz 1983, Niemela et al. 1979, Wratten et al. 1984). Similar chemical changes occur in leaves both distal and proximal to damaged leaves (Haukioja and Niemela 1979, Wratten et al. 1984). Foliage of *Pinus sylvestris* L. attacked by the European pine sawfly, *Neodiprion sertifer* (Geoff), has elevated concentrations of polyphenols, compared to foliage of nonattacked trees (Thielges 1968). Similarly, *Pinus radiata* D. Don trees wounded by the wasp, *Sirex noctilio* (F.), produce more phenols at a higher rate than unwounded trees (Shain and Hillis 1972).

Damaged sugar maple and poplar trees elevate leaf phenol levels in their own leaves and cause increased phenol production in leaves of adjacent, nonconnected trees of the same species, suggesting aerial communication between trees regarding leaf wounding (Baldwin and Schultz 1983, Perry and Pitman 1983, Rhoades 1983). Young leaves are more easily induced to increase phenol content after reception of airborne stimuli, and damage to young leaves stimulates greater phenol production in neighboring foliage than damage to older leaves (Baldwin and Schultz 1984).

Tuomi et al. (1984) hypothesize that the responses induced by mechanical damage are actually by-products of plant metabolic events which rearrange plant carbon/nitrogen ratios, due to the nutrient stress caused by defoliation. When nutrients from *B. pubescens* trees are removed by defoliation in nutrient-poor soils, higher concentrations of carbon-based allelochemicals (phenols) are produced. Depending on the rate of nutrient stress, the C/N ratio returns to predamage levels over a period of one to several years (Haukioja 1982). The return may require several years in trees with poor nutrient uptake or in trees growing in nutrient poor soils.

2.2 Increased Plant Susceptibility

Generally, insect defoliation to crop plants has more adverse effects on yield when it occurs during the reproductive stage than if inflicted during the vegetative stage. Studies with tomato (Wolk et al. 1983), sugar beet (Dunning and Winder 1972, Capinera 1979), apple (Howell 1978), rice (Bowling 1978, Navas 1976), soybean (Camery and Weber 1953), and wheat (Capinera and Roltsch 1980) indicate that crop yields are greater when defoliation occurs during the vegetative stage than during the fruiting or reproductive stage of development. For insects feeding on vegetative tissues, heavy early season defoliation may be resource limiting, as in the case of armyworms feeding on various Graminae, whereas lighter infestations will ensure development of later generations. For insects feeding on reproductive tissues, defoliation of the host plant during the vegetative stage can be highly detrimental, because their food supply is reduced or eliminated.

The type of defoliation incurred by plant tissues also affects arthropod populations. Severing the main veins of apple leaves causes greater reduction in photosynthesis than damage to interleaf veins (Hall and Ferree 1975). The net result of such defoliation would be a reduction of two-spotted spider mite populations on apple, since this arthropod utilizes leaf photosynthate during feeding.

Mechanical damage to plant tissue increases plant susceptibility to some insects. Lewis (1984) observed perferential feeding by the grasshopper, *Melanoplus differentialis* (Thomas), for damaged leaves or inflorescences of the wild sunflower, *Helianthus annuus* L. Defoliation of grand fir, *Abies grandis* Douglas (Lindley), elicits reductions in tree monoterpene content which predisposes trees to attack by the fir engraver beetle, *Scolytus ventralis* LeConte (Wright et al. 1979). Root pruning of loblolly pine by mechanical damage during timber harvest, or rooting damage by wild pigs, temporarily increase susceptibility to the southern pine beetle (Hetrick 1949). Blanche et al. (1985) indicate that root wounding reduces the total resin flow in pine trees, making them less able to impede beetle attacks. Similarly, extreme damage to the bark of *P. radiata* causes enhanced susceptibility to attack by *S. noctilio* (Coutts and Dolezal 1966). Defoliation to sugar maple trees causes changes in root chemical content which favors development of the root fungus, *Armillaria mellea* (Wargo 1972).

3. EFFECTS OF PLANT THIGMOMORPHOGENESIS ON INSECT POPULATIONS

Jaffee (1973) defined thigmomorphogenesis as plant growth retardation resulting from rubbing or bending of plant parts. Plants with rubbed inter-

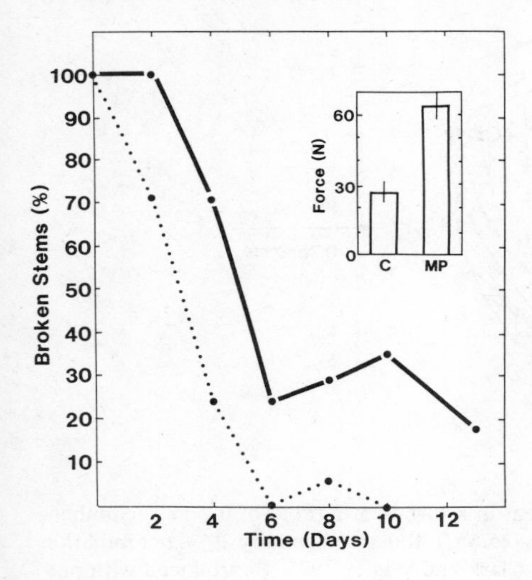

Figure 3. Effects of previous mechanical perturbation on mechanical stress-induced stem rupture in beans (main graph) and pine (inset). Main graph: Amount of stem breakage due to lateral force of 1.76 N of the thigmostimulator. Plants rubbed (···) or not rubbed (——) for up to 13 days. Inset graph: Lateral force necessary to rupture pine stems mechanically perturbed (MP) or not (C) for one growing season (about 7 months). (From Jaffee and Telewski 1984. Reproduced with permission of Plenum Press.)

nodes exhibit reduced internode elongation and increased radial growth, resulting in thicker stems and leaves (Jaffee 1976).

Thigmomorphogenesis is similar to wound induced phytochemical production, as it is translocatable between internodes but occurs instantaneously, presumably due to changes in cell membrane permeability (Jaffee 1980). Endogenous ethylene is thought to mediate thigmomorphogenesis, because concentrations increase twofold after plants are stimulated (Jaffee and Telewski 1984). Ethylene production has also been observed in sweet potato roots wounded by the sweet potato weevil (Uritani et al. 1975), in *Pinus radiata* trees wounded by the wasp, *S. noctilio* (Shain and Hillis 1972), and in cotton shoots fed on by the cotton fleahopper, *Pseudatomoscelis seriatus* (Duffey and Powell 1979, Grisham et al. 1987).

In *Phaseolus* plants, thickening of stem cell walls occurs within four days after stimulation (Jaffee and Biro 1979), causing stem cuticular ridges to be broader and more twisted. This same reaction occurs on stems of bean (Jaffee and Biro 1979), celery (Venning 1949, Walker 1960), and corn (Whitehead and Luti 1962). The effects of stem thickening on insect populations is unknown, but plants with thickened stems can withstand drought and frost stress better than control plants (Jaffee and Biro 1979). It would be interesting to determine if stem thickening adversely affects stem-feeding Homoptera, as stimulated plants are less susceptible to stem rupture than nonstimulated plants (Fig. 3) (Jaffee and Telewski 1984).

Both positive and negative effects of thigmomorphogenesis occur in plants. Excessive handling of cotton plants reduces internode length, number, and plant height (Frizzell et al. 1974), while stroking of apple stems (Ferree and Hall 1981) and tomato plants (Mitchell et al. 1977) reduces net photosynthetic capacity. However, shaking plants of a wheat variety resis-

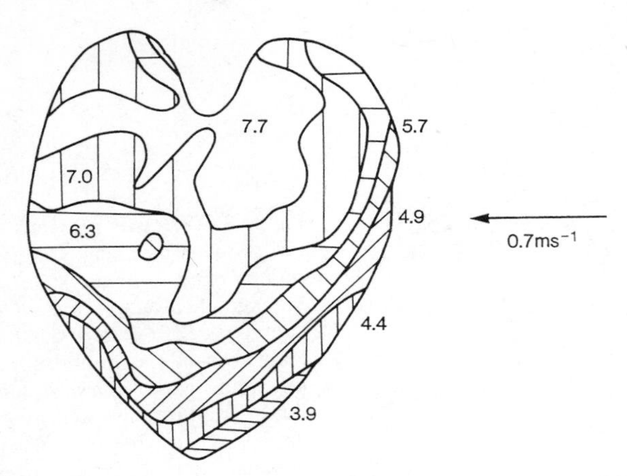

Figure 4. Temperature variation over a bean leaf in a turbulent airflow of 0.7 m/s. Numbers denote surface temperature in excess of ambient (25.6°C). Relative humidity 29%; net radiation 150 W/m^2; leaf resistance 400–1300 s/m. (From Clark and Wigley 1975. Reproduced with permission of Hemisphere Publ. Co.)

tant to lodging reduces shoot length, suggesting a beneficial effect of thigmomorphogenesis (Steucek and Gordon 1975). Shaking chrysanthemum plants reduces elongation, creating a favorable plant type (Hammer et al. 1974), and shaking *Liquidambar* trees stimulates development of terminal buds (Neel and Harris 1971).

It is difficult to generalize about the effects of thigmomorphogenesis on insect growth and development, although adverse effects occur in insects exposed to changes in plant height (Tingey and Leigh 1974), and stem thickness (Djamin and Pathak 1967). The role of thigmomorphogenesis in plant–insect interactions awaits investigation to determine its actual effects on insect population dynamics.

4. EFFECTS OF CLIMATIC STRESS TO PLANTS ON INSECT POPULATIONS

4.1 Wind

Wind adversely affects plant growth and development. Decreases in plant growth occur at wind speeds moderately above normal (4.0 m/s) in *Brassica napus* (Wadsworth 1960) and alfalfa, clover, and tomato (Morse and Evans 1962). Wind also causes a turbulence around and within the plant canopy which affects leaf surface temperature. Temperatures of bean leaves vary from 4.0°C above ambient on the leaf edge to 7.7°C above ambient near the petiole junction at only 0.7 m/s (Fig. 4) (Clark and Wigley 1975). The effects of these differences on insect behavior and development have not been intensively studied, but changes in temperature directly affect insect-feeding activity.

Wind damage to plant leaves occurs due to leaf folding and surface abrasion (Thompson 1974), edge burning (Hogg 1964), or leaf tearing (Taylor and Sexton 1972). This damage has direct negative effects on the growth and development of foliage-feeding insects, because the quantity and quality of foliar materials available for insect consumption is decreased. Leaf tearing and abrasion from wind damage induce changes in the immune responses of plants similar to other types of mechanical damage. Wind-related damage to leaf surface cells also alters leaf surface wax composition, which has been shown to serve as a defense against some insects (Atkin and Hamilton 1982, Littledyke and Cherrett 1978, Way and Murdie 1965, Woodhead 1982).

Direct effects of wind on trees do affect changes in insect populations. Outbreaks of the geometrid *Selidosema suavis* Butler are greater on the leeward sides of *Pinus radiata* trees and, in general, on ''suppressed'' trees in which growth is reduced by drought stress, root breakage, and windthrow (White 1974). Engelman spruce trees blown down by high winds are preferably infested by Engelman spruce beetles (Wygant 1958), as downed trees apparently afford more protection from predation and low winter temperatures. Wind-thrown loblolly pine trees are also more readily attacked by southern pine beetles (St. George 1930).

4.2 Lightning

Lightning strikes to coniferous trees are a major factor in the population dynamics of several species of bark beetles. Trees struck by lightning have reduced water content, decreased oleoresin pressure, and decreased nonreducing sugar content (Hodges and Pickard 1971), resulting in increased attack by the southern pine beetle, western pine beetle, *Dendroctonus brevicomis* LeConte, small southern pine engraver, *Ips avulsus* (Eichh.), *Ips grandicollis* (Eichh.), and *Ips calligraphus* (Germ.) (Hetrick 1949, Hodges and Pickard 1971, Johnson 1966, St. George 1930). The two most plausible theories for increased tree susceptibility deal with attraction of beetles to volatile allelochemicals (Blanche et al. 1983). Evidence for direct tree attraction exists because *Ips* spp. are attracted to volatiles emanating from lightning struck trees (All and Anderson 1974, Werner 1979). Secondary attraction to black turpentine beetles, *Dendroctonus terebrans* (Oliver), infesting lightning-struck trees also occurs in *D. frontalis*, *I. grandicollis*, and *I. avulsus* (Hodges and Pickard 1971, Merkel 1981). Trees adjacent to lightning-struck trees are also more susceptible to insect attack (Schmitz and Taylor 1969, Minko 1966).

4.3 Dust Particles

Both abrasive and nonabrasive dusts have insecticidal properties (Ebling 1971). Abrasive dusts, such as those derived from alumina, carborundum, flint, and quartz, have varying degrees of insecticidal activity that are correlated to abrasiveness (Alexander et al. 1944, Briscoe 1943). Nonabrasive (sorptive) dusts from nonabrasive clays, charcoal, diatomaceous earth, and silica differ in insect toxicity, depending on particle surface characteristics (Krishnakumar and Majumder 1962, Chiu 1939). Though the two types of dust have differing modes of action (sorptive-removal of cuticular lipids; abrasive-desiccation), no information exists regarding the detrimental effects, if any, of dusts on insect feeding and the effects of dust intake on digestion and metabolism.

Inert road dust particles deposited on plants are generally known to make them less attractive for insect feeding. Dust produced by plowing agricultural land reduces populations of mite predators and parasites (DeBach 1974). Similarly, washing dust from citrus trees increases survival of California red scale parasites (DeBach 1974). Cultures of hymenopterous parasites are especially susceptible to the insecticidal actions of inert dust particles (Ebling 1971). Philogene (1972) noted that volcanic dust is toxic to *Periplaneta americana* (L.) placed in contact with the dust. Bentonite, a clay of volcanic origin composed of aluminum, iron, magnesium, and silica, is also toxic to the bean weevil, *Acanthoscelides obtectus* (Say) (Chiu 1939).

5. SUMMARY AND CONCLUSIONS

Plant wounding due to mechanical tissue destruction from both abiotic and biotic forces has both negative and positive effects on insects associated with the wounded plant. From the previous review a general pattern of plant and insect responses to plant wounding is apparent (Fig. 5). Damage to plant tissues, whether mechanical or insect related, adversely affects insect populations feeding on damaged plants. Root, stem, and foliar damage to several species of plants elicits a wound response that may affect insects feeding on these plants in as little as four to six hours. In a few cases plant wounding increases susceptibility to insects, either directly, due to a specific plant response, or indirectly, by reducing the total resources available to the insect population.

Rubbing or bending plants results in decreased apical growth and increased radial growth (thigmomorphogenesis) which may also adversely affect insects feeding on these plants. These effects may be manifested physically, in the strengthening of plant stem tissues for protection against stem-feeding insects, or chemically, since ethylene production mediates thigmomorphogenesis and is part of the general plant wounding response. Indirect adverse effects on insects may also result from alterations of plant height or canopy density due to thigmomorphogenesis, but these possibilities have not been investigated.

Wounded plants exhibit changes in chemistry linked to allelochemicals

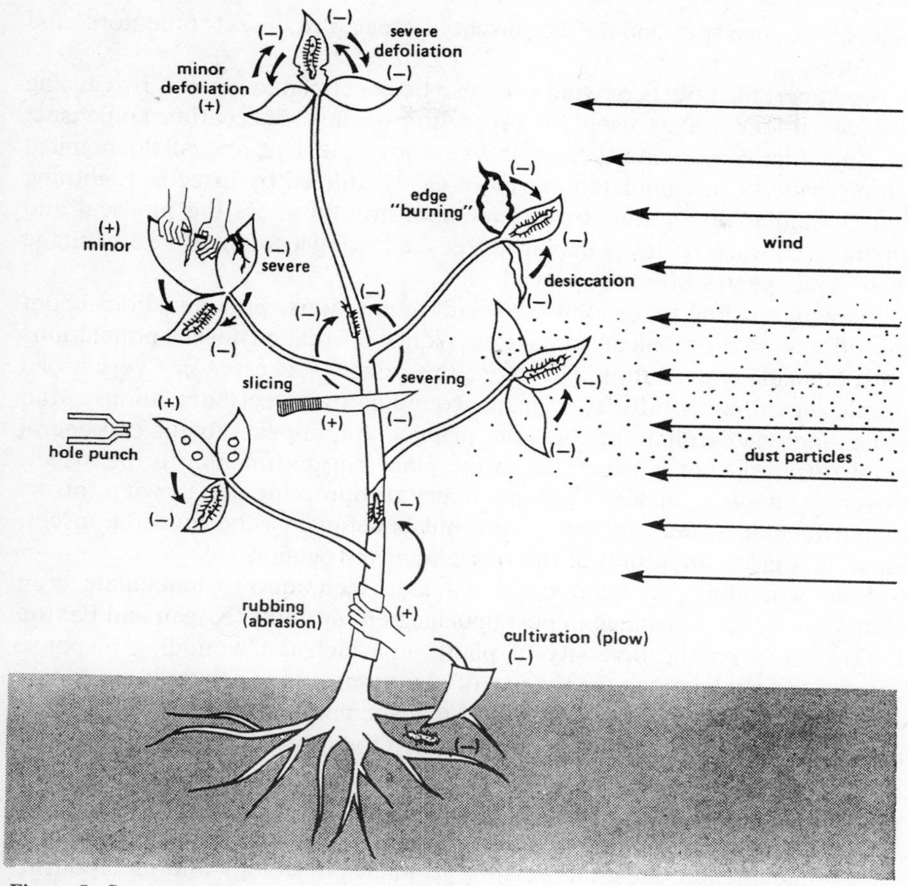

Figure 5. Summary diagram of plant and insect responses to mechanical wounding of plants. Arrows indicate the direction of the effect (+) or (−) of induction to insects on leaves, stems, or roots adjacent to tissues induced. Symbols (+) and (−) at the base of arrows denote the effect of the induction on the originating tissue. (Minor defoliation elicits positive effects on the induced tissue but adverse effects on the inducing insect; severe defoliation induces negative effects on both organisms.)

involved in resistance to insects. One of the most exciting recent developments is the discovery that damage to poplar and sugar maple tree foliage increases the total phenol content of foliage of adjacent, nonconnected trees, suggesting that plants are capable of pheromonal communication regarding wounding.

Climatic stresses to plants, resulting from the effects of wind, dust, and lightning also affect insect populations. Foliar tearing and abrasion related to the effects of wind are similar to general mechanical damage, causing negative effects on insects. Dust particles carried by wind and deposited on plants are also detrimental to insects, especially insect predators and parasites.

However, the effects of wind may also be beneficial to insects, by causing loss of surface waxes used for protection against desiccation and insect feeding. Similarly, wind-thrown trees, whose nutrient and allelochemical supply have been eliminated are more easily utilized by insects. Lightning causes similar disruptions to tree metabolism and alters the physical and chemical characteristics of damaged trees, allowing a competitive advantage to several species of insect pests.

As with most all types of insect–plant interactions, generalizations about the effects of mechanical damage to plants on related insect populations must be made with caution. However, this review suggests some very broad trends regarding the effects of plant wounding on insect populations. Mild wounding, via defoliation, abrasion, or injection, appears to elicit a general plant response that is beneficial to the plant but detrimental to the insect. Severe wounding, such as lightning injury or uprooting due to wind throw, is detrimental to the tree population but beneficial to the invading insect, because a gross alteration of the tree chemistry occurs.

Mild wounding has been suggested as a technique to innoculate crop plants for insect resistance in pest management systems (Kogan and Paxton 1983). Based on the diversity of plants in which the wounding response occurs, the implications for the use of plant innoculation by mild wounding are very strong. If practiced in insect management systems, it will be very interesting to determine the extent to which the general wounding response will function synergistically with more specific physical and chemical plant defenses.

The entire process of the response of insects to damaged plant tissue also raises fundamental questions relating to plant physiology and metabolism. How rapidly do plants induce an immune response to wounding? Most past research has studied induction effects 24 to 48 hours after wounding. In one instance (Carroll and Hoffman 1980) induction occurred within 40 minutes. Is the wound response instantaneous as is thigmomorphogenesis? Is there a general resistance phenomenon in plants, and if so, how does it differ from the specific resistance that has been bred into much of the cultivated crop acreage of the world? Can this general resistance or the primitive immune response of plants be increased or enhanced by breeding as specific resistance factors have?

These questions, in addition to those posed previously in this chapter, indicate that there is much to learn about the effects of mechanical damage on plants caused by biotic and abiotic stresses, and the resulting effects on phytophagous insects. Answers to these questions will require solutions derived by the cooperative efforts of plant physiologists, plant breeders, and entomologists, and should provide much-needed additional information about plant–insect interactions.

REFERENCES

Akazawa, T., I. Uritani, and H. Kubota. 1960. Isolation of ipomeamarone and two coumarin derivatives from sweet potato roots injured by the weevil, *Cylas formicarius elegantus*. *Arch. Biochem. Biophys.* 88: 150–156.

Alexander, P., J. A. Kitchener, and H. V. A. Briscoe. 1944. Inert dust insecticides. III. The effect of dusts on stored products pests other than *Calandra granaria*. *Ann. Appl. Biol.* 31: 156–159.

All, J. N., and R. F. Anderson. 1974. The influence of various odors on host selection by pioneer beetles of *Ips grandicollis*. *J. Ga. Entomol. Soc.* 9: 223–228.

Atkin, D. S. J., and R. J. Hamilton. 1982. The effects of plant waxes on insects. *J. Nat. Prod.* 45: 694–696.

Baldwin, I. T., and J. C. Schultz. 1983. Rapid changes in tree leaf chemistry induced by damage: evidence for communication between plants. *Science* 221: 277–279.

Baldwin, I. T., and J. C. Schultz. 1984. Damage- and communication-induced changes in yellow birch leaf phenolics. *In* Lenner, R. (ed.), *Proc. 8th Annu. For. Biol. Workshop*. Utah State Univ., Logan. pp. 25–33.

Benz, G. 1974. Negative Rückkoppelung durch reumund Nährungskonkurrenz sowie zyklische Veranderung der Nährungsgrundlage als Regelprinzip in der Populations Dynamik des grauen Arch Einwicklers, *Zeiraphera diniana* (Guenee) (Lep. Tortricidae). *Z. Angew. Entomol.* 76: 196–228.

Bernays, E. A., R. F. Chapman, E. M. Leather, A. R. McCaffery, and W. W. D. Modder. 1977. The relationship of *Zonocerus variegatus* (L.) (Acridoidea: Pyrgomorphidae) with cassava (*Manihot esculenta*). *Bull. Entomol. Res.* 67: 391–404.

Blanche, C. A., J. D. Hodges, T. E. Nebeker, and D. M. Moehring. 1983. Southern Pine Beetle: The Host Dimension. *Miss. Agric. & Forest. Exp. Sta. Bull.* 917.

Blanche, C. A., T. E. Nebeker, J. D. Hodges, B. L. Karr, and J. J. Schmitt. 1985. Effect of thinning damage on bark beetle susceptibility indicators in loblolly pine. *In* Shoulders, E. (ed.), *Proc. 3d Bienn. Southern Silvicultural Res. Conf. USDA For. Serv. Gen. Tech. Rep.* 50–54, pp. 471–479.

Bowling, C. C. 1978. Simulated insect damage to rice: effects of leaf removal. *J. Econ. Entomol.* 71: 377–378.

Bramstedt, F. 1938. Der nachweis der blutlausunanfalligkeit der apfelsorten auf histologischer grundlage. *Z. Pflanzenkrank Pflanzenschutz.* 48: 480–488.

Briscoe, H. V. A. 1943. Some properties of inorganic dusts. *J. Royal Soc. Arts* 91: 593–607.

Broadway, R. M., S. S. Duffey, G. Pearce, and C. A. Ryan. 1986. Plant proteinase inhibitors: a defense against herbivorous insects? *Entomol. Exp. Appl.* 41: 33–38.

Camery, M. P., and R. C. Weber. 1953. Effects of certain components of simulated hail injury on soybeans and corn. *Agric. Exp. Sta., Iowa St. Coll. Res. Bull.* 400: 465–504.

Capinera, J. L. 1979. Effect of simulated insect herbivory on sugarbeet yield in Colorado. *J. Kansas Entomol. Soc.* 52: 712–718.

Capinera, J. L., and W. W. J. Roltsch. 1980. Response of wheat seedlings to actual and simulated migratory grasshopper defoliation. *J. Econ. Entomol.* 73: 258–261.

Carroll, C. R., and C. A. Hoffman. 1980. Chemical feeding deterrent mobilized in response to insect herbivory and counter adaptation by *Epilachna tredecimnotata*. *Science* 209: 414–416.

Chiu, S. F. 1939. Toxicity studies of so-called "inert" materials with the bean weevil, *Acanthoscelides obtectus* (Say). *J. Econ. Entomol.* 32: 240–248.

Clark, J. A., and G. Wigley. 1975. Heat and mass transfer from real and model leaves. *In* de

Vries, D. A. and N. H. Afgan (eds.), *Heat and Mass Transfer in the Biosphere. Part 1. Transfer Processes in the Plant Environment*. Hemisphere, Washington, DC, pp. 413–422.

Coutts, M. P., and J. E. Dolezal. 1966. Some effects of bark cincturing on the physiology on *Pinus radiata* and on *Sirex* attack. *Austral. For. Res.* 2: 17–28.

Crute, I. R., P. J. G. M. De Wit, and M. Wade. 1985. Mechanisms by which genetically controlled resistance and virulence influence host colonization by fungal and bacterial parasites. *In* Fraser, R. S. S. (ed.), *Mechanisms of Resistance to Plant Diseases*. Martinus Nijhoff/Dr. Junk Publ., Boston, pp. 197–309.

Davies, E., and A. Schuster. 1981. Intercellular communication in plants: evidence for a rapidly generated, bidirectionally transmitted wound signal. *Proc. Natl. Acad. Sci. USA*. 78: 2422–2426.

DeBach, P. 1974. *Biological Control by Natural Enemies*. Cambridge Univ. P., Cambridge.

De Wit, P. J. G. M. 1985. Induced resistance to fungal and bacterial diseases. *In* Fraser, R. S. S. (ed.), *Mechanisms of Resistance to Plant Diseases*. Martinus Nijhoff/Dr. Junk Publ., Boston, pp. 405–424.

Djamin, A., and M. D. Pathak. 1967. Role of silica in resistance to Asiatic rice borer, *Chilo supressalis* (Walker), in rice varieties, *J. Econ. Entomol.* 60: 347–351.

Duffey, J. E., and R. D. Powell. 1979. Microbial-induced ethylene synthesis as a possible factor in square abscission in cotton infested by cotton fleahopper. *Ann. Entomol. Soc. Am.* 72: 599–601.

Dunning, R. A., and G. H. Winder. 1972. Some effects, especially on yield, of artificially defoliating sugar beet. *Ann. Appl. Biol.* 70: 89–98.

Ebling, W. 1971. Sorptive dusts for pest control. *Annu. Rev. Entomol.* 46: 123–158.

Edwards, P. J., and S. D. Wratten. 1983. Wound induced defenses in plants and their consequences for patterns of insect grazing. *Oecologia* 59: 88–93.

Edwards, P. J., S. D. Wratten, and H. Cox. 1985. Wound-induced changes in the acceptability of tomato to larvae of *Spodoptera littoralis*: a laboratory bioassay. *Ecol. Entomol.* 10: 155–158.

English, J., and J. Bonner. 1937. The wound hormones of plants. I. Traumatin, the active principle of the bean test. *J. Biol. Chem.* 121: 791–799.

Ferree, D. C., and F. R. Hall. 1981. Influence of physical stress on photosynthesis and transpiration of apple leaves. *J. Am. Soc. Hortic. Sci.* 106: 348–351.

Frizell, J. L., L. C. Brown, and B. A. Waddle. 1974. Some effects of handling on the growth and development of cotton. *Agron. J.* 70: 69–71.

Galliard, T. 1978. Lipolytic and lipoxygenase enzymes in plants and their action in wounded tissues. In G. Kahl (ed.), *Biochemistry of Wounded Plant Tissues*. de Gruyter, Berlin, pp. 155–202.

Green, T. R., and C. A. Ryan. 1972. Wound-induced proteinase inhibitor in plant leaves: a possible defense mechanism against insects. *Science* 175: 776–777.

Grisham, M. P., W. L. Sterling, R. D. Powell, and P. W. Morgan. 1987. Characterization of the induction of stress ethylene synthesis in cotton caused by the cotton fleahopper (Hemiptera: Miridae) and its microorganisms. *Ann. Entomol. Soc. Am.* 80: 411–416.

Guerra, D. J. 1981. Natural and *Heliothis zea* (Boddie)-induced levels of specific phenolic compounds in *Gossypium hirsutum* L. M.S. Thesis. University of Arkansas, Fayetteville, 85 p.

Hall, F. R., and D. C. Ferree. 1975. Influence of two spotted spider mite populations on photosynthesis of apple leaves. *J. Econ. Entomol.* 68: 517–520.

Hammer, P. A., C. A. Mitchell, and T. C. Weiler. 1974. Height control in greenhouse chrysanthemum by mechanical stress. *HortScience*. 9: 474–475.

Haukioja, E. 1982. Inducible defenses of white birch to a geometrid defoliator, *Epirrita autumnata*. *In Proc. 5th Intl. Symp. Insect–Plant Relationships*. Pudoc, Wageningen, The Netherlands, pp. 199–203.

Haukioja, E., and P. Niemela. 1977. Retarded growth of a geometrid larva after mechanical damage to leaves of its host tree. *Ann. Zool. Fennici*. 14: 48–52.

Haukioja, E., and P. Niemela. 1979. Birch leaves as a resource for herbivores: seasonal occurrence of increased resistance in foliage after mechanical damage of adjacent leaves. *Oecologia* 39: 151–159.

Henke, O. 1963. Uber den stoffwechsel reblausanfalliger und-unanfalliger Reben. *Phytopathol. Z.* 47: 314–326.

Hetrick, L. A. 1949. Susceptibility of pine trees to bark beetle attack. *Arborist's News* 14: 149–151.

Hodges, J. D., and L. S. Pickard. 1971. Lightning in the ecology of the southern pine beetle, *Dendroctonus frontalis* (Coleoptera: Scolytidae). *Can. Entomol.* 103: 44–51.

Hogg, W. H. 1964. Measurements of the shelter effect of landforms and other topographical features, and of artificial wind-breaks. *Sci. Hortic.* 17: 20–30.

Hori, K. 1973. Studies on the feeding habits of *Lygus disponsi* Linnavuori (Hemiptera: Miridae) and the injury to its host plant. III. Phenolic compounds, acid phosphatase and oxidative enzymes in the injured tissue of sugar beet leaf. *Appl. Entomol. Zool.* 8: 103–112.

Hori, K., and R. Atalay. 1980. Biochemical changes in the tissue of Chinese cabbage injured by the bug, *Lygus disponsi*. *Appl. Entomol. Zool.* 15: 234–241.

Howell, J. F. 1978. Spotted cutworm: simulated damage to apples. *J. Econ. Entomol.* 71: 437–439.

Jaffee, M. J. 1973. Thigmomorphogenesis: the response of plant growth and development to mechanical stimulation. *Planta* 114: 143–157.

Jaffee, M. J. 1976. Thigmomorphogenesis: a detailed characterization of the response of beans (*Phaseolus vulgaris* L.) to mechanical stimulation. *Z. Pflanzenphysiol.* 77: 437–453.

Jaffee, M. J. 1980. Morphogenetic responses of plants to mechanical stimuli or stress. *BioScience* 30: 239–243.

Jaffee, M. J., and R. Biro. 1979. Thigmomorphogenesis: the effect of mechanical perturbation on the growth of plants, with special reference to anatomical changes, the role of ethylene, and interaction with other environmental stresses. *In* Mussell, H. and R. C. Staples (eds.), *Stress Physiology in Crop Plants*. Wiley, New York, pp. 25–59.

Jaffee, M. J., and F. W. Telewski. 1984. Thigmomorphogenesis: callose and ethylene in the hardening of mechanical stressed plants. *In* Timmermann, B. N., C. Steelink, and F. A. Loewus (eds.), *Phytochemical Adaptations to Plant Stress. Rec. Adv. Phytochem.*, Vol. 18. Plenum, New York, pp. 79–95.

Johnson, P. C. 1966. Attractiveness of lightning-struck ponderosa pine trees to *Dendroctonus brevicomis* (Coleoptera: Scolytidae). *Ann. Entomol. Soc. Am.* 59: 615.

Karban, R. 1985. Resistance against spider mites in cotton induced by mechanical abrasion. *Entomol. Exp. Appl.* 37: 137–141.

Karban, R., and J. R. Carey. 1984. Induced resistance of cotton seedlings to mites. *Science* 225: 53–54.

Klun, J. A., C. L. Tipton, and T. A. Brindley. 1967. 2,4-dihydroxy-7-methoxy-1,4-benzoxacin, 3-one (DIMBOA), an active agent in the resistance of maize to the European corn borer. *J. Econ. Entomol.* 60: 1529–1533.

Kogan, M., and J. Paxton. 1983. Natural inducers of plant resistance to insects. In P. A. Hedin (ed.), *Plant Resistance to Insects. Am. Chem. Soc. Symp. Series* 208, American Chemical Society, Washington, DC, pp. 153–171.

Krishnakumar, M. K., and S. K. Majumder. 1962. Modes of insecticidal action of active carbon and clay on *Tribolium castaneum* (Hbst.) *Nature* 193: 1310–1311.

Leszczynski, B. 1985. Changes in phenols content and metabolism in leaves of susceptible and resistant winter wheat cultivars infested by *Rhopalosiphum padi* (L.) (Hom., Aphididae). *Z. Angew. Entomol.* 100: 343–348.

Leszczynski, B., J. Warchol, and S. Naraz. 1985. The influence of phenolic compounds on the preference of winter wheat cultivars by cereal aphids. *Insect Sci. Applic.* 6: 157–158.

Lewis, A. C. 1984. Plant quality and grasshopper feeding: effects of sunflower condition on preference and performance in *Melanoplus differentialis*. *Ecology* 65: 836–843.

Littledyke, M., and J. M. Cherrett. 1978. Defense mechanisms in young and old leaves against cutting by the leaf-cutting ants *Atta cephalotes* (L.) and *Acromyrmex octospinosus* (Reich.) (Hymenoptera: Formicidae). *Bull. Entomol. Res.* 68: 263–271.

Merkel, E. P. 1981. Control of the black turpentine beetle. *Ga. For. Res. Pap.* 15, 4 p.

Miles, P. W. 1968. Insect secretions in plants. *Annu. Rev. Phytopathology* 6: 137–164.

Minko, G. 1966. Lightning in radiata pine stands in northeastern Victoria. *Austral. For.* 30: 257–267.

Mitchell, C. A., H. C. Dostal, and T. M. Seipel. 1977. Dry weight reduction in mechanically-dwarfed tomato plants. *J. Am. Soc. Hortic. Sci.* 102: 605–608.

Morse, R. N., and L. T. Evans. 1962. Design and development of CERES-an Australian phytotron. *J. Agric. Engr. Res.* 7: 128–140.

Navas, D. 1976. Fall armyworm in rice. In *Proc. Tall Timbers Conf. Ecol. Anim. Control Habitat Manag.* 1974 Meeting, pp. 99–106.

Nebeker, T. E. and J. D. Hodges. 1983. Influence of forestry practices on host-susceptibility to bark beetles. *Z. Angew. Entomol.* 96: 194–208.

Neel, P. L., and R. W. Harris. 1971. Motion-induced inhibition of elongation and induction of dormancy in *Liquidambar*. *Science* 173: 58–59.

Niemela, P., E. M. Aro, and E. Haukioja. 1979. Birch leaves as a resource for herbivores. Damaged-induced increase in leaf phenols with trypsin-inhibiting effects. *Rep. Kevo Subarctic Res. Stat.* 15: 37–40.

Perry, D. A., and G. B. Pitman. 1983. Genetic and environmental influence in host resistance to herbivory: Douglas-fir and the western spruce budworm. *Z. Angew. Entomol.* 96: 217–228.

Philogene, B. J. R. 1972. Volcanic ash for insect control. *Can. Entomol.* 104: 1487.

Raupp, M. J., and R. F. Denno. 1984. The suitability of damaged willow leaves as food for the leaf beetle, *Plagiodera versicolora*. *Ecol. Entomol.* 9: 443–448.

Reynolds, G. W., and C. M. Smith. 1985. Effects of leaf position, leaf wounding, and plant age of two soybean genotypes on soybean looper (Lepidoptera: Noctuidae) growth. *Environ. Entomol.* 14: 475–478.

Rhoades, D. F. 1979. Evolution of plant chemical defense against herbivores. *In* Rosenthal, G. A. and D. Janzen (eds.), *Herbivores. Their Interaction with Secondary Plant Metabolites*. Academic Press, New York, pp. 3–54.

Rhoades, D. F. 1983. Responses of alder and willow to attack by tent caterpillars and webworms: evidence for pheromonal sensitivity of willows. *In* Hedin, P. A. (ed.), *Plant Resistance to Insects. Am. Chem. Soc. Symposium Series* 208, Washington, DC, pp. 56–68.

Rhodes, J. M., and L. S. C. Wooltorton. 1978. The bio-synthesis of phenolic compounds in wounded plant storage tissues. *In* Kahl, G. (ed.), *Biochemistry of Wounded Plant Tissues*. de Gruyter, Berlin, pp. 243–286.

Rohfritsch, O. 1981. A "defense mechanism" of *Picea excelsa* L. against the gall former *Chermes abietis* L. (Homoptera, Adelgidae). *Z. Angew. Entomol.* 92: 18–26.

Rottger, U., and F. Klingauf. 1976. Anderung im stoffwechsel von Zuckerruben blattern durch befall mit *Pegomya betae* Curt. (Muscidae: Anthomyidae). *Z. Angew. Entomol.* 82: 220–227.

Ryan, C. A. 1974. Assay and biochemical properties of the proteinase inhibitor-inducing factor, a wound hormone. *Plant Physiol.* 54: 328–332.

Schmitz, R. F., and A. R. Taylor. 1969. An instance of lightning damage and infestation of ponderosa pines by the pine engraver beetle in Montana. *USDA For. Serv. Res. Note INT-88*, 8 p.

Schultz, J. C., and I. T. Baldwin. 1982. Oak leaf quality declines in response to defoliation by gypsy moth larvae. *Science* 221: 149–151.

Scott, F. M., B. G. Bystrom, and V. Sjaholm. 1961. Anatomy of traumatic acid-treated internodes of *Ricinus communis*. *Bot. Gaz.* 122: 311–314.

Shain, L., and W. E. Hillis. 1972. Ethylene production in *Pinus radiata* in response to *Sirex-Amylostereum* attack. *Phytopathology* 62: 1407–1409.

Smith, C. M. 1985. Expression, mechanisms and chemistry of resistance in soybean, *Glycine max* L. (Merr.) to the soybean looper, *Pseudoplusia includens* (Walker). *Insect Sci. Applic.* 6: 243–248.

Steucek, G. L., and L. K. Gordon. 1975. Response of wheat (*Triticum aestivum*) seedlings to mechanical stress. *Bot. Gaz.* 136: 17–19.

St. George, R. A. 1930. Drought-affected and injured trees attractive to bark beetles. *J. Econ. Entomol.* 23: 825–828.

Sucoff, E., H. Ratsch, and D. D. Hook. 1967. Early development of wound-initiated discoloration in *Populus tremuloides* Michx. *Can. J. Bot.* 45: 649–656.

Taylor, S. E., and O. J. Sexton. 1972. Some implications of leaf tearing in Musaceae. *Ecology* 53: 143–149.

Thielges, B. A. 1968. Altered polyphenol metabolism in the foliage of *Pinus sylvestris* associated with European pine sawfly attack. *Can. J. Bot.* 46: 724–725.

Thompson, J. R. 1974. The effect of wind on grasses. II. Mechanical damage in *Festuca arundinacea* Schreb. *J. Exp. Bot.* 25: 965–972.

Tingey, W. M., and T. F. Leigh. 1974. Height preference of *Lygus* bugs for oviposition on caged cotton plants. *Environ. Entomol.* 3: 350–351.

Tjia, B., and D. B. Houston. 1975. Phenolic constituents of Norway spruce resistant or susceptible to the eastern spruce gall aphid. *For. Sci.* 211: 180–184.

Treshow, M. 1955. The etiology, development, and control of tomato fruit tumor. *Phytopathology* 45: 132–137.

Tuomi, J., P. Niemala, E. Haukioja, S. Siren, and S. Neuvonen. 1984. Nutrient stress: an explanation for anti-herbivore responses to defoliation. *Oecologia* 61: 208–210.

Uritani, I., T. Saito, H. Honda, and W. K. Kim. 1975. Induction of furano-terpenoids in sweet potato roots by the larval components of the sweet potato weevils. *Agr. Biol. Chem.* 39: 1857–1862.

Venning, F. D. 1949. Stimulation by wind motion of collenchyma formation in celery petioles. *Bot. Gaz.* 110: 511–514.

Wadsworth, R. M. 1960. The effect of artificial wind on the growth-rate of plants in water culture. *Ann. Bot.* 24: 200–211.

Wahlroos, V., and A. I. Virtanen. 1959. The precursors of 6 MBOA in maize and wheat plants, their isolations and some of their properties. *Acta Chem. Scand.* 13: 1906–1908.

Walker, W. S. 1960. The effects of mechanical stimulation and etiolation on the collenchyma of *Datura stramonium*. *Am. J. Bot.* 47: 717–724.

Walker-Simmons, M., and C. A. Ryan. 1977. Wound-induced accumulation of trypsin inhibitor activities in plant leaves. Survey of several plant genera. *Plant Physiol.* 59: 437–439.

Walker-Simmons, M., C. H. Hollander, J. K. Anderson, and C. A. Ryan. 1984. Wound signals in plants: a systematic plant wound signal alters plasma membrane integrity. *Proc. Natl. Acad. Sci. USA* 81: 3737–3741.

Wallner, W. E., and G. S. Walton. 1979. Host defoliation: a possible determinant of gypsy moth population quality. *Ann. Entomol. Soc. Am.* 72: 62–67.

Wargo, P. M. 1972. Defoliation-induced chemical changes in sugar maple roots stimulate growth of *Armillaria mellea*. *Phytopathology* 62: 1278–1283.

Way, M. J., and G. Murdie. 1965. An example of varietal variations in resistance of Brussels sprouts. *Ann. Appl. Biol.* 56: 326–328.

Werner, R. A. 1979. Influence of host foliage on development, survival, fecundity, and oviposition of the spear-marked black moth, *Rheumaptera hastata* (Lepidoptera: Geometridae). *Can. Entomol.* 111: 317–322.

West, C. 1985. Factors underlying the late seasonal appearance of the lepidopterous leaf-mining guild on oak. *Ecol. Entomol.* 10: 111–120.

Westphal, E., R. Bonner, and M. LeRet. 1981. Changes in leaves of susceptible and resistant *Solanum dulcamara* infested by the gall mite *Eriophyes cladophthirus* (Acarina, Eriophyoidea). *Can. J. Bot.* 59: 875–882.

White, T. C. R. 1974. A hypothesis to explain outbreaks of looper caterpillars, with special reference to populations of *Selidosema suavis* in a plantation of *Pinus radiata* in New Zealand. *Oecologia* 16: 279–301.

Whitehead, F. H., and R. Luti. 1962. Experimental studies of the effect of wind on plant growth. I. *Zea mays*. *New Phytol.* 61: 56–58.

Wolk, J. O., D. W. Kretchman, and D. G. Ortega, Jr. 1983. Response of tomato to defoliation. *J. Am. Soc. Hortic. Sci.* 108: 536–540.

Woodhead, S. 1982. p-Hydroxybenzaldehyde in the surface wax of sorghum: its importance in seedling resistance in acridids. *Entomol. Exp. Appl.* 31: 296–302.

Woodhead, S., and E. A. Bernays. 1977. Changes in release rates of cyanide in relation to palatability of *Sorghum* to insects. *Nature* 270: 235–236.

Wratten, S. D., P. J. Edwards, and I. Dunn. 1984. Wound-induced changes in the palatability of *Betula pubescens* and *B. pendula*. *Oecologia* 61: 372–375.

Wright, L. C., A. A. Berryman, and S. Gurusiddaiah. 1979. Host resistance to the fir engraver beetle, *Scolytus ventralis* (Coleoptera: Scolytidae). IV. Effect of defoliation on wound monoterpene and inner bark carbohydrate concentrations. *Can. Entomol.* 111: 1255–1262.

Wygant, N. D. 1958. Engelmann spruce beetle control in Colorado. *In Proc. 10th Intl. Cong. Entomol.*, Vol. 4, 1956, pp. 181–184.

10

SENSITIVITY OF INSECT-DAMAGED PLANTS TO ENVIRONMENTAL STRESSES

Dale M. Norris

Department of Entomology
University of Wisconsin
Madison, Wisconsin

1. INTRODUCTION

The effects of insect-induced damage (as caused by herbivory, ovipositional punctures, microbial transmissions, excretions, etc.) on a plant's sensitivity to subsequent environmental stresses (as measured especially by crop yield losses) are interpreted here. The discussion is based on the premise, if not the fact, that plant species survive evolutionarily by developing a successful strategy of overall stress management which, rarely if ever, focuses functionally on a specific stress (e.g., insect herbivory, moisture deficiency, or nitrogen shortage). Thus plants usually do not specifically evolve resistance to insects, pathogens, or given abiotic factors. Our best knowledge tells us that some genetic lines of a plant species simply have a greater inherent potential for specific types of primary and secondary growth and development (including differentiation) which, when realized (expressed) in the phenotype, enable those individual plants not only to survive but also to maintain "normal" yields despite exposures to a variety of biotic or abiotic stress factors.

Though relatively stress-resistant plants possess both constitutive (e.g., stress-independent) and inducible (e.g., stress-dependent) characteristics that jointly account for such resistance, these relatively durable (hardy) plants usually excel in their genetic-based potential for the inducible traits. Regardless of the stressful agent(s) presented in their environment, these plants are especially able to unleash a dynamic secondary biochemical metabolism, and associated tissue differentiation and development, which not only adversely affect the immediate agent of stress but also may significantly reduce the negative effects (e.g., on crop yield) of subsequent "attacks" by other varied agents.

This chapter discusses the ability of different plant species and cultivars to maintain "normal" crop yields despite insect attacks and subsequent "attacks" by chemical, physical, biological, and mechanical agents. Regarding various plants' abilities to maintain their activities despite the various environmental stresses, four tenets are frequently considered: (1) individual plants (e.g., cultivars) within a species differ in their genetic potential for reacting to environmental stress (e.g., some individuals within a species are capable of vigorous self-preserving reactions to a given stress, whereas others lack such significant defensive responses); (2) various plant species employ some common principles (strategies) in coping with a variety of environmental stresses and do not usually seem to have evolved defenses against specific agents (e.g., herbivorous insects or microbial pathogens); (3) some plant species do employ unique traits to survive environmental stress; and (4) a defense against environmental stress for one plant species may possibly be a submission to such stress for another.

Selected research findings are presented as contributions toward an understanding of the interdependencies that may exist among the effects of various biotic and abiotic stresses on crop plants' performances as measured especially by crop yield. Data on food legumes and cereals are especially discussed because of their international importance and the greater relevant information available on such crops.

For purely didactic reasons, environmental stresses affecting the plant are classified as chemical, physical, biological, and mechanical. Cited specific potential agents of plant stress may involve more than one of these classified forms (e.g., slugs [Phylum Mollusca] during their feeding may *mechanically* disrupt plant cells, excrete [regurgitate] digestive enzymes which may constitute a *chemical* stress, and introduce microbes into the created wounds which may function as a *biological* stress).

2. SIGNIFICANCE OF TIMING OF THE STRESS

Beyond a plant having the necessary genetic potential for inducible resistance, timing of an initial stressful attack may be the next most important parameter in determining the effects of the plant's elicited responses to subsequent secondary environmental stresses. For example, the tobacco budworm, *Heliothis virescens* (F.), attack on cotton, *Gossypium hirsutum* L., during the earlier stages of fruit formation and development normally has the greatest impact on the plant's stress physiology and yield (McCarty et al. 1986). During such fruit formation (FF), especially under secondary drought conditions, insect-caused defoliation of more than about 20% may significantly reduce the cotton yield (Reynolds et al. 1982). In contrast, a 50% or greater defoliation of cotton plants during either the plant establishment (PE) or fruit maturation (FM) stage may not affect the lint yield even under dry soil conditions. In the case of the pink bollworm, *Pectinophora gossypiella* (Saunders), multiple attacks per boll during the late boll period of cotton slows or prevents boll opening especially through accelerated dehydration, and this reduces the harvested yield of lint (Kittock et al. 1983). The tarnished plant bug (TPB), *Lygus lineolaris* (Palisot de Beauvois), only reduces cotton yield in late planted cotton (Gaylor et al. 1983). Cotton apparently can compensate, through growth, for TPB damage to early planted cotton, but this is not usually possible with the late crop.

With food legumes (e.g., soybeans), any insect attack that shortens or disrupts the seed-filling period (SFP) is likely to reduce yield (Smith and Nelson 1986). In addition, a high correlation ($r = +0.8$) exists between seed weight and leaf area at each soybean node; thus in some grain (food) legumes a set fruit together with the leaf attached at the same node constitutes a functional unit as regards fruit (seed) development (Sinha and Savithri 1978). Any insect attack that significantly disrupts this functional unit of the plant is thus likely to reduce yield. Insect disruption of legume flowering and fruit (pod) set is not likely to reduce yield because these plants usually produce such a large (excessive) number of flower buds and flowers (Sinha and Savithri 1978). Most of the flowers (e.g., 80 to 90%) normally abscise; thus only a few fruits are ever set.

In the specific cases of the legumes, mungbeans and cowpeas, dry matter production, and thus crop yields, is limited by the plant growth which occurs during a very brief (e.g., two- to three-week) period six to nine weeks after sowing (Sinha and Savithri 1978). Insect attack during this critical period thus especially decreases the realization of the sink potential and therefore lowers the yield.

Among the grasses, *Zea mays* L. is especially susceptible to the southwestern corn borer, *Diatraea grandiosella* (Dyar), during the whorl stage of plant growth (Williams et al. 1983). Yield loss from attack at this stage was 20%, whereas, at anthesis such loss was only 9%. Removal of increasing percentages (e.g., 33, 50, 67, and 100%) of leaf area from grain sorghum (*Sorghum bicolor* L.) during boot and anthesis stages decreased mean yield by 23, 35, 43, and 95%, respectively (Stickler and Pauli 1961). Thus insect defoliation of *S. bicolor* during these stages of plant growth can have a striking effect on crop yield.

With deciduous tree fruits such as apples, insects (e.g., *Rhagoletis pomonella* [Walsh], apple maggot) that directly attack the developing fruit are especially harmful because the quality of the crop must be very high to be marketable; greater than 1% insect-infested or insect-damaged fruit represents a real reduction in marketable yield.

In summary, assuming that the plant is genetically competent to respond defensively to environmental stresses, such as insect herbivory, then the timing of the initial stress in relation to the stage of plant growth, differentiation, and development especially determines the plant's responses to subsequent environmental stresses.

3. CHEMICAL STRESSES

Several chemical parameters in the environment that may function as stresses to plant growth, differentiation, and development are discussed in Chapters 2, 6, 7, and 8. Such stressful factors are divided into (1) chemical properties of soils, (2) growth regulators and pesticides, and (3) pollutants. Some effects of prior insect attack (e.g., herbivory) upon a plant's subsequent response to such chemical environmental stresses are now discussed.

3.1. Chemical Properties of Soils

Plants may suffer injury if any of the essential chemical parameters of soil are not adequately available to the plant in the root medium (Epstein and Läuchli 1983). Insect damage to the plant's root system can variously impair absorption of essential or nutrient elements even though these soil factors are present in normal amounts and physical or chemical forms. This deficiency (injury) may occur especially if insects, such as northern corn rootworms, *Diabrotica longicornis* (Say), destroy a critical portion of the ''root-hair'' zone of the root system. Such insect damage thus may result in plant stress (disease) attributable to insufficient total ''root-hair'' volume for normal element absorption or inadequate distribution of functional ''hair-root'' regions to gain ready access to elements as distributed in the soil.

Insect infestations and associated feeding on leguminous (e.g., soybean) plants may significantly reduce the level of nitrogen fixation. Specifically, feeding by the three-cornered alfalfa hopper, *Spissistilus festinus* (Say), on early vegetative soybean plants, *Glycine max* (L.) Merrill, disrupted the movement of assimilates and photosynthates in the phloem (Hicks et al. 1984). Significant reductions in nodule number, nodule dry weight, root dry weight, leaf area, and nitrogen fixation resulted. The mean nitrogen fixed per soybean plant decreased with increasing *S. festinus* population. The major direct cause of the observed lowered N_2 fixation was reduction in the amount of photosynthates available to the soybean roots (Hicks et al. 1984). Pea aphid, *Acyrthosiphon pisum* (Harris), feeding on young vegetative peas, *Pisum sativum* (L.), likewise reduced plant growth and nitrogen fixation (Maiteki and Lamb 1985). Such feeding specifically reduced plant dry matter and the weight of nitrogen-fixing nodules.

In a stable ecological system, metabolically useful nitrogen is supplied to leguminous plants by the action of nitrogenase on nitrogen in the atmosphere (Oaks and Hirel 1985). With plant injury and tissue death, for example, as caused by cutworms and grubs feeding on roots, such nitrogen fixation may be disrupted. Then nitrogen is lost from the plant nutrient cycle through its conversion into amino acids, purines, and pyrimidines, leading to its release as ammonia. This series of conversions involving nitrogen constitutes a major component in the stress physiology of most plant species, especially leguminous ones.

Insect herbivory also may significantly alter a plant's efficiency of utilization of potassium and phosphorus otherwise available in the soil substrate. This can occur by such insect feeding reducing the amount of photosynthates available to the roots, and thus reducing the root growth (extension) rate which has the greatest influence on nutrient (e.g., phosphate) extraction from a given volume of soil (Clarkson 1985).

3.2 Growth Regulators and Pesticides

It is now well established that insect herbivory and other plant injuries (e.g., ovipositional punctures) caused by animals alter the phenolic metabolism of both locally involved (e.g., physically disrupted) and systemically remote (e.g., physically nondisrupted) plant cells (Ryan 1983, Kogan and Paxton 1983, Norris 1986, Chiang et al. 1986a). Phenolics participate in a diverse array of plant physiological processes including the stimulation and the inhibition of auxin activities (Misaghi 1982). Thus insect herbivory and other injuries certainly may alter the responses of plants to growth regulators. A major phytohormone, the phenolic indoleacetic acid (IAA), has been isolated from the salivary glands of phytophagous coccids and aphids and the ovipositional glands of plant gall-forming tenthredinids (Hymenoptera) (Norris 1979). Some phytophagous insects are directly equipped with chemicals, such as IAA, that function as hormones and growth regulators in plants. Presence of such insects on a plant more or less ensures that it is conditioned to respond uniquely to growth regulators otherwise presented in its environment. Such insect-induced, or conditioned, plant growth and differentiation constitute a unique spectrum of form and function (Norris 1979). At one end of this spectrum are the amorphous changes in plant growth patterns. These may be usefully divided into self-limiting overgrowths (e.g., galls) and non-self-limiting overgrowths (e.g., tumors) which involve hyperplasia, hypertrophy, or both. Several species of gall-forming insects, even when occurring on the same leaves of the same species of plant, each elicit a unique self-limiting overgrowth (e.g., zoocecidium) in the plant. Insect-altered plant responses to cytokinin phytohormones also apparently contribute to galling and overgrowth in plants (Misaghi 1982).

At the other end of this spectrum of insect-induced change in plant responses to growth regulators are the relatively harmonious (e.g., proportional) alterations in plant growth, whether increases or decreases. Examples include (1) cercopid-induced reductions in the length and diameter of sugarcanes, and increases in the development of adventitious buds on such canes, and (2) spittle bug-induced rosetting and shortening of internodes on alfalfa and clover (Norris 1979).

Wounding of plants, as during insect herbivory, elicits additional ethylene production. This chemical is an ubiquitous hormone capable of influencing a wide range of plant responses to various subsequent stressful factors in the environment (Misaghi 1982). Insect attack can, for example, alter plant growth, chlorophyll content, foliar senescence, epinasty, adventitious roots, ripening of fruit, and yield, through changed ethylene interactions.

3.3 Air Pollution

Light, temperature, humidity, soil moisture, and nutrition are a few of the identified variables that alter plant responses to air pollutants (Pell 1979). Insect herbivory can alter each of these environmental parameters with regard to plant growth and reproduction (Denno and McClure 1983), and thus it can indirectly influence plant responses to air pollutants. Symptoms of air pollutants commonly appear on plants at certain times but not at others. The timing of symptoms frequently involves the interaction of pollutants with other environmental stresses, such as insect herbivory, to exceed the particular plant's abilities to remain healthy. The abilities of viruses, fungi, and bacteria to influence the sensitivity of plants to air pollutants have been documented (Heagle 1973). Because hundreds of insect species also transmit such organisms into plants (Harris and Maramorosch 1980), the overall importance of these specific insect activities on plants in influencing their responses to air pollutants is very great.

Though a plant is composed of many organs (leaves, stems, roots, fruits, flowers, seeds, etc.) that are exposed to pollutants, the leaves are the primary receptors of air pollutants, and much of the pertinent literature concerns the impact of such substances on foliage (Pell 1979). Though insects also attack most organs of plants, insect herbivory on foliage can have a major impact on the plant (Denno and McClure 1983). In fact, both insects and air pollutants can affect plant health through foliage, and interactions between these two types of environmental stresses to plants can be expected. Air pollutant effects on leaves can entail stomata, cell surfaces, plasma membranes, chloroplasts, and mitochondria (Pell 1979), all of which represent sites along the path which an air pollutant follows in reaching a plant cell.

Stomata are the major portals by which air pollutants (e.g., O_3) reach plant cells; thus the effects of other stresses on stomatal aperture are of major interest. There are many reports that abscisic acid (ABA) hormonally induces stomatal closures (Mansfield 1976) and that foliar applications of ABA reduce O_3 injury to plants (Adedipe et al. 1973). Insect herbivory stimulates ABA levels in leaves (Visscher-Neumann 1982); thus such stress may reduce the susceptibility of plants to air pollutants, such as O_3, by mechanistically closing stomata.

The mechanisms by which the pollutant SO_2 damages leaves are unclear; however, subsidiary cells around guard cells, not the guard cells, seem to be directly affected (Pell 1979). These affected cells lose turgor, and associated guard cells then collapse with stomata in the open position permitting greater conductance and SO_2 uptake. Insect herbivory-increased ABA levels in leaves may reduce the net opening effects of pollutant SO_2 on stomata; however, herbivory also may directly cause dehydration of affected leaf tissues. Such H_2O loss may add to moisture loss in plant cells resulting from SO_2 toxicity. Thus insect herbivory may have either subtractive or additive effects on the toxicity of air pollutants to plants.

Leaf cell type influences the degree of leaf exposure to acid precipitation and thus the degree of injury (Pell 1979). Epidermal cells are normally contacted first in the intact plant; however, insect herbivory can injure other cells, such as mesophyll cells, that are more exposed to the acid. In the intact leaf, mesophyll cells are only exposed via the stomata.

Lignification of plant cell walls has been associated with reduced cell damage by air pollutants (Pell 1979). Insect herbivory clearly can stimulate phenylpropanoid biosynthesis in some plants, and this can increase lignification processes (Chiang et al. 1986a). Thus insect herbivory may reduce plant susceptibility to air pollutants through promoted lignification of cell walls.

The cell membrane appears to be the major site of irreversible and potentially injurious plant plastic strain, injury, and damage caused by environmental stresses, including air pollutants (Levitt 1980). Thus the plasma membrane is the especially obvious common site where the effects of various environmental stresses may be mechanistically interactive. At this level in the plant's cell structure, insect herbivory may alter the the plant's responsiveness to other environmental stresses such as air pollutants. The simplest kind of cell-membrane damage is, of course, membrane laceration. Insect herbivory may lacerate dozens and hundreds of cell membranes per leaf. Cells that do survive the herbivory may be more susceptible to subsequent environmental stresses such as air pollutants.

Besides simple laceration of cell membranes, the next sequential form of membrane damage apparently is a dysfunction of membrane pumps, and therefore of the active transport process (Levitt 1980). This effect may be initiated by protein denaturation or by a dehydration strain leading to an extension of the membrane-pump proteins. A major effect of insect herbivory is plant cell dehydration (moisture loss); thus herbivory can have a drastic influence at this membrane level on the disrupted cell's or the whole plant's ability to respond subsequently to air pollutants.

Another kind of plant injury common to different environmental stresses involves metabolic disturbances. These may be subclassed as (1) deficiencies of an intermediate or terminal metabolite or (2) toxicities (Levitt 1980). End product deficiencies (e.g., of carbohydrate or protein) are especially common effects of several stresses including insect herbivory and air pollutants.

In summary, current knowledge indicates that insect attack (e.g., herbivory) on a plant can render it more susceptible to some air pollutants but less susceptible to others. These opposite effects are explainable because different environmental stresses are now known to affect plants through the common mechanisms which may be either stimulated or inhibited; thus a dichotomy of effects may result from two or more forms of stress.

4. PHYSICAL STRESSES

Physical parameters of the environment that may function as stresses to plant growth, differentiation, and development include (1) physical properties of soils, (2) extreme temperature, and (3) electromagnetic energy and sound. Some effects of prior insect attack (e.g., herbivory) upon a plant's subsequent responses to such physical environmental stresses are now discussed.

As presented in Section 3, a plant employs some common mechanisms (e.g., lignification) in defending itself against several different types of environmental stress. Thus a prior exposure to insect herbivory may be expected to alter a plant's subsequent responses to physical, as well as chemical, environmental conditions. Though defense is a significant energy expense to plants (Denno and McClure 1983), through the use of common principles and mechanisms of defense they may obtain significant protection from several stresses for a given (same) energy expense. As an example, lignification can increase a plant's protection against (1) herbivory, (2) drought, (3) pathogens, (4) wind, (5) temperature extreme, (6) hail, and (7) flooding. Thus most of the mechanisms that plants use for defense against chemical stresses also benefit it against physical factors in the environment.

Insect defoliation of a plant can increase its subsequent susceptibility to high soil temperature by reducing the plant's effective leaf-surface area available for transpiration. High transpiration rate is a major plant mechanism for tolerating a high soil temperature (Levitt 1980).

Leaf feeding as done by the two-spotted mite, *Tetranychus urticae* Koch, on peppermint, *Mentha piperita* L., may improve the plant's short-term abilities to cope with a moisture deficiency in the soil (DeAngelis et al. 1983a). This resultant improved peppermint ability to cope with the stress of deficient soil moisture was attributed to an increased amount of soluble carbohydrates in the water-stressed leaves fed on by the mites. The added carbohydrates may help such leaves maintain cell turgor under drought conditions through an osmotic adjustment (osmoregulatory) mechanism (DeAngelis et al. 1983b). However, such mite feeding also will increase water loss (dehydration) of peppermint leaves, and if such plants are stressed by a soil moisture deficiency, an extraordinary closure of stomata will result. This will cause a major inhibition of CO_2 influx (gaseous exchange) in the injured leaves which is the predominant factor determining photosyntheis and finally crop yield.

Greenbug, *Schizaphis graminum* (Rondani), feeding on drought-resistant winter wheat, *Triticum aestivum* L., increased cell membrane injury which rendered the plants more susceptible to soil moisture deficiency (Dorschner et al. 1986). Without greenbug feeding, the wheat cell membranes were much more stable against solute leakage when subjected to the osmotic stress associated with a soil moisture deficiency. Pink bollworm, *P. gossypiella*, infestation of cotton (*G. hirsutum*) resulted in the bolls opening three days earlier than noninfected ones under soil moisture-deficient conditions, but this difference in opening was only 0.5 day under an adequate soil moisture situation (Kittock et al. 1983).

In summary, very limited information is available on insect attacks affecting subsequent plant responses to physical environmental stresses. Existing studies have yielded findings that fit well into the idea that many stresses affect plants through common mechanisms such as cell dehydration and alterations of cell-membrane stability and permeability.

5. BIOLOGICAL STRESSES

It is now clear that plants do not usually respond specifically to a particular environmental stress such as insect herbivory. It appears instead that evolutionarily successful plants have a strategy of defense against a spectrum of stressful events more or less regardless of the environmental agent. Thus some individuals of plant species or cultivars become more resistant to some insect herbivores as well as some other phytopathogens when attacked by a specific microbial pathogen (Kogan and Paxton 1983).

5.1 Insect-Inducible Plant Defenses against Insects

Inducible plant factors that yield resistance to insects include those stimulated by various environmental stresses as well as prior insect attacks on the plant. Gypsy moth, *Lymantria dispar* (L.), larval feeding on gray birch and black oak leaves can cause changes in those plants (trees) that may reduce the survival and growth rates of other larvae feeding on the foliage several days after the initial herbivory (Wallner and Walton 1979). Pea aphids, *Acyrthosiphon pisum* (Harris), feeding on alfalfa (*Medicago sativa* L.) stimulate the biosynthesis of coumestrol, which makes the plants more resistant to subsequent aphids (Loper 1968). European pine sawfly, *Neodiprion sertifer* (Geoffroy), larval feeding on pine needles increases the polyphenolic content of the needles which makes those tissues more resistant to the insect (Thielges 1968). Haukioja and Niemelä (1977) reported that lepidopterous larval feeding on birch foliage increased the phenolic content of the leaves which proved deleterious to the larvae. Cotton bollworm, *Heliothis armigera* (Hubner), feeding on cotton (*G. hirsutum*) increased phenolics in the plant which were detrimental to the insect (Kogan and Paxton 1983). Lygus bug feeding on sugar beets resulted in increased quinones, which inhibited subsequent bug feeding (Hori 1973), and likewise such initial feeding on Chinese cabbage increased the phenol content in such plants, which also decreased insect feeding at a later time (Hori and Atalay 1980). Feeding by the striped cucumber beetle, *Acalyma vittata* (Fabr.), on squash plants resulted in amounts of cucurbitacins in the plants which are deterrent to the squash beetle, *Epilachna tredecimnotata* (Carroll and Hoffman 1980).

5.1.1 Proteinase Inhibitors

Insect-induced chemicals in given plants may include proteinase inhibitors (Green and Ryan 1972, Ryan 1983). Such substances are polypeptides and proteins that complex tightly with proteolytic enzymes to inhibit their catalytic activity. The majority of such inhibitors in plants are specific for the serine class of proteinases, which are common as major food protein-digesting enzymes in insects (Ryan 1983).

The understanding that induced proteinase inhibitors in plants may function as protective substances against subsequent insect attack (e.g., herbivory) originated with studies by Mikel and Standish (1947). Applebaum and associates (e.g., Birk and Applebaum 1960) expanded on the initial proteinase-inhibitor studies, and showed the presence of two such fractions in soybean meal that inhibited both the growth of *Tribolium* larvae and the proteolytic activity of their midgut proteinases. Ryan and colleagues (Ryan 1983) clearly elucidated that insect (Colorado potato beetle) herbivory on tomato (*Lycopersicon esculentum*) induced accumulation of proteinase inhibitors in those plants. Inhibitors accumulate as protein globules or membraneless bodies in the central vacuole, or lysosomal compartment, of the plant cells. In such compartments the inhibitors can survive and protect the tissues from subsequent insect attack over long periods.

5.1.2 Elicitor Chemicals

In the situation where insect herbivory induces proteinase inhibitors in genetically competent plants, a chemical signal or wound hormone called a proteinase-inhibitor-inducing factor (PIIF) is released at, or near, the wound site (Green and Ryan 1972). The magnitude of the induced signal depends on the cultivar, the stage of plant growth, and the location and severity of the insect-caused wound. Transport of such an elicitor occurs within a few hours, and it apparently moves in the phloem (Ryan 1983). The released factors are informational macromolecules that communicate with cells in nearby or distant tissues of the attacked plant by binding with receptor sites on each cell's membrane. The first isolated PIIF, from tomato leaves, was a highly methylated polysaccharide containing galacturonic acid, rhamnose, galactose, arabinose, and fucose. A second potent inhibitor, PIIF II, was soon isolated from "Russet Burbank" potatoes (Ryan 1983). Recent research indicates that smaller oligosaccharide fragments also may function as signals. Some pectic polysaccharides do not exhibit PIIF activity.

Active pectic oligosaccharide fragments (e.g., β-glucans), released at the site of initial insect attack, are produced by degradation of the plant cell wall. This is accomplished by hydrolytic enzymes that either are compartmentalized in the intact plant and mixed with the cell wall during wounding, or are introduced into the plant by the invading insect (Bishop et al. 1981).

5.1.3 Insect-Antifeedant Phytoalexins

Wound (stress)-induced elicitor chemicals may stimulate plant production of defensive chemicals other than proteinase inhibitors (Kogan and Paxton 1983; Chiang et al. 1986a, b) (Fig. 1). These compounds may include phytoalexins, some of which are antifeedants to some insect species or biotypes as well as antimicrobial agents. Sweet potato weevil, *Cylas formicarius elegantulus* (Summers), feeding and tunneling in potatoes induce the biosynthesis of the furanoterpenoid phytoalexin, ipomeamarone, which is allelochemic to the insects (Akazawa et al. 1960). The phytoalexins, phoretin in *Malus pumila* and 2′,6′-dihydroxy-4′-methoxy dihydrochalcone from *Populus deltoides* Bartr., inhibited smaller European elm bark beetle, *Scolytus multistriatus* (Marsh.), feeding (Norris 1977). Mexican bean beetle (MBB) (*Epilachna varivestis* Mulsant) consumption of about 30% of the middle trifoliolate leaflet on the second leaf (first trifoliolate) of V3 stage PI 227687

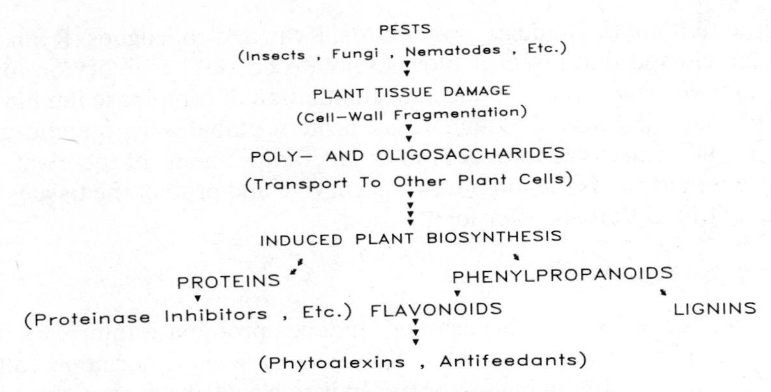

Figure 1. Pest-induced plant resistances to pests.

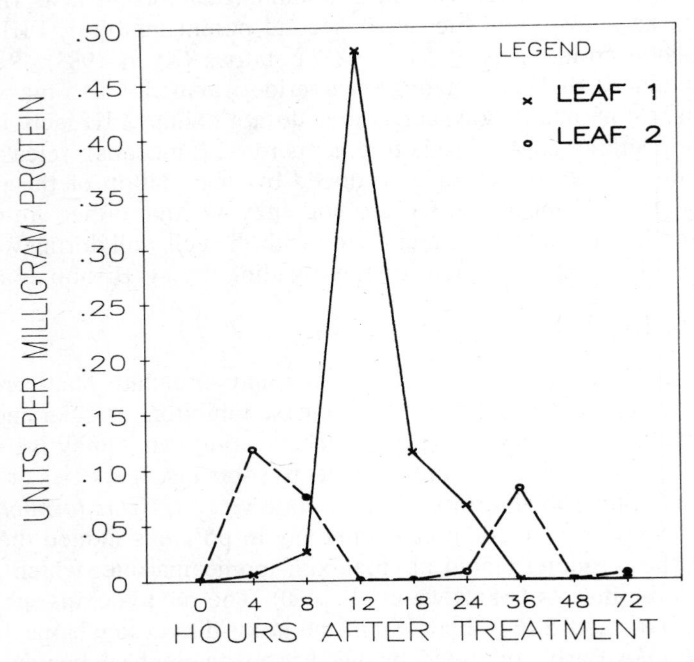

Figure 2. L-Phenylalanine ammonia-lyase (PAL) activity in the indicated leaf of healthy (non-stressed) PI 227687 soybean plants, at the given time after treatment, growing in a biotron room. Leaf 2(M) refers to the middle leaflet of the first trifoliolate leaf; and Leaf 2(O), to the outside (lateral) leaflet(s) of such a leaf. (From Chiang et al. 1986a)

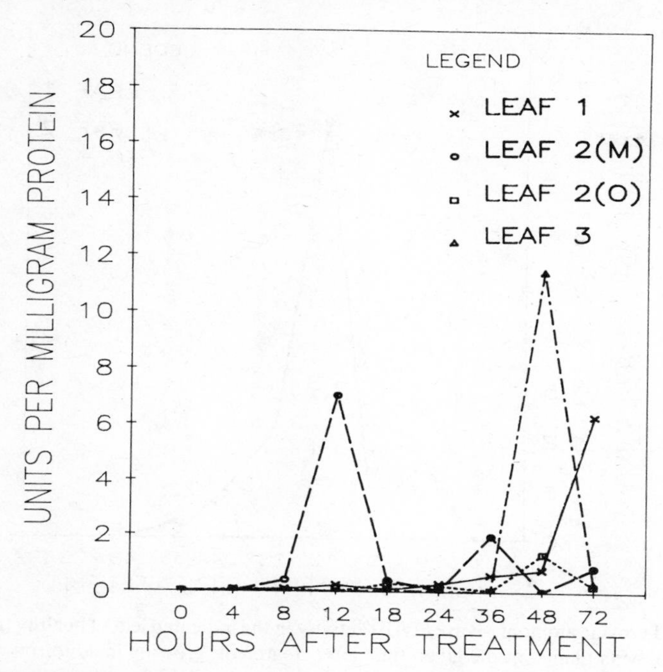

Figure 3. L-Phenylalanine ammonia-lyase (PAL) activity in the indicated leaf of experimentally stressed PI 227687 soybean plants, at the given time after the herbivory (stress) treatment, growing in a biotron room. (From Chiang et al. 1986a)

soybean plants resulted, within 72 hours, in increased antifeedants in leaves throughout these plants (Chiang et al. 1986a). The chemicals responsible for such inducible resistance to MBB and *Trichoplusia ni* apparently are phenylpropanoid compounds (Chiang et al. 1986a, Khan et al. 1986). This induced resistance in PI 227687 soybean plants also was positively correlated with temporally unique, e.g., characteristic, increased L-phenylalanine ammonialyase (PAL) and L-tyrosine ammonialyase (TAL) enzyme activities (Figs. 2–5) (Chiang et al. 1986a, b). The induced chemicals responsible for the increased resistance of PI 227687 to insects thus may be flavonoids (e.g., phytoalexins). This possibility is also supported by their extractability from plant tissues and their thin-layer chromatographic mobilities in several solvent systems (Khan et al. 1986).

The observed temporal aspects of the enzymatic (PAL and TAL) changes associated with greater MBB resistance in PI 227687 are compatible with the reported timetables for similar enzymatic events associated with microorganismal induction of flavonoid phytoalexins in soybean plants (Sequeira 1983, Darvill and Albersheim 1984). Initial mite feeding on the cotyledons of cotton seedlings resulted in subsequent mites not growing as rapidly on such "exposed" seedlings as on "nonexposed" cotton plants

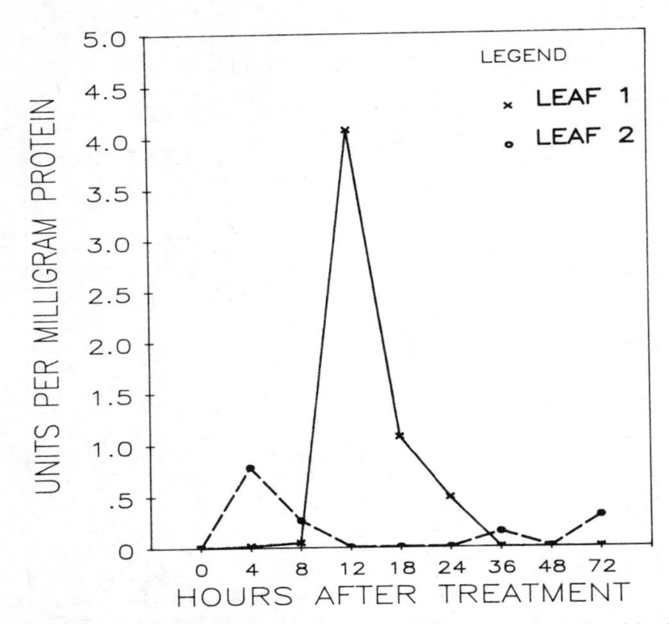

Figure 4. L-Tyrosine ammonia-lyase (TAL) activity in the indicated leaf of healthy (nonstressed) PI 227687 soybean plants, at the given time after treatment, growing in a biotron room. (From Chiang et al. 1986a)

(Karban and Carey 1984); the systemically induced differences probably involve phytoalexin chemicals.

5.2 Insect-Inducible Plant Defenses against Insect-Transmitted and Other Phytopathogens

Initial attacks on genetically competent individuals within a plant (tree) species or cultivar by insects and their ectosymbiotic microbes (Norris 1979) may result in those plants being more resistant to subsequent invasions by insect-transmitted plant pathogens and other plant pathogens (Reid et al. 1967; Hanover 1975; Wong and Berryman 1977; Raffa and Berryman 1982, 1983; Raffa et al. 1985; Shigo 1984). Such elicited defense consists of at least two components, for example, (1) local lesion or barrier formation (compartmentalization; Shigo 1984) and (2) chemical (e.g., monoterpene) synthesis and translocation (Raffa and Berryman 1982, 1983; Raffa et al. 1985). The lesion formation limits the initially attacking agents and the chemical synthesis and translocation especially render the overall plant more resistant to subsequent microbial stresses, for instance, for at least several days or weeks.

Specific more toxic and repellent chemicals induced initially in trees attacked by a scolytid beetle and an ectosymbiotic fungus include limonene,

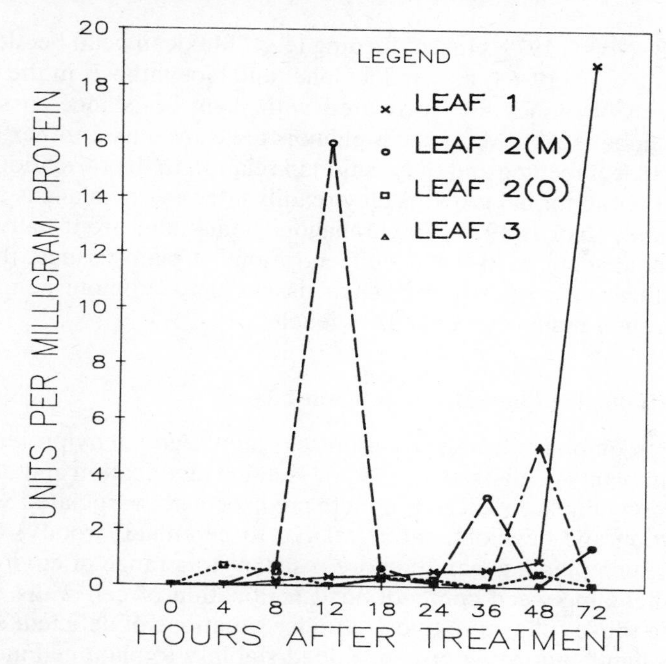

Figure 5. L-Tyrosine ammonia-lyase (TAL) activity in the indicated leaf of experimentally stressed PI 227687 soybean plants, at the given time after the herbivory (stress) treatment, growing in a biotron room. Leaf 2(M) refers to the middle leaflet of the first trifoliolate leaf; and Leaf 2(O), to the outside (lateral) leaflet(s) of such a leaf. (From Chiang et al. 1986a)

myrcene, and Δ-3-carene (Raffa and Berryman 1982, 1983; Raffa et al. 1985). These terpenoid compounds strongly inhibit phytopathogenic microorganisms, including the ectosymbiotic fungi of scolytid insects, and the beetles themselves. In addition, initial insect attack on genetically competent plant species, and cultivars may elicit the production of classical flavonoidal phytoalexins which render the plants more resistant to some phytopathogenic microorganisms (Kogan and Paxton 1983).

5.3 Insect-Inducible Plant Defenses against Phytophagous Nematodes

A major plant parameter that is involved in their defense against phytophagous nematodes, such as *Aphelenchoides ritzemabosi*, is tissue hypersensitivity (acute necrosis) (Giebel 1982). Such rapid plant cell necrosis can readily trap nematodes physically and nutritionally. The feeding of many species of insects, especially the piercing-sucking Homoptera and Hemiptera, may cause such necrotic areas in plant tissues which may increase their resistance to some nematodes. The capsid bugs (Hemiptera: Miridae) present an especially strong example of such pertinent severe phytotoxemia

(Miles 1968, Norris 1979). Insect feeding (e.g., Mexican bean beetle on soybean, Chiang et al. 1986a) may induce phenolic biosynthesis in the attacked tissues. Such phenolics are correlated with plant resistance to some nematodes (Giebel 1982). Many such phenols are stored as a rather nontoxic glycoside. Insect feeding and oviposition in relation to the formation of plant galls and other abnormal growths may readily alter auxin balances and titers in such tissues (Norris 1979). Phytophagous nematodes are usually favored by high indole acetic acid (IAA) concentrations in plant tissues; thus some insect feeding and oviposition in such tissues may promote nematode invasions through inducing higher IAA levels.

5.4 Insect-Inducible Plant Defenses against Weeds

Insect attack on plants frequently promotes increased phenylpropanoid biosynthesis in plants (Chiang et al. 1986a), yielding lignification and deposition of other polyphenolics. These products are especially associated with plant differentiation and development as related to perennial (woody) traits and may make such plants generally more resistant to a range of environmental stresses, including weed competition. Lignification of cell walls can especially make those cells more tolerant under situations of deficient soil moisture. Such lignification can provide added stability to plant cell membranes which are mechanistically a major site of stress expression irrespective of its biotic or abiotic nature (Levitt 1980).

5.5 Summary

Consideration of the marked effects that an initial insect attack on a plant may have on the latter's responses to subsequent attacks by stressful biotic agents reveals conclusively that plants have evolved a strategy of defense against such stresses irrespective of the involved agent. Plants thus do not usually express resistance specifically against a given biotic agent. Stress in plants ultimately is expressed at the cell-membrane and cell-metabolism levels; thus a plant's defense strategy seeks to stabilize these levels of function and activity regardless of the specific stressful agent.

6. MECHANICAL STRESSES

Perhaps the most dynamic and useful discussion of the effects of an initial insect attack upon a plant's responses to subsequent mechanical stresses is one that focuses attention at the major, common sites (levels) of stress expression regardless of the inflicting mechanical agent. Levitt (1980) has effectively developed a scientifically based rationale for such interpretations. Following his organized facts from the scientific literature, one may conclude that mechanical forces largely stress plants at the following mechanistic levels (1) cell-membrane stability (semipermeability) and (2) cell-metabolism regulation (e.g., gene expression). Prior discussions in this chapter clearly indicated that insect attack on plants may also stress them at especially these two mechanistic levels, and thus by so doing, insects surely may alter a plant's responses to subsequent mechanical environmental stresses.

The documented (Chiang et al. 1986; Chiang and Norris 1984, 1985) induction of lignification processes in plant cells by an initial insect attack readily alters (e.g., increases) the involved plant's resistance to many subsequent mechanical agents. Lignification variously alters indirectly the chances of a plant's cell membranes to retain their stability (semipermeability) when subsequently exposed to mechanical stresses (Levitt 1980). If cell-membrane semipermeability is affected by such initial stress, then the involved cells' subsequent responses at the metabolism-regulation and gene-expression levels to mechanical stresses also are likely to be affected. This is true because changes in cell-membrane permeability may significantly alter the ionic balance and charge distribution in the cell cytoplasm with resultant effects on gene regulation and metabolism.

Insect-induced lignification in a plant may render it phenotypically more resistant to subsequent mechanical stresses through numerous altered characteristics. Lignification is a process that is centered in plant cell walls and especially strengthens them physically (mechanically). Thus insect-induced lignification should increase a plant's subsequent tolerances of (1) wind and its shaking effects, (2) cultivation damages to roots, including associated reduction in the volume of roots available to physically anchor the plant, and (3) the abrasive actions of volcanic ash, sand, and dust particles. Lignification also renders plants more tolerant of lightning strikes. Most highly lignified, older, large plants (trees) have survived one or more lightning strike, whereas herbaceous plants generally are killed by a single strike.

In summary, the frequent, if not universal, induction of plant lignification processes by insect attack usually will increase the plant's tolerances of subsequent mechanical stresses such as wind, abrasive particles, and lightning. This is true because lignification physically (mechanically) strengthens cell walls, and this usually will improve the involved plant cells' abilities to maintain the required semipermeability of the cell membrane and regulate the gene expression and cell metabolism.

7. SUMMARY AND CONCLUSIONS

Insect attack apparently triggers, in genetically competent plant species and cultivars, the implementation of an evolved defense strategy. This strategy seems to have evolved as a generalized defense. It does not usually seem to be directed specifically at a given stressful environmental agent such as an insect attacker. Thus initial exposure to limited insect herbivory, oviposition, microbial-transmission, regurgitation, excretion, and so on, usually renders the genetically competent plant more resistant to the variety of other environmental stressful factors, whether physical, chemical, biological, or mechanical. Obviously, if the initial insect attack "devours" or otherwise "destroys" the plant, then secondary considerations regarding plant resistance become trivial.

A more generalized strategy of inducible plant defense against the diverse stressful environmental factors seems evolutionarily acceptable because all such agents ultimately "take their toll" on a plant at the common levels of gene expression, cell metabolism, and cell physiology. All agents must affect the semipermeability (stability) of the cell membrane to effect stress on the plant. Efforts to improve plant resistance to insect pests can and should benefit from all efforts to improve plant resistance to stress. Entomologists, plant pathologists, physiologists, plant breeders, geneticists, and plant nutritionists, for example, should not proceed in this field as if they possessed a unique cause.

REFERENCES

Adedipe, N. O., H. Khatamian, and D. P. Ormrod. 1973. Stomatal regulation of ozone phytotoxicity in tomato. *Z. Pflanzenphysiol* 68: 323–328.

Akazawa, T., I. Uritani, and H. Kubota. 1960. Isolation of ipomeamarone and two coumarin derivatives from sweet potato roots injured by the weevil, *Cylas formicaris elegantulus*. *Arch. Biochem. Biophys.* 88: 150–156.

Birk, Y., and S. W. Applebaum. 1960. Effect of soybean trypsin inhibitors on the development and midgut proteolytic activity of *Tribolium castaneum* larvae. *Enzymologia* 22: 318–326.

Bishop, P. D., D. J. Makus, G. Pearce, and C. A. Ryan. 1981. Proteinase inhibitor-inducing factor activity in tomato leaves resides in oligosaccharides enzymically released from cell walls. *Proc. Natl. Acad. Sci. USA* 78: 3536–3540.

Carroll, C. R., and C. A. Hoffman. 1980. Chemical feeding deterrent mobilized in response to insect herbivory and counter adaptation by *Epilachna tredecimnota*. *Science* 209: 414–416.

Chiang, H. S., and D. M. Norris. 1984. "Purple stem," a new indicator of soybean stem resistance to bean flies (Diptera: Agromyzidae). *J. Econ. Entomol.* 77: 121–125.

Chiang, H. S., and D. M. Norris. 1985. Expression and stability of soybean resistance to agromyzid beanflies. *Insect Sci. Appl.* 6: 265–270.

Chiang, H. S., D. M. Norris, A., Ciepiela, P. Shapiro, and A. Oosterwyk. 1986a. Inducible versus constitutive PI 227687 soybean resistance to Mexican bean beetle, *Epilachna varivestis*. *J. Chem. Ecol.* 13: 741–749.

Chiang, H. S., D. M. Norris, A. Ciepiela, A. Oosterwyk, P. Shapiro, and M. Jackson. 1986b. Comparative constitutive resistance in soybean lines to Mexican bean beetle. *Entomol. Exp. Appl.* 42: 19–26.

Clarkson, D. T. 1985. Factors affecting mineral nutrient acquisition by plants. *Annu. Rev. Plant Physiol.* 36: 77–115.

Darvill, A. G., and P. Albersheim. 1984. Phytoalexins and their elicitors—a defense against microbial infection in plants. *Annu. Rev. Plant Physiol.* 35: 243–275.

DeAngelis, J. D., R. E. Berry, and G. W. Krantz. 1983a. Evidence for spider mite (Acari:

Tetranychidae) injury-induced leaf water deficits and osmotic adjustment in peppermint. *Environ. Entomol.* 12: 336–339.

DeAngelis, J. D., R. E. Berry, and G. W. Krantz. 1983b. Photosynthesis, leaf conductance, and leaf chlorophyll content in spider mite (Acari: Tetranychidae)-injured peppermint leaves. *Environ. Entomol.* 12: 345–348.

Denno, R. F., and M. S. McClure. 1983. *Variable Plants and Herbivores in Natural and Managed Systems.* Academic Press, New York.

Dorschner, K. W., R. C. Johnson, R. D. Eikenbary, and J. D. Ryan. 1986. Insect–plant interactions: greenbugs (Homoptera: Aphididae) disrupt acclimation of winter wheat to drought stress. *Environ. Entomol.* 15: 118–121.

Epstein, E., and A. Läuchli. 1983. Mineral deficiencies and excesses, *In* Kommedahl, T. and P. H. Williams, (eds.), *Challenging Problems in Plant Health.* American Phytopathological Society, St. Paul, MN, pp. 196–205.

Gaylor, M. J., G. A. Buchanan, F. R. Gilliland, and R. L. Davis. 1983. Interactions among a herbicide program, nitrogen fertilization, tarnished plant bugs, and planting dates for yield and maturity of cotton. *Agron. J.* 75: 903–907.

Giebel, J. 1982. Mechanism of resistance to plant nematodes. *Annu. Rev. Phytopathol.* 20: 257–279.

Green, T. R., and C. A. Ryan. 1972. Wound-induced proteinase inhibitor in plant leaves: a possible defense mechanism against insects. *Science* 175: 776–777.

Hanover, J. W. 1975. Physiology of tree resistance to insects. *Annu. Rev. Entomol.* 20: 75–95.

Harris, K. F., and K. Maramorosch. 1980. *Vectors of Plant Pathogens.* Acad. Press, New York.

Haukioja, E., and P. Niemalä. 1977. Retarded growth of a geometrid larva after mechanical damage to leaves of its host tree. *Ann. Zool. Fenn.* 14: 48–52.

Heagle, A. S. 1973. Interactions between air pollutants and plant parasites. *Annu. Rev. Phytopathol.* 11: 365–388.

Hicks, P. M., P. L. Mitchell, E. P. Dunigan, L. D. Newsom, and P. K. Bollich. 1984. Effect of threecornered alfalfa hopper (Homoptera: Membracidae) feeding on, translocation and nitrogen fixation in soybeans. *J. Econ. Entomol.* 77: 1275–1277.

Hori, K. 1973. Studies on the feeding habits of *Lygus disponsi* Linnavuori (Hemiptera: Miridae) and the injury to its host plant. III. Phenolic compounds, acid phosphatase and oxidative enzymes in the injured tissue of sugar beet leaf. *Appl. Ent. Zool.* 8: 103–112.

Hori, K., and R. Atalay. 1980. Biochemical changes in the tissue of Chinese cabbage injured by the bug *Lygus disponsi. Appl. Ent. Zool.* 15: 234–241.

Karban, R., and J. R. Carey. 1984. Induced resistance of cotton seedlings to mites. *Science* 225: 53–54.

Khan, Z. R., D. M. Norris, H. S. Chiang, N. E. Weiss, and A. S. Oosterwyk. 1986. Light-induced susceptibility in soybean to cabbage looper, *Trichoplusia ni* (Lepidoptera: Noctuidae). *Environ. Entomol.* 15: 803–808.

Kittock, D. L., T. J. Henneberry, L. A. Bariola, B. B. Taylor, and W. C. Hofmann. 1983. Cotton boll period response to water stress and pink bollworm. *Agron. J.* 75: 17–20.

Kogan, M., and J. Paxton. 1983. Natural inducers of plant resistance to insects. *In* Hedin, P. A. (ed.), *Plant Resistance to Insects. ACS Symp. Ser.* 208. Washington, DC, pp. 153–171.

Levitt, J. 1980. *Responses of Plants to Environmental Stresses. Water, Radiation, Salt, and Other Stresses*, Vol. 2. Academic Press, New York.

Loper, G. M. 1968. Effects of aphid infestation on the coumestrol content of alfalfa varieties differing in aphid resistance. *Crop Sci.* 8: 104–106.

Maiteki, G. A., and R. J. Lamb. 1985. Growth stages of field peas sensitive to damage by the pea aphid, *Acyrthosiphon pisum* (Homoptera: Aphididae). *J. Econ. Entomol.* 78: 1442–1448.

Mansfield, T. A. 1976. Stomatal behavior. Chemical control of stomatal movements. *Philos. Trans. R. Soc. Lond.* (Ser. B) 273: 541–550.

McCarty, J. C., Jr., J. N. Jenkins, and W. L. Parrott. 1986. Yield response of two cotton cultivars to tobacco budworm infestation. *Crop Sci.* 26: 136–139.

Mikel, C. E., and J. Standish. 1947. Susceptibility of processed soy flour and soy grits in storage to attack by *Tribolium castaneum* (Herbst). *Minn. Agric. Exp. Sta., Tech. Bull.* 178: 1–20.

Miles, P. W. 1968. Insect secretions in plants, *Annu. Rev. Phytopathol.* 6: 137–164.

Misaghi, I. J. 1982. *Physiology and Biochemistry of Plant–Pathogen Interactions*. Plenum, New York.

Norris, D. M. 1977. Role of repellents and deterrents in feeding of *Scolytus multistriatus*. *In* Hedin, P. A. (ed.), *Host Plant Resistance to Pests. ACS Symp. Ser.* 62. Washington, DC, pp. 215–230.

Norris, D. M. 1979. How insects induce disease. *In* Horsfall, J. G. and E. B. Cowling, (eds.), *Plant Disease: an Advanced Treatise, IV, How Pathogens Induce Disease*. Academic Press, New York, pp. 239–255.

Norris, D. M. 1986. Anti-feeding compounds. *In* Haug, G., and H. Hoffmann, (eds.), *Chemistry of Plant Protection, 1, Sterol Biosynthesis, Inhibitors and Anti-Feeding Compounds*. Springer-Verlag, Berlin, pp. 97–146.

Oaks, A., and B. Hirel. 1985. Nitrogen metabolism in roots. *Annu. Rev. Plant Physiol.* 36: 345–365.

Pell, E. J. 1979. How air pollutants induce disease. *In* Horsfall, J. G. and E. B. Cowling, (eds.), *Plant Disease: An Advanced Treatise. IV. How Pathogens Induce Disease*. Academic Press, New York, pp. 273–292.

Raffa, K. F., and A. A. Berryman. 1982. Accumulation of monoterpenes and associated volatiles following inoculation of grand fir with a fungus transmitted by the fir engraver, *Scolytus ventralis* (Coleoptera: Scolytidae). *Can. Entomol.* 114: 797–810.

Raffa, K. F., and A. A. Berryman. 1983. Physiological aspects of lodgepole pine wound responses to a fungal symbiont of the mountain pine beetle, *Dendroctonus ponderosae* (Coleoptera: Scolytidae). *Can. Entomol.* 115: 723–734.

Raffa, K. F., A. A. Berryman, J. Simasko, W. Teal, and B. L. Wong. 1985. Effects of grand fir monoterpenes on the fir engraver, *Scolytus ventralis* (Coleoptera: Scolytidae) and its symbiotic fungus. *Environ. Entomol.* 14: 552–556.

Reid, R. W., H. S. Whitney, and J. A. Watson. 1967. Reactions of lodgepole pine to attack by *Dendroctonus ponderosae* Hopkins and blue stain fungi. *Can. J. Bot.* 45: 1115–1116.

Reynolds, H. T., P. L. Adkisson, R. F. Smith, and R. E. Frisbie. 1982. Cotton insect pest management. *In* Metcalf, R. L. and W. H. Luckmann, (eds.), *Introduction to Insect Pest Management*. 2d ed. Wiley-Interscience, New York, pp. 375–441.

Ryan, C. A. 1983. Insect-induced chemical signals regulating natural plant protection responses. *In* Denno, R. F. and M. S. McClure, (eds.), *Variable Plants and Herbivores in Natural and Managed Systems*. Academic Press, New York, pp. 43–60.

Sequeira, L. 1983. Mechanisms of induced resistance in plants. *Annu. Rev. Microbiol.* 37: 51–79.

Shigo, A. L. 1984. Compartmentalization: a conceptual framework for understanding how trees grow and defend themselves. *Annu. Rev. Phytopathol.* 22: 189–214.

Sinha, S. K., and K. S. Savithri. 1978. Biology of yield in food legumes. *In* Singh, S. R., H. F. van Emden, and T. A. Taylor, (eds.), *Pest of Grain Legumes: Ecology and Control*. Academic Press, New York, pp. 233–240.

Smith, J. R., and R. L. Nelson. 1986. Relationship between seed-filling period and yield among soybean breeding lines. *Crop Sci.* 26: 469–472.

Stickler, F. C., and A. W. Pauli. 1961. Leaf removal in grain sorghum. I. Effects of certain defoliation treatments on yield and components of yield. *Agron. J.* 53: 99–102.

Thielges, B. A. 1968. Altered polyphenol metabolism in the foliage of *Pinus sylvestris* associated with European pine sawfly attack. *Can. J. Bot.* 46: 724–725.

Visscher-Neumann, S. 1982. Plant growth hormones affect grasshopper growth and reproduction. *In* Visser, J. H. and A. K. Minks, (eds.), *Proc. 5th Intl. Symp. Insect–Plant Relationships*. Centre for Agric. Publish. and Documents, Wageningen, The Netherlands. pp. 57–62.

Wallner, W. E., and G. S. Walton. 1979. Host defoliation: a possible determinant of gypsy moth population quality. *Ann. Entomol. Soc. Am.* 72: 62–67.

Williams, W. P., F. M. Davis and G. E. Scott. 1983. Second-brood southeastern corn borer infestation levels and their effect on corn. *Agron. J.* 75: 132–134.

Wong, B. L., and A. A. Berryman. 1977. Host resistance to the fir engraver, *3*, Lesion development and containment of infection by resistant *Abies grandis* innoculated with *Trichosporium symbioticum*. *Can. J. Bot.* 55: 2358–2365.

11

PLANT-INDUCED STRESSES AS FACTORS IN NATURAL ENEMY EFFICACY

Merle Shepard

Department of Entomology
International Rice Research Institute
Manila, Philippines

D. L. Dahlman

Department of Entomology
University of Kentucky
Lexington, Kentucky

I. INTRODUCTION

The action of certain biological control agents may be enhanced or retarded by the type of plant upon which their host feeds. Invertebrate predators, parasitoids, or diseases may be stressed directly by plant chemicals or structure, indirectly by chemicals that are obtained from plants by their hosts or by lack of food quality for hosts that feed on unsuitable plants (Duffey 1980, Duffey et al. 1986). The term stress, as used in this chapter, refers to "any adverse force or condition, strain, pressure, or inconvenience to which the insect is subjected" (Steinhaus 1956). Thus any one of numerous factors in the chemical or physical environment of the insect has the potential for causing stress.

The tritrophic interactions between host plant, herbivore, and natural enemies are poorly understood, and almost no effort has been expended toward developing breeding programs that consider the effects of the host plant on natural enemies. Although Wilson and Huffaker (1976) pointed out that "biological control, together with plant resistance, form nature's principal means of keeping phytophagous insects within bounds in environments which otherwise favor them," most studies of bi- or tritrophic interactions involving biological control agents have been conducted in the laboratory. Results from these studies may provide insights into factors that influence the regulation of insect pest populations in natural and agricultural ecosystems, but success under field conditions remains to be demonstrated. Duffey and Bloem (1986) discussed the interactions between plant defenses, herbivores, and parasitoids as they relate to biological control.

Insects that feed on resistant plants normally require longer to develop, experience higher mortality and decreased fecundity, and usually undergo some behavioral changes. The net result of these factors is an overall reduction in the equilibrium density of the potential host (pest). Although this is the ultimate objective of the control program, these low host densities may result in the decimation of beneficial insect populations (Pimentel and Wheeler 1973, Kennedy et al. 1975, Casagrande and Haynes 1976, Schuster et al. 1976). It is likely that in cases where resistance is strongly expressed, pest biotypes that can overcome resistance will develop pest status more rapidly than will those feeding on moderately resistant plants. As Bergman and Tingey (1979) suggested, "development of pest biotypes capable of colonizing a resistant host might be slowed by diversification in selection pressure associated with partial control by natural enemies."

The importance of allelochemicals to biological control has been suggested by several authors, including Barbosa et al. (1982), Campbell and Duffey (1979), Greenblatt and Barbosa (1981), Schultz (1983), and Smith (1957, 1978). Although most of this work dealt with parasitoids, there are reports of plant-mediated effects on predators either directly or via their arthropod host (Rothschild et al. 1973, Dodd 1973, Rogers and Sullivan 1986, Orr and Boethel 1986). In addition, the action of certain entomopathogens may be retarded by plant chemicals (Felton and Dahlman 1984).

A general overview of interactions of resistant plants with biological control agents was provided by Bergman and Tingey (1979) and more recently by Boethel and Eikenbary (1986). There are only a few examples where resistant plants compliment the action of natural enemies (Starks et al. 1972, Obrycki et al. 1983). In most cases results from interactions of pests and their natural enemies that colonize resistant plants are unknown, and the need for more research in this area has been emphasized by Herzog et al. (1984) and Herzog and Funderburk (1985).

Some possible stresses placed on natural enemy populations by resistant plants were suggested by Painter (1951). Prey populations may be reduced to levels where predators and parasites cannot effectively find them; host plants may induce changes in the behavior or physiology of the insect herbivore which could impact adversely on natural enemy populations; or the suitability of the host may be altered through the presence of plant chemicals. Plant height, canopy development, and presence of trichomes, for example, may influence communities of natural enemies directly or through changes induced in their hosts.

This review will focus on those plant–herbivore–natural enemy (predators, parasitoids, and entomopathogens) interactions found primarily in agricultural ecosystems. An in-depth review of theory related to the interactions of plants, herbivores, and natural enemies was published by Price et al. (1980). Vinson (1976) provided a review of the host selection process by parasitoids, including the influence of plant volatiles and morphology. Plant-mediated variation in host suitability for parasitoids has been reviewed by Barbosa et al. (1982).

2. PLANT STRESSES TO PARASITOIDS

2.1 Plant-Induced Stresses via the Insect Host

There are several examples of how the nutrition of host plants or presence of toxic substances can influence parasitoids of insects that feed on these plants. Because of the difficulty in separating and studying direct and indirect interactions in the field, most studies have been conducted in the laboratory, glasshouse, or in field cages. Thus the overall effects of plant–herbivore–parasitoid interactions on parasitoid and host insect population dynamics in the field can only be hypothesized.

The examples to follow will illustrate that parasitoids can be stressed when their hosts feed on specific plants that may or may not be resistant to the herbivore. Food plants of the California red scale, *Aonidiella aurantii* (Mask.), significantly affected the biology of its parasitoids (Smith 1957). The percentage of females of the parasitoid *Comperiella bifasciata* How. ranged from about 45 when hosts were reared on orange to 84 when yucca was provided as food. Oviposition, survival, and adult longevity of *C. bifasciata* and *Habrolepis rouxi* Comp. reared from scales on different hosts were significantly affected.

The braconid *Microplitis demolitor* Wilkinson often failed to develop in its soybean looper host, *Pseudoplusia includens* (Walker), when the host fed on resistant soybean (Yanes and Boethel 1983). Powell and Lambert (1984) recorded that development of another braconid, *M. crociepes* (Cresson), was retarded when its hosts, *Heliothis zea* (Boddie) and *Heliothis virescens* (F.), were reared on resistant soybean. Mueller (1983) observed the host plant to be an important factor in determining both the likelihood of parasitoid attack and the probability of successful parasitism by *M. Crociepes* in both *H. virescens* and *H. zea*. Mortality of a South American population of a tachinid parasitoid, *Voria ruralis* (Fallen), was higher when the soybean cultivar ED73-377 was used for food than when a susceptible line was fed to its soybean looper host (Grant and Shepard 1985). Another recent example of the impact of plant genotype on the parasitoid of a leaf-feeding insect was reported by Orr and Boethel (1985) in a study in which they fed *P. includens* leaves of either resistant or susceptible soybean. The encyrtid parasitoid *Copidosoma truncatellum* (Dalman) parasitizes the eggs of *P. includens* and emerges from "mummified" late stage larvae. Although mortality of the herbivore was higher on resistant plants, they reported no significant differences in numbers of parasitoids that emerged from individual hosts reared on susceptible or resistant plants. However, the parasitoid pupation period was significantly longer in hosts reared on resistant plants. Results of later research revealed that few *C. truncatellum* emerged from *P. includens* reared on resistant plants (Beach and Todd 1986). Likewise, emergence of the parasitoid *Pediobius foveolatus* Crawford was reduced when its host, the Mexican bean beetle (*Epilachna varivestis* Mulsant), fed on leaves from resistant soybean plants (Dover 1984) (Fig. 1).

Resistant soybean varieties have also been shown to influence the biology of egg parasitoids at both the third and fourth trophic levels. The thelyotokous egg parasitoid *Telenomus chloropus* Thomson exhibited reduced adult emergence and almost a 50% reduction in progeny production when reared from eggs of *Nezara viridula* (L.) that had been reared on resistant soybean (Orr et al. 1985). Orr and Boethel (1986) also reported a reduction in fecundity of *Telenomus podisi* Ashmead when pre-imaginal development occurred in eggs of the spined soldier bug, *Podisus maculiventris* Say, which had been reared on *P. includens* larvae that had been reared on resistant soybean foliage.

Campbell and Duffey (1979) reported that α-tomatine acquired by *H. zea* from artificial diet provided protection from the parasitoid *Hyposoter exiguae* (Viereck). However, they subsequently showed that α-tomatine toxicity was reduced by various dietary phytosterols. Therefore only tomato plants with low total sterol-α-tomatine ratios were most toxic to *H. exiguae* (Campbell and Duffey 1981).

Cardiac glycosides found in higher concentrations in certain host plants utilized by the butterfly *Danaus chrysippus* (L.) caused a reduction in parasitism by tachinids. The mechanism of protection afforded by the plants containing the cardenolide was not understood (Smith 1978). Transfer of a

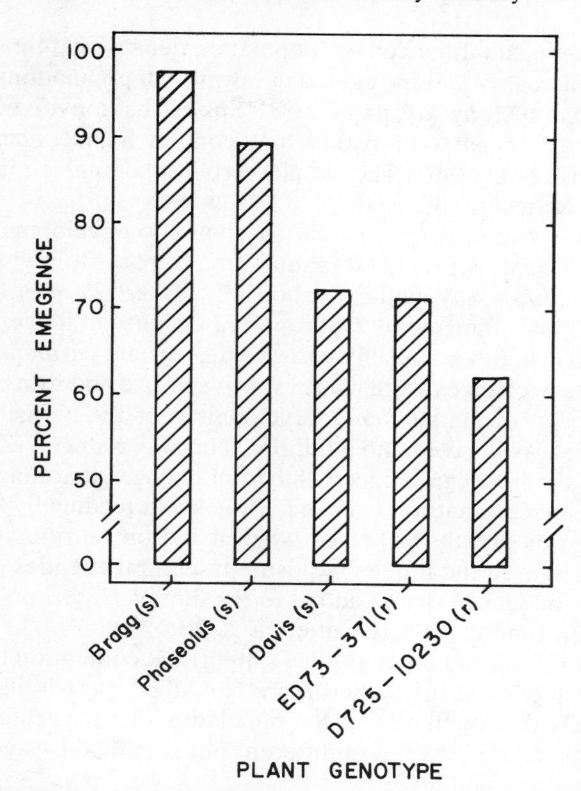

Figure 1. Emergence of the parasitoid *Pediobius foveolatus* from Mexican bean beetle larvae fed foliage from *Phaseolus* sp. and susceptible (*s*) and resistant (*r*) soybean. (Modified after Dover 1984)

plant alkaloid through a herbivorous larva, *Nyctemera annulata* Boisduval, to a braconid was demonstrated by Benn et al. (1979).

Resistant sorghum and barley varieties complimented the action of the parasite *Lysiphlebus testaceipes* (Cresson) in regulating populations of biotype "C" of the greenbug, *Schizaphis graminum* (Rodani). However, fewer and smaller "mummies" (parasitized hosts) were produced (Starks et al. 1972).

Successful development and emergence of the gregarious parasitoid *Apanteles congregatus* (Say) on tobacco hornworm, *Manduca sexta* (L.), was severely retarded when 0.1% nicotine was added to the host diet (Thurston and Fox 1972). When tested under field conditions using two varieties of tobacco that had about a 10-fold difference in nicotine content, the number of parasitoids reaching adulthood on the low nicotine treatment was nearly twice that on the high nicotine treatment (Thorpe and Barbosa 1986). Because tobacco is a favorite host of the hornworm, it is clear that the presence of this plant chemical influenced the population density and thus the overall efficiency of this parasitoid in regulating hornworm populations in nature.

Some insects, such as *Myzus persicae* Sulzer, have evolved the ability to avoid feeding on parts of plants that contain high concentrations of nicotine (Guthrie et al. 1962) This would certainly minimize nicotine stress on parasitoids attacking this host.

Host nutrition can affect parasitoid survival and development. Changes in the quality of host diet resulted in inhibition of parasitoid embryogenesis of *Aphaereta pallipes* (Say) within the housefly, *Musca domestica* L. (Lange and Bronskill 1964). Pimentel (1966) reported changes in longevity and percent parasitism in another housefly parasitoid, *Nasonia vitripennis* (Walker), when host dietary changes were made. When sucrose and iron content were altered in the diet of *M. persicae*, development of the parasite *Aphelinus asychis* Walker was delayed and adult life span was reduced (Zhody 1976).

The diet of the host can induce behavioral changes that may increase or decrease the impact of natural enemies. Time spent feeding by the imported cabbageworm was greatly increased when it fed on nutritionally deficient plants. This increased the time of exposure to natural enemies (Slansky and Feeny 1977). Condensed tannin added to the diet of *H. virescens* increased its development time by 21% (Chan et al. 1978).

Results from these and other studies support the contention that the quality and quantity of food of the herbivore can affect parasitoids that attack them. It is likely that regulation of the population of a particular insect host by a parasitoid can be affected in different but significant ways, depending on the plant upon which it feeds.

2.2 Plant Stresses to Parasitoids Directly

2.2.1 Stress by Plant Morphology

Many species of parasitoids are affected by morphological features of the plant upon which their host lives. *Bracon mellitor* (Say), a native parasitoid of the boll weevil, *Anthonomus grandis* (Boh.), was able to parasitize 50% of the immature weevils in cotton with frego bract fruit compared to only 7% in cotton with normal fruit (McGovern and Cross 1976). However, *Heteroloccus grandis* (Burks) oviposited more frequently in weevil larvae in normal bract fruit. These examples illustrate the importance of identifying those parasitoid species that may be playing a major role in regulating a pest population and suggest that plant structures may be altered to favor their population buildup for the purpose of biological control.

The presence of dense trichomes on cotton leaves stressed *Trichogramma pretiosum* (Riley) such that fewer eggs of *H. zea* were parasitized (Treacy et al. 1985a). May (1951), Turner (1983), and Treacy et al. (1986) suggested that pubescence on the leaves reduced the mobility of parasitoids and increased the time required to locate host eggs, thereby rendering them less

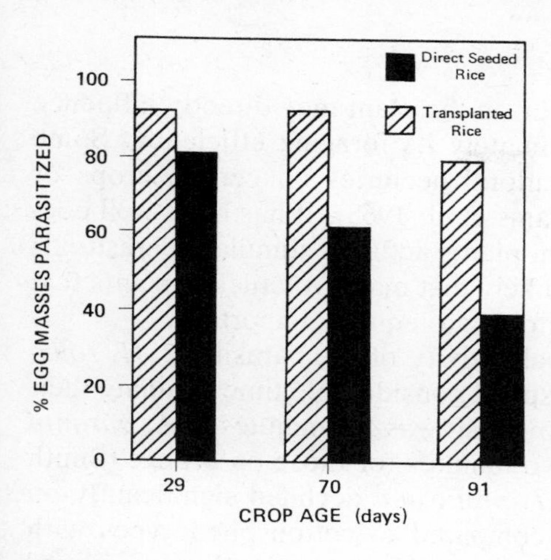

Figure 2. Percent parasitism of yellow stem borer, *Scirpophaga incertulas*, egg masses at 29, 70, and 91 days after seeding in direct seeded and transplanted rice; Laguna Province, Philippines. (Modified after Shepard and Arida 1986)

effective. The overall negative effect of pubescent leaves on the efficacy of parasitoids was reported in tobacco (Rabb and Bradley 1968, Elsey and Chaplin 1978) and soybean (Turner 1983). Obrycki and Tauber (1984) reported that glandular pubescence of potato plants grown in the greenhouse reduced the effectiveness of 11 aphidophagous species. However, this reduction did not carry over to the field.

There are many examples of herbivore hosts that are attacked by parasitoids on one food plant but not on another (Smith 1957, Walker 1940, Zwolfer and Krous 1957). For example, the tachninid parasitoid, *Gonia capitata* (De Geer), preferentially searches native grasses for potential hosts even though its cutworm host, *Perosagrotis* spp., is found in greater abundance on cultivated grains (Strickland 1923). Similarly, survival of *M. sexta* was much poorer when eggs were laid on Jimson weed than when laid on tobacco. These differences resulted primarily from differences in egg parasitism, which was high on Jimson weed and very low on tobacco (Katanyukul and Thurston 1979). Ullyett (1953) observed that the parasitoid *Chelonus texanus* Cresson preferred to search for its host, *Loxostege frustralis* Walker, at a particular height from the ground.

Percent parasitism of yellow stem borer, *Scirpophaga incertulas* (Walker), egg masses was lower in direct-seeded than in transplanted rice (Fig. 2). (Shepard and Arida 1986). The explanation given for this was that the denser canopy of the direct-seeded rice made discovery of the egg masses by the parasitoids more difficult. Overall, parasitism in rice established by both techniques declined as the season progressed. Parasitization by *Trichogramma* spp. was greater on hosts that were found in soybean with weeds or interplanted with corn than in soybean grown in monoculture (Altieri et al. 1981). Thus the relative plant species mix may influence host habitat location by the parasitoid, and searching may be directed preferentially in a particular crop due to the physical or chemical makeup of the crop.

2.2.2 *Stress by Plant Chemicals*

The presence of certain chemicals in or on the plant may directly influence a parasitoid's search pattern and ultimately its foraging efficiency. Some chemicals, namely, those from extrafloral nectaries on certain crops or weeds, may enhance parasitism (Streams et al. 1968). It has been well documented that volatile synomones from plants actually stimulate parasitoids to search (Nordlund et al. 1985). It is likely that many volatiles in plants that retard or prevent searching by parasitoids are equally important.

Sago palm leaves induced abnormal activity of the parasitoids *H. rouxi* and particularly *C. bifasciata* which spend considerable time cleaning their legs and antennae. Time required to oviposit was 5.3 minutes in *A. aurantii* hosts reared on palm compared to two minutes for those on orange (Smith 1957). Parasitism of *H. zea* eggs by *T. pretiosum* declined significantly on cotton without extrafloral nectaries compared to cotton phenotypes with nectaries (Treacy et al. 1985a). It was suggested that extrafloral nectar served as an essential food source for the parasitoid. In many cases it was not clear if plant chemicals, plant morphology, or both, were involved in differential parasitism.

3. PLANT STRESSES TO PREDATORS

As with parasitoids, influences of plants on predators may be due to the chemistry or morphology of the plant. Stresses may impact on predators directly from the plant or indirectly via the prey. However, it is difficult to separate influences emanating from the plant itself from those which are sequestered by the prey and later taken up by the predator. The reason for this difficulty is that predators require several hosts to develop to maturity, some of which may or may not have fed on toxic plants. Many predators, such as mirids and lygaeids, feed to some extent on the plant in addition to being predatory. Thus in most cases we can only surmise the influence of plant chemistry and/or structure on predators.

The growth rate of *Geocoris punctipes* (Say), an important predator in soybean, was reduced and mortality increased when its prey, the velvetbean caterpillar, *Anticarsia gemmatalis* Hübner, was reared on resistant soybean genotypes (Rogers and Sullivan 1986) (Fig. 3). The predator also feeds to some extent on the host plant. Direct effects from feeding by *G. punctipes* on resistant plants increased nymphal mortality.

Orr and Boethel (1986) also reported that *P. maculiventris* was affected by soybean antibiosis in a manner similar to that of its prey, *P. includens*. Pre-imaginal development time and mortality were increased, whereas cumulative weight gain tended to be reduced.

Glabrous cotton contained fewer predatory arthropods than did pilose genotypes, with the exception of *Orius insidiosus* (Say) (Shepard et al. 1972).

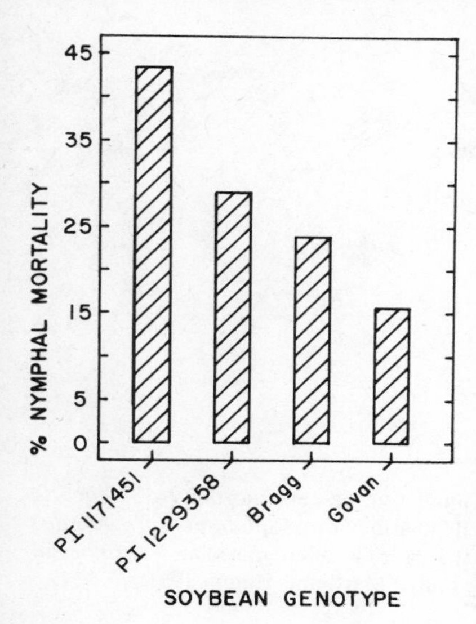

Figure 3. Nymphal mortality of the predator *Geocoris punctipes* when its host, the velvetbean caterpillar *Anticarsia gemmatalis*, was fed foliage from susceptible and resistant (PIs) soybean. (Modified after Rogers and Sullivan 1986)

Mussett et al. (1979) later reported that predatory arthropods were reduced by 68% on resistant varieties of cotton. It was not clear whether this reduction was due to a decline in the prey items, particularly *Pseudatomoscelis seriatus* (Reuter), the cotton fleahopper, or to other factors associated with the resistant cotton that impact directly on the predators. Early instars of the predators *Chrysopa carnea* Stephens and *Coleomegilla maculata* (De Geer) searched cotton faster than tobacco because of the higher trichome density on tobacco (Elsey 1974). Likewise, successful attacks by *Chrysopa rufrilabris* (Burmeister) on *H. zea* eggs were reduced with increased density of trichomes on cotton leaves (Treacy et al. 1985b). Nymphal mortality and rate of development of the predatory insects *P. maculiventris* (Say) and *Perillus bioculatus* F. were affected by both the host insect species and the type of plant upon which the host was reared (Landis 1937).

A preferred height for oviposition was found for syrphids (Chandler 1968). Interestingly this height did not necessarily represent the preferred searching zone of the predators. Although examples of influences of plants on predators are not numerous, it is likely that these influences exist for a wide array of predatory arthropods, and further research will undoubtedly uncover many other cases.

An interesting example of an insect herbivore that sequesters chemicals from host plants and uses them for defense was reported by Eisner et al. (1974). Larvae of the European pine sawfly, *Neodiprion sertifer* (Geoff.), collect resin from pine trees (*Pinus* sp.) and store it in diverticular pouches of the foregut. When predators such as ants and spiders disturb the larvae, the resin is discharged orally and acts as a strong deterrent.

Some predatory insects may actually be stressed by feeding on insects

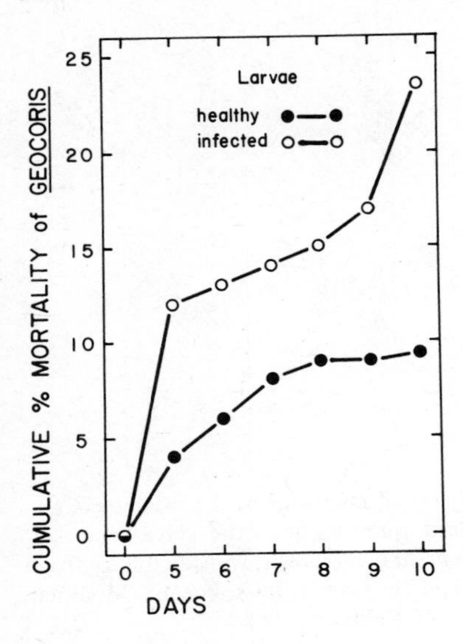

Figure 4. Cumulative percent mortality of *Geocoris punctipes* fed fall armyworm (*Spodoptera frugiperda*) larvae infected with the microsporidian *Vairimorpha* sp. (Modified after Marti and Hamm 1985)

that are infected by disease. When the predatory lygaeid, *G. punctipes*, fed on fall armyworm, *Spodoptera frugiperda* (J. E. Smith), infected with the microsporidian *Vairimorpha* sp., survival of the predator was significantly reduced (Fig. 4) (Marti and Hamm 1985). Decreased survival was attributed to the non-nutritive tissue of the infected host. Thus, though disease epizootics by entomopathogens may reduce pest populations, their effects on other natural enemies may be attributed directly to nutritional deficiencies in diseased hosts as well as to host numbers that are too low to maintain a viable population of parasitoids and predators.

4. PLANT–HERBIVORE–PATHOGEN INTERACTIONS

In a review paper published in 1983, Kogan and Paxton stated that "it is thus apparent that a plant challenged by herbivores, by plant pathogens, or otherwise stressed by specific chemicals does not remain biochemically indifferent." The main focus of their paper was on postinfestation induction of resistance by arthropods in a manner similar to that known to result from pathogen infections. Nevertheless, it is important to remember that the interaction between the insect and the plant is dynamic and that there are a number of ways in which the plant can influence the response of the insects to both naturally occurring and artificially disseminated pathogens.

As long ago as 1962 it was reported that antibacterial substances in the leaves insects eat may play a direct part in protecting them against ingested bacteria (Kushner and Harvey 1962). However, these insects were not protected against injected bacteria. The presumption was that these antibacterial substances became incorporated into the gut contents and, as the insects encountered pathogenic bacteria, they were destroyed before they could render harm to the insect. It was also demonstrated that increased age of foliage helped to improve the resistance of insects to bacterial attack, especially by species of *Bacillus*. A decade later Morris (1972) reported on the antibiotic effects on *B. thuringiensis* Berliner of a number of terpenes extracted from various species of forest trees. Subsequently, Merdan et al. (1975) showed that mortality due to *Bacillus* infection was significantly reduced in lepidopterous larvae feeding on certain plants containing antibacterial substances. However, the response was species specific both with the insect and the host plant. About this same time Iizuka et al. (1974) reported that protocatechuic acid and *p*-hydroxybenzoic acid found in *Bombyx mori* (L.) feces had antibacterial activity. A few years later, Koike et al. (1979) demonstrated that caffeic acid was responsible for the antibacterial activity. Caffeic acid was a metabolite of chlorogenic acid, a component of mulberry leaves. Silkworm larvae raised on artificial diet that did not contain chlorogenic acid were susceptible to infection by *Streptococcus* species. Jones (1984) reviewed the literature on plant phenolics, terpenoids, and other plant allelochemicals that reduce susceptibility of many Lepidoptera to entomopathogenic bacteria. Hayashiya (1978) isolated a red fluorescent protein from the digestive juice of the silkworm larvae that had antiviral activity against silkworm NPV (see Hayashiya 1978 for other references). Each of these papers dealt in rather general terms with effects that enhanced the resistance of the insect to various bacterial and viral pathogens.

More recently, Felton et al. (1986) have shown that rutin and chlorogenic acid significantly inhibited the infectivity of NPV *in vivo* and *in vitro* in larvae of *H. zea*. Both of these compounds have been implicated in resistance of tomato to insects. Elleman and Entwistle (1985a, b) have demonstrated that leaf glands of certain cotton varieties produce a "dust" containing calcium and magnesium that rapidly inactivates the NPV of *Spodoptera littoralis* (Boisduval). The use of EDTA and/or a low pH reduced the rate of inactivation, whereas additional $MgCl_2$ reversed the effect of EDTA. The possibility of Mg^{2+} inhibition of alkaline protease was eliminated because the same effect was observed with heat-killed alkaline protease. However, the specific mode of action remains to be elucidated.

Examples of potentiation of pathogens by various plant components are more rare and more recent. Felton and Dahlman (1984) reported that the LC_{50} of *B. thuringiensis* in *M. sexta* was reduced in larvae that had fed upon an artificial diet containing the nonprotein amino acid, L-canavanine, an analogue of L-arginine. Another interesting interaction between insects, plants, and pathogens was reported by Hare and fellow workers. They initially observed that tobacco infected with tobacco mosaic virus seemed to be protected against attack by both a fungus and a bacterium. They also noted that the reproductive rate of the green peach aphid was reduced when feeding on infected plants. This suggested that plant stress resulting from disease rendered the plant more resistant to other pathogens and predators (McIntyre et al. 1981). Hare and Andreadis (1983) followed up on this work and showed that the susceptibility of the Colorado potato beetle *Leptinotarsa decemlineata* (Say) to the fungal pathogen *Beauveria bassiana* (Balsamo) was greater when the larvae were reared on the least suitable species of *Solanum*. Greenhouse-grown plants appeared to be less suitable than field-grown plants, and host plant suitability declined with host plant age. In another paper Hare (1983) reported that the growth rate of *M. sexta* was reduced 27% when the larvae were reared on leaves infected with tobacco mosaic virus.

Enhancement of pathogen activity via plant structures is poorly documented. One example was reported by Wellso (1973) who found that greater numbers of trichomes increased development time and lowered adult weight and survivorship of the cereal leaf beetle. The small larvae needed to bite the trichomes repeatedly before getting to the nutritious adaxial cells. Larger larvae ingested the siliceous trichomes, which were undigestible and capable of lacerating the midgut, permitting entry of bacteria into the hemocoel where it produces a septicemic condition.

An interesting area of research that has recently emerged deals with interactions between insects and various plants infected with endophytic fungi (see Chapter 12). Little is presently known about the third trophic level effect of consumption of the infected plant tissue by insect herbivores. However, because infected plants possess qualities that render them resistant to insect attack, it seems likely that the reduced viability of the herbivore would have at least an indirect detrimental effect on the parasitoids and predators of these herbivores.

5. SUMMARY AND CONCLUSIONS

It is clear that plant-induced stresses are factors that influence the effectiveness of natural populations of insect parasitoids and predators. There is considerable evidence that the success of parasitoids is indirectly related to the nutritional quality of the food of the host and to the types of secondary chemicals produced by specific host plants. In addition, parasitoids are directly affected by plant stress in that stressed plants often produce larger amounts of secondary metabolites that may influence the search patterns of parasitoids. These metabolites, in the form of trichome secretions, for example, may literally capture the small adult parasitoids.

Less information is available on the effects of plant stress on predators, but it is likely that predator fitness can be altered as a result of feeding on less fit prey taken from stressed plants. These prey may be of poorer nutritional quality and/or may be suffering from chronic disease.

There are numerous documented cases where consumption of certain plant materials directly protect the herbivore from bacterial and/or viral attack. In a few cases plant allelochemicals have been shown to enhance the effectiveness of a pathogen.

The net impact of plant-induced stresses on populations or communities of biological control agents has been the subject of only fairly recent studies. As our understanding of these tritrophic interactions increase, we will be able to plan more long-term insect control strategies, including the direct and indirect influences of plant structure and chemicals on natural enemy communities as well as on the target pest. Examples of this type of approach might include breeding plants that are preferred for feeding by mirid and lygaeid predators, or breeding for high concentrations of plant volatiles that may enhance the searching activity of parasitoids. Biotechnological approaches may be employed to expedite breeding programs and to produce plant materials once specific chemicals and their metabolic pathways are identified. These genetically engineered plants should possess characteristics that improve the economically beneficial interactions between insect predators/parasitoids and their potential prey/hosts. It is suggested that host plant resistance programs should try to achieve only medium levels of resistance to insect pests, thus allowing subthreshold level reservoirs of pests to provide food for biocontrol agents. This would be expected to reduce the rate of selection for insect biotypes that can overcome resistance.

REFERENCES

Altieri, M. A., W. J. Lewis, D. A. Nordlund, R. C. Gueldner, and J. W. Todd. 1981. Chemical interactions between plants and *Trichogramma* wasps in Georgia soybean fields. *Prot. Ecol.* 3: 259–263.

Barbosa, P., J. A. Saunders, and M. Waldvogel. 1982. Plant-mediated variation in herbivore suitability and parasitoid fitness. *In Proc. Fifth Intl. Symp. Insect-Plant Relationships*, Wageningen, pp. 63–71.

Beach, R. M., and J. W. Todd. 1986. Foliage consumption and larval development of parasitized and unparasitized soybean looper, *Pseudoplusia includens* (Lep.: Noctuidae), reared on a resistant soybean genotype and effects on an associated parasitoid, *Copidosoma truncatellum* (Hym: Encyrtidae). *Entomophaga* 31: 237–242.

Benn, M., J. DeGrave, C. Cnanasunderam, and R. Hutchins. 1979. Host-plant pyrrolizidine alkaloids in *Nyctemera annulata* Boisduval: Their persistence through the life cycle and transfer to a parasite. *Experientia* 35: 731–732.

Bergman, J. M., and W. M. Tingey. 1979. Aspects of interaction between plant genotypes and biological control. *Bull. Entomol. Soc. Am.* 25: 275–279.

Boethel, D. J. and R. D. Eikenbarry (eds.). 1986. *Interactions of Plant Resistance and Parasitoids and Predators of Insects*. Ellis Horwood Ltd., West Sussex, England.

Campbell, B. C., and S. S. Duffey. 1979. Tomatine and parasitic wasps: potential incompatibility of plant antibiosis with biological control. *Science* 205: 700–702.

Campbell, B. C., and S. S. Duffey. 1981. Alleviation of α-tomatine-induced toxicity to the parasitoid, *Hyposoter exiguae*, by phytosterols in the diet of the host, *Heliothis zea*. *J. Chem. Ecol.* 7: 927–946.

Casagrande, R. A., and D. L. Haynes. 1976. The impact of pubescent wheat on the population dynamics of the cereal leaf beetle. *Environ. Entomol.* 5: 153–159.

Chan, B. G., A. C. Waiss, and M. Lukefahr. 1978. Condensed tannin, an antibiotic chemical from *Gossypium hirsutum*. *J. Insect. Physiol.* 24: 113–118.

Chandler, A. E. F. 1968. Height preferences for oviposition of aphidophagous Syphidae (Diptera). *Entomophaga* 13: 187–195.

Dodd, G. D. 1973. Integrated control of the cabbage aphid *Breviocoryne brassicae* (L.). Ph.D. dissertation. University of Reading.

Dover, B. A. 1984. Development of the parasitoid, *Pediobius foveolatus* in Mexican bean beetles reared on different dietary sources. M.S. Thesis. Clemson University, Clemson, SC.

Duffey, S. S. 1980. Sequestration of plant natural products by insects. *Annu. Rev. Entomol.* 25: 447–477.

Duffey, S. S., and K. A. Bloem. 1986. Plant defense, herbivore/parasite interactions and biological control. *In* Kogan, M. (ed.), *Ecological Theory and Integrated Pest Management* Environ. Sci. & Tech. Series. Wiley, New York, pp. 135–183.

Duffey, S. S., K. A. Bloem, and B. C. Campbell. 1986. Consequences of sequestration of plant natural products in plant–insect–parasitoid interactions. *In* Boethel, D. J. and R. D. Eikenbary, (eds.), *Interactions of Plant Resistance and Parasitoids and Predators of Insects*. Ellis Howard, West Sussex, England, pp. 31–60.

Eisner, T., J. S. Johnessee, and J. Carrel. 1974. Defensive use by an insect of a plant resin. *Science* 184: 996–999.

Elleman, C. J., and P. F. Entwistle. 1985a. Inactivation of a nuclear polyhedrosis virus on cotton by the substances produced by the cotton leaf surface glands. *Ann. Appl. Biol.* 106: 83–92.

Elleman, C. J., and P. F. Entwistle. 1985b. The effect of magnesium ions on the solubility of polyhedral inclusion bodies and its possible role in the inactivation of the nuclear polyhedrosis virus of *Spodoptera littoralis* by the cotton leaf gland exudate. *Ann. Appl. Biol.* 106: 93–100.

Elsey, K. D. 1974. Influence of plant host on searching speed of two predators. *Entomophaga* 19: 3–6.

Elsey, K. D., and J. F. Chaplin. 1978. Resistance of tobacco introduction 1112 to the tobacco budworm and green peach aphid. *J. Econ. Entomol.* 71: 723–725.

Felton, G. W., and D. L. Dahlman. 1984. Allelochemical induced stress: effects of L-canavanine on the pathogenicity of *Bacillus thuringiensis* in *Manduca sexta*. *J. Invertebr. Pathol.* 44: 187–191.

Felton, G. W., S. S. Duffey, P. V. Vail. H. K. Kaya, and J. Manning. 1986. Interaction of nuclear polyhedrosis virus with catechols: potential incompatibility for host-plant resistance against noctuid larvae. *J. Chem. Ecol.* 13: 947–957.

Grant, J. F., and M. Shepard. 1985. Influence of three soybean genotypes on development of *Voria ruralis* and foliage consumption by its soybean looper host. *Florida Entomol.* 68: 672–677.

Greenblatt, J. A., and P. Barbosa. 1981. Effects of host's diet on two pupal parasitoids of the gypsy moth: *Brachymeria intermedia* (Nees) and *Coccygomimus turionellae* (L.). *J. Appl. Ecol.* 18: 1–10.

Guthrie, F. E., W. V. Campbell, and R. L. Baron. 1962. Feeding sites of the green peach aphid with respect to its adaptation to tobacco. *Ann. Entomol. Soc. Am.* 55: 42–46.

Hare, J. D. 1983. Manipulation of host suitability for herbivore pest management. *In* Denno, R. F. and M. S. McClure (eds.), *Variable Plants and Herbivores in Natural and Managed Systems*. Academic Press, New York, pp. 655–680.

Hare, J. D., and T. G. Andreadis. 1983. Variation in the susceptibility of *Lepinotarsa decem-*

lineata (Coleoptera: Chrysomelidae) when reared on different host plants to the fungal pathogen, *Beauveria bassiana* in the field and laboratory. *Environ. Entomol.* 12: 1892–1896.

Hayashiya, K. 1978. Red fluorescent protein in the digestive juice of the silkworm larvae on host-plant mulberry leaves. *Entomol. Exp. Appl.* 24: 228–236.

Herzog, D. C., and J. E. Funderburk. 1985. Plant resistance and cultural practice interactions with biological control. *In* Hoy, M. A. and D. C. Herzog (eds.), *Biological Control in Agricultural IPM systems*. Academic Press, New York, pp. 67–88.

Herzog, D. C., J. L. Stimac, D. G. Boucias, and V. H. Waddill. 1984. Compatibility of biological control in soybean insect management. *In* Adkisson, P. L. and Shijun Ma (eds.), *Proc. Chinese Acad. Sci./U.S. Natl. Acad. Sci. Joint Symp. Biol. Control of Insects*. Science Press, Beijing, China (1984), pp. 37–60.

Iizuka, T., S. Koike, and J. Mizutani. 1974. Antibacterial substances in feces of silkworm larvae reared on mulberry leaves. *Agr. Biol. Chem.* 38: 1549–1550.

Jones, C. G. 1984. Microorganisms as mediators of plant resource exploitation by insect herbivores. *In* Price, P. W., C. N. Slobodchikoff and W. S. Gaud, (eds.), *A New Ecology: Novel Approaches to Interactive Systems*. Wiley, New York, pp. 53–99.

Katanyukul, W., and R. Thurston. 1979. Mortality of eggs and larvae of the tobacco hornworm on Jimson weed and various tobacco cultivars in Kentucky. *Environ. Entomol.* 8: 802–807.

Kennedy, G. G., A. N. Kishaba, and G. W. Bohn. 1975. Response of several pest species to *Cucumis melo* L. lines resistant to *Aphis gossypii* Glover. *Environ. Entomol.* 4: 653–657.

Kogan, M., and J. Paxton. 1983. Natural inducers of plant resistance to insects. *In* Hedin, P. A. (ed.), *Plant Resistance to Insects*. American Chemical Society, Washington, DC, pp. 153–171.

Koike, S., T. Iizuka, and J. Mizutani. 1979. Determination of caffeic acid in the digestive juice of silkworm larvae and its antibacterial activity against the pathogenic *Streptococcus faecalis* AD-4. *Agric. Biol. Chem.* 43: 1727–1731.

Kushner, D. J., and G. T. Harvey. 1962. Antibacterial substances in leaves: their possible role in insect resistance to disease. *J. Insect Pathol.* 4: 155–184.

Landis, B. J. 1937. Insect hosts and nymphal development of *Podisus maculiventris* Say and *Perillus bioculatus* F. (Hemiptera: Pentatomidae). *Ohio J. Sci.* 37: 252–259.

Lange, R., and J. F. Bronskill. 1964. Reactions of *Musca domestica* to parasitism by *Aphaereta pallipes* with special reference to host diet and parasitoid toxin. *Z. Parasitologie* 25: 193–210.

Marti, O. J., Jr., and J. J. Hamm. 1985. Effect of *Vairimorpha* sp. on survival of *Geocoris punctipes* in the laboratory. *J. Entomol. Sci.* 20: 354–358.

May, A. W. S. 1951. Jassid resistance of the cotton plant. *Queensland J. Agric. Sci.* 8: 43–68.

McGovern, W. L., and W. H. Cross. 1976. Effects of two cotton varieties on levels of boll weevil parasitism (Coleoptera: Curculionidae). *Entomophaga* 21: 123–125.

McIntyre, J. L., J. A. Dodds, and J. D. Hare. 1981. Effects of localized infections of *Nicotiana tabacum* by tobacco mosaic virus on systemic resistance against diverse pathogens and an insect. *Phytopathology* 71: 297–301.

Merdan, A., H. Abdel-Rahman, and A. Soliman. 1975. On the influence of host plants on insect resistance to bacterial diseases. *Z. Angew. Entomol.* 78: 280–285.

Morris, O. N. 1972. Inhibitory effects of foliage extracts of some forest trees on commercial *Bacillus thuringiensis*. *Can. Entomol.* 104: 1357–1361.

Mueller, T. F. 1983. The effect of plants on the host relations of a specialist parasitoid of *Heliothis* larvae. *Ent. Exp. Appl.* 34: 78–84.

Mussett, K. S., J. H. Young, R. G. Price, and R. D. Morrison. 1979. Predatory arthropods and their relationship to fleahoppers on *Heliothis*-resistant cotton varieties in southwestern Oklahoma. *Southwestern Entomol.* 4: 35–39.

Nordlund, D. A., R. B. Chalfant, and W. J. Lewis. 1985. Response of *Trichogramma pretiosum* females to volatile synomones from tomato plants. *J. Entomol. Sci.* 20: 372–376.

Obrycki, J. J., and M. J. Tauber. 1984. Natural enemy activity on glandular pubescent potato plants in the green house: an unreliable predictor of effects in the field. *Environ. Entomol.* 13: 679–683.

Obrycki, J. J., M. J. Tauber, and W. M. Tingey. 1983. Predator and parasitoid interaction with aphid-resistant potatoes to reduce aphid densities: a two year field study. *J. Econ. Entomol.* 76: 456–462.

Orr, D. B., and D. J. Boethel. 1985. Comparative development of *Copidosoma truncatellum* (Hymenoptera: Encyrtidae) and its host *Pseudoplusia includens* (Lepidoptera: Noctuidae), on resistant and susceptible soybean genotypes. *Environ. Entomol.* 14: 612–616.

Orr, D. B., and D. J. Boethel. 1986. Influence of plant antibiosis through four trophic levels. *Oecologia* 70: 242–249.

Orr, D. B., D. J. Boethel, and W. A. Jones. 1985. Biology of *Telenomus chloropus* (Hymenoptera: Scelionidae) from eggs of *Nezara viridula* (Hemiptera: Pentatomidae) reared on resistant and susceptible soybean genotypes. *Can. Entomol.* 117: 1137–1142.

Painter, R. H. 1951. *Insect Resistance in Crop Plants*. Univ. Kansas P., Lawrence.

Pimentel, D. 1966. Wasp parasite (*Nasonia vitripennis*) survival on its house fly host (*Musca domestica*) reared on various foods. *Ann. Entomol. Soc. Am.* 59: 1031–1038.

Pimentel, D., and A. G. Wheeler, Jr. 1973. Influence of alfalfa resistance on a pea aphid population and its associated parasites, predators and competitors. *Environ. Entomol.* 2: 1–11.

Powell, J. E., and L. Lambert. 1984. Effects of three resistant soybean genotypes on development of *Microplitis croceipes* and leaf consumption by its *Heliothis* spp. hosts. *J. Agr. Entomol.* 1: 169–176.

Price, P. W., C. E. Bouton, P. Gross, B. A. McPheron, J. N. Thompson, and W. E. Weis. 1980. Interactions among three trophic levels: influence of plants on interactions between herbivores and natural enemies. *Annu. Rev. Ecol. Syst.* 11: 41–65.

Rabb, R. L., and J. R. Bradley. 1968. The influence of host plants on parasitism of eggs of the tobacco hornworm. *J. Econ. Entomol.* 61: 1249–1252.

Rogers, D. J., and M. Sullivan. 1986. Nymphal performance of *Geocoris punctipes* (Say) Hemiptera: Lygaeidae) on pest resistant soybeans *Environ. Entomol.* 15: 1032–1036.

Rothschild, M., J. von Euw, and T. Reichstein. 1973. Cardiac glycosides in a scale insect (*Aspidiotus*), a ladybird (*Coccinella*) and a lacewing (*Chrysopa*). *J. Entomol.* 48: 89–90.

Schultz, J. C. 1983. Impact of variable plant defensive chemistry on susceptibility of insects to natural enemies. *In* Hedin, P. A. (ed.), *Plant Resistance to Insects*. American Chemical Society, Washington, DC, pp. 37–55.

Schuster, M. F., M. J. Lukefahr, and F. G. Maxwell. 1976. Impact of nectariless cotton on plant bugs and natural enemies. *J. Econ. Entomol.* 69: 400–402.

Shepard, M., and G. S. Arida. 1986. Parasitism and predation of yellow stem borer, *Scirpophaga incertulas* (Walker) (Lepidoptera: Pyralidae), eggs in transplanted and direct seeded rice. *J. Entomol. Sci.* 21: 26–32.

Shepard, M., W. Sterling, and J. K. Walker, Jr. 1972. Abundance of beneficial arthropods on cotton genotypes. *Environ. Entomol.* 1: 117–121.

Slansky, F., and P. Feeny. 1977. Stabilization of the rate of nitrogen accumulation by larvae of the cabbage butterfly on wild and cultivated food plants. *Ecol. Monogr.* 47: 209–228.

Smith, D. A. S. 1978. Cardiac glycosides in *Danaus chrysippus* (L.) provide some protection against an insect parasitoid. *Experientia* 34: 844–846.

Smith, J. M. 1957. Effects of the food of California red scale, *Aonidiella aurantii* (Mask.) on reproduction of its hymenopterous parasites. *Can. Entomol.* 89: 219–230.

Starks, K. J., R. Muniappan, and R. D. Eikenbary. 1972. Interaction between plant resistance and parasitism against the greenbug on barley and sorghum. *Ann. Entomol. Soc. Am.* 65: 650–655.

Steinhaus, E. A. 1956. Stress as a factor in insect disease. *Proc. 10th Intl. Cong. Entomol.* 4: 725–730.

Streams, F. A., M. Shahjahan, and H. G. LeMasurier. 1968. Influence of plants on the parasitization of the tarnished plant bug by *Leiophron pallipes*. *J. Econ. Entomol.* 61: 996–999.

Strickland, E. H. 1923. Biological notes on parasites of prairie cutworms. *Can. Dept. Agr. Entomol. Br. Bull.* 22: 1–40.

Thorpe, K. W. and P. Barbosa. 1986. Effects of consumption of high and low nicotine tobacco by *Manduca sexta* (Lepidoptera: Sphingidae) on the survival of the gregarious endoparasitoid *Cotesia congregatus* (Hymenoptera: Braconidae). *J. Chem. Ecol.* 12: 1329–1337.

Thurston, R., and P. M. Fox. 1972. Inhibition by nicotine of emergence of *Apanteles congregatus* from its host, the tobacco hornworm. *Ann. Entomol. Soc. Am.* 65: 547–550.

Treacy, M. F., J. H. Benedict, and M. H. Walmsley. 1985a. Effect of nectariless cotton on a parasite of *Heliothis zea* (Boddie) eggs. *In Proc. Beltwide Cotton Prod. Res. Conf.*, Memphis, pp. 394–395.

Treacy, M. F., G. R. Zummo, and J. H. Benedict. 1985b. Interactions of host-plant resistance in cotton with predators and parasites. *Agric. Ecosys. Environ.* 13: 151–157.

Treacy, M. F., J. H. Benedict, J. C. Segers, R. K. Morrison, and J. D. Lopez. 1986. Role of cotton trichome density in bollworm (Lepidoptera: Noctuidae) egg parasitism. *Environ. Entomol.* 15: 365–368.

Turner, J. W. 1983. Influence of plant species on the movement of *Trissolcus basalis* Wollaston (Hymenoptera: Scelionidae)—a parasite of *Nezara viridula* L. *J. Aust. Entomol. Soc.* 22: 271–272.

Ullyett, G. C. 1953. Biomathematics and insect population problems. *Entomol. Soc. S. Africa Mem.* 2: 1–89.

Vinson, S. B. 1976. Host selection by insect parasitoids. *Annu. Rev. Entomol.* 21: 109–133.

Walker, M. G. 1940. Notes on the disruption of *Cephus pygmaeus* Linn. and of its parasite, *Collyria caleitrator* Grav. *Bull. Entomol. Res.* 30: 551–573.

Wellso, S. G. 1973. Cereal leaf beetle: larval feeding, orientation, development, and survival on four small-grain cultivars in the laboratory. *Ann. Entomol. Soc. Am.* 66: 1201–1208.

Wilson, F., and C. B. Huffaker. 1976. The philosophy, scope, and importance of biological control. In Huffaker, C. B. and P. S. Messenger (eds.). *Theory and Practice of Biological Control.* Academic Press, New York, pp. 3–15.

Yanes, J., Jr., and D. J. Boethel. 1983. Effect of a resistant soybean genotype on the development of the soybean looper (Lepidoptera: Noctuidae) and an introduced parasitoid, *Microplitis demolitor* Wilkinson (Hymenoptera: Braconidae). *Environ. Entomol.* 12: 1270–1274.

Zhody, N. 1976. On the effect of the food of *Myzus persicae* Sulz. on the hymenopterous parasite *Aphelinus asychis* Walker. *Oecologia* 26: 185–191.

Zwolfer, H., and M. Krous. 1957. Bioclimatic studies on the parasites of two fir- and oak tortricids. *Entomophaga* 2: 173–196.

12

QUALITY OF DISEASED PLANTS AS HOSTS FOR INSECTS

Abner M. Hammond and Tad N. Hardy

Department of Entomology
Louisiana State University Agricultural Center
Baton Rouge, Louisiana

1. INTRODUCTION

The prevailing tendency in current empirical and theoretical examination of plant–insect interactions is toward consideration of the abiotic and biotic

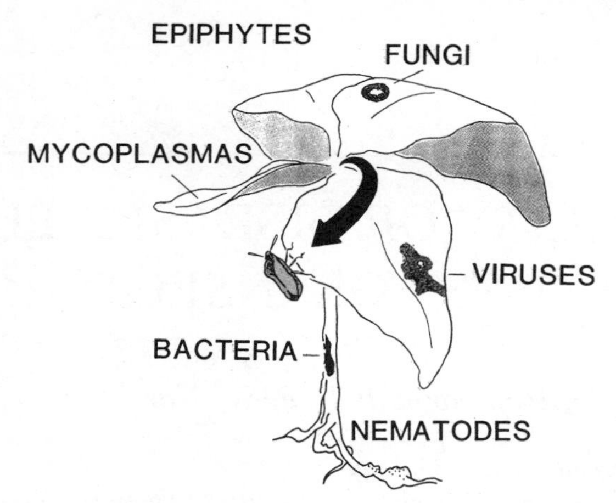

Figure 1. The interaction of insects, host plants, and plant disease agents creates a complex, multitrophic system. Phytopathogens may act directly upon the insect herbivore or may act indirectly by altering characteristics of the plant host.

factors that influence those interactions. This holistic approach incorporates the conceptual development of trophic interrelationships involving insect herbivores, their host plants, and the factors that regulate their association, including chemical, environmental, predatory, parasitic, pathogenic, and microorganismal stresses (e.g., Beck and Reese 1976, Rhoades and Cates 1976, Price et al. 1980, Scriber and Slansky 1981, Rhoades 1979, 1983). Although a large volume of information exists regarding specific, one-on-one interactions of these factors with either the plant or the insect, few studies have sought to delineate the multitrophic relationships of plant–insect complexes within their physical/biological environments (Price et al. 1980, Jones 1984).

One combination of interacting trophic levels that has received only anecdotal attention is that of insect–plant–plant pathogen (Fig. 1). Often reference is made to this interaction solely as an area lacking critical research, devoid of hard and fast generalizations but warranting study (Purcell 1982a, Jones 1984).

The frequency with which an insect herbivore encounters a diseased plant in natural or monoculture plant populations would presumably be quite high. For example, there are over 60 well-described diseases and an approximately equivalent number of known insect pests of rice, an economic crop of international distribution (Ou 1980, Bowling 1980). Thus the probability of interaction between synchronous insect and disease agents is great. A dated, but applicable, estimate of annual major crop losses on a global scale due to insect invaders and plant pathogens is 14 and 12%, respectively (Cramer 1967), further emphasizing the practical significance of potential insect–pathogen associations on mutual host plants. The relative dearth of observational or experimental data on diseased plant–insect interactions appears to be as great as the potential for its occurrence.

The complexity of such a three-dimensional biotic association is inherent in any tritrophic system and generally is compounded by abiotic effects. Environmental stresses may weaken a plant, predisposing it to phytopathogen (or insect) attack (Schoeneweiss 1975, Rhoades 1983); conversely, abiotic factors may in some way exert an influence over induction of plant defenses to these invaders (Kuc 1965). Additionally, consideration must be given to the nature of the phytopathogen (often microbial), its effect on the plant and the beneficial, neutral, or harmful effects of the diseased plant on the phytophagous insect. It is this latter, biotic (phytopathogen) stress variable and its role in modifying plant–insect interactions that is the primary focus of this chapter.

Phytopathogenic infection of an insect host plant presents a problematical situation to the insect herbivore. On the one hand, the plant–pathogen complex may be viewed by the insect as an exploitable resource, representing a unique and readily accessible source of refuge (Carter 1939) or nutrition (Langdon and Champ 1954, Markkula and Laurema 1964), or an opportunity to colonize a previously unsuitable host (Maramorosch 1958, Raffa and Berryman 1982). Alternatively, morphological (Bald et al. 1946) or physiological changes in the diseased plant, such as nutritional modifications (Misaghi 1982), accelerated production of induced chemical defenses against pathogen invasion (Ayres 1981), or production of defensive chemicals by the pathogen itself may render the plant undesirable, suboptimal, unsuitable, or even lethal (Ahmad et al. 1985). Harris and Frederiksen (1984), in reviewing host plant resistance to plant pathogens and arthropods, and Jones (1984), in examining microorganismal mediation of plant resource exploitation, concluded that plant response to damage, regardless of the source, may have a broad spectrum of efficacy against both disease agents and potential insect herbivores. Thus the diseased host plant engages in an ambiguous, often contrasting relationship with the phytophagous insect, resulting in effects that lie on a continuum ranging from highly beneficial to acutely detrimental to the insect.

The consequences of insect exposure to or utilization of a diseased host plant are diverse and may depend upon the extent of *vis-à-vis* interaction with the phytopathogen proper. Effects attributable to direct insect–plant pathogen association are not easily isolated and discerned (i.e., Sections 2 and 3). Some phytopathogenic organisms are capable of utilizing the insect as an alternative host (i.e., many viruses and mycoplasmas) and in such instances the influence of the pathogen apart from the host plant emerges (Nakasuji and Kiritani 1970, Granados and Meehan 1975). Some insects that are facultative mycophages probably derive nutritional benefit from incidental fungus consumption on infested hosts (Yarwood 1943, Lewis 1979). However, most plant disease agents appear to interact indirectly with the insect by modifying host plant attractiveness (Whitcomb and Williamson 1979; Hardy et al. 1985, 1986), digestibility and food assimilation (Kingsley et al. 1983), defenses (Francke-Grosmann 1967, Harris and Frederiksen 1984), and nutritional value (Laurema et al. 1966, Ajayi and Dewar 1982).

The role of the host plant in this multitrophic menage may also be to functionally modify the plant pathogen–insect relationship. At one extreme, the plant may serve in a limited capacity as liaison, simply providing a station for insect–pathogen interaction. Such an interaction might be better defined as a two-dimensional trophic system involving only the insect and the microbial pathogen. At the other extreme, the plant may be transformed by the disease to such a degree that it can no longer function as a host to biotrophic insects (i.e., extensive tissue necrosis). Most plants lie on the gradient bounded by these extremes and exist as intermediaries or altered hosts.

Plants considered intermediary hosts have active but limited involvement in the plant–insect–phytopathogen interrelationship, and the effects of the pathogen on the insect are direct. Symbiotic interactions of plant and pathogen creating a singular complex which the insect encounters (e.g., possibly fungal endophytes, Clay 1986, Carroll 1986) may also constitute somewhat of an intermediary role. Altered host plants, however, are those that are qualitatively or quantitatively modified by pathogen association. These plants are altered by invading phytopathogens that indirectly affect the insect through changes in plant morphology (e.g., cell wall lignification, Ayres 1984) or metabolism (Section 6). Obviously, no host plant may be strictly assigned to one or the other category and will usually display characteristics of each, depending on the particular pathogen involved.

In this chapter plant disease is defined as the disruption of orderly processes that operate within a living plant by causal microorganisms ("phytopathogens") encountered by that plant (broadly modified from Bateman 1978). This working definition excludes pathogenic effects of insect feeding ("damage") and limits discussion of diseases to those of viral, mycoplasmal, bacterial, or fungal origin, although generalizations may apply to additional organisms often considered as agents of plant disease such as epiphytes and nematodes. Certain organisms (i.e., a few plant viruses and mycoplasmas, and fungal endophytes) that may interact with the host plant in a symbiotic manner and have maintained an ambiguous pathogenic status are also included.

Excluded from discussions of plant–insect–phytopathogen interactions are saprophytic bacteria and fungi that are almost exclusively associated with nonliving plants, and insects considered strict detritivores. Symbiotic relationships of microbes (irrespective of plant pathogenicity) with insects on plants have recently been reviewed (Jones 1984) and are not covered here, although several of Jones's (1984) postulates for adaptation in insect–microbe associations and consequences of microbial mediation are directly applicable to insect–plant pathogen systems.

Insect vectors of plant disease constitute a special case of phytopathogen mediation of plant–insect interaction. Although some plant pathogens (i.e., bacteria and many fungi) utilize insects as secondary agents of transmission or dispersal (Carter 1973), their continued transmission rarely is dependent upon a relationship with a specific vector. Obligate plant-pathogenic parasites that pass freely between their plant hosts and insect (host) vectors, however, rely entirely upon the continued success of a vector and its maintenance of a plant–insect interaction with appropriate plant hosts. These relationships are frequently complex, often involving vector specificity (Purcell 1982a). The evolutionary and ecological implications have been discussed elsewhere (Maramorosch and Jensen 1963; Purcell 1982a, b) and are referred to only briefly in this chapter.

2. VIRUSES

Phytopathogenic viruses, in conjunction with mycoplasmas (Section 3), constitute a major portion of the causal organisms responsible for plant diseases. More than 700 viral pathogens are known to utilize as hosts ornamental, cereal, vegetable, forage, and orchard crops of considerable economic importance (Fraenkel-Conrat 1985). Viral associations with agricultural systems, and the often intimate reliance upon arthropods for dispersal, have directed much attention to the host plant–virus–insect relationship.

Viruses are, by definition, obligate intracellular parasites incapable of existing beyond the confines of living host tissue. Only rarely are they transmitted by natural mechanical means, depending largely on biological agents for dissemination. Approximately 99% of all known arthropod vectors of plant viruses are insects (Harris 1981). Aphids account for the majority of vector species, but other important vectors are leaf- and planthoppers (Harris 1980), whiteflies, mealybugs, thrips, and beetles (Walkey 1985). Thus it comes as no surprise that observational and experimental studies of plant-pathogenic viruses and insects focus almost exclusively on vectors. As a result the "flying needle" concept of virus vectors acting simply as passive intermediaries has been modified (Maramorosch and Jensen 1963) to include the dynamic nature of viral phytopathogen–insect vector interaction.

2.1 Vectored Viruses

The effects of vectored plant viruses on their respective insect vectors, and in several instances, on nonvectors, lie on a continuum ranging from highly beneficial to lethal (Maramorosch and Jensen 1963, Jensen 1969). An understanding of the underlying mechanisms responsible for these effects is complicated by the nature of the tritrophic interaction. Direct effects of the virus on its vector, and indirect effects ascribed to the diseased host plant or the plant–pathogen complex on vectors often are not easily or justifiably separated.

One unique aspect of most vector-borne viral (and mycoplasmal) plant diseases is that dispersal is only possible via transport within the vector. The insect harbors the pathogenic agent for a variable period of time (the duration of which is dependent on the specific insect–virus combination) and, quite literally, plays host throughout that period. Many viruses are nonpersistent transients, acquired by the vector during protracted feeding bouts on an infected plant. Such stylet-borne viruses do not circulate throughout the vector, and within hours after vacating the infected plant, the vector can no longer transmit the virus. Other viruses are semipersistent, transmissible for days but noncirculating in the vector (Harris 1981). Persistent viruses, however, pervade their insect host, are transstadially transmitted, and frequently confer permanent virulence to the vectoring individual. Some propagate within their vector; a few, such as sowthistle yellow vein virus (Sylvester 1973), are passed transovarially and are maintained in this manner for generations. The biology of a given virus following acquisition by an insect may partially determine proximate changes in individual vector biology and may contribute directly to extended changes in vector population structure.

Several studies have indicated deleterious effects on insect host vectors directly attributable to the plant virus proper, separate from the host plant. Rice stripe disease, a persistent-propagative, transovarially transmitted virus, reduced longevity and egg production of its leafhopper vector, *Delphacodes striatella* (Fallen), when maintained in the population by serial transfer (Okuyama 1962). The transovarially transmitted rice dwarf virus also decreased adult longevity and fecundity and adversely affected nymphal survival and development of the leafhopper, *Nephotettix cincticeps* Uhler (Nakasuji and Kiritani 1970). Hoja blanca virus of rice was responsible for similar, negative effects in viruliferous colonies of its leafhopper reservoir, *Sogatodes oryzicola* (Muir) (Jennings and Pineda 1971).

Cytological and metabolic changes associated with persistent plant virus (and mycoplasma) diseases have been reported in numerous vectors and were reviewed previously (Maramorosch and Jensen 1963, Jensen 1969, Shikata 1979). Inclusion bodies in various tissues, increased vacuole formation, nuclear abberrations, histological changes in mycetomes, and altered oxygen consumption rates and cellular carbohydrate levels are all known from virus-infected vectors. A few studies confirm the correlation between assumed cytopathogenicity, or abnormal physiology, and reduction in longevity or fecundity (Okuyama 1962), but most are strictly cytological or histochemical in nature. Often no differences exist between tissues of virus-free and infected individuals, and when deviations from normal morphology or metabolism do occur, the assignment of overt pathogenicity without behavioral or developmental verification may be unwarranted.

Occasionally, morphological anomalies of the host plant resulting from infection are deemed exclusively responsible for differential vector response. Thrips, for example, exploit the curled leaves of infected *Emilia* as refugia (Carter 1939). Bald et al. (1946) indicated that nonpreference by

Macrosiphum gei Koch for potato plants afflicted with potato leaf roll virus was due to curled leaves and stunting. In contrast, the reversion disease of black currant produces a more glabrous plant surface, enhancing population growth of the mite, *Phytoptus ribis* Nal. (Thresh 1964, 1967). However, consequences of viral contact by a vector via choice and utilization of infected host plants as feeding sites frequently are considered the collective result of the plant-pathogen influence. Table 1 is a summary of selected phytopathogens, their respective host plants, and their net effect on the multiplicity of behavioral and developmental parameters of vector species.

Demonstration of a preference for either healthy or infected host plants as feeding or colonization sites by insects has been indicated for several viral diseases. *Myzus persicae* (Sulz.) appeared to disregard hosts harboring mosaic infections when provided a clear choice (Kuan and Wang 1965) or during colony establishment (Lowe and Strong 1963). Nonpreference was associated with a reduction in levels of host plant nitrogen. Other aphids are known to prefer diseased plants over healthy ones (Baker 1960, Garran and Gibbs 1982, Ajayi and Dewar 1983).

Additional effects of plant infection on vectors, such as survival, longevity, fecundity, development and/or population fluctuations, typically are attributed to viral intervention once inside the insect host or to modifications of host plant processes that indirectly affect the vector. Rice dwarf virus was responsible for premature mortality of transovarially infected progeny of viruliferous leafhoppers (Shinkai 1960), but perpetual access to diseased plants did not consistently reinforce the effect (Shinkai 1960, Satomi, Nasu, and Suenaga in Maramorosch and Jensen 1963). Decreased longevity was demonstrated for the aphid vector of sowthistle yellow vein virus feeding on plants inoculated with source material obtained by serial transfer in the insect host (Sylvester 1973).

A positive correlation between aphid fecundity and higher levels of amino acids in barley yellow dwarf (BYDV) and bean yellow mosaic (BYMV) infected plants indicates the potential importance of indirect, host plant-mediated effects for certain insect-diseased host combinations (Markkula and Laurema 1964). The aphids *Acyrthosiphon pisum* Harris (Fig. 2) and *Rhopalosiphum padi* (L.) produce more young on moderately diseased red clover and oats, respectively, than on healthy or severely infected plants. Infected plants demonstrated higher levels of free amino acids as infection progressed, followed by a reduction in amino acid concentration in severely diseased plants to a level below that of healthy plants (e.g., in BYMV-infected peas). Mean *A. pisum* progeny production paralleled the disease-induced changes in amino acid content. Progeny production in *Macrosiphum avenae* F. and *Acyrthosiphon dirhodum* (Wlk.) on oats and *A. pisum* on peas was unaffected by disease status of the host, however. Markkula and Laurema concluded that aphid reproduction response and amino acid levels are dependent upon plant age, viral agent, and stage of viral infection. Ajayi and Dewar (1982) suggested that reduced frequency of honeydew excretion

Table 1. Net Effect of Phytopathogenic Viral Infection on Insect Vectors via Natural Acquisition from Host Plant

Pathogen (Type of Transmission)	Host Plant	Insect Vector	Net Effect[a]	Reference
Beet mosaic (Nonpersistent)	Sugar beet	*Aphis fabae* Scopoli	+ fecundity	Kennedy 1951
Turnip mosaic (Nonpersistent)	Chinese cabbage	*Myzus persicae* (Sulzer)	− feeding preference	Kuan and Wang 1965
Cucumber mosaic (Nonpersistent)	Cucumber	*M. persicae*	− colonization	Lowe and Strong 1963
Alfalfa mosaic (Nonpersistent)	Lucerne	*Therioaphis trifolii* f. *maculata* (Buckton)	+ feeding preference on Siriver lucerne +/− feeding preference on Hunter River lucerne	Garran and Gibbs 1982
Beet yellow mosaic (Nonpersistent)	Pea	*Acyrthosiphon pisum* (Harris)	+/− progeny production	Markkula and Laurema 1964
	Red clover	*A. pisum*	+ progeny production (moderate symptoms) − progeny production (severe symptoms)	Markkula and Laurema 1964
Sugar beet yellows (Semipersistent)	Sugar beet	*A. fabae, M. persicae, Myzus ascalonicus* Doncaster, *Myzus solani* (Kalt.)	+ feeding preference, survival, fecundity	Baker 1960

Barley yellow dwarf (Persistent)	Oats	*Acyrthosiphon dirhodum* (Wlk.), *Macrosiphum avenae* (Fabricius), *Rhopalosiphum padi* (L.)	+/− progeny production	Markkula and Laurema 1964
		Sitobion avenae (Fabricius)	+ alate production	Gildow 1984
		Macrosiphum granarium (Kirby)	+ longevity, fecundity	Miller and Coon 1964
	Wheat, barley	*S. avenae*, *Metopolophium dirhodum* (Walker)	+ digestibility, dispersal preference	Ajayi and Dewar 1982, 1983
Potato leaf roll (Persistent, transovarial)	Potato	*M. persicae*	+/− feeding preference	Bald et al. 1946
		Macrosiphum gei Koch	− feeding preference	Bald et al. 1946
European wheat striate mosaic (Persistent?, propagative, transovarial)	Wheat	*Delphacodes pellucida* Fabricius	− progeny production +/− longevity	Watson and Sinha 1959
Yellow spot of pineapple (Persistent)	*Emilia*	*Thrips tabaci* Lind.	+ population growth	Carter 1939
Unidentified oat virus (Persistent)	Oats	*Graminella nigrifrons* (Forbes)	− egg and early instar survival	Jedlinski, in Maramorosch and Jensen 1963
Rice dwarf (Persistent, propagative, transovarial)	Rice	*Recilia dorsalis* (Motschulsky)	+/− nymph survival	Shinkai 1960
		Nephotettix cincticeps (Uhler)	− fecundity, nymph survival	Satomi et al., in Maramorosch and Jensen 1963
		Nephotettix bipunctatus (Uhler)	− survival, longevity	Suenaga 1962
Sowthistle yellow vein (Persistent, propagative, transovarial)	Sowthistle	*Hyperomyzus lactucae* (L.)	− longevity +/− larviposition and excretion	Sylvester 1973

[a] A + symbol indicates a beneficial effect, a − symbol a detrimental effect, and a +/− symbol indicates no apparent effect.

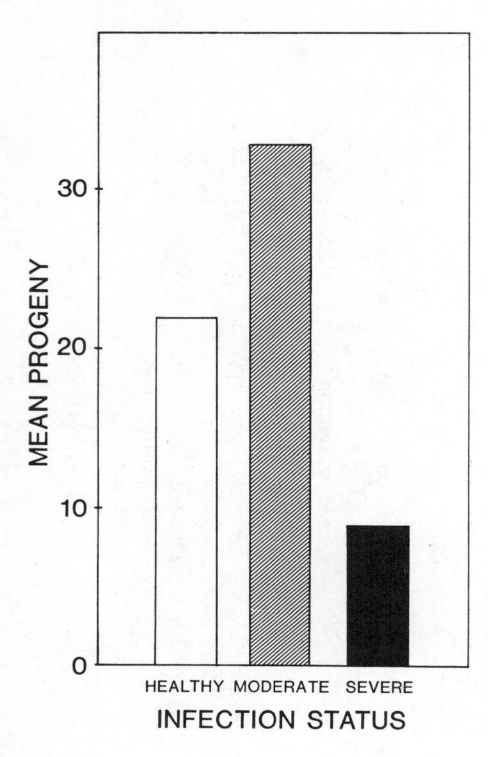

Figure 2. Mean progeny production of the aphid *Acyrthosiphon pisum* Harris on healthy red clover and on red clover supporting moderate or severe infections of bean yellow mosaic virus. (Modified from Markkula and Laurema 1964)

by aphid vectors on cereals infected with barley yellow dwarf virus implies nutritional superiority of infected over healthy plants. Pathological symptoms of viral infections that physiologically mimic leaf senescence (Kennedy 1951, Gildow 1984) corroborate the indirect role of plant mediation. Gildow (1984) reported a twofold increase in alatae production in *Sitobion avenae* F. reared on barley yellow dwarf virus-infected as opposed to healthy oats. Detached leaves and senescing plants also increased alatae incidence, suggesting that common factors influencing aphid development may exist between infected and aging plants.

In an innovative study Laurema et al. (1966) excluded direct pathogenic effects on the insect by caging known species of aphid vectors on oats infected with a virus (or mycoplasma) transmissible only by leafhoppers. Longevity of *R. padi* and *M. avenae* was reduced significantly on plants infected with European wheat striate mosaic virus, and as a result fewer progeny were produced (Fig. 3). Chromatographic examination of free amino acids in healthy and virus-diseased oats revealed consistent, negative correlations of aphid longevity and reproduction to plant infection and increased amino acid concentration. These results are contrary to possible amino acid effects on these aphids reported earlier (Markkula and Laurema 1964). Apparently, virus-induced changes in plant metabolism alone can alter the likelihood that

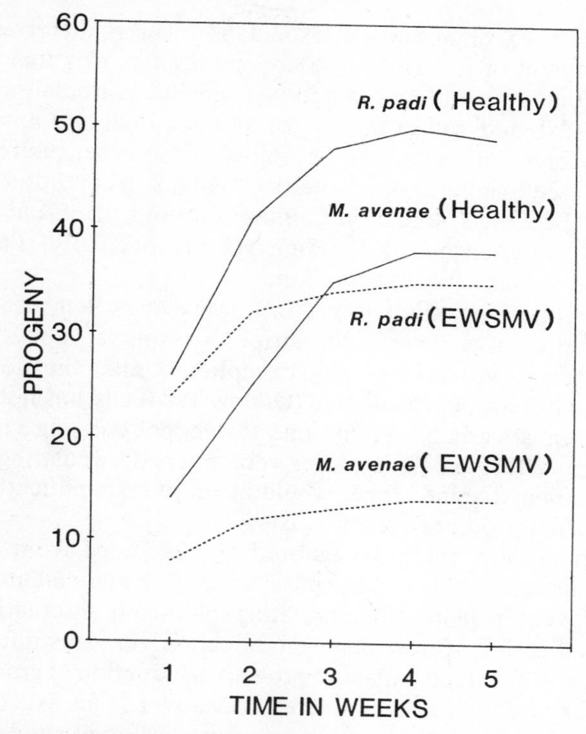

Figure 3. Progeny production on healthy and European wheat striate mosaic virus (EWSMV)-infected oats by the aphids *Rhopalosiphum padi* L. and *Macrosiphum avenae* (F.). (Modified from Laurema et al. 1966)

a diseased plant will be utilized or suitable as a host by phytophagous insects; amino acid content, however, is only one factor to be considered.

Few generalizations concerning qualitative or quantitative aspects of vector–plant–virus interaction can be drawn from existing experimental data. Preferential feeding (or lack thereof) by insect vectors on both healthy and diseased plants has been demonstrated, and varies with the phytopathogen, vector, and host plant involved. The influence of viral plant infection on vector biology likewise provides no clear consensus that phytopathogens are categorically beneficial or detrimental.

However, certain transmission (persistence) characteristics (direct effect) and altered metabolic processes of infected plants (plant-mediated, or indirect effects) are of interest. The association of noncirculating (non- or semipersistent) viruses with the plant often favor or have no affect upon the vector. In instances of negative net effect, a lack of preference rather than physiological detriment seems a likely possibility. It is unfortunate that few studies convincingly eliminate host preference as a contributory factor when differences in vector survival or development exist between healthy and diseased plants (i.e., Saini and Peterson 1965). The relatively short period and limited amount of exposure to a noncirculating virus that a vector receives argues that direct effects should be negligible. Circulating (persistent) viruses are distributed within the vectoring individual and have ample opportunity to strongly influence vector biology. Over time, such relationships might be expected to emerge as mutually beneficial, a condition of viral (and mycoplasmal, Section 3) infection common in both plant and insect hosts (Maramorosch and Jensen 1963). High vector specificity (Purcell 1982a) would seem to support this expectation.

Persistent viral associations in vectors, however, often result in marked pathogenicity (i.e., rice dwarf, rice stripe, sowthistle yellow vein virus). Barley yellow dwarf virus is a notable exception. Unlike the aforementioned phytopathogens, it is nonpropagative (Gildow 1984) and has not been shown to be passed transovarially. To assume that noncirculating viruses have a positive/neutral effect on their insect vectors, and circulating/propagative viruses have a negative influence, would be an oversimplification, although evidence for such a generalization exists.

Indirect effects are not to be ignored nor excluded from the potential influence of characteristics of transmission. Both beneficial and deleterious outcomes of vector–plant–noncirculating pathogen interactions are undoubtedly a reflection of host plant modification via virus infection (Baker 1960). The opposing effects on aphid progeny production of graded symptom severity in beet yellow mosaic-infected red clover is an excellent example (Markkula and Laurema 1964). Evidence for nutritional similarity between plant senescence (amino acid accumulation) and the occurrence of phytopathogen infection (barley yellow dwarf, persistent in its vector, Gildow 1984) and similarly high levels of amino acids in virus-diseased oats (Laurema et al. 1966) appear to establish a trend. But biochemical or physiological deviations in diseased host plants vary with the unique virus–host combination and therefore are best considered on an individual basis.

2.2 Nonvectored Viruses

It is difficult to estimate the frequency with which insect phytophages encounter nonvectored plant viruses (or for that matter, how often nonvectors contact vectored plant virus diseases). As stated previously, the insect–virus interaction is, with few exceptions, studied only with insect vectors. One would assume that such nonvector encounters occur with a relative degree of regularity, given the abundance of plant pathogens and insect herbivores.

Plant viruses that do not rely on arthropods as vectors for primary infection would not be expected to exercise a direct influence over insects in natural or managed ecosystems. Insect–virus associations would be totally dependent upon plant mediation and thus invariably indirect. Relatively few studies have examined the nature of these nonvectored plant virus–insect relationships and their indirect mechanisms.

Tobacco mosaic virus (TMV), well known for its broad host plant distribution, economic impact, and nearly exclusive mechanical mode of transmission (Lucas et al. 1985), seems a logical candidate for probing these relationships. Three studies dealing with induced immunity to viral infection using this disease agent provide insight into the effect of a nonvectored virus on insects.

Asymptomatic and symptomatic infections of tomato plants inoculated with TMV significantly increased survival of the Colorado potato beetle, *Leptinotarsa decemlineata* (Say), relative to that on virus-free tomato plants (Hare and Dodds 1978). In contrast, reproduction of *M. persicae* was reduced when held on inoculated tobacco (McIntyre et al. 1981). Higher nitrogen content of diseased plants was suggested to be a factor in enhanced beetle survival; lowered aphid reproduction was attributed to the translocation of allelochemic metabolites in infected tobacco plants. In both instances, virus infection induced changes in host plant constituents that affected the insect. Hare, McIntyre, and Dodds (in Hare 1983) also present data demonstrating reductions in growth rate of fourth instar tobacco hornworms, *Manduca sexta* (L.), fed on foliage from tobacco previously infected with TMV. They concluded that growth rate reduction on TMV-infected leaves (local lesions) was due to reduced ingestion by larvae, whereas growth rate reductions observed in larvae fed nonlesion leaves from TMV-infected plants were due to reduced efficiency of plant tissue assimilation.

Over the last decade approximately 12 subviral phytopathogens (viroids) have been recognized as naturally occurring plant disease agents (Diener and Prusiner 1985). These viroids, capable of autonomous replication in host cells, have not been demonstrated to be transmissible by arthropod vectors (Diener 1979). Little information regarding insect–host plant–viroid relationships is available.

3. MYCOPLASMAS

It is only within the past two decades that mycoplasmas were discovered, described, and designated as entities separate from viruses and as agents of disease in vascular plants (Doi et al. 1967, Ishiie et al. 1967). The plant diseases with which these diverse prokaryotes are now associated previously were considered viral in origin due to physical properties and epidemiological characteristics common to both groups (Maramorosch et al. 1975). Over 100 plant diseases formerly attributed to viruses are now believed to be of mycoplasmal etiology (Maramorosch 1981).

The term "mycoplasma" typically has been utilized as a generic term in a trivial sense and as a recognized genus of microorganisms lacking cell walls. Mycoplasmas are taxonomically categorized in the Class Mollicutes, Order Mycoplasmatales under which mycoplasmalike organisms (MLOs) and spiroplasmas are also classified (Freundt 1981). MLOs are pleomorphic organisms that resemble mycoplasmas but await final classification. Spiroplasmas, however, have a distinctive helical form and are placed in a taxonomic family separate from their nonhelical counterparts (Freundt 1981). As recently as 1979, the few plant-pathogenic mycoplasmas that had been isolated and identified were spiroplasmas (McCoy 1979).

An additional group of obligate parasites, the rickettsialike organisms (RLOs), are occasionally included in discussions of mycoplasmal phytopathogenicity. These organisms superficially resemble mycoplasmas in epidemiology and have been implicated as etiological agents of plant disease but, unlike other Mollicutes, possess a cell wall (Tsai 1979). For our purposes, however, RLOs will be considered in greater detail in Section 4.

With rare exception, all known plant-pathogenic MLOs and spiroplasmas are transmitted by leafhopper or planthopper vectors (Banttari and Zeyen 1979, Harris 1980, Tsai 1979), and, like viruses, the focus has been on vector–plant-pathogen relationships. Most mycoplasmas are presumed to be transmitted in a persistent-propagative manner (Harris 1980), an additional trait shared with leafhopper-borne viral phytopaths. Thus the recurrent dilemma of separating direct from indirect (plant-derived) effects of plant disease agents on their vectors also exists for the mycoplasmas.

Table 2 is a selected listing of plant mycoplasmas and their observed effects on the biology of the insect vector. Only those studies utilizing natural means of mycoplasmal acquisition or exposure (via feeding on the host plant) are included.

Aster yellows-diseased plants appear to have a net beneficial influence on the insects that contract and harbor the causal disease agent (Severin 1946, Kunkel 1954). Indeed, pathogen infection in the plant was shown to be a prerequisite for vector survival in most instances. Maramorosch (1958) demonstrated that previous feeding on infected host plants by *Dalbulus maidis* (DeLong and Wolcott) (which harbors but does not transmit the disease) is responsible for subsequent survival on nonhost plant species and on healthy host plants incapable of supporting individuals that had not first been confined on a diseased plant. He indicated that direct mycoplasmal intervention contributed to increased leafhopper survival; and though Severin (1946) did not eliminate the potential for direct effects on vectors, he provided adequate evidence to implicate indirect effects of the disease as beneficial. Higher survival of two insects not vectoring aster yellows, the aphids *R. padi* and *M. persicae*, on aster yellows-infected China aster (Saini and Peterson 1965) lends further support to plant-moderated influence.

Many insect–plant–mycoplasma associations reduce longevity or fecundity of insect vectors (Table 2). Corn stunt spiroplasma, for example, caused premature mortality in all but one of its leafhopper vectors (Granados and Meehan 1975, Madden and Nault 1983, Madden et al. 1984, Nault et al. 1984) as a result of a one-week acquisition access period to infected plants followed by transferral to healthy plants. This mycoplasma apparently has direct, detrimental effects on its vectors (i.e., cytopathological changes in

Table 2. Net Effect of Phytopathogenic Mycoplasmal Infection on Insects Encountering or Acquiring the Disease Agent via Host Plant

Pathogen	Host Plant	Insect[a]	Net Effect[b]	Reference
		Mycoplasma (MLO)		
Aster yellows	Aster	*Dalbulus maidis* (Delong & Walcott)*	+ survival	Maramorosch 1958
		Cloanthonus irroratus (Van Duzee), *Eusceles maculipenis* DeLong, *Collandonus geminatus* (Van Duzee)	+ survival, longevity	Severin 1946
		Collandonus montanus (Van Duzee)	+ longevity	Severin 1946
	Celery	*Texanus denticulus* Osborn & Lathrop, *Texanus pergradus* DeLong,	+ survival, longevity	Severin 1946
		Texanus spatulatus Van Duzee	+ survival, longevity, nymphal development	Severin 1946
		Texanus lathropi Baker	+ nymphal development	Severin 1946
	Aster and celery	*Idiodonus kirkaldyi* (Ball)	+ survival, longevity	Severin 1946
		Macrosteles fascifrons (Stål)	+/− longevity	Kunkel 1954
	China aster	*Rhopalosiphum padi* (L.)*, *Myzus persicae* (Sulzer)	+ survival	Saini and Peterson 1965
Maize bushy stunt	Maize	*Dalbulus longulus* DeLong, *Dalbulus guevarai* DeLong	− longevity	Nault et al. 1984
		D. guevarai, *Dalbulus tripsacoides* DeLong & Nault, *Dalbulus quinquenotatus* DeLong & Nault, *D. maidis Baldulus tripsaci* Kramer & Whitcomb	− longevity	Madden and Nault 1983
		Dalbulus elimatus Ball, *Dalbulus gelbus* DeLong	+/− longevity	Madden and Nault 1983

Table 2. (*continued*)

Pathogen	Host Plant	Insect[a]	Net Effect[b]	Reference
Western (peach) X-disease	Celery	*C. montanus*	− fecundity	Jensen 1962
Clover witches'-broom	Red clover	*Euscelis plebejus* (Fallén)	− longevity, development	Posnette and Ellenberger 1963
Potato purple top roll	Potato	*Alebroides nigroscutellatus* (Dist.)	− longevity, fecundity	Singh et al. 1983
Oat sterile dwarf	Oats	*R. padi**	+/− longevity − fecundity	Laurema et al. 1966
		Macrosiphum avenae (Fabricius)*	+ longevity	Laurema et al. 1966
		Spiroplasma		
Citrus stubborn (*Spiroplasma citri*)	?	*E. plebejus*	+/− longevity	Markham and Townsend 1974
	?	*D. elimatus*	+/− longevity	Whitcomb et al. 1973
	?	*M. fascifrons*	− survival, longevity	Whitcomb et al. 1974, Whitcomb and Williamson 1975
Corn stunt	Maize	*D. longulus, D. guevarai*	− longevity	Nault et al. 1984
		D. elimatus	− longevity, fecundity	Granados and Meehan 1975
		D. maidis	− longevity	Granados and Meehan 1975
			+/− longevity	Madden and Nault 1983
		D. guevarai, D. quinquenotatus, D. tripsacoides, B. tripsaci	+/− longevity	Madden and Nault 1983
		D. gelbus, D. elimatus	− longevity, fecundity	Madden et al. 1984
		Rickettsialike Organism (RLO)		
Sugar beet witches'-broom	Sugarbeet	*Piesma quadratum* Fieb.	− longevity	Fronsch 1983

[a] An * symbol indicates that the insect is not known to vector the mycoplasma.
[b] A + symbol indicates a beneficial effect, a − symbol a detrimental effect, and a +/− symbol indicates no apparent effect.

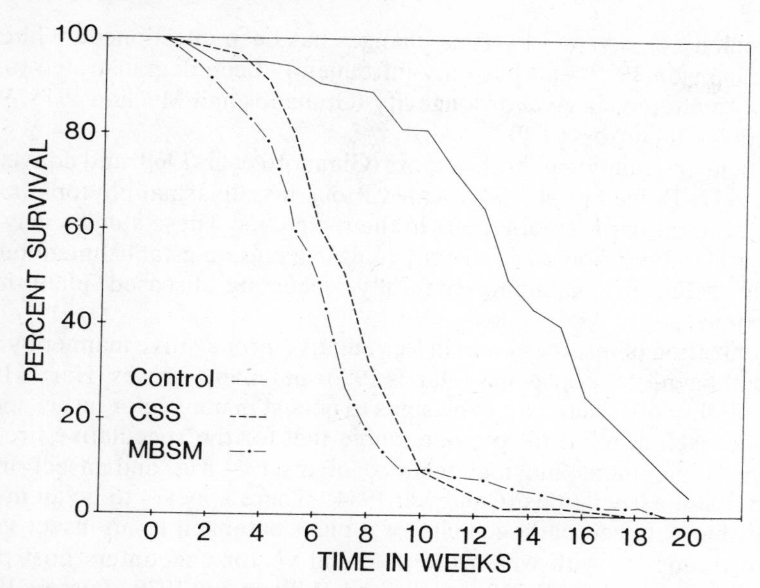

Figure 4. Survival curves of adult *Dalbulus longulus* DeLong leafhoppers held on healthy maize plants following a one-week acquisition access period on healthy (control) or mycoplasma-infected plants (CSS = corn stunt spiroplasma; MBSM = maize bushy stunt mycoplasma). (Modified from Nault et al. 1984)

neural and glandular tissues were observed in *Dalbulus elimatus* [Ball], Granados and Meehan 1975). Maize bushy stunt mycoplasma had a similar negative effect on *Dalbulus longulus* DeLong (Fig. 4) and other leafhopper vectors (Nault et al. 1984). Laurema et al. (1966) found that oat sterile dwarf virus-diseased oats reduced longevity and fecundity of *R. padi,* yet increased fecundity and reproductive period duration of *M. avenae* when aphids were held for over a month on the infected plants. Evidently, increased progeny production of *M. avenae* corresponded to an extended growing period of diseased over healthy oat plants.

Citrus stubborn disease, *Spiroplasma citri,* is an exception to the general trend of harmful mycoplasmal effects. There are few if any reports of pathogenicity of this plant disease agent to its natural leafhopper vectors (Daniels 1979). However, *S. citri* multiplied in and reduced longevity of a nonvector, *Macrosteles fascifrons* (Stål) (Whitcomb et al. 1974, Whitcomb and Williamson 1975).

As with viruses, mycoplasmal infection in insect vectors is associated with modifications of cellular structure (Maramorosch and Jensen 1963, Jensen 1969). Many are considered cytopathological, such as lesions in midgut and epithelial cells of leafhoppers infected with clover phyllody (Giannotti 1969, Sinha and Paliwal 1970), and irregular cell division in leafhoppers harboring Western X-mycoplasma (Whitcomb and Jensen 1968). However, the pathogenicity of several observed changes has been questioned (Whitcomb and Williamson 1979) and has only infrequently been demonstrated to correlate with altered survival or longevity (Granados and Meehan 1975, Whitcomb and Williamson 1979).

Artificial inoculation of both vectors (Giannotti et al. 1968) and nonvectors (Clark 1977, Dowell et al. 1981) with various mycoplasmal phytopathogens produced harmful or lethal effects in the recipients. These studies may provide useful information on *in vivo* mycoplasma culture establishment but are of little value in examining naturally occurring diseased plant–insect interactions.

Colonization of insect vectors in a circulative-propagative manner by most plant pathogenic mycoplasmas (Harris 1980) and many viruses (Harris 1981), and the ability of selected mycoplasmas to persist in nonvector insect species (Maramorosch et al. 1975) provide ample fuel for the speculative fire concerning development and maintenance of insect–virus and insect–mycoplasma–plant associations (Caudwell 1984). There appears to be an inverse relationship between pathogenicity of a plant pathogen to an insect vector and the frequency with which the potential vector encounters host plants harboring that pathogen (Whitcomb and Williamson 1979, Purcell 1982b, Nault et al. 1984). Thus insects possessing vector capabilities that constantly encounter transmissible phytopathogens would be expected to develop a certain degree of tolerance to or symbiotic commensalism or mutualism with the disease agent (Whitcomb and Williamson 1979, Maramorosch 1981, Madden and Nault 1983).

Madden and Nault (1983) provided evidence to support this mutualistic association using the interrelationships of *Dalbulus* leafhoppers, native host plants and two Mollicutes phytopathogens. All *Dalbulus* spp. used in their study were capable of vectoring corn stunt and maize bushy stunt diseases; however, vector–pathogen compatibility (i.e., lack of pathogenicity to the vector) was observed only for those species known to consistently encounter the pathogens on disease-susceptible plant hosts. On the contrary, the results of a similar study using corn stunt and *D. elimatus* do not lend support to this concept (Granados and Meehan 1975).

It has also been suggested that spiroplasmas and MLOs may have originated as insect pathogens, allowing for a gradual loss of pathogenicity to the insect host over time (Maramorosch 1981). Generalizations of low pathogenicity of these mycoplasmas (and persistent viruses as well) to known vectors (Maramorosch 1981, Purcell 1982b), however, are only valid when variables such as vector ecology and biotype, pathogen strain, and likelihood of interaction are considered (Carter 1973, Granados and Meehan 1975, Madden and Nault 1983, and Tables 1 and 2 this chapter). The consequences of long-term association of insect vectors of obligate pathogens with diseased plants are difficult to assess (Purcell 1982a) and are dependent upon specific insect–pathogen combinations and the highly adaptable nature of many pathogens (Whitcomb 1975) and their vectors (Madden and Nault 1983).

4. BACTERIA

Plant bacterial diseases are much less damaging to crop plants than the diseases caused by viruses, mycoplasmas, and fungi. However, all major agricultural crops including maize, rice, potato, cotton, tobacco, and pome fruits are attacked by bacterial pathogens. Bacterial taxonomy is in a transitional state, but most authors suggest that there are about 200 species of plant pathogenic bacteria. These pathogenic bacteria are found in five main genera: *Agrobacterium, Corynebacterium, Erwinia, Pseudomonas,* and *Xanthomonas*. Recently, Raju and Wells (1986) proposed a new genus, *Xylemella,* for prokaryotic plant pathogens referred to as "xylem-limited bacteria." These fastidious bacteria are transmitted by xylem-feeding leafhoppers and spittlebugs. When first associated with plant diseases these agents were referred to as "rickettsialike bacteria" (RLOs) (Goheen et al. 1973, Hopkins and Mollenhaver 1973).

Bacterial diseases are not totally dependent on insects for their dissemination or for their inoculation into the plant tissue. Plant pathogens can disperse to new hosts in a variety of ways: (1) via seed or propagules of the host plant, (2) by wind or moving water, or (3) as a "hitchhiker" on some other organism that serves as an active vector (Purcell 1982b). These pathogens, however, are unable to penetrate plant tissue without a court of entry. Natural plant entry occurs through stomata, lenticels, water pores, and flower nectaries. Insects contribute to plant infection through the production of feeding and oviposition wounds, as a mechanical carrier of the organisms, and in some cases, by virtue of a mutualistic relationship between the organism and the insect which ensures an association between pathogen, insect, and host plant (Carter 1973).

In a few cases close biological relationships appear to have developed between insects and plant pathogenic bacteria which are highly beneficial to one or both of the organisms (Harrison et al. 1980). For example, Stewart's bacterial wilt, caused by *Erwinia stewartii* (E. F. Sm.) Dye, can be seedborne or transmitted by moving water to fresh wounds, but transmission by insect vectors is the most frequent mode of dissemination (Pepper 1967). The larval and adult stages of the 12-spotted cucumber beetle, *Diabrotica undecimpunctata howardi* Barb., the northern and western corn rootworms, *Diabrotica* spp., and the corn flea beetle, *Chaetocnema pulicaria* Melsh., provide a means to spread the pathogen to new hosts, to enter the plant, and perhaps to overwinter from one year to the next. Pepper (1967) in his excellent monograph emphasized that only the corn flea beetle is important in overwintering of *E. stewartii;* the pathogen overwinters in the alimentary tract of hibernating adults.

According to Harrison and Brewer (1982), a large number of insects, particularly Diptera, have been shown to be important vectors of soft rot bacteria, *Erwinia* spp. The insects spread soft rot bacteria within already infested fields, but they are probably most important in the initial introduction of the pathogens into disease-free areas. Leach (1940) suggested a symbiotic mutualism between the seedcorn maggot, *Hylemya platura* (Meigen), and *Erwinia carotovora* (Jones) Bergey et al. These bacteria were reported to be normal inhabitants of the insect gut and, according to Leach (1926, 1931) and Huff (1928), to aid the insect nutritionally by reducing the plant tissues into a form more usable by the larvae. A similar relationship might exist with *Hylemya brassicae* (Weidemann), the cabbage maggot, and soft rot in the Brassicaceae (Johnson 1930).

Doane (1953) demonstrated that the onion maggot, *Hylemya* (= *Delia*) *antiqua* Meigen, was important to the survival of *E. carotovora,* the cause of soft rot of onion. The feeding activities of the onion maggot aid the bacterium by providing wounds for entry into the plant. In turn, the bacterium aids the insects nutritionally. Friend et al. (1959) reported that larvae of onion maggot could not be reared to adults on sterilized onions. The bacterial pathogen apparently reduces onion tissue to a more suitable form for the larvae, and it is possible that the ingestion of bacteria provides certain nutritional requirements of the insects. The fact that adult onion maggots and some other insects are attracted to rotting onions (Doane 1953) suggests that infected tissue may be important to the survival of the insects.

Recent studies on insect relationships with *E. carotovora* (which is also the causal organism of heart rot of celery) and possible insect vectors are lacking. Leach (1927) reared *Scaptomyza graminum* (Fallen) and *Elachiptera costata* (Loew) (Diptera: Drosophilidae and Chloropidae, respectfully) on rotting celery leaves. From this early report it is not possible to determine if the insects were vectors or just provided entry courts for bacteria already on leaf surfaces. Indeed, Leach (1940) suggested that experimental evidence for a mutualistic relationship was lacking.

Harrison et al. (1980) offer a good discussion of bacterial wilt of cucurbits, incited by *Erwinia tracheiphila* (Smith) Bergey et al., and the possible association of the phytopathogen with the striped cucumber beetle *Acalymma vittata* (F.) and the spotted cucumber beetle *D. undecimpunctata howardi*. After carefully reviewing several papers on the subject of beetle–pathogen relationships involving overwintering, inoculation, and dissemination, the authors concluded that little evidence exists to support the conclusion that the pathogen is dependent upon the insect species. Leach (1964) even suggested that bacteria-contaminated beetles do not survive as well as healthy ones during overwintering.

Fire blight of apple, pear, and related species, caused by *Erwinia amylovora* (Burrill) Winslow et al., was the first bacterial disease shown to be vectored by insects (Waite 1891). Today more than 100 insect species are suspected of being involved in transmission, survival, and inoculation of this bacterium into host plants. However, it has not been demonstrated that the bacterium must spend part of its life cycle in insect species. More research is needed to establish if the numerous insect vectors are adversely affected by the bacterium via the host plant. Although *E. amylovora* can infect plants through insect-induced wounds, wounding by a vector is not absolutely necessary for infection. The bacterium can overwinter in cankers on its perennial hosts. Rain or moisture may transport the pathogen to new sites. Insect-proof netting, however, significantly reduces blossom infections (see Harrison et al. 1980); furthermore, considerable numbers of new infections occur during prolonged dry periods (Miller and Schroth 1972) when insect vectors undoubtedly play a significant role in the tree-to-tree dissemination. Caution is called for by Harrison et al. (1980) in their recent review of insect involvement in phytobacterial diseases when assessing the importance of insects in the overwinter survival and long-distance dispersal of inoculum of our most-studied diseases such as fire blight, Stewart's wilt of corn, and cucurbit wilt.

Bacterial rot of apple, caused by *Pseudomonas melophthora* Allen and Riker, is associated with the apple maggot, *Rhagoletis pomonella* (Walsh). In a detailed study Allen et al. (1934) reported that two stages of this insect appeared to assist in the dissemination of the bacteria: the adult fly, by depositing contaminated eggs beneath the cuticle of the apple, and the larvae, by aiding the movement of bacteria through the fruit by their burrowing habit. The pathogen may be transmitted exclusively by the apple maggot. Harrison et al. (1980), in reviewing the data concerning this insect–vector relationship, found evidence to suggest that the bacterium may be necessary for normal development of the larvae, serving as an obligate extracellular symbiote that renders the apple habitable for the apple maggot larvae. Miyazaki et al. (1968) suggested that the bacterium may provide essential amino acids that are absent in apple tissue. They showed that *P. melophthora* synthesizes methionine and cystine which are not present in apple tissue and suggested that these amino acids allow *Rhagoletis* larvae to develop. These authors recognized, however, that because the nutritional requirements of *R. pomonella* were not known, it was not possible to be unequivocal about the conclusions regarding their study.

Similarly, the symbiotic relationships between intestinal bacteria, including *Pseudomonas savastanoi* (Smith) Stevens (which causes olive knot), and the olive fruit fly *Dacus oleae* (Gmelin), have been the subject of numerous research papers in recent years (for complete references see Harrison et al. 1980). Investigations in which the bacterial microflora of the intestine have been eliminated by incorporating an antibiotic, usually streptomycin, in the diet of adult flies have consistently shown that the adult females oviposit normally but that the larvae are unable to survive in olive fruits (Hagen et al. 1963, Hagen 1966, Fytizas and Tzanakakis 1966, Rey 1969). Hagen (1966) suggested that *P. savastanoi* hydrolyzes protein in the olive tissue and makes it available to olive fly larvae and that the bacterium may also synthesize methionine and threonine, essential amino acids not present in the olive fruit. The effects of streptomycin treatment on additional bacteria or other organisms is often overlooked in studies of this type. Considerably more study is needed to fully elucidate the role of prokaryotic bacterial pathogens and other possible symbiotes in the life history of the olive fly and the role of the fly in the etiology of olive knot.

The term ‘‘xylem-limited bacteria’’ is now being used to describe prokaryotic plant pathogens difficult to isolate by standard bacteriologic procedures. These fastidious organisms require complex media for growth, occur only in the xylem of infected plants, and are transmitted by xylem-feeding leafhoppers and spittlebugs. Purcell (1979) discusses some of the leafhopper vectors of these xylem-borne plant pathogens. Freitag (1951) identified at least 75 symptomless hosts of Pierce’s disease bacterium and determined that the leafhopper vectors usually overwintered on the symptomless wild hosts, spreading the pathogen to economically important crops the following spring. Most of the xylem-limited bacteria that cause plant diseases in the United States have been isolated and maintained in pure culture, yet their taxonomic status has not been defined (Raju and Wells 1986). Since the nutritional requirements of these bacteria are not well understood, it takes little courage to suggest that the vector–pathogen–host relationships need intensive research.

Unlike viruses and mycoplasmas, there is little evidence that the direct effects of bacterial pathogens on insect carriers or the effects on insects via the host plant have been seriously considered. This lack of concern may be partially due to the concept that insects are only casually involved in transmission of most phytopathogenic bacteria.

5. FUNGI

Fungi constitute an immensely diverse group of organisms, spanning a broad ecological continuum of niche occupation. Many are saprophytes; others have established symbiotic relationships to varying degrees. Still others are parasitic, dependent upon living hosts for proliferation. Nearly 8000 fungal species are phytoparasites known to cause plant disease (Lucas et al. 1985). These include economically devastating pathogenic rusts and smuts, downy and powdery mildews, and blights that attack a heterogeneous cross section of hosts and host tissues in varied habitats.

Insect–fungus relationships are as multiform as the fungi themselves. Both obligate and facultative mycophagy are common among several insect orders, Diptera and Coleoptera in particular (Benick 1952, Fogel 1975, Crowson 1984). For example, a majority of mycophagous insects is known to utilize sporocarps of polyporaceous bracket and shelf fungi as a primary food resource (Weiss and West 1920, Fogel 1975). These comparatively long-lived fungi employ a variety of substrates and, although living trees are among those utilized as hosts in a parasitic sense, it is the macrofungal structure alone rather than a plant–fungus complex that interacts with the insect. Ambrosia beetles (Scolytidae) subsist on domesticated fungi that are transported to and flourish in the beetle galleries but apparently are not regarded as pathogenic to living trees (Francke-Grosmann 1967). Such strict mycophagous relationships will not be included in this section and discussion of insect–fungal pathogen–host plant interactions will be limited to those explicitly involving all three components.

Most plants, when infected with a pathogenic fungus, function in an altered host capacity (see Sections 1 and 6), becoming more or less favorable to the insect (Williams 1965, Graham 1967, Cobb et al. 1974, Carruthers et al. 1986). Others serve as intermediaries by providing a dynamic interface for insect–fungal phytopathogen encounters. Notable examples include insect associations with rusts and ergot on cereals (Langdon and Champ 1954, Gilbertson 1984). Occasionally, suspected mutualistic microbial–plant or microbial–insect associations develop (Wright 1935, Francke-Grosmann 1967, Nebeker et al. 1984, Clay et al. 1985a), creating a more complex and intriguing multitrophic system.

Fungal plant pathogens, unlike viral or prokaryotic disease agents, are not solely dependent upon specific biological vectors for internal transport or primary inoculation. They are frequently wind-borne or may be carried as external contaminants on arthropods (e.g., ectosymbionts of bark beetles, Barras 1973). Pathogenic fungi are ubiquitous in their plant host distribution, and insect interaction with these disease agents is known throughout the vascular plants. This interaction is discussed in the following sections, arbitrarily separated into tree-associated, and shrub-, herb-, and grass-associated relationships.

5.1 Tree-Associated Fungi

Few insect–phytopathogenic fungus–plant interrelationships have been as actively studied as that of the bark beetle-bluestain (and related fungi, Francke-Grosmann 1967) and root disease interactions (Cobb et al. 1974, Barras and Perry 1975). The southern pine beetle, *Dendroctonus frontalis* Zimmermann, has long been associated with the blue-stain fungus, *Ceratocystis minor* (Hedgcock) in pine (Craighead 1928). The relationship between the insect and the fungus has often been described as facultative mutualism (Francke-Grosmann 1967, Barras 1973) in which the fungus derives benefit through phoretic transport by the beetle and the beetle benefits through fungus-induced changes in tree physiology. Phytopathogenic effects of the fungus are varied, but include xylem disruption and subsequent reduction in tree water content, believed to be essential for beetle brood production (Nelson 1934).

There is mounting evidence, however, that the mutualistic relationship may not occur as frequently as previously assumed (Bridges et al. 1985) and is tempered by certain nonbeneficial effects of the diseased tissue on the beetle participants. Several studies (Barras 1970, Franklin 1970) have determined that fungal presence in the phloem and inner bark is detrimental to larval development and larval and adult feeding. Reductions in total sugar content and abnormally low and high levels of free amino acids and protein-bound amino acids, respectively, in tree tissues proximate to the fungus (Hodges et al. 1968, Barras and Hodges 1969) may be responsible.

The persistence of the beetle–fungus association despite its apparent negative influence on beetle brood attests to the complexity of the facultative relationship. It is interesting that brood development proceeds normally in the absence of the fungus (Hetrick 1949, Bridges et al. 1985). According to Bridges et al. (1985), "when there is an abundance of highly susceptible host trees or when beetle populations are very large, *C. minor* may not be required" for successful beetle infestation. Apparently, tree morbidity resulting from adverse effects of bluestain inoculum on water relations (Hodges et al. 1985a, b) may more than compensate for nutritional microhabitat deficiencies encountered by individual beetles. Thus beetle abundance, extent of fungal infection, and tree susceptibility to both are only a few of the factors demanding consideration in attempts to understand this interaction.

A related fungal pathogen and causative agent of Dutch elm disease, *Ceratocystis ulmi* (Buism.) C. Mor., exists in an apparent mutualistic symbiosis with its primary vector, *Scolytus multistriatus* (Marsham), similar to that of other bark beetle–bluestain associations. Epidemiology and etiology of the disease have been studied extensively (i.e., Finnegan 1964), and phytopathogenic effects appear to be due to additive blockage of internal water transport (Francke-Grosmann 1967).

Associations of other wood-inhabiting beetles and wood-staining fungi in coniferous systems have been examined to a lesser degree. The presence of blue-staining fungi in Scots pine, *Pinus sylvestris* L., sapwood corresponds to delayed larval development in the common furniture beetle, *Anobium punctatum* DeG. (Bletchly 1969). Yearian et al. (1972) reported that the bluestain fungus *Ceratocystis ips* (Rumb.) C. Moreau inhibited oviposition by *Ips* spp. beetles, but brood size, pupal weights, and fecundity were unaffected. *Scolytus ventralis* LeC. appeared to rely upon a facultative relationship with the brown-staining fungus *Trichosporium symbioticum* Wright in fir (Wright 1935), analogous to that of bluestain fungus with *D. frontalis*. Nonstaining fungi that readily colonize sapwood adjacent to fungus-stained sections are also known to escalate tree decline and invite bark beetle attack (Whitney and Cobb 1972).

Root pathogens of various conifers frequently co-occur with bark beetle infestation (Cobb et al. 1974). Unlike the intricate mutualistic bluestain relationship, these disease agents are credited principally with predisposing their host to bark beetle colonization.

Moeck et al. (1981), using a quantitative sampling method for estimating the severity of infection by *Verticicladiella wagenerii* (Kendrick), a root pathogen in ponderosa pine, *Pinus ponderosa* Laws., found that 77% of root-diseased trees were killed by scolytid infestation or scolytid-buprestid attack over a three-year period. Only 10% of the healthy control trees were affected. They also found that primary attraction of five scolytid species to host trees was unrelated to the presence or severity of disease. Similar results were reported as unpublished data in Wood (1982) for landing rates of *Dendroctonus brevicomis* LeConte and *Dendroctonus ponderosae* Hopkins on pine harboring *Ceratocystis wageneri* Goheen and Cobb (the imperfect state of *V. wagenerii*). Thus dispersing bark beetles are not preferentially attracted to root-rotted trees but are almost certain to colonize them following incidental encounter.

A direct relationship exists between bark beetle infestation and severity of *C. wageneri* infection in ponderosa pine (Goheen and Cobb 1980). Phytopathogenic invasion of host roots typically results in beetle establishment once the pathogen has fully advanced to the root crown. Other root pathogens invading pine and fir, such as *Fomes annosus* (Fr.) Karst. and *Armillaria mellea* (Fr.) Quel., also invite secondary infestation by scolytid and buprestid beetles (Partridge and Miller 1972, Cobb et al. 1974, Hertert et al. 1975, Goheen and Cobb 1980), with nearly 80% of all beetle-infested trees constituting a plant–pathogen complex system. Cobb et al. (1974) suggested that such consistent predisposition of diseased trees to beetle attack may dictate geographic and quantitative fluctuations in bark beetle population dynamics.

A recent detailed study of beetle–fungus–host plant interaction (Kulhavy et al. 1984) suggests that endemic bark beetle populations are maintained in stands of western white pine, *Pinus monticola* Douglas, hosting a multiple complement of pathogenic fungi. Trees weakened by the crown-defoliating blister rust *Cronartium ribicola* Fisch. were susceptible to root pathogen invasion which subsequently incited bark beetle establishment. Beetle activity (primarily *D. ponderosae* and *Pityogenes fossifrons* (LeConte) was associated with the presence of one or more root pathogens (primarily *A. mellea*) in over 90% of all diseased trees, and greater than 60% of host trees sustaining blister rust damage also suffered bark beetle infestation. The extent of phytopathogenic infection determined the probability of beetle attack (Fig. 5).

Although bark beetles are capable of successfully rearing brood in apparently pathogen-free trees (i.e., Bridges et al. 1985) they rarely do so. Instead, they rely upon microorganismal augmentation to overcome host tree defenses. Hodges et al. (1985b) recently proposed a working hypothesis of beetle, fungus, and host tree interaction, whereby an invading phytopathogenic fungus (*C. wageneri*) renders the host susceptible to beetle infestation by reducing the efficiency of the oleoresin defense system normally enlisted against beetle attack (Fig. 6). The ability of the tree to "pitch out" invading beetles is impaired and, ultimately, physiological changes lower oleoresin flow rates and exudation pressure (OEP) and the tree succumbs to bark beetle colonization. Trees resistant to beetle attack display a more rapid, elevated, and localized mobilization of the oleoresin system in direct response to fungus inoculation compared to susceptible cohorts (Raffa and Berryman 1982). The specific mechanisms responsible for initiation and out-

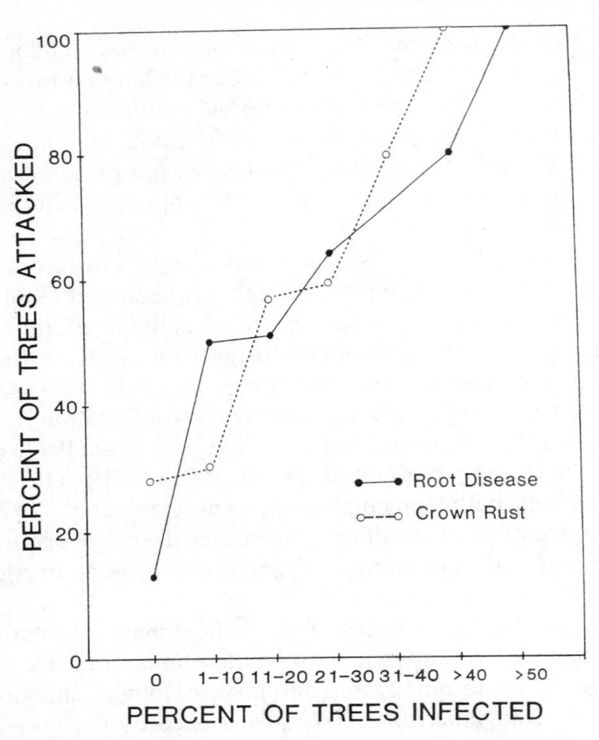

Figure 5. Relationship between the incidence of fungal phytopathogen infection and subsequent attack by bark beetles. (Modified from Kulhavy et al. 1984)

come of beetle–fungus–host interactions, however, and the generalized nature of the interaction among fungi as diverse as wood-staining and root rot pathogens, awaits elucidation.

Although the relationship between insects and phytopathogenic fungi in arboreal ecosystems has been examined in some detail for damaging, bark beetle-associated diseases such as bluestain and Dutch elm, other associations have received scant attention or are poorly understood. For example, an interesting interaction between *Scolytus* spp. that vector Dutch elm disease and a fungal endophyte of elms has been discussed (Webber 1981). The presence of *Phomopsis oblonga* (Desm.) Trav., an elm-colonizing endophyte, is responsible for declines in beetle populations. Infected wood deterred bark beetle gallery initiation and feeding (Claydon et al. 1985). Carroll (1986) and Stone (1985) demonstrated a marked increase in larval mortality of gall midges (*Contarinia* spp.) in Douglas fir harboring endophytic infections of *Rhabdocline parkeri* Sherwood-Pike et Carroll. The phytopathogenic status of such endophytic fungal associations often is uncertain, and the relationship may be more appropriately described as mutualistic (Carroll 1986).

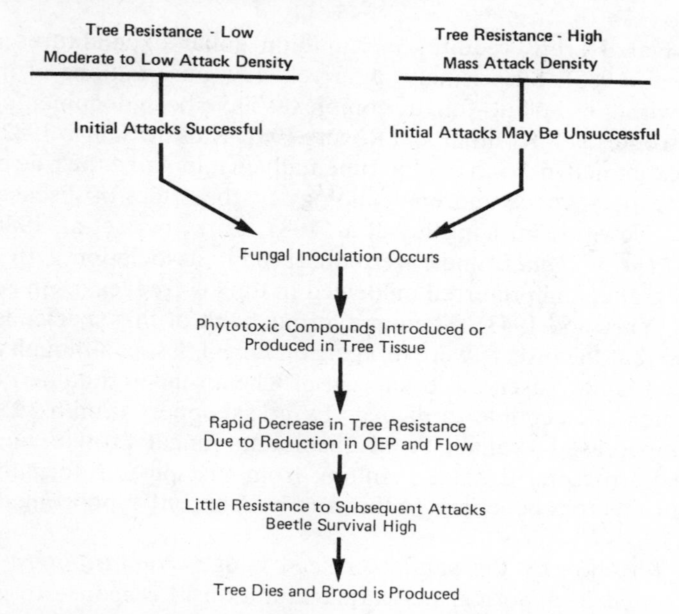

Figure 6. Observed and hypothesized sequence of events involving interactions between beetle, fungi, and host following bark beetle attack. (From Hodges et al. 1985b)

Termite interactions with fungi in dead, decaying wood have long been acknowledged (i.e., Gilbertson 1984); however, less frequent termite infestations in living tree heartwood may also be in complement with or dependent upon fungal preparation of the wood. Williams (1965) found that *Coptotermes niger* Snyder damage in standing pitch pine was restricted to heartwood infected with brown rot. Subsequent laboratory studies suggested that fungal conversion of resinous wood components allowed for secondary infestation by the termite. Disease-free wood successfully repelled termite attack and significantly reduced worker longevity. An identical relationship has been established for fungus-affected heartwood and *Coptotermes acinaciformis raffrayi* Wasmann in Australian eucalypt and pine forests (Perry et al. 1985). In both instances, fire damage precedes fungal invasion which, in turn, preconditions the wood, inviting termite attack. The insect-diseased wood interrelationship is one of obligate commensalism in which the termites depend wholly upon the fungus to convert sound wood to a utilizable resource.

5.2 Shrub-, Herb-, and Grass-Associated Fungi

The broad range of macrophytes collectively exploited by phytopathogenic fungi encompasses many plant taxa. Numbered among potential hosts are native plant species and their cultivated counterparts. Control of disease in these cultivated crops requires multimillion dollar expenditures annually (Lucas et al. 1985). Surprisingly, relatively few observations of insect association with these plant–fungus complexes have been documented (Smith 1939, Yarwood 1943, Kreitner and Rogers 1981, Rios de Saluso 1982). Fewer still have examined in depth or experimentally manipulated the role of insects in pathogen persistence and epidemiology or the effect of diseased plants on insects (Lewis 1979, Kingsley et al. 1983, Carruthers et al. 1986).

Thrips (*Thrips tabaci* Lind.) were observed in association with powdery mildew of grape, and preferred mildewed to fungus-free leaves in controlled situations (Yarwood 1943). The omnivorous habit of this species led to the conclusion that the insects were foraging on fungal tissue, although ingestion of hyphae was not observed. Stem rust of wheat rendered numerous wheat varieties more susceptible to damage by grasshoppers (Smith 1939). Plant maturity, increased availability of sugar and fungal protein on infected plants, and structural damage resulting from rust pustule formation were suggested as factors beneficial to the grasshopper, but supporting data were lacking.

Seed infestation by the sunflower seed midge, *Neolasioptera helianthi* (Felt), resulted in abnormal pericarp maturation in response to accompanying fungal phytopathogen invasion in the sunflower *Helianthus annuus* L. (Kreitner and Rogers 1981). Successful completion of larval development appeared to depend upon fungal presence, with the fungus serving as a supplemental source of nutrition or possibly as an agent that preconditions seed tissue for assimilation by larvae. Lewis (1979) noted preferential feeding of the grasshopper, *Melanoplus differentialis* (Thomas), on rust-infected (*Puccinia helianthii* Schw.) leaves and downy mildew-infected leaves (A. C. Lewis, personal communication) of both native and cultivated sunflower. Feeding was confined to rust pustules and necrotic tissue immediately surrounding them, indicating a highly localized phenomenon. Preference for diseased over healthy leaves on individual plants supported this finding. Analogous results were reported for *Zonocerus variegatus* (L.) on fungus-infected cassava (Bernays et al. 1977). Differential preference for abiotically or artificially stressed sunflower (Lewis 1984) by *M. differentialis*, however, suggests that localized changes in host plant physiology as a consequence of pathogenic stress may further focus grasshopper feeding at the site of rust infection.

Only a few recent studies have quantitatively examined the influence of pathogen-infected crop plant consumption on insect developmental traits. Alfalfa infected by *Verticillium* wilt had no effect on growth rate of penultimate instars of the southern armyworm (*Spodoptera eridania* (Cram) (Kingsley et al. 1983), although efficiency of conversion of ingested material was significantly lower on infected plants. Larvae were able to compensate by increasing consumption rates, but the principal factor(s) responsible for diminished conversion of digested host plant was not determined. Carruthers et al. (1986) found that European corn borer *Ostrinia nubilalis* (Hübner) larvae exhibited accelerated growth rates when reared on maize infected with the pathogenic anthracnose fungus, *Colletotrichum graminicola* (Ces.) Wils. They reported a 20% decrease in larval developmental time on infected tissue and postulated that fungal activity may catabolize complex carbohydrates to more easily assimilated sugars. Interestingly, *O. nubilalis* and *C. graminicola* are frequently found in tandem on maize under natural field conditions, an association not unlike that of this insect with *Fusarium graminearum* Schw., a stalk rot which also was observed to favorably influence larval development (Chiang and Wilcoxon 1961).

Karban et al. (1987) recently reported that cotton plants inoculated with *Verticillium dahliae* (SS-4 fungus strain) supported smaller populations of spider mites, *Tetranychus urticae,* than uninoculated control plants. They suggest that reduced leaf area of infected seedlings may account for the negative effect on mite population growth.

Vesicular-arbuscular mycorrhizal (VAM) fungi, although not plant pathogens, can induce metabolic deviations in host plant chemistry mimicking those of a pathogen-burdened plant (Bowen 1978). VAM-infected soybean plants were shown to reduce larval and pupal weights and delay development of two polyphagous noctuids relative to larvae reared simultaneously on uninfected control plants (Rabin and Pacovsky 1985). Degree of root colonization was directly correlated with the severity of observed effects. No obvious differences in nitrogen, micronutrient, soluble sugar or phenolic content could be detected between infected and uninfected soybeans, and negative effects were attributed to underlying antixenotic, antibiotic, or dietary constraints.

Perhaps one of the most intensively studied insect–fungus–host plant multitrophic systems of late is that of fungal endophyte associations with grasses. Grass endophytes, like those persisting in woody hosts (Carroll 1986), penetrate the intercellular spaces of host tissue and may complete their life cycle without producing overt symptoms of infection in the plant (Clay 1986). Lack of apparent pathogenic effects, and the potential for increased vegetative vigor in infected grasses (Latch et al. 1985) has reinforced current thought regarding endophytes as plant mutualists (Clay et al. 1985b, Siegel et al. 1985, Carroll 1986). Many, however, sterilize the host by aborting the developing infloresence and replacing it with fungal fruiting structures (Clay et al. 1985b). Because of the large number of studies on this model system, the differing response of phytophagus insects to infected and uninfected plants, and the potential modifications in host plant metabolism to accommodate these microbes, we have chosen to include a discussion of endophyte fungi–insect interactions here, even though these fungi are largely symbiotic and infrequently are considered plant disease agents.

The presence of endophyte infection in grasses has consistently proved to be detrimental to a broad group of insect herbivores utilizing these grasses as a food resource. This phenomenon of "endophyte-mediated resistance" to insects is well established in perennial ryegrass (*Lolium perenne* L.) and tall fescue (*Festuca arundinaceae* Schreb.) infected by clavicipitaceous fungi.

Crickets, *Acheta domesticus* L., feeding on endophyte-infected perennial ryegrass suffered 100% mortality in less than four days (Ahmad et al. 1985). Abnormal distension of the foregut and pathological changes in proventriculus tissue effectively halted the alimentary process, culminating in prostration and death of adults maintained on *Acremonium loliae* Latch, Christensen and Samuels-infected plants. Two aphid species, *R. padi* and *Schizaphis graminum* (Rondani), and the large milkweed bug, *Oncopeltus fasciatus* (Dallas), avoided *Acremonium coenophialum* Morgan-Jones and Gams-infected tall fescue leaf sheaths or their methanolic extract and neither aphid species was able to survive when confined on individual infected plants (Latch et al. 1985, Johnson et al. 1985).

Mazur et al. (1981) observed differential responses of selected perennial ryegrass cultivars to damage by sod webworm (*Crambus* spp.), which later was confirmed to be due to the presence of the "*Lolium* endophyte" (*A. loliae*) (Funk et al. 1983). Reduced damage to certain cultivars by the bluegrass billbug, *Sphenophorus parvulus* Gyllenhal, also suggested fungal endophyte involvement (Ahmad and Funk 1983). Similar resistance qualities of infected ryegrass to feeding and oviposition by the Argentine stem weevil, *Listronotus bonariensis* (Kuschel), have been reported in New Zealand (Kain et al. 1982; Prestidge et al. 1982; Mortimer and diMenna 1983). More recent studies using this insect have demonstrated adult nonpreference for leaves from ryegrass and tall fescue containing *Acremonium* endophytes (Barker et al. 1983, 1984), reduced oviposition and larval damage in infected stands (Gaynor and Hunt 1983), and deterred adult-feeding response to extracts of infected foliage (Gaynor et al. 1983, Rowan and Gaynor 1986) and broth cultures of the fungus (Prestidge et al. 1985).

Deleterious effects of endophyte infection on sustained feeding, survival and developmental parameters of the fall armyworm, *Spodoptera frugiperda* (J. E. Smith), have been quantified on perennial ryegrass, tall fescue and naturally infected stands of several grasses and sedges common to pastures (Fig. 7). Larvae preferred to feed on uninfected over infected ryegrass leaves of mixed age in controlled choice studies (Hardy et al. 1985) and preferentially selected uninfected fescue when provided equivalent age leaves from infected plants (Hardy et al. 1986). The magnitude of nonpreference for leaves from an infected host was directly related to the relative age of the leaf (Fig. 8). Survival was halved for larvae reared on infected ryegrass foliage from the field (Clay et al. 1985a). An expansion of the fall armyworm developmental studies revealed that fungal infection consistently hampered larval development, resulting in delayed growth on five grass and two sedge hosts (Clay et al. 1985a, b; Hardy et al. 1985).

The intimate relationship of a grass and its fungal inhabitant and the enigmatic extent of their individual and collective contribution to functional insect resistance suggest that several mechanisms may be operating that neg-

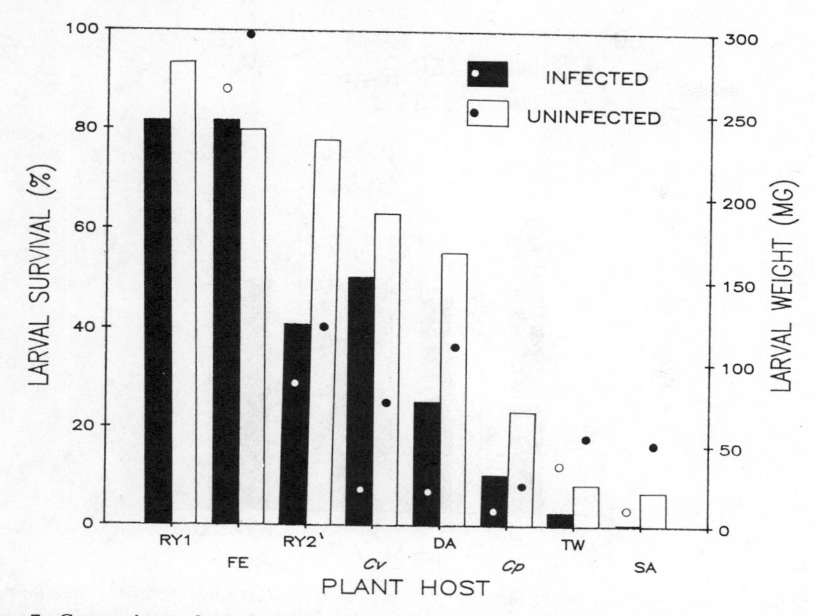

Figure 7. Comparison of survival to pupation (bars) and weights (dots) of 12–14-day-old fall armyworm larvae reared on endophyte-infected and uninfected grass and sedge hosts. RY1 and RY2 = greenhouse-raised and field-collected perennial ryegrass, respectively; FE = tall fescue; *Cv* = *Cyperus virens* Michx. (sedge); DA = dallisgrass; *CP* = *C. pseudovegetus* Steud. (sedge); TW = Texas wintergrass; SA = sandbur. (From Clay et al. 1985a, b)

atively influence insect herbivores. No differences in crude nitrogen content, amino acid composition, water-soluble carbohydrate or forage fiber characteristics were found between infected and fungus-free fescue foliage (Bond et al. 1984, Belesky et al. 1985, Hardy et al. 1986). However, biologically active alkaloids exclusively associated with fungal infection have been implicated as potential allelochemic agents. Tall fescue forage with high incidence of fungal infection contained substantial quantities of clavine ergot alkaloids (Lyons et al. 1986). *N*-formyl and *N*-acetyl lolines are also present in relatively high levels in infected fescue (Bush et al. 1982) and are associated with aphid-deterring extracts (Johnson et al. 1985), although loline content had no influence on fall armyworm larval choice or development (Hardy et al. 1986). Prestidge et al. (1985) reported fungal culture extracts from ryegrass endophytes contained nonloline weevil-deterring compounds that have been partially characterized as indolic alkaloids (Rowan and Gaynor 1986). Clearly, these alkaloidal compounds may hold profound feeding deterrency and possibly antibiotic effects for insects encountering infected grasses.

A final insect–fungus–host plant interaction receiving meager attention is that of nectar-feeding insects and their utilization of ephemeral fungal

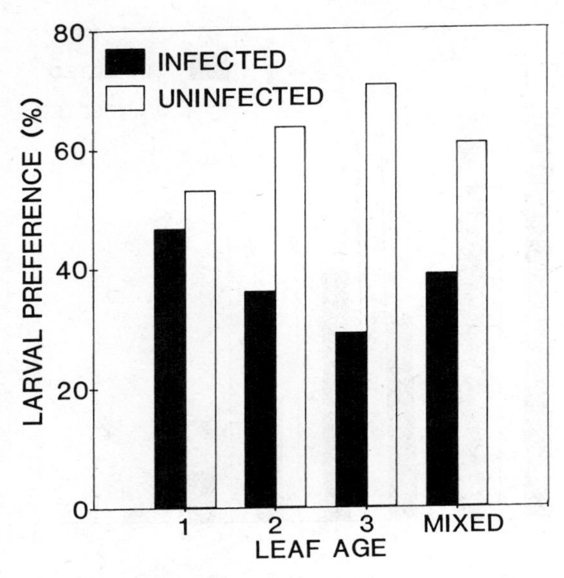

Figure 8. Fall armyworm neonate larval preference for endophyte-infected (*Acremonium coenophialum* Morgan-Jones and Gams) and uninfected tall fescue when provided infected and uninfected leaf blades of equivalent age. Leaf age 1 = new growth; leaf ages 2 and 3 = consecutively older growth; mixed = mixture of leaf ages and represents larval response on both tall fescue and perennial ryegrass (infected by *Acremonium loliae* Latch, Christensen and Samuels). (From Hardy et al. 1985, 1986)

exudates as supplemental food resources. These associations are often seasonally dependent upon phytopathogen life history and may represent casual exploitation of plant disease by the insect, but may also exemplify fungal manipulation of potential insect vectors. Infecting pathogenic fungi often benefit from this association via fortuitous dispersal and consequent secondary infection by insects visiting their fruiting structures.

Vegetative structures of blueberries and huckleberries infected by *Monilinia* spp. fungi secrete sugars at the site of infection, attracting a host of natural pollinators of these plants (Batra and Batra 1985). Infected parts are thought to mimic the ultraviolet-reflective patterns of inflorescences. Rust fungi were reported to attract honeybees to feed on exudates from a diseased South African grass (Lundie 1938). Ergot (*Claviceps* spp.) infection of grasses attracts a multitude of insects to its sticky secretion, primarily flies, beetles, bees, moths, and occasionally aphids and thrips (Atanasoff 1920, Langdon and Champ 1954, Sharma et al. 1983). Karr (1976) postulates that ergot–insect association in tropical climates is a necessity for fungal dispersal in the absence of wind. Ergot and related fungal honeydew is a reservoir for infrequently encountered oligosaccharides (Mower et al. 1973) and thus undoubtedly functions as a unique source of nutrients for insects.

6. THE DISEASED PLANT AS AN ALTERED HOST

The physical and chemical characteristics of a given plant species or an individual plant define its suitability as a host for a variety of biotic agents. Insect herbivores and microbial plant pathogens commonly exploit the plant as a resource, utilizing it to varying degrees as a primary source of nutrition (Jones 1984). Abiotic factors influence the progression of plant–insect or plant–pathogen interactions and frequently predispose host plants to successful phytopathogen or insect attack (Schoeneweiss 1975, Ayres 1984, other chapters in this book).

The biotic agents exploiting the plant resource also constitute a considerable source of highly variable stress on the plant. Generally, plant pathogens are randomly dispersed, establish close association with their host, and remain on the host for an extended period of time (often completing their life cycle on the same individual). Phytophagous insects, on the other hand, participate in active and directed dispersal, seeking and selecting a host and often establishing a temporary association (Harris and Frederiksen 1984). Thus a plant must be prepared to contend with a diversity of exploitation strategies.

There is an enormous amount of literature concerning theory and mechanisms of plant–disease and plant–insect interaction, much of it addressing various aspects of nutrition and plant defense (e.g., Feeny 1975, Beck and Reese 1976, Levin 1976, Rhoades and Cates 1976, Harborne 1978, Rosenthal and Janzen 1979, Scriber and Slansky 1981, Bailey and Mansfield 1982, Denno and McClure 1983). However, little is known about the specific changes occurring in plants under phytopathogenic stress and their subsequent effects on insect herbivores utilizing the plants as hosts. As discussed in Section 1, most diseased plants probably function in a capacity best defined as an altered host, and any influence of the pathogen on the insect herbivore is exerted indirectly through modifications in host plant morphology, physiology, or possibly plant population dynamics (i.e., Bridges et al. 1985). Only a few studies (e.g., Laurema et al. 1966, McIntyre et al. 1981, Rabin and Pacovsky 1985, Hardy et al. 1986) have quantitatively examined the relationship between microbe-induced plant characteristics and the resulting insect response. Therefore, any discussion of host plant-dependent mechanisms responsible for mediating diseased plant–insect interactions must be restricted to plausible corollaries based upon assumptions of insect behavioral response to and developmental requirements derived from the host plant.

Prediction or inference of pathogen-infected host effects on insects is extremely difficult. First, generalizations regarding phytopathogenic symptoms produced by a given group of disease agents are not without numerous exceptions. For example, plant infection may be somewhat localized (certain viruses, bacteria, and fungi) or may be systemic (many viruses and mycoplasmas, some fungi). An individual plant may, as a result of disease, represent a mosaic of nutritional attractiveness to an insect (Denno and McClure 1983). Second, although most insects have basic nutritional requirements in common (Beck and Reese 1976, Mattson 1980, Scriber and Slansky 1981), they are quite variable in their response to many "secondary" plant metabolites, particularly those functioning as allelochemics (Levin 1976, Rhoades and Cates 1976). There are some relatively consistent changes in structure and metabolism that do occur in plants experiencing pathogenic stresses, however, and a certain degree of speculation as to how an insect herbivore might be affected by such an altered host plant is possible. The brief discussion of inferred effects that follows points out the potential for significant impact of plant disease on insect–host plant relations.

6.1 Morphology

Phytopathogenic diseases can produce an array of overt symptoms that modify or disrupt the existing morphology of an infected plant. The most obvious signs of disease include leaf chlorosis, wilting, reduction in vegetative proliferation, defoliation, stunting, gall formation, and necrosis (Dickinson and Lucas 1982). Any one of these abnormalities could affect basic insect–host plant interaction.

Changes in color of available vegetation could render a plant more or less attractive to an approaching insect herbivore. Leaf necrosis accompanying aster yellows infection, for example, was a suggested factor in attraction of aphids to diseased plants (Whitcomb and Williamson 1979). Certain biotrophic phytopathogen infections are responsible for the formation of "green islands," localized sites of photosynthetically active tissues surrounded by areas of chlorotic vegetation (Misaghi 1982). These sites might also act as nutritional islands, selectively drawing in foraging insect herbivores.

Foliage-limiting diseases causing outright defoliation or leaf wilt could also influence host suitability. Defoliated plants offer fewer feeding sites, although defoliation-induced changes in metabolism may outweigh site availability. Wilted leaves are more palatable to some insects, particularly those with reduced water preferences or requirements (Lewis 1979); however, chemical factors associated with wilting are important, and are not easily separated from the physical effects of turgidity loss in insect interaction. Symptoms of growth-altering diseases (i.e., stunting or gall formation) modify insect response to the host (Bald et al. 1946). Speculation on the degree or direction of response hinges on the unique life history traits of the host and insect involved (e.g., gall-associated species may exploit tissue proliferation of diseased hosts).

6.2 Physiology

6.2.1 Nutritional Factors

Qualitative and quantitative changes in nutritional factors and allelochemic compounds of diseased plants are likely to have profound behavioral and developmental effects on insects utilizing them as hosts. Nutritional requirements of primary importance to most phytophagous insects normally are met by healthy, acceptable host plants. Alterations in food quality—such as nutrient depletion, conversion to an inaccessible form, or nutritional shunting within diseased plants creating optimal feeding sites—could greatly influence insect performance. Basically, insects require an adequate dietary source of water, nitrogen, carbohydrate, and certain lipids and vitamins (Dadd 1973, Rockstein 1978, Mattson 1980, Scriber and Slansky 1981). Phytophagous insects must rely upon appropriate host plants to provide acceptable amounts of necessary primary plant metabolites. Carbohydrate requirements may often be fulfilled by simple sugars, which also function as stimuli for initiation and maintenance of feeding (Schoonhoven 1968). Nitrogen, obtainable in the form of proteins or amino acids, appears to be a crucial, limiting factor in insect growth and survival (Mattson 1980, Scriber and Slansky 1981). Thus any disease-induced increases in availability of nutritional components of insect host plants should enhance host suitability, but reductions in nutrient availability should render plant hosts less suitable for insect herbivores.

Plant pathogens can cause a myriad of changes in host plant metabolism that potentially are beneficial or detrimental to insects. Some changes effectively parallel those caused by abiotic stress conditions (i.e., drought) which have been postulated to enhance proximate nutritional quality of insect host plants (Rhoades 1983, Holtzer et al. Chapter 3 this volume) or resemble wound response mobilization of energy and nutrient sources resulting in metabolic sinks (Kosuge and Kimpel 1981) favorable to both pathogens and insects. As a rule, viruses (and probably other plant disease agents as well) interfere with normal photosynthetic activity in infected plants. Notable reductions in soluble carbohydrates may be observed with systemic viral infection and certain fungal infections because of impaired rates of photosynthesis, or possibly due to alternative metabolism of photosynthates (Buchanan et al. 1981). If these changes also reduce phagostimulant concentrations or accessibility of usable carbohydrate for insects, the infected plant may be less desirable than its healthy counterpart. In contrast, accumulations of starch and soluble carbohydrates at sites of infection are not uncommon in plants afflicted with diseases of viral or fungal origin (Misaghi 1982). Singh and Smalley (1969) reported sevenfold increases in free sugar concentrations in elms inoculated with *C. ulmi,* the Dutch elm disease agent. Mycoplasmas and biotrophic bacteria also are known to induce accumulation of soluble carbohydrates in infected tissues (Whipps and Lewis 1981). These tissues might serve as attractive energy reserve sources for phytophagous insects, until extensive pathogenic damage or subsequent invasion by necrotrophic organisms occurs.

Apparent nitrogen deficiency is a common accompaniment of plant disease. The pathogen may remove nitrogenous compounds from a plant for its own maintenance or growth, and thus produce symptomatic changes in the host such as stunting or chlorosis (Russell 1981). Frequently, shifts in carbohydrate metabolism divert photosynthetic production from sugars to amino acids (Buchanan et al. 1981). This seems especially true for obligate parasite infections, including viruses and fungal smuts (although exceptions have been demonstrated; see Hodges and Robinson 1977). Accumulation of selected amino acids has been shown to closely correlate with symptom severity (Markkula and Laurema 1964, Tu and Ford 1970). Increased amino acid content in a plant host might be expected to favor insect response and performance, but the limited information referring to this association reveals ambiguous results (Markkula and Laurema 1964, Laurema et al. 1966, Gildow 1984). Overabundance of some amino acids can even be detrimental (Auclair 1965). Obviously, insect foraging and life history traits (i.e., phloem-sucking versus defoliation) and specific qualitative or quantitative amino acid requirements in insect–plant (and disease) interactions are only a few of the biological factors that prevent generalizations regarding the effects of altered plant nitrogen metabolism on insects.

Wilting is a common symptom of phytopathogen invasion of a host plant and is characteristic of deteriorating water relations within most diseased hosts. Water stress may be induced by physical blockage of xylem tissue, morphological abnormalities, or the release of high molecular weight compounds into water-conducting tissue by certain bacteria and fungi (Russell 1981, Misaghi 1982). Most plant diseases, regardless of etiological origin, negatively affect water uptake or translocation. Failure to maintain adequate water relations has been associated with other metabolic anomalies in diseased plants, including altered carbohydrate levels (Singh and Smalley 1969) and nitrogen concentration (Mattson 1980). Disrupted water movement and wilting of plant tissues has also been correlated to increased insect populations on affected host plants. As mentioned in Section 5, reduced water content of fungus-infected trees allows successful bark beetle colonization (Nelson 1934). Grasshoppers prefer flaccid to turgid sunflower leaves (Lewis 1979). Often, any stress factors that negatively alter water relations in a plant can preferentially predispose the plant to insect attack, partially due to favorable changes in concentrations of primary metabolites utilized by insects (Rhoades 1983). However, there are many insects, particularly larvae of foliage-chewing species, that require relatively high water content in host plants for optimal performance (Scriber and Slansky 1981). Clearly, nutritional requirements of insects are to some degree dependent upon the species and developmental stage, type of host plant utilized, and insect life history. The magnitude of effectiveness of phytopathogenically induced changes in plant nutrient factors on an insect will vary with each unique insect–plant–pathogen combination. More important, the altered nutritional state of a diseased host plant is only a partial reflection of the physiological and biochemical changes that can potentially influence food plant suitability for the insect. Allelochemical changes resulting from plant disease may be the determining factor in the outcome of insect–diseased host interactions.

6.2.2 Allelochemicals

Although an adequate complement of commonly recognized nutritional factors is essential to insect survival, growth, and development on a host plant, it is usually the interaction of plant allelochemicals with the insect that determines the successful and continued host–insect relationship (Beck and Reese 1976, Rosenthal and Janzen 1979). Phytopathogenic stress is responsible for induction and/or elevated levels of numerous allelochemic compounds (e.g., Kuc 1972, Harris and Frederiksen 1984), many of which may similarly possess biological activity against insects (Levin 1976). Research efforts in the field of plant interaction with and defense against pathogens and herbivores has produced an enormous amount of information. It is beyond the scope of this chapter to attempt a categorical review of potential effects of specific disease-induced changes in host plant chemistry on insects.

Levin (1976) has stated that host plant response to most plant pathogens and nematodes consists of translocation or modification of existing metabolites following plant–pathogen interaction ("induced resistance"), whereas defense against herbivores centers upon inhibitors or intrinsic deterrents ("constitutive resistance") (although recent evidence indicates that herbivory also triggers induced responses in plants; Ryan 1983, Kogan 1986). Although both systems utilize similar chemical components, including phenolics, tannins, alkaloids, glycosides, saponins, and terpenes (Rhoades and Cates 1976, Beck and Reese 1976, Rosenthal and Janzen 1979), and modifications in plant growth regulator production (Ayres 1981, Misaghi 1982), different functional mechanisms are implicated. Cross protection, or acquired immunity to phytopathogen invasion subsequent to exposure to the same or occasionally an unrelated plant disease agent, is well documented (Horsfall and Cowling 1980). These induced changes in diseased host plants could, following Levin's classification, provide constitutive defense against potential insect herbivores. The phytoalexins, antifungal compounds produced by plants in response to fungal pathogens (Bailey and Mansfield 1982), are one example. Hart et al. (1983) offered phytoalexin-rich and phytoalexin-poor soybean cotyledons to adult Mexican bean beetles, *Epilachna varivestis* Mulsant, in dual-choice tests and found that phytoalexin-rich tissues strongly deterred feeding. Additional studies provide mounting support for phytoalexins as pathogen-induced metabolites active against insect herbivores (McIntyre et al. 1981, Kogan and Paxton 1983, Kogan 1986, M. Kogan pers. comm.), but their effects are probably selective for certain insect species (Kogan and Paxton 1983).

Conversely, the altered host may represent an optimal find for insects that are selectively adapted to unique allelochemic profiles (Feeny 1975, Rhoades and Cates 1976, Kogan and Paxton 1983), or the insect may rely to a varying degree upon the etiological agent to weaken or overcome plant defenses that might otherwise deter successful insect utilization of the host (e.g., oleoresin and bark beetle invasion of pines; Nebeker et al. 1984). Finally, in some cases pathogen infection may have no bearing on subsequent insect herbivory. Phytoalexin content had no effect on feeding preferences of lepidopteran larvae for host plant tissue or artificial diets (Hart et al. 1983, Kogan and Paxton 1983).

A multiplicity of factors interplay to shape the complex and highly variable interactions expected with tritrophic level allelochemic associations. Abiotic stresses, microbial pathogen requirements and toxin production, plant and insect age, plant tissue involved, health of the insect, insect biotype, specific isolate of the phytopathogen and plant cultivar, are only some of the variables known to mediate insect performance on diseased plants (Kennedy 1951, Maramorosch and Jensen 1963, Kuc 1972, Garran and Gibbs 1982, Gildow 1984, Hardy et al. 1986). Host plant susceptibility or resistance to plant disease (and insect herbivores) frequently is attributable to relatively low or high levels of certain allelochemicals, respectively. By understanding allelochemic differences in host plant defensive capabilities, we may gain insight into the underlying mechanisms regulating host–pathogen relationships and concomitant insect interactions (Garran and Gibbs 1982, Gildow 1984, Bridges et al. 1985).

7. CONCLUSIONS

The factors that influence insect–host plant–plant pathogen interactions are dynamic, complex, and highly variable. Although ample opportunity exists in both agricultural and naturally occurring plant ecosystems for phytopathogen mediation of insect–host plant interactions, relatively little is known about its ecological significance or practical importance. Future avenues of research in this area are legion and hold the potential for substantial contribution to our current understanding of fundamental mechanisms and concepts of insect herbivory as it relates to plant stress.

We recommend that future research in insect–plant–pathogen relationships address the following as areas in need of critical examination:

1. Direct interaction with phytopathogens may enable an insect to utilize previously undesirable or unsuitable hosts (Maramorosch 1958). This ability would allow for expansion of the host range of an insect, especially in instances where the disease agent could be maintained in its insect host (harbored) and/or transmitted (vectored) to the new host plant, allowing continual exposure. This insect–plant disease system could serve as a working model for evaluating mechanisms of host range expansion and its ecological and evolutionary significance.
2. Are certain diseased plants sought out by insect herbivores or simply

utilized as hosts in space or time following random encounter? Preferential feeding on infected plants could indicate movement toward a permanent association. Thus an insect might narrow its ecological niche by exclusive use of diseased plants or might be diversifying its host options by opportunistic interaction with phytopathogen-harboring hosts [see (1) above].

3. Do certain diseased plants that normally evoke negative responses from insects become relatively more acceptable to or tolerable for the insect during periods of adverse abiotic conditions and/or poor vegetation quality?
4. Are host-limited (oligophagous) insects forced to subsist on nonpreferred hosts when the primary host plant(s) is afflicted with phytopathogens deleterious to those insects?
5. What characteristics of insects (i.e., monophagy vs. polyphagy) and host plants (i.e., annuals vs. perennials) are most important in insect–host plant–disease relationships? And what are the long-term ramifications of transient versus sustained (Rhoades 1983) insect association with stress (disease) in the plant host?
6. Many insects sequester plant products for use in defense, and some obtain antimicrobial protection by ingestion of selected host substances (Merdan et al. 1975). Could changes in host plant metabolism due to disease enhance the potential for such protection to insects?
7. How do phytopathogenic infection and its multidimensional effects alter insect interaction with predators and parasitoids?
8. Multiple-resistant cultivars (resistant to both insects and diseases) and their cross-resistance status are excellent sources of research material for exploring heterogeneous mechanisms of plant–insect interactions, including both nutritional and allelochemic factors.

In addition, Jones (1984) has proposed testable hypotheses and practical approaches to examining microbial mediation of insect–plant relationships. Although his suggestions and conclusions are structured around endosymbiotic relationships of microbial inhabitants of insect hosts, the concepts are directly applicable to plant pathogen–insect associations. Jones suggests that microbial intervention in insect–plant interactions, such as microbial detoxification of plant allelochemicals and modification of plant nutrients, supports a "general ecological phenomenon" of microbially mediated insect–plant interactions.

A fundamental grasp of insect-diseased plant relationships is also an essential first step in assessing possible manipulation strategies for improving crop production in managed agricultural systems. One functional approach would involve the incorporation of phytopathogen strains that actually benefit the plant host with respect to resistance or tolerance to insect herbivores. For example, introduction of forage grasses infected by fungal endophyte strains selected for high insect resistance and low mammalian toxicity could provide large acreages of insect-resistant pasture that would not induce livestock toxicoses. Endophyte-infected, insect-resistant turfgrass cultivars are already available but are unsafe for cattle comsumption. Integration of plant disease incidence data into established insect monitoring and management programs could provide insight into seasonal or host growth stage-related fluctuations in pest population dynamics (e.g., bark beetle research). The role of infected wild host plants as overwintering sites of insect vectors or as seasonal disease sinks (Caudwell 1984) also is extremely important but remains undetermined.

In conclusion, increased knowledge of the impact of plant pathogens on host plant suitability to insects will require considerable effort and multidisciplinary research, but promises beneficial and intriguing insight and application to general herbivore theory, coevolutionary relationships, and practical applications in insect pest management strategies.

8. SUMMARY

Phytopathogenic mediation of insect–host plant interactions is a frequent but little understood ecological phenomenon. The complexity of such multitrophic interactions is great and is compounded by abiotic influence on each insect, host plant, and phytopathogen component. The insect–plant disease agent association may be one of direct interaction in which the plant functions primarily as an intermediary; more commonly, the interaction is indirect, and the plant functions as an altered host via disease-induced changes in plant morphology or metabolism. Occasionally, a symbiotic relationship between the insect and pathogen (beneficial to the insect) or between disease agent and host (usually beneficial to the host) exists.

Diseased host plants represent a unique yet problematical feeding niche for phytophagus insects, affecting reproductive biology and population dynamics in ways ranging from highly beneficial to acutely detrimental. Based on our current understanding of these intriguing relationships, few generalizations with regard to diseased plant effects on insects can be made. Some plant pathogens are directly responsible for reductions in insect survival, longevity, and fecundity, particularly the obligately parasitic viruses and mycoplasmas that rely upon insect vector transmission and persistence in their insect host. Others may indirectly benefit the insect by enhancing host plant attractiveness or suitability through increased nutritional quality or by weakening existing host plant defense systems (i.e., allelochemical response). Specific examples from observational or experimental studies examining insect–plant–pathogen interactions are discussed in this chapter for viral, mycoplasmal, bacterial, and fungal phytopathogens, and general considerations for future research are presented.

ACKNOWLEDGMENTS

We thank J. H. Benedict, M. Kogan, and D. Waller for critical comments and suggestions on earlier drafts of this manuscript. Stuart LeBas assisted with technical aspects of manuscript preparation, and Debbie Woolf provided clerical assistance during preparation of this chapter.

REFERENCES

Ahmad, S., and C. R. Funk. 1983. Bluegrass billbug (Coleoptera: Curculionidae) tolerance of ryegrass cultivars and selections. *J. Econ. Entomol.* 76: 414–416.

Ahmad, S., S. Govindarajan, C. R. Funk, and J. M. Johnson-Cicalese. 1985. Fatality of house crickets on perennial ryegrasses infected with a fungal endophyte. *Entomol. Exp. Appl.* 39: 183–190.

Ajayi, O., and A. M. Dewar. 1982. The effect of barley yellow dwarf virus on honeydew production by the cereal aphids, *Sitobion avenae* and *Metopolophium dirhodum*. *Ann. Appl. Biol.* 100: 203–212.

Ajayi, O., and A. M. Dewar. 1983. The effect of barley yellow dwarf virus on field populations of the cereal aphids, *Sitobion avenae* and *Metopolophium dirhodum*. *Ann. Appl. Biol.* 103: 1–11.

Allen, T. C., J. A. Pinckard, and A. J. Riker. 1934. Frequent association of *Phytomonas melophthora* with various stages in the life cycle of the apple maggot, *Rhagoletis pomonella*. *Phytopathology* 24: 228–238.

Atanasoff, D. 1920. Ergot of grains and grasses. *USDA Bur. Plant Ind.,* 127 pp.

Auclair, J. L. 1965. Feeding and nutrition of the pea aphid, *Acyrthosiphum pisum* (Homoptera: Aphidae), on chemically defined diets of various pH and nutrient levels. *Ann. Entomol. Soc. Am.* 58: 855–875.

Ayres, P. G. 1981. Effects of disease on plant water relations. *In* Ayres, P. G. (ed.), *Effects of Disease on the Physiology of the Growing Plant*. Cambridge Univ. P., Cambridge, pp. 131–148.

Ayres P. G. 1984. The interaction between environmental stress injury and biotic disease physiology. *Annu. Rev. Phytopathol.* 22: 53–75.

Bailey, J. A., and J. W. Mansfield (eds.). 1982. *Phytoalexins*. Blackie, Glasgow.

Baker, R. F. 1960. Aphid behaviour on healthy and on yellows–virus-infected sugar beet. *Ann. Appl. Biol.* 48: 384–391.

Bald, J. G., D. O. Norris, and G. A. H. Helson. 1946. Transmission of potato virus diseases. V. Aphid populations, resistance, and tolerance of potato varieties to leaf roll. *Commonwealth Australia, Council Sci. and Ind. Res. Bull.* 196: 1–32.

Banttari, E. E., and R. J. Zeyen. 1979. Interactions of mycoplasmalike organisms and viruses in dually infected leafhoppers, planthoppers, and plants. *In* Maramorosch, K. and K. F. Harris (eds.), *Leafhopper Vectors and Plant Disease Agents*. Academic Press, New York, pp. 327–347.

Barker, G. M., R. P. Pottinger, and P. J. Addison. 1983. Effect of tall fescue and ryegrass endophytes on Argentine stem weevil. *In Proc. 36th N. Z. Weed and Pest Cont. Conf.,* pp. 216–219.

Barker, G. M., R. P. Pottinger, P. J. Addison, and R. A. Prestidge. 1984. Effect of *Lolium* endophyte fungus infections on behaviour of adult Argentine stem weevil. *N. Z. J. Agric. Res.* 27: 271–277.

Barras, S. J. 1970. Antagonism between *Dendroctonus frontalis* and the fungus *Ceratocystis minor*. *Ann. Entomol. Soc. Am.* 63: 1187–1190.

Barras, S. J. 1973. Reduction of progeny and development in the southern pine beetle following removal of symbiotic fungi. *Can. Entomol.* 105: 1295–1299.

Barras, S. J. and J. D. Hodges. 1969. Carbohydrates of inner bark of *Pinus taeda* as affected by *Dendroctonus frontalis* and associated microorganisms. *Can. Entomol.* 101: 489–493.

Barras, S. J. and T. J. Perry. 1975. Interrelationships among microorganisms, bark or ambrosia beetles, and woody host tissue: an annotated bibliography, 1965–1974. *USDA For. Ser. Gen. Tech. Rep. S0-10*. New Orleans, LA.

Bateman, D. F. 1978. The dynamic nature of disease. *In* Horsfall, J. G. and E. B. Cowling (eds.), *Plant Disease: An Advanced Treatise,* Vol. 3. Academic Press, New York, pp. 53–83.

Batra, L. R. and S. W. T. Batra. 1985. Floral mimicry induced by mummy-berry fungus exploits host's pollinators as vectors. *Science* 228: 1011–1013.

Beck, S. D., and J. C. Reese. 1976. Insect plant interactions: nutrition and metabolism. *Rec. Adv. Phytochem.* 10: 41–92.

Belesky, D. P., J. J. Evans, and S. R. Wilkinson. 1985. Amino acid composition of tall fescue seed produced from fungal endophyte (*Acremonium coenophialum*)-free and infected plants. *Agron. J.* 77: 796–798.

Benick, L. 1952. Pilzkäfer und Käferpilze. *Acta Zool. Fenn.* 70: 1–250.

Bernays, E. A., R. F. Chapman, E. M. Leather, A. R. McCaffery, and W. W. D. Modder. 1977. The relationship of *Zonocerus variegatus* (L.) (Acridoidea: Pyrgomorphidae) with cassava (*Manihot esculenta*). *Bull. Entomol. Res.* 67: 391–404.

Bletchly, J. D. 1969. Effect of staining fungi in Scots pine sapwood (*Pinus sylvestris*) on the development of the larvae of the common furniture beetle (*Anobium punctatum* De G.). *J. Inst. Wood Sci.* 22: 41–42.

Bond, J., J. B. Powell, D. J. Undersander, P. W. Moe, H. F. Tyrrell, and R. R. Oltjen. 1984. Forage composition and growth and physiological characteristics of cattle grazing several varieties of tall fescue during summer conditions. *J. Anim. Sci.* 59: 584–593.

Bowen, G. D. 1978. Disfunction and shortfalls in symbiotic responses. *In* Horsfall, J. G. and E. B. Cowling (eds.), *Plant Disease: An Advanced Treatise,* Vol. 3. Academic Press, New York, pp. 231–256.

Bowling, C. C. 1980. Insect pests of the rice plant. *In* Luh, B. S. (ed.), *Rice: Production and Utilization*. AVI, Westport, CN, pp. 260–288.

Bridges, J. R., W. A. Nettleton, and M. D. Connor. 1985. Southern pine beetle (Coleoptera: Scolytidae) infestations without the bluestain fungus, *Ceratocystis minor*. *J. Econ. Entomol.* 78: 325–327.

Buchanan, B. B., S. W. Hutcheson, A. C. Magyarosy, and P. Montalbini. 1981. Photosynthesis in healthy and diseased plants. *In* P. G. Ayres (ed.). *Effects of Disease on the Physiology of the Growing Plant*. Cambridge Univ. P., Cambridge, pp. 13–28.

Bush, L. P., P. L. Cornelius, R. C. Buckner, D. R. Varney, R. A. Chapman, P. B. Burrus, II, C. W. Kennedy, T. A. Jones, and M. J. Saunders. 1982. Association of *N*-acetyl loline and *N*-formyl loline with *Epichloe typhina* in tall fescue. *Crop Sci.* 22: 941–943.

Carroll, G. C. 1986. The biology of endophytism in plants with particular reference to woody perennials. *In* Fokkema, N. J., and J. van den Heuvel (eds.), *Microbiology of the Phyllosphere*. Cambridge Univ. P., Cambridge, pp. 205–222.

Carruthers, R. I., G. C. Bergstrom, and P. A. Haynes. 1986. Accelerated development of the European corn borer, *Ostrinia nubilalis* (Lepidoptera: Pyralidae), induced by interactions with *Colletotrichum graminicola* (Melanconiales: Melanconiaceae), the causal fungus of maize anthracnose. *Ann. Entomol. Soc. Am.* 79: 385–389.

Carter, W. 1939. Populations of *Thrips tabaci,* with special reference to virus transmission. *J. Anim. Ecol.* 8: 261–276.

Carter, W. 1973. *Insects in Relation to Plant Disease*. 2d ed. Wiley, New York.

Caudwell, A. 1984. Mycoplasma-like organisms (MLO), pathogens of the plant yellows diseases, as a model of coevolution between prokaryotes, insects and plants. *Isr. J. Med. Sci.* 20: 1025–1027.

Chiang, H. C., and R. D. Wilcoxson. 1961. Interactions of the European corn borer and stalk rots in corn. *J. Econ. Entomol.* 54: 779–782.

Clark, T. B. 1977. *Spiroplasma* sp., a new pathogen in honey bees. *J. Invert. Pathol.* 29: 112–113.

Clay, K. 1986. Grass endophytes. *In* Fokkema, N. J., and J. van den Heuvel (eds.), *Microbiology of the Phyllosphere*. Cambridge Univ. P., Cambridge, pp. 188–204.

Clay, K., T. N. Hardy, and A. M. Hammond, Jr. 1985a. Fungal endophytes of grasses and their effects on an insect herbivore. *Oecologia* 66: 1–5.

Clay, K., T. N. Hardy, and A. M. Hammond, Jr. 1985b. Fungal endophytes of *Cyperus* and their effect on an insect herbivore. *Am. J. Bot.* 72: 1284–1289.

Claydon, N., J. F. Grove, and M. Pople. 1985. Elm bark beetle boring and feeding deterrents from *Phomopsis oblonga*. *Phytochemistry* 24: 937–943.

Cobb, F. W., Jr., J. R. Parmeter, Jr., D. L. Wood, and R. W. Stark. 1974. Root pathogens as agents predisposing ponderosa pine and white fir to bark beetles. *In* Kuhlman, E. G. (ed.), *Proc. 4th Intl. Conf. on Fomes annosus*. September 17–22, 1973, Athens, GA, pp. 8–15.

Craighead, F. C. 1928. Interrelation of tree-killing bark beetles (*Dendroctonus*) and bluestain. *J. For.* 26: 886–887.

Cramer, H. H. 1967. *Plant Protection and World Crop Production*. Leverkusen: Bayer Pflanzenschutz.

Crowson, R. A. 1984. The associations of Coleoptera with Ascomycetes. *In* Wheeler, Q., and M. Blackwell (eds.), *Fungus-Insect Relationships*. Columbia Univ. P., New York, pp. 256–285.

Dadd, R. H. 1973. Insect nutrition: current developments and metabolic implications. *Annu. Rev. Entomol.* 18: 381–420.

Daniels, M. J. 1979. Mechanisms of spiroplasma pathogenicity. *In* Whitcomb, R. F., and J. G. Tully (eds.), *The Mycoplasmas,* Vol. 3. Academic Press, New York, pp. 209–228.

Denno, R. F., and M. S. McClure (eds.). 1983. *Variable Plants and Herbivores in Natural and Managed Systems*. Academic Press, New York.

Dickinson, C. H., and J. A. Lucas. 1982. *Plant Pathology and Plant Pathogens*. 2d ed. Blackwell Scientific, Oxford.

Diener, T. O. 1979. *Viroids and Viroid Diseases*. Wiley, New York.

Diener, T. O. and S. B. Prusiner. 1985. The recognition of subviral pathogens. *In* Maramorosch, K., and J. J. McKelvey, Jr. (eds.), *Subviral Pathogens of Plants and Animals: Viroids and Prions*. Academic Press, New York, pp. 3–20.

Doane, C. C. 1953. The onion maggot in Wisconsin and its relation to rot in onions. Ph.D. dissertation. University Wisconsin, Madison.

Doi, Y., M. Terenaka, K. Yora, and H. Asuyama. 1967. Mycoplasma or PLT group-like microorganisms found in the phloem elements of plants infected with mulberry dwarf, potato witches' broom, aster yellows, or paulownia witches' broom. *Ann. Phytopathol. Soc. Jpn.* 33: 223–226.

Dowell, R. V., H. G. Basham, and R. E. McCoy. 1981. Influence of five spiroplasma strains on growth rate and survival of *Galleria mellonella* (Lepidoptera: Pyralidae) larvae. *J. Invert. Pathol.* 37: 231–235.

Feeny, P. P. 1975. Biochemical coevolution between plants and their insect herbivores. *In* Gilbert, L. E., and P. H. Raven (eds.), *Coevolution of Animals and Plants*. University of Texas Press, Austin, pp. 3–18.

Finnegan, R. J. 1964. Rôle des insectes dans la transmission de la maladie hollandaise de l'Orme, *Ceratocystis ulmi* (Buism.) C. Moreau. *Phytoprotection* 45: 117–124.

Fogel, R. 1975. Insect mycophagy: a preliminary bibliography. *USDA For. Ser. Gen. Tech. Rep.* PNW-36. Portland, OR.

Fraenkel-Conrat, H. 1985. *The Viruses: Catalogue, Characterization, and Classification.* Plenum, New York.

Francke-Grosmann, H. 1967. Ectosymbiosis in wood-inhabiting insects. *In* Henry, S. M. (ed.), *Symbiosis,* Vol. 3. Academic Press, New York, pp. 141–205.

Franklin, R. T. 1970. Observations on the blue stain-southern pine beetle relationship. *J. Ga. Entomol. Soc.* 5: 53–57.

Freundt, E. A. 1981. Isolation, characterization, and identification of spiroplasmas and MLOs. *In* Maramorosch, K., and S. P. Rayachaudhuri (eds.), *Mycoplasma Diseases of Trees and Shrubs*. Academic Press, New York, pp. 1–34.

Friend, W. G., E. H. Salkeld, and I. L. Stevenson. 1959. Nutrition of onion maggots, larvae of *Hylemya antiqua* (Meig.) with reference to other members of the genus *Hylemya*. *Ann. N. Y. Acad. Sci.* 77: 384–393.

Frietag, J. H. 1951. Host range of the Pierce's disease virus of grapes as determined by insect transmission. *Phytopathology* 41: 920–934.

Fronsch, M. 1983. Occurrence and distribution of the age of the latent rosette (witches-broom) disease of the sugar beet (*Beta vulgaris*) in its insect vector *Piesma quadratum* Fieb (Heteroptera, Piesmidae). *Z. Angew. Entomol.* 95: 310–318.

Funk, C. R., P. M. Halisky, M. C. Johnson, M. R. Siegel, A. V. Stewart, S. Ahmad, R. H. Hurley, and I. C. Harvey. 1983. An endophytic fungus and resistance to sod webworms: association in *Lolium perenne* L. *Biotechnology* 1: 189–191.

Fytizas, E., and M. E. Tzanakakis. 1966. Some effects of streptomycin, when added to the adult food, on the adults of *Dacus oleae* (Diptera: Tephritidae) and their progeny. *Ann. Entomol. Soc. Am.* 59: 269–273.

Garran, J., and A. Gibbs. 1982. Studies on alfalfa mosaic virus and alfalfa aphids. *Aust. J. Agric. Res.* 33: 657–664.

Gaynor, D. L. and W. F. Hunt. 1983. The relationship between nitrogen supply, endophytic fungus, and Argentine stem weevil resistance in ryegrasses. *Proc. N. Z. Grassl. Assoc.* 44: 257–263.

Gaynor, D. L., D. D. Rowan, G. C. M. Latch, and S. Pilkington. 1983. Preliminary results on the biochemical relationship between adult Argentine stem weevil and two endophytes in ryegrass. *In Proc. 36th N. Z. Weed and Pest Cont. Conf.*, pp. 220–224.

Giannotti, J. 1969. Lésions cellulaires chez deux cicadelles vectrices de la phyllodie du trèfle. *Ann. Soc. Entomol. Fr.* 5: 155–160.

Giannotti, J., C. Vago, G. Marchoux, G. Devauchelle, and J. L. Duthoit. 1968. Recherches sur les microorganismes de type mycoplasma dans les cicadelles vectrices et dans les vegetaux atteints de jaunisses. *Entomol. Exp. Appl.* 11: 470–474.

Gilbertson, R. L. 1984. Relationships between insects and wood-rotting Basidiomycetes. *In* Wheeler, Q., and M. Blackwell (eds.), *Fungus–Insect Relationships*. Columbia Univ. P., New York, pp. 130–165.

Gildow, F. E. 1984. Biology of aphid vectors of barley yellow dwarf virus and the effect of BYDV on aphids. *In Barley Yellow Dwarf: A Proceedings of the Workshop,* December 6–8, 1983. CIMMYT, Mexico, pp. 28–35.

Goheen, A. C., G. Nyland, and S. K. Lowe. 1973. Association of rickettsialike organism with

Pierce's disease of grapevines and alfalfa dwarf and heat therapy of the disease in grapevines. *Phytopathology* 63: 341–345.

Goheen, D. J. and F. W. Cobb, Jr. 1980. Infestation of *Ceratocystis wageneri*-infected ponderosa pines by bark beetles (Coleoptera: Scolytidae) in the Central Sierra Nevada. *Can. Entomol.* 112: 725–730.

Graham, K. 1967. Fungal-insect mutualism in trees and timber. *Annu. Rev. Entomol.* 12: 105–126.

Granados, R. R., and D. J. Meehan. 1975. Pathogenicity of the corn stunt agent to an insect vector, *Dalbulus elimatus*. *J. Invert. Pathol.* 26: 313–320.

Hagen, K. S. 1966. Dependence of the olive fly, *Dacus oleae,* larvae on symbiosis with *Pseudomonas savastanoi* for the utilization of olive. *Nature* (London) 209: 423–424.

Hagen, K. S., L. Santas, and A. Tsecouras. 1963. A technique of culturing the olive fly *Dacus oleae* Gmel. on synthetic media under xenic conditions. *In Radiation and Radioisotopes Applied to Insects of Agricultural Importance*. Proc. Symp. Athens, April 22–26, 1963. Intl. At. Energy Agency, Vienna, pp. 333–356.

Harborne, J. B. (ed.). 1978. *Biochemical Aspects of Plant and Animal Coevolution*. Academic Press, New York.

Hardy, T. N., K. Clay, and A. M. Hammond, Jr. 1985. Fall armyworm (Lepidoptera: Noctuidae): a laboratory bioassay and larval preference study for the fungal endophyte of perennial ryegrass. *J. Econ. Entomol.* 78: 571–575.

Hardy, T. N., K. Clay, and A. M. Hammond, Jr. 1986. Leaf age and related factors affecting endophyte-mediated resistance to fall armyworm (Lepidoptera: Noctuidae) in tall fescue. *Environ. Entomol.* 15: 1083–1089.

Hare, J. D. 1983. Manipulation of host suitability for herbivore pest management. *In* Denno, R. F., and M. S. McClure (eds.), *Variable Plants and Herbivores in Natural and Managed Systems*. Academic Press, New York, pp. 655–680.

Hare, J. D. and J. A. Dodds. 1978. Changes in food quality of an insect's marginal host species associated with a plant virus. *J. N.Y. Entomol. Soc.* 86: 292.

Harris, K. F. 1980. Aphids, leafhoppers, and planthoppers. *In* Harris, K. F., and K. Maramorosch (eds.), *Vectors of Plant Pathogens*. Academic Press, New York, pp. 1–13.

Harris, K. F. 1981. Arthropod and nematode vectors of plant viruses. *Annu. Rev. Phytopathol.* 19: 391–426.

Harris, M. K., and R. A. Frederiksen. 1984. Concepts and methods regarding host plant resistance to arthropods and pathogens. *Annu. Rev. Phytopathol.* 22: 247–272.

Harrison, M. D., and J. W. Brewer. 1982. Field dispersal of soft rot bacteria. *In* Mount, M. S., and G. H. Lacy (eds.), *Phytopathogenic Prokaryotes,* Vol. 2. Academic Press, New York, pp. 31–69.

Harrison, M. D., J. W. Brewer, and L. D. Merrill. 1980. Insect development in the transmission of bacterial pathogens. *In* Harris, K. F., and K. Maramorosch (eds.), *Vectors of Plant Pathogens*. Academic Press, New York, pp. 201–292.

Hart, S. V., M. Kogan, and J. D. Paxton. 1983. Effect of soybean phytoalexins on the herbivorous insects Mexican bean beetle and soybean looper. *J. Chem. Ecol.* 9: 657–672.

Hertert, H. D., D. L. Miller, and A. D. Partridge. 1975. Interaction of bark beetles (Coleoptera: Scolytidae) and root-rot pathogens in grand fir in northern Idaho. *Can. Entomol.* 107: 899–904.

Hetrick, L. A. 1949. Some overlooked relationships of southern pine beetle. *J. Econ. Entomol.* 42: 466–469.

Hodges, C. F., and P. W. Robinson. 1977. Sugar and amino acid content of *Poa pratensis* infected with *Ustilago striiformis* and *Urocystis agropyri*. *Physiol. Plant.* 41: 25–28.

Hodges, J. D., S. J. Barras, and J. K. Mauldin. 1968. Amino acids in inner bark of loblolly pine, as affected by the southern pine beetle and associated microorganisms. *Can. J. Bot.* 46: 1467–1472.

Hodges, J. D., T. E. Nebeker, J. D. DeAngelis, and C. A. Blanche. 1985a. Host/beetle interactions: influence of associated microorganisms, tree disturbance, and host vigor. *USDA For. Ser. Gen. Tech. Rep.* SO-56. New Orleans, LA.

Hodges, J. D., T. E. Nebeker, J. D. DeAngelis, B. L. Karr, and C. A. Blanche. 1985b. Host resistance and mortality: a hypothesis based on the southern pine beetle–microorganism–host interactions. *Bull. Entomol. Soc. Am.* 311: 31–35.

Hopkins, D. L. and H. H. Mollenhaver. 1973. Rickettsia like bacterium associated with Pierce's disease of grapes. *Science* 179: 298–300.

Horsfall, J. G., and E. B. Cowling (eds.). 1980. *Plant Disease: An Advanced Treatise,* Vol. 5. Academic Press, New York.

Huff, C. G. 1928. Nutritional studies on the seed-corn maggot *Hylemyia cilicrura* Rondani. *J. Agri. Res.* 36: 625–630.

Ishiie, T., Y. Doi, K. Yora, and H. Asuyama. 1967. Suppressive effects of antibiotics of tetracycline group on symptom development of mulberry dwarf disease. *Ann. Phytopathol. Soc. Jpn.* 33: 267–275. (In Japanese).

Jennings, P. R., and A. T. Pineda. 1971. The effect of the hoja blanca virus on its insect vector. *Phytopathology* 61: 142–143.

Jensen, D. D. 1962. Pathogenicity of western X-disease virus of stone fruits to its leafhopper vector, *Colladonus montanus* (Van Duzee). *In Proc. 11th Intl. Cong. Entomol.,* Vienna, August 17–25, 1960, pp. 790–791.

Jensen, D. D. 1969. Insect diseases induced by plant-pathogenic viruses. In Maramorosch, K., (ed.), *Viruses, Vectors, and Vegetation.* Wiley, New York, pp. 505–525.

Johnson, D. E. 1930. The relation of the cabbage maggot and other insects to the spread and development of soft rot of Cruciferae. *Phytopathology* 20: 857–872.

Johnson, M. C., D. L. Dahlman, M. R. Siegel, L. P. Bush, G. C. M. Latch, D. A. Potter, and D. R. Varney. 1985. Insect feeding deterrents in endophyte-infected tall fescue. *Appl. Environ. Microbiol.* 49: 568–571.

Jones, C. G. 1984. Microorganisms as mediators of plant resource exploitation by insect herbivores. *In* Price, P. W., C. N. Slobodchikoff, and W. S. Gaud (eds.), *A New Ecology: Novel Approaches to Interactive Systems.* Wiley, New York, pp. 53–99.

Kain, W. M., T. W. Wyeth, D. L. Gaynor, and M. W. Slay. 1982. Argentine stem weevil (*Hyperodes bonariensis* Kuschel) resistance in perennial ryegrass (*Lolium perenne* L.). *N. Z. J. Agric. Res.* 25: 255–259.

Karban, R., R. Adamchak, and W. C. Schnathorst. 1987. Induced resistance and interspecific competition between spider mites and a vascular wilt fungus. *Science* 235: 678–680.

Karr, J. R. 1976. An association between a grass (*Paspalum virgatum*) and moths. *Biotropica* 8: 284–285.

Kennedy, J. S. 1951. Benefits to aphids from feeding on galled and virus-infected leaves. *Nature* 168: 825–826.

Kingsley, P., J. M. Scriber, C. R. Grau, and P. A. Delwiche. 1983. Feeding and growth performance of *Spodoptera eridania* (Noctuidae: Lepidoptera) on "vernal" alfalfa, as influenced by *Verticillium* wilt. *Protection Ecol.* 5: 127–134.

Kogan, M. 1986. Plant defense strategies and host-plant resistance. *In* Kogan, M. (ed.), *Ecological Theory and Integrated Pest Management Practice.* Wiley, New York, pp. 83–134.

Kogan, M., and J. Paxton. 1983. Natural inducers of plant resistance to insects. *In* Hedin, P. (ed.), *Plant Resistance to Insects. ACS Symp. Ser.* 208. American Chemical Society, Washington, DC, pp. 153–171.

Kosuge, T. and J. A. Kimpel. 1981. Energy use and metabolic regulation in plant-pathogen interactions. *In* Ayres, P. G. (ed.), *Effects of Disease on the Physiology of the Growing Plant*. Cambridge Univ. P., Cambridge, pp. 29–45.

Kreitner, G. and C. E. Rogers. 1981. Sunflower seed midge: effects of larval infestation on pericarp development in sunflower. *Ann. Entomol. Soc. Am.* 74: 431–435.

Kuan, C., and S. Wang. 1965. On some physiological changes of Chinese cabbage infected by the Kwuting strain of turnip mosaic virus in relation to the development of *Myzus persicae* (Sulzer). *Acta Phytophyl. Sinica* 4: 27–33.

Kuc, J. 1965. Resistance of plants to infectious agents. *Annu. Rev. Microbiol.* 20: 337–370.

Kuc, J. 1972. Phytoalexins. *Annu. Rev. Phytopathol.* 10: 207–232.

Kulhavy, D. L., A. D. Partridge, and R. W. Stark. 1984. Root diseases and blister rust associated with bark beetles (Coleoptera: Scolytidae) in western white pine in Idaho. *Environ. Entomol.* 13: 813–817.

Kunkel, L. O. 1954. Maintenance of yellows-type viruses in plant and insect reservoirs. *In* Hartman, F. W., F. L. Horsfall, and J. G. Kidd (eds.), *The Dynamics of Virus and Rickettsial Infections*. McGraw-Hill, New York, pp. 150–163.

Langdon, R. F. N., and B. R. Champ. 1954. The insect vectors of *Claviceps paspali* in Queensland. *J. Aust. Inst. Agric. Sci.* 20: 115–118.

Latch, G. C. M., M. J. Christensen, and D. L. Gaynor. 1985. Aphid detection of endophyte infection in tall fescue. *N. Z. J. Agric. Res.* 28: 129–132.

Laurema, S., M. Markkula, and M. Raatikainen. 1966. The effect of virus diseases transmitted by the leafhopper *Javesella pellucida* (F.) on the concentration of free amino acids in oats and on the reproduction of aphids. *Ann. Agric. Fenn.* 5: 94–99.

Leach, J. G. 1926. The relation of the seed-corn maggot (*Phorbia fusciceps* Zett.) to the spread and development of potato blackleg in Minnesota. *Phytopathology* 16: 149–176.

Leach, J. G. 1927. The relation of insects and weather to the development of heart rot of celery. *Phytopathology* 17: 663–667.

Leach, J. G. 1931. Further studies on the seed-corn maggot and bacteria with special reference to potato blackleg. *Phytopathology* 21: 387–406.

Leach, J. G. 1940. *Insect Transmission of Plant Diseases*. McGraw-Hill, New York.

Leach, J. C. 1964. Observations on cucumber beetles as vectors of cucurbit wilt. *Phytopathology* 54: 606–607.

Levin, D. A. 1976. The chemical defenses of plants to pathogens and herbivores. *Annu. Rev. Ecol. Syst.* 7: 121–159.

Lewis, A. C. 1979. Feeding preference for diseased and wilted sunflower in the grasshopper, *Melanoplus differentialis*. *Ent. Exp. Appl.* 26: 202–207.

Lewis, A. C. 1984. Plant quality and grasshopper feeding: effects of sunflower condition on preference and performance in *Melanoplus differentialis*. *Ecology* 65: 836–843.

Lowe, S., and F. A. Strong. 1963. The unsuitability of some viruliferous plants as hosts for the green peach aphid, *Myzus persicae*. *J. Econ. Entomol.* 56: 307–309.

Lucas, G. B., C. L. Campbell, and L. T. Lucas. 1985. *Introduction to Plant Diseases: Identification and Management*. AVI, Westport, CN.

Lundie, A. E. 1938. Honeybees working a fungus growth (*Hyparrhenia filipendula* var. *pilosa*). *S. Afr. Bee J.* 13: 19.

Lyons, P. C., R. D. Plattner, and C. W. Bacon. 1986. Occurrence of peptide and clavine ergot alkaloids in tall fescue grass. *Science* 232: 487–489.

Madden, L. V., and L. R. Nault. 1983. Differential pathogenicity of corn stunting mollicutes to leafhopper vectors in *Dalbulus* and *Baldulus* species. *Phytopathology* 73: 1608–1614.

Madden, L. V., L. R. Nault, S. E. Heady, and W. E. Styer. 1984. Effect of maize stunting mollicutes on survival and fecundity of *Dalbulus* leafhopper vectors. *Ann. Appl. Biol.* 105: 431–441.

Maramorosch, K. 1958. Cross-protection between two strains of corn-stunt virus in an insect vector. *Virology* 6: 448–459.

Maramorosch, K. 1981. Spiroplasmas: agents of animal and plant diseases. *Bioscience* 31: 374–380.

Maramorosch, K., and D. D. Jensen. 1963. Harmful and beneficial effects of plant viruses in insects. *Annu. Rev. Microbiol.* 17: 495–530.

Maramorosch, K., H. Hirumi, M. Kimura, and J. Bird. 1975. Mollicutes and *Rickettsia*-like plant disease agents (Zoophytomicrobes) in insects. *Ann. N. Y. Acad. Sci.* 266: 276–292.

Markham, P. G., and R. Townsend. 1974. Transmission of *Spiroplasma citri* to plants. *Colloq. Inst. Natl. Sante Rech. Med.* 33: 201–206.

Markkula, M., and S. Laurema. 1964. Changes in the concentration of free amino acids in plants induced by virus diseases and the reproduction of aphids. *Annu. Agric. Fenn.* 3: 265–271.

Mattson, W. J., Jr. 1980. Herbivory in relation to plant nitrogen content. *Annu. Rev. Ecol. Syst.* 11: 119–161.

Mazur, G., C. R. Funk, W. K. Dickson, R. F. Bara, and J. M. Johnson-Cicalese. 1981. Reactions of perennial ryegrass varieties to sod webworm larvae. *Rutgers Turfgr. Proc.* 12: 85–91.

McCoy, R. E. 1979. Mycoplasmas and yellows diseases. *In* Whitcomb, R. F., and J. G. Tully (eds.), *The Mycoplasmas,* Vol. 3. Academic Press, New York, pp. 229–264.

McIntyre, J. L., J. A. Dodds, and J. D. Hare. 1981. Effects of localized infections of *Nicotiana tabacum* by tobacco mosaic virus on systemic resistance against diverse pathogens and an insect. *Phytopathology* 71: 297–301.

Merdan, A., H. Abdel-Rahman, and A. Soliman. 1975. On the influence of host plants on insect resistance to bacterial disease. *Z. Angew. Entomol.* 78: 280–286.

Miller, J., and B. F. Coon. 1964. The effect of barley yellow dwarf virus on the biology of its vector the English grain aphid, *Macrosiphum granarium*. *J. Econ. Entomol.* 57: 970–974.

Miller, T. D., and M. N. Schroth. 1972. Monitoring the epiphytic population of *Erwinia amylovora* on pear with a selective medium. *Phytopathology* 62: 1175–1182.

Misaghi, I. J. 1982. *Physiology and Biochemistry of Plant-Pathogen Interactions*. Plenum, New York.

Miyazaki, S., G. M. Boush, and R. J. Baerwald. 1968. Amino acid synthesis by *Pseudomonas melophthora,* bacterial symbiote of *Rhagoletis pomonella* (Diptera). *J. Insect Physiol.* 14: 513–518.

Moeck, H. A., D. L. Wood, and K. Q. Lindahl, Jr. 1981. Host selection behavior of bark beetles (Coleoptera: Scolytidae) attacking *Pinus ponderosa,* with special emphasis on the western pine beetle, *Dendroctonus brevicomis*. *J. Chem. Ecol.* 7: 49–83.

Mortimer, P. H., and M. E. diMenna. 1983. Ryegrass staggers: further substantiation of a *Lolium* endophyte aetiology and the discovery of weevil resistance of ryegrass pastures infected with *Lolium* endophyte. *Proc. N. Z. Grassl. Assoc.* 44: 240–243.

Mower, R. L., G. R. Gray, and C. E. Ballou. 1973. Sugars from *Sphacelia sorghi* honeydew. *Carbohydrate Res.* 27: 119–134.

Nakasuji, F. and K. Kiritani. 1970. Ill-effects of rice dwarf virus upon its vector, *Nephotettix cincticeps* Uhler (Hemiptera: Deltocephalidae) and its significance for changes in relative abundance of infected individuals among vector populations. *Appl. Entomol. Zool.* 5: 1–12.

Nault, L. R., L. V. Madden, W. E. Styer, B. W. Triplehorn, G. F. Shambaugh, and S. E. Heady. 1984. Pathogenicity of corn stunt spiroplasma and maize bushy stunt mycoplasma to their vector, *Dalbulus longulus*. *Phytopathology* 74: 977–979.

Nebeker, T. E., C. A. Blanche, and J. DeAngelis. 1984. Host, bark beetle, and microorganism interactions. *In* Payne, T. L., R. F. Billings, R. N. Coulson, and D. L. Kulhavy (eds.), *History, Status, and Future Needs for Entomology Research in Southern Forests*. Texas A&M Misc. Publ. 1553. Texas A&M University, College Station, pp. 19–23.

Nelson, R. M. 1934. Effect of bluestain fungi on southern pines attacked by bark beetles. *Phytopathol. Ztschr.* 7: 327–353.

Okuyama, S. 1962. The propagation of the rice stripe virus in the body of the vector. Symp. on Vectors of Plant Viruses, Hokkaido Univ., Sapporo, Japan, September 20, 1962, pp. 8–10. (Abst. in Japanese).

Ou, S. H. 1980. Rice plant diseases. *In* B. S. Luh (ed.), *Rice: Production and Utilization*. AVI, Westport, CN, pp. 235–259.

Partridge, A. D., and D. L. Miller. 1972. Bark beetles and root rots related in Idaho conifers. *Plant Dis. Rep.* 56: 498–500.

Pepper, E. H. 1967. Stewart's bacterial wilt of corn. Monograph no. 4. Am. Phytopathol. Soc. Heffernan Press, Worcester, MA.

Perry, D. H., M. Lenz, and J. A. L. Watson. 1985. Relationships between fire, fungal rots and termite damage in Australian forest trees. *Aust. For.* 48: 46–53.

Posnette, A. F., and C. E. Ellenberger. 1963. Further studies of green petal and other leafhopper-transmitted viruses infecting strawberry and clover. *Ann. Appl. Biol.* 51: 69–83.

Prestidge, R. A., D. R. Lauren, S. G. Van Der Zijpp, and M. E. diMenna. 1985. Isolation of feeding deterrents to Argentine stem weevil in cultures of endophytes of perennial ryegrass and tall fescue. *N. Z. J. Agric. Res.* 28: 87–92.

Prestidge, R. A., R. P. Pottinger, and G. M. Barker. 1982. An association of *Lolium* endophyte with ryegrass resistance to Argentine stem weevil. *Proc. N. Z. Weed Pest Contr. Conf.* 35: 119–122.

Price, P. W., C. E. Bouton, P. Gross, B. A. McPheron, J. N. Thompson, and A. E. Weis. 1980. Interactions among three trophic levels: influence of plants on interactions between insect herbivores and natural enemies. *Annu. Rev. Ecol. Syst.* 11: 41–65.

Purcell, A. H. 1979. Leafhopper vectors of xylem-borne plant pathogens. *In* Maramorosch, K., and K. F. Harris (eds.), *Leafhopper Vectors and Plant Disease Agents*. Academic Press, New York, pp. 603–625.

Purcell, A. H. 1982a. Insect vector relationships with procaryotic plant pathogens. *Annu. Rev. Phytopathol.* 20: 397–417.

Purcell, A. H. 1982b. Evolution of the insect vector relationship. *In* Mount, M. S., and G. H. Lacy (eds.), *Phytopathogenic Procaryotes,* Vol. 1. Academic Press, New York, pp. 121–156.

Rabin, L. B. and R. S. Pacovsky. 1985. Reduced larva growth of two Lepidoptera (Noctuidae) on excised leaves of soybean infected with a mycorrhizal fungus. *J. Econ. Entomol.* 78: 1358–1363.

Raffa, K. F., and A. A. Berryman. 1982. Physiological differences between lodgepole pines resistant and susceptible to mountain pine beetle and associated microorganisms. *Environ. Entomol.* 11: 486–492.

Raju, B. C., and J. M. Wells. 1986. Diseases caused by fastidious xylem-limited bacteria and strategies for management. *Plant Dis.* 70: 182–186.

Rey, J. M. 1969. Development of a larval diet for the rearing of *Dacus oleae* Gmel. (Diptera: Tripetidae). *Graellsia* 24: 253–259.

Rhoades, D. F. 1979. Evolution of plant chemical defense against herbivores. *In* Rosenthal, G. A., and D. H. Janzen (eds.), *Herbivores: Their Interaction with Secondary Plant Metabolites*. Academic Press, New York, pp. 3–54.

Rhoades, D. F. 1983. Herbivore population dynamics and plant chemistry. *In* Denno, R. F., and M. S. McClure (eds.), *Variable Plants and Herbivores in Natural and Managed Systems*. Academic Press, New York, pp. 155–220.

Rhoades, D. F., and R. G. Cates. 1976. Toward a general theory of plant antiherbivore chemistry. *Rec. Adv. Phytochem.* 10: 168–213.

Rios de Saluso, L. A. 1982. Discovery of a dipteran of the genus *Mycodiplosis* on blades of grass attacked by rust. *INTA Estac. Exp. Agropecu. Concordia ser. Notas Tec.* 0: 1–5.

Rockstein, M. (ed.). 1978. *Biochemistry of Insects*. Academic Press, New York.

Rosenthal, G. A. and D. H. Janzen (eds.). 1979. *Herbivores: Their Interaction with Secondary Plant Metabolites*. Academic Press, New York.

Rowan, D. D., and D. L. Gaynor. 1986. Isolation of feeding deterrents against Argentine stem weevil from ryegrass infected with the endophyte *Acremonium loliae*. *J. Chem. Ecol.* 12: 647–658.

Russell, G. E. 1981. Disease and crop yield: the problems and prospects for agriculture. *In* Ayres, P. G. (ed.), *Effects of Disease on the Physiology of the Growing Plant*. Cambridge Univ. P., Cambridge, pp. 1–11.

Ryan, C. A. 1983. Insect-induced chemical signals regulating natural plant protection responses. *In* Denno, R. F., and M. S. McClure (eds.), *Variable Plants and Herbivores in Natural and Managed Systems*. Academic Press, New York, pp. 43–60.

Saini, R. S., and A. G. Peterson. 1965. Colonization of yellows-infected asters by two species of aphids. *J. Econ. Entomol.* 58: 537–539.

Schoeneweiss, D. F. 1975. Predisposition, stress, and plant disease. *Annu. Rev. Phytopathol.* 13: 193–211.

Schoonhoven, L. M. 1968. Chemosensory bases of host plant selection. *Annu. Rev. Entomol.* 13: 115–136.

Scriber, J. M., and F. Slansky, Jr. 1981. The nutritional ecology of immature insects. *Annu. Rev. Entomol.* 26: 183–211.

Severin, H. H. P. 1946. Longevity, or life histories, of leafhopper species on virus-infected and on healthy plants. *Hilgardia* 17: 121–133.

Sharma, Y. P., R. S. Singh, and R. K. Tripathi. 1983. Role of insects in secondary spread of the ergot disease of pearl millet (*Pennisetum americanum*). *Indian Phytopathol.* 36: 131–133.

Shikata, E. 1979. Cytopathological changes in leafhopper vectors of plant viruses. *In* Maramorosch, K., and K. F. Harris (eds.), *Leafhopper Vectors and Plant Disease Agents*. Academic Press, New York, pp. 309–325.

Shinkai, A. 1960. Virus transmission by leafhoppers infected with rice dwarf disease. *Ann. Phytopathol. Soc. Japan* 25: 42.

Siegel, M. R., G. C. M. Latch, and M. C. Johnson. 1985. *Acremonium* fungal endophytes of tall fescue and perennial ryegrass: significance and control. *Plant Dis.* 69: 179–183.

Singh, D., and E. B. Smalley. 1969. Changes in amino acid and sugar constituents of the xylem sap of American elm following inoculation with *Ceratocystis ulmi*. *Phytopathology* 59: 891–896.

Singh, S., K. S. Bhargava, and B. B. Nagaich. 1983. Lethality of the potato purple top roll pathogen (mycoplasma) to its vector *Alebroides nigroscutellatus*. *Indian Phytopathol.* 36: 646–650.

Sinha, R. C., and Y. S. Paliwal. 1970. Localization of a mycoplasma-like organism in tissues of a leafhopper carrying clover phyllody agent. *Virology* 40: 665–672.

Smith, R. W. 1939. Grasshopper injury in relation to stem rust in spring wheat varieties. *J. Am. Soc. Agron.* 31: 818–821.

Stone, J. K. 1985. Foliar endophytes of *Pseudotsuga menziesii* (Mirb.) Franco. Cytology and physiology of the host-endophyte relationship. Ph.D dissertation. University of Oregon, Eugene.

Suenaga, H. 1962. The occurrence of the rice dwarf disease and the ecology of the green rice leafhopper, 1. Symp. on Vectors of Plant Viruses, Hokkaido Univ., Sapporo, Japan, September 20, 1962. (Abst. in Japanese).

Sylvester, E. S. 1973. Reduction of excretion, reproduction, and survival in *Hyperomyzus lactucae* fed on plants infected with isolates of sowthistle yellow vein virus. *Virology* 56: 632–635.

Thresh, J. M. 1964. Association between black currant reversion virus and its gall mite vector (*Phytoptus ribis* Nal.). *Nature* (London) 202: 1085–1087.

Thresh, J. M. 1967. Increased susceptibility of black-currant bushes to the gall-mite vector (*Phytoptus ribis* Nal.) following infection with reversion virus. *Ann. Appl. Biol.* 60: 455–467.

Tsai, J. H. 1979. Vector transmission of mycoplasmal agents of plant diseases. *In* Whitcomb, R. F., and J. G. Tully (eds.), *The Mycoplasmas,* Vol. 3. Academic Press, New York, pp. 266–309.

Tu, J. C., and R. E. Ford. 1970. Free amino acids in soybeans infected with soybean mosaic virus, bean pod mottle virus, or both. *Phytopathology* 60: 660–664.

Waite, M. B. 1891. Results from recent investigations in pear blight (abst.). *Bot. Gaz.* 16: 259.

Walkey, D. G. A. 1985. *Applied Plant Virology*. Wiley, New York.

Watson, M. A., and R. C. Sinha. 1959. Studies on the transmission of European wheat striate mosaic virus by *Delphacodes pellucida* Fabricius. *Virology* 8: 139–163.

Webber, J. 1981. A natural control of Dutch elm disease. *Nature* (London) 292: 449–451.

Weiss, H. B. and E. West. 1920. Fungous insects and their hosts. *Biol. Soc. Wash. Proc.* 33: 1–20.

Whipps, J. M., and D. H. Lewis. 1981. Patterns of translocation, storage, and interconversion of carbohydrates. *In* Ayres, P. G. (ed.), *Effects of Disease on the Physiology of the Growing Plant*. Cambridge Univ. P., Cambridge, pp. 47–83.

Whitcomb, R. F. 1975. Pathogenic effects of plant disease agents on vector insects: an introduction. *Ann. N. Y. Acad. Sci.* 266: 259.

Whitcomb, R. F., and D. D. Jensen. 1968. Proliferative symptoms in leafhoppers infected with western X-disease virus. *Virology* 35: 174–177.

Whitcomb, R. F., and D. L. Williamson. 1975. Helical wall-free prokaryotes in insects: multiplication and pathogenicity. *Ann. N. Y. Acad. Sci.* 266: 260–275.

Whitcomb, R. F., and D. L. Williamson. 1979. Pathogenicity of mycoplasmas for arthropods. *Zbl. Bakt. Hyg., I. Abt. Orig. A* 245: 200–221.

Whitcomb, R. F., J. G. Tully, J. M. Bove, and P. Saglio. 1973. Spiroplasmas and acholeplasmas: multiplication in insects. *Science* 182: 1251–1253.

Whitcomb, R. F., D. L. Williamson, J. Rosen, and M. Coan. 1974. Relationship of infection and pathogenicity on the infection of insects by wall-free prokaryotes. *Colloq. Inst. Natl. Sante Rech. Med.* 33: 275–282.

Whitney, H. S., and F. W. Cobb, Jr. 1972. Non-staining fungi associated with the bark beetle, *Dendroctonus brevicomis,* (Coleoptera: Scolytidae) on *Pinus ponderosa*. *Can. J. Bot.* 50: 1943–1945.

Williams, R. M. C. 1965. Infestation of *Pinus caribaea* by the termite *Coptotermes niger* Snyder. *In Proc. 12th Intl. Cong. Entomol.,* London, pp. 675–676.

Wood, D. L. 1982. The role of pheromones, kairomones, and allomones in the host selection and colonization behavior of bark beetles. *Annu. Rev. Entomol.* 27: 411–446.

Wright, E. 1935. *Trichosporium symbioticum* n. sp., a woodstaining fungus associated with *Scolytus ventralis*. *J. Agric. Res.* 50: 525–538.

Yarwood, C. E. 1943. Association of thrips with powdery mildews. *Mycologia* 35: 189–191.

Yearian, W. C., R. J. Gouger, and R. C. Wilkinson. 1972. Effects of the bluestain fungus, *Ceratocystis ips,* on development of *Ips* bark beetles in pine bolts. *USDA For. Ser. Gen. Tech. Rep.* SO-10 New Orleans, LA.

13

THE DYNAMICS OF INSECT POPULATIONS IN CROP SYSTEMS SUBJECT TO WEED INTERFERENCE

Miguel A. Altieri

Agricultural Experiment Station
Division of Biological Control
University of California, Berkeley
Albany, California

1. INTRODUCTION

Through several adaptations (wide environmental tolerance, high seed production and longevity, rapid growth, etc.) weed species have managed successfully to persist in agroecosystems. The presence of weeds within or around crop fields influences the dynamics of the crop and associated biotic communities. Studies over the past 15 years have produced a great deal of evidence that the manipulation of a specific weed species, a particular weed control practice, or a cropping system can affect the ecology of insect pests and associated natural enemies (van Emden 1965, Altieri et al. 1977, Altieri and Whitcomb 1979, Thresh 1981, William 1981, Norris 1982, Andow 1983). Weeds exert a direct biotic stress by competing for sunlight, moisture, and some nutrients, and thus reducing crop yields. Weeds indirectly affect crop plants through positive and/or negative effects on insect herbivores and also on the natural enemies of herbivores (Price et al. 1980). Herbivore–natural enemy interactions occurring in a crop system can be influenced by the presence of associated weeds, or by the presence of herbivores on associated weed plants (Altieri and Letourneau 1982). On the other hand, herbivores can mediate the interaction between crops and weeds, as in a natural community where the competitiveness of two-plant species was altered substantially by the selective feeding of a foliage-consuming beetle (Bentley and Whittaker 1979). Such relationships have been little explored in agricultural systems, however.

In this chapter the multiple interactions among crops, weeds, herbivores, and natural enemies are discussed, in particular weed ecology and management that affect the dynamics of insect populations and thus crop health. The stress imposed on crops by weeds is viewed beyond the mere competitive interaction, incorporating three trophic level system interactions.

2. CROP–WEED INTERFERENCE

Agronomists have traditionally considered that the presence of weeds in crop systems lowers crop yields through competition for limited resources, contending that increasing weed density decreases yields proportionally. A more careful analysis of the literature suggests that the weed density–crop yield relationship is far from linear, but rather sigmoidal (Zimdahl 1980). A low weed density does not usually affect yields, and under some circumstances certain weeds even stimulate crop growth. For example, in the Indian desert, presence of the leguminous weed *Indigofera cordifolia* Heyne enhanced the growth and yields of *Pennisetum typhoideum* L. Rich. and *Sesamum indicum* L. (Bhandari and Sen 1979). In many traditional farming systems, small farmers seem aware of these relationships and encourage the growth of weeds that benefit the crop (Chacon and Gliessman 1982).

However, not all suppression of plant growth in agroecosystems can be explained by resource competition. Some weeds supplement aggressiveness by the release of toxic or growth-inhibiting substances into the soil as root exudates or leachates of their dead and decaying vegetative parts, thus reducing the growth of associated crops. Such biochemical interaction is called "allelopathy" (Rice 1984). Many scientists postulate that allelopathy is an important mechanism by which weeds and crops affect each other, and are engaged in research designed to exploit allelopathy as a viable means of weed management in agroecosystems (Altieri and Doll 1978a, Putnam and Duke 1982, Gliessman 1983). The term "interference" is widely used today to include resource competition and allelopathy.

Crops and weeds interact in a manner determined by physical (climate, soil), biological (crop–weed density, species composition, etc.), and cultural management factors (fertilizer use, tillage, cropping patterns, etc.). A crop–weed balance results that is determined by the crop and weed reproductive cycles (Bantilan et al. 1974). Weed management is the understanding of the factors affecting this balance and the shifting of the balance to favor the crop. The outcome of such crop–weed interference depends on a number of factors, including crop type, weed species composition and abundance, soil fertility and moisture levels, and the period of weed growth in relation to crop emergence. Consequently several management strategies aimed at changing the spatial and temporal arrangement of crops have been proposed to maximize or reduce the impact of these factors in agroecosystems. Some of these proposed techniques are crop rotations, crop densities and row spacing patterns, date of planting, and selection of crop genotypes and crop mixtures (Walker and Buchanan 1982). In third world countries where farmers lack capital or credit to purchase herbicides, a considerable amount of multiple-cropping systems research is being conducted principally on the manipulation of biological factors that promote intercrop dominance over weeds (Moody 1977, Shetty and Rao 1981, Altieri and Liebman 1986). For example, substantial weed suppression was obtained in intercropping low growing species of melons, sweet potatoes, or legumes such as *Centrosema pubescens* Benth. and *Psophocarpus palustris* Desv. between rows of maize and cassava (Akobundu 1980). The smother crops not only served as a labor-saving means of weed control but also provided erosion control through increased soil coverage.

3. WEEDS AS SOURCES OF INSECT PESTS IN AGROECOSYSTEMS

Weeds are important hosts of insect pests and pathogens in agroecosystems. Van Emden (1965) cites 442 references relating to weeds as reservoirs of pests. One hundred such references concern cereals. A series of publications concerning weeds as reservoirs for organisms affecting crops has been published by the Ohio Agricultural Research and Development Center. More than 70 families of arthropods affecting crops were reported as being primarily weed associated (Bendixen and Horn 1981). Many pest outbreaks can be traced to locally abundant weeds belonging to the same family as the affected crop plants. Detailed examples of the role of weeds in the epidemiology of insect pests and plant diseases can be found in Thresh (1981), especially for crop diseases transmitted from weeds to adjacent crop plants by insect vectors.

Often the domesticated version of a plant species is the result of an intensive breeding program that results in, among other factors, a reduction in the concentration of secondary substances in various plant parts, thus producing plants with simpler and less stable defenses against herbivores (Harlan 1975).

The presence of wild relatives in crop borders may have local effects on the population genetics of a crop pest. Presumably the wild plants can contribute to the genetic diversity of pest populations, at least if migration is limited. Thus maintenance of a variety of plants that possess different compliments of plant defenses in crop border flora could result in the preservation of genetic diversity in the local pest population, and decrease the rate of selection for new biotypes that have the ability to overcome host plant resistance or withstand the application of pesticides (Thresh 1981).

Weedy plants near crop fields can provide the requisites for pest outbreaks. The presence of *Urtica dioica* L. in the host layer of noncrop habitats surrounding carrot fields was the most important factor determining high levels of carrot fly larval damage to adjacent carrots (Wainhouse and Coaker 1981). Adult leafhoppers invade peach orchards from edge vegetation and subsequently colonize trees whose ground cover is composed of preferred wild hosts (McClure 1982). Plantains (*Plantago* spp.) and docks (*Rumex* spp.) provide alternative food for the rosy apple aphid *Dysaphis plantaginea* (Passerini) and dock sawfly *Ametastegia glabrata* (Fallén), important pests of apple in England (Altieri and Letourneau 1982).

4. THE ROLE OF WEEDS IN THE ECOLOGY OF NATURAL ENEMIES

Certain weeds are important components of agroecosystems because they positively affect the biology and dynamics of beneficial insects. Weeds offer many important requisites for natural enemies such as alternative prey/hosts, pollen, or nectar as well as microhabitats that are not available in weed-free monocultures (van Emden 1965). The beneficial entomofauna associated with weeds has been surveyed for many species, including the perennial stinging nettle (*U. dioica*), Mexican tea (*Chenopodium ambrosioides* L.), camphorweed [*Heterotheca subaxillaris* (Lam.)], and a number of ragweed species (Altieri and Whitcomb 1979).

In the last 20 years research has shown that outbreaks of certain types of crop pests are less likely to occur in weed-diversified crop systems than in weed-free fields, mainly due to increased mortality imposed by natural enemies (Pimentel 1961, Adams and Drew 1965, Dempster 1969, Flaherty 1969, Root 1973, Smith 1969, Altieri et al. 1977). Crop fields with a dense weed cover and high diversity usually have more predaceous arthropods than do weed-free fields (Pimentel 1961, Dempster 1969, Flaherty 1969, Root 1973, Perrin 1975, Smith 1969, Speight and Lawton 1976). The successful establishment of several parasitoids has depended on the presence of weeds that provided nectar for the adult female wasps. Relevant examples of cropping systems in which the presence of specific weeds has enhanced the biological control of particular pests are given in Altieri and Letourneau (1982).

Van Emden (1962) demonstrated that certain Ichneumonidae, such as *Mesochorus* spp., must feed on nectar for egg maturation, and Leius (1967) reported that carbohydrates from the nectar of certain Umbelliferae are essential in normal fecundity and longevity in three Ichneumonid species. In studies of the parasitoids of the European pine shoot moth, *Rhyacionia buoliana* (Denis and Schiffermüller), Syme (1975) showed that fecundity and longevity of the wasps, *Exeristes comstockii* (Cresson) and *Hyssopus thymus* Girault were significantly increased with the presence of several flowering weeds.

Spectacular parasitism increase has been observed in annual crops and orchards with rich undergrowths of wild flowers. In apple, parasitism of tent caterpillar eggs and larvae and codling moth larvae was 18 times greater in those orchards with floral undergrowths than in orchards with sparse floral undergrowth (Leius 1967).

Soviet researchers at the Tashkent Laboratory (Telenga 1958) cited lack of adult food supply as a reason for the inability of *Aphytis proclia* (Walker) to control its host, the San Jose scale (*Quadraspidiotus perniciosus* (Comstock). The effectiveness of the parasitoid improved as a result of planting a *Phacelia* sp. cover crop in the orchards. Three successive plantings of *Phacelia* increased scale parasitization from 5% in clean cultivated orchards to 75% where these nectar producing plants were grown. These Russian researchers also noted that *Apanteles glomeratus* (L.), a parasite of two cabbageworm species (*Pieris* spp.) on crucifer crops, obtained nectar from wild mustard flowers. The parasites lived longer and laid more eggs when these weeds were present. When quick-flowering mustards were actually planted in the fields with cole crops, parasitization of the host increased from 10 to 60% (Telenga 1958).

Weed flowers are also important food sources for various insect predators (Van Emden 1965). Pollen appears to be instrumental in egg production of many syrphid flies and is reported to be a significant food source for many predaceous Coccinellidae. Lacewings seem to prefer several composite flowers that supply nectar, thus satisfying their sugar requirements.

Weeds may increase populations of nonpestiferous herbivorous insects in crop fields. Such insects serve as hosts to entomophagous insects, thus improving the survival and reproduction of these beneficial insects in the agroecosystem. For example, the effectiveness of the tachinid *Lydella grisecens* Robineau-Desvoidy, a parasite of the European corn borer, *Ostrinia nubilalis* (Hübner), can be increased in the presence of an alternate host, *Papaipema nebris* (Guenée), a stalk borer on giant ragweed (*Ambrosia* spp.) (Syme 1975).

Several other authors have reported that the presence of alternate hosts

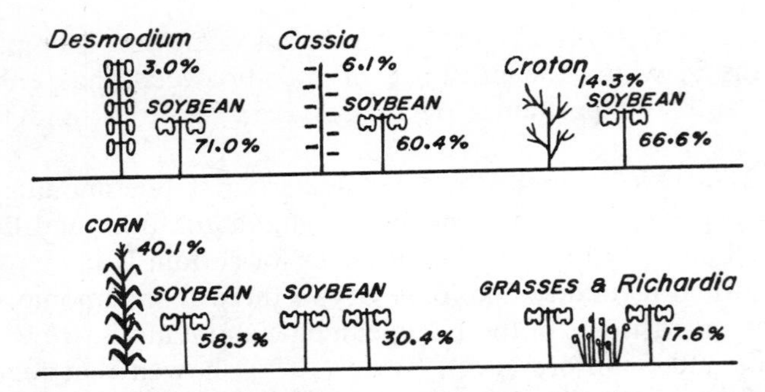

Figure 1. The effect of weed diversity on percent parasitization of corn earworm (*Heliothis zea*) eggs by parasitic wasps (*Trichogramma* spp.) in Georgia soybean fields. (From Altieri et al. 1981a)

on ragweeds near crop fields increased parasitism of specific crop pests. Examples include *Eurytoma tylodermatis* Ashmead against the boll weevil, *Anthonomus grandis grandis* Boheman, and *Macrocentrus delicatus* (Cresson) against the oriental fruit moth, *Grapholita molesta* (Bosch). The parasite *Horogenes* spp. uses the caterpillar of *Swammerdamia lutarea* on the weed *Crataegus* sp. to overwinter each year after emergence from the diamondback moth, *Plutella maculipennis* (Curt.) (van Emden 1965). A similar situation occurs with the egg parasitoid *Anagrus epos* Girault, whose effectiveness in regulating the grape leafhopper, *Erythroneura elegantula* Osborn, is greatly increased in vineyards near areas invaded by wild blackberry (*Rubus* sp.) This plant hosts an alternate leafhopper *Dikrella cruentata* Gillette that breeds in its leaves during winter (Doutt and Nakata 1973).

Some entomophagous insects are attracted to particular plants, even in the absence of host or prey, by chemicals released by the herbivore's host plant or other associated plants (Altieri et al. 1981a). For example, the parasitic fly *Eucelatoria* sp. prefers okra to cotton, and the wasp *Peristenus pseudopallipes* (Loan), which attacks the tarnished plant bug, prefers *Erigeron* to other weed species (Monteith 1960, Nettles 1979). Parasitism by *Diaeretiella rapae* (McIntosh) was much higher when the aphid *Myzus persicae* (Sulzer) was on collard then when it was on beet, a plant lacking attactive mustard oil (Read et al. 1970).

Of significant practical interest are the findings of Altieri et al. (1981a), which showed that parasitization rates of *Heliothis zea* (Boddie) eggs by *Trichogramma* sp. were greater when the eggs were placed on soybeans next to corn and the weeds *Desmodium* sp., *Cassia* sp., and *Croton* sp., than on soybeans grown alone (Fig. 1). Although the same number of eggs was placed on the associated plants, few of these eggs were parasitized, suggesting that these plants were not actively searched by *Trichogramma* sp. but nevertheless enhanced the efficiency of parasitization on the associated soybean plants. It is possible that they emitted volatiles with kairomonal action. Further tests showed that application of water extracts of some of these associated plants (especially *Amaranthus* sp.) to soybean enhanced parasitization of *H. zea* eggs by *Trichogramma* spp. wasps. The authors stated that a stronger attraction and retention of wasps in the extract treated plots may be responsible for the higher parasitization levels. The possibility that vegetationally complex plots are more chemically diverse than monocultures, and therefore more acceptable and attractive to parasitic wasps, opens new dimensions for biological control through weed management and behavior modification.

In general, most beneficial insects present on weeds tend to disperse to crops, but in a few instances the prey found on weeds prevent or delay this dispersal. In such cases allowing weeds to grow to ensure concentrations of insects, and then cutting them regularly to force movement, could be an effective strategy. For example, by cutting patches of stinging nettle (*U. dioica*) in May or June, predators (mainly Coccinellidae) were forced to move into crop fields (Perrin 1975). Similarly, cutting the grass weed cover drove Coccinellidae into orchard trees in southeastern Czechoslovakia (Hodek 1973). By cutting hedges of *Ambrosia trifida* infested with the weevil *Lixus scrobicollis* Boheman, a 10% increase of boll weevil parasitization by *E. tylodermatis* was obtained in two test plots of cotton adjacent to the hedgerow (Pierce et al. 1912). These practices must be carefully timed and based on the biology of beneficial insects. For example, in California the annual cleanup of weeds along the edges of alfalfa fields should be delayed until after mid-March, when aggregations of dormant Coccinellidae have largely dispersed (van den Bosch and Telford 1964).

5. INSECT DYNAMICS IN WEED-DIVERSIFIED CROP SYSTEMS

Field experiments show that careful diversification of the weedy component of agricultural systems often lowers pest populations significantly (Altieri et al. 1977, Risch et al. 1983). Researchers have found at least two underlying mechanisms involved in these pest reductions. In some cases plant dispersion and diversity appear to influence herbivore density, primarily by altering herbivore movement or searching behavior (Risch 1981, Bach 1980, Kareiva 1983). In other cases predators and parasites encounter a greater array of alternative resources and microhabitats in weedy crops, reach greater abundance and diversity levels, and impose greater mortality on pests (Root 1973, Letourneau and Altieri 1983).

Many studies indicate that insect pest dynamics are affected by the lower concentration and/or greater dispersion of crops intermingled with weeds. For example, adult and nymph densities of *Empoasca kraemeri* Ross and Moore, the main bean pest of the Latin American tropics, were reduced significantly as weed density increased in bean plots. Conversely, the chrysomelid *Diabrotica balteata* (LeConte) was more abundant in weedy bean habitats than in bean monocultures; bean yields were not affected because feeding on weeds diluted the injury to beans. In other experiments *E. kraemeri* populations were reduced significantly in weedy habitats, especially in bean plots with grass weeds [*Eleusine indica* (L.) Gaertn. and *Leptochloa filiformis* (Lam.)]. *D. balteata* densities fell by 14% in these systems. When grass-weed borders one meter wide surrounded bean monocultures, populations of adults and nymphs of *E. kraemeri* fell drastically. If bean plots were sprayed with a water homogenate of fresh grass-weed leaves, adult leafhoppers were repelled. Continuous applications affected the reproduction of leafhoppers, as evinced by a reduction in the number of nymphs (Altieri et al. 1977). Their regulatory effect was greater than that of broadleaf weeds such as *Amaranthus dubius* Mart. Also pure stands of *L. filiformis* reduced adult leafhopper populations significantly more than *E. indica* (Schoonhoven et al. 1981); this effect ceased when the weed was killed with paraquat.

Weeds within a crop system can reduce pest incidence by enticing pest insects away from the crop. For example, flea beetles, *Phyllotreta cruciferae* (Goeze) concentrate their feeding more on the intermingled *Brassica campestris* L. plants than on collards (Altieri and Gliessman 1983). The weed species has significantly higher concentrations of allylisothiocyanate (a powerful attractant of flea beetle adults) than collards, thus diverting the beetles from the crops. Similarly, in Tlaxcala, Mexico, the presence of flowering *Lupine* spp. in tasseling corn fields often diverts the attack of the scarab beetle, *Macrodactylus* sp., from female corn flowers to lupine flowers (Trujillo, pers. comm.).

A couple of studies have documented pest reduction due to an increase of natural enemies in weedy crop fields. Fall armyworm *Spodoptera frugiperda* (J. E. Smith) incidence was consistently higher in weed-free corn plots than in corn plots containing natural weed complexes or selected weed associations. Corn earworm (*H. zea*) damage was similar in all weed-free and weedy treatments, suggesting that this insect is not affected greatly by weed diversity. Experimental design has been shown to be a crucial factor in demonstrating the effect of weeds on predator populations. In an experiment conducted by Altieri and Whitcomb (1980), field plots were close together and predators moved freely between habitats. As a result it was difficult to identify between-treatment differences in the composition of predator communities. In another experiment increased distances between plots minimized such migrations, resulting in greater population densities and diversity of common foliage insect predators in the weed-manipulated corn systems than in the weed-free plots. Trophic relationships in the weedy habitats were more complex than food webs in monocultures.

Similarly, spring-planted alfalfa plots infested with weeds had a less diverse substrate predator complex but a greater foliage predator complex than did weed-free plots (Barney et al. 1984). The carabid *Harpalus pennsylvanicus* De Geer and the foliage predators [i.e., *Orius insidiosus* (Say) and Nabidae] were more abundant in alfalfa fields where grass weeds were dominant.

One problem with the above experiments is that they do not isolate crop–weed diversity as an independent variable. In most cases weed diversity could only have reduced herbivore abundance because it reduced the size or quality of crop plants (Kareiva 1983). Weed density, diversity, plot or patch size, are all interacting factors that may influence crop quality and herbivore densities.

Although all hypotheses explaining herbivore population dynamics in mixed cropping systems attribute a role to the physical structure of the plant canopies in achieving decreased herbivore abundance and/or increased natural enemy densities, few experiments have removed the confounding influence of plant species richness to assess the effects of plant architecture and density per se (Altieri and Letourneau 1984). The presence of weeds in crops affects both plant density and spacing patterns, factors known to significantly influence insect populations (Mayse 1983). In fact, many herbivores respond specifically to plant density; some proliferate in close plantings, whereas others reach high numbers in open canopy crops. Predator and parasite populations tend to be greater in high density plantings. Mayse suggests that the microclimate associated with canopy closure, which occurs earlier in dense plantings, may increase development rates of some predators and possibly facilitate prey capture.

Careful consideration of the units to be used in expressing population numbers is crucial for meaningful interpretations of results and for determining general patterns among various research findings. For example, Mayse and Price (1978) found that numbers of certain arthropod species sampled in different soybean row spacing treatments were significantly different on a per plant basis, but those same population values converted to a square meter of soil area basis were not significantly different.

Based on the preceding considerations, in addressing the effects of crop–weed density/spacing on insect populations, we should fully consider (1) the effects on the growth, development and nutritional status of the crop plants and weeds, (2) effects on the microclimate and microhabitats available for the life processes of herbivores and their natural enemies, and (3) effects of potentially different levels of herbivores in the population dynamics of predators and parasites.

Andow (1983) purposely designed an experiment to separate the effects of decreased pest attack [from *Epilachna varivestis* Mulsant and *Empoasca fabae* (Harris)] and increased plant competition that often occur simultaneously in weed-diversified bean systems (Fig. 2). In this system weeds directly affected bean herbivores by reducing their population densities and intensity of attack, thereby indirectly benefiting the bean plants. At the same time the weeds directly competed with the beans. There is a negative correlation between the intensity of insect attack on beans and the intensity of competition. At one extreme, in the monoculture, there was very high pest attack

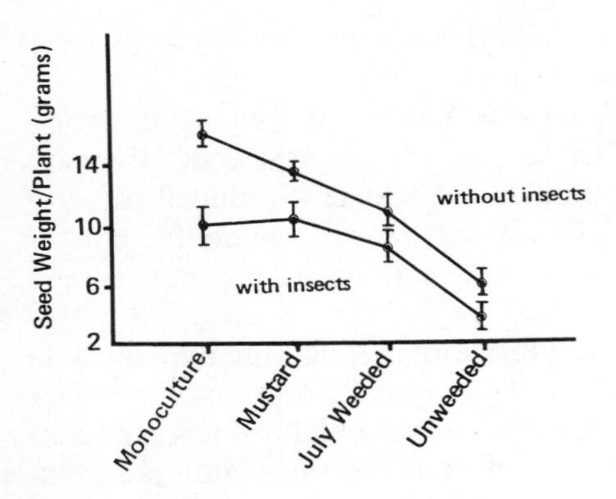

Figure 2. Yields of beans (as seed weight per plant) with and without herbivores and with and without weeds. Insects were eliminated with insecticides. The four levels of weediness were (1) no weeds (monoculture), (2) interplanted with 8200 clumps of wild mustards (*Brassica kaber*)/ha, midway between alternate bean rows, (3) natural populations of weeds for the first 35 days after planting and weeded from then on (July weeded), and (4) natural populations of weeds for the entire growing season (unweeded). (From Andow 1983)

but no interspecific competition, whereas at the other, the unweeded treatment, there was very intense competition but very little insect attack. Without insects, competition reduced the yield of beans in these treatments. However, when insects were present there was no difference in yield among the three treatments. Thus at low levels of crop–weed competition the effects of reduced insect pest damage were large enough to balance out the effects of increased plant competition (Fig. 3).

In contrast, there was no significant interaction between yields and insect damage in the unweeded treatment. This was the treatment with the greatest reduction in insect pest populations and herbivore damage. Despite this great decrease in the intensity of herbivory, there was no detectable yield response. Apparently the reduction in yield from the intense weed competition was so large that the positive effect of reduced herbivory was swamped out. Andow (1983) concluded that three-way interactions among beans, weeds, and bean herbivores were important when bean–weed competition was not very intense.

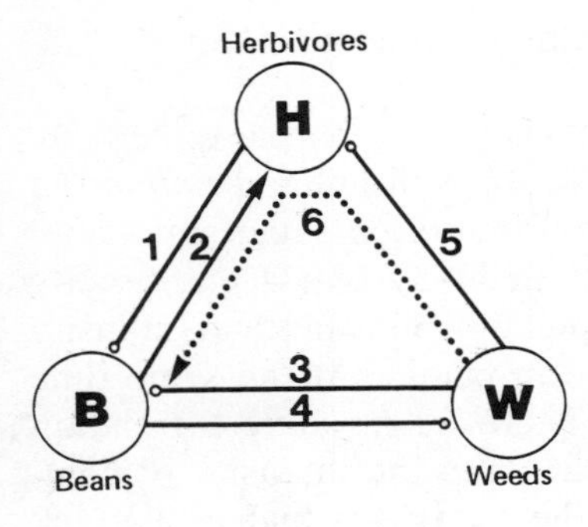

Figure 3. Interactions among beans (B), weeds (W), and bean herbivores (H). The interactions are denoted by lines; interactions that are beneficial for the recipient end in arrows, and those that are detrimental to the recipient end in circles. The following interactions are depicted: (1) herbivory on beans, (2) feeding by herbivores, (3) competition by weeds that affects beans, (4) competition by beans that affects weeds, (5) interference-type effects of weeds on bean herbivores, and (6) indirect beneficial effect that weeds have on beans, the product of the negative interactions 1 and 5. (From Andow 1983)

6. CROP–WEED MANAGEMENT CONSIDERATIONS

As indicated by the studies discussed in this chapter, much evidence suggests that encouragement of specific weeds in crop fields may improve the regulation of certain insect pests (Altieri and Whitcomb 1979). Naturally, careful manipulation strategies need to be defined in order to avoid weed competition with crops and interference with certain cultural practices. In other words, economic thresholds of weed populations need to be defined, as well as factors affecting crop–weed balance within a crop season (Bantilan et al. 1974).

Shifting the crop–weed balance so that insect regulation is achieved and crop yields are not economically reduced may be accomplished by using herbicides or selecting cultural practices that favor the crop cover over weeds. Suitable levels of desirable weeds that support populations of beneficial insects can be attained within fields by (1) designing competitive crop mixtures, (2) allowing weed growth as alternate rows or in field margins only, (3) use of cover crops, (4) adoption of close row spacings, (5) providing weed-free periods (i.e., keeping crops free of weeds during the first third of their growth cycle from sprouting to harvest), (6) mulching, and (7) cultivation regimes.

In addition to minimizing the competitive interference of weeds, changes in the species of weed communities are desirable to ensure the presence of plants that affect insects. Manipulation of weed species can be achieved by several means, such as changing levels of key chemical constituents in the soil, use of herbicides that suppress certain weeds while encouraging others, direct sowing of weed seeds, and timing soil disturbances (Altieri and Whitcomb 1979).

Perhaps one of the most useful concepts in managing weeds for insect regulation within fields is the "critical period." This is the maximum period that weeds can be tolerated without affecting final crop yields, or the point after which weed growth does not affect final yield. Duration of weed competition data for particular crops has been compiled by Zimdahl (1980). In studies in southern Georgia, Altieri et al. (1981b) observed that populations of the velvetbean caterpillar (*Anticarsia gemmatalis* Hübner) and of the southern green stink bug (*Nezara viridula* L.) were greater in weed-free soybeans than in either soybeans left weedy for two or four weeks after crop emergence, or for the whole season. Soybeans maintained weed-free for two or four weeks after emergence required no further weed control to produce optimum yield (Walker et al. 1984).

In another experiment, allowing weed growth during selected periods of the collard crop cycle (two or four weeks weed-free or weedy all season) resulted in lower flea beetle (*P. cruciferae*) densities in the weedy monocultures than in the weed-free monocultures. Lowest densities occurred in systems allowed to remain weedy all season. No differences in the abundance of beetles were observed between collards kept weed-free for two or four weeks after transplanting (Altieri and Gliessman 1983).

In collard systems with "relaxed" weeding regimes, flea beetle densities were at least five times greater on a per plant basis on *B. campestris* (the dominant plant of the weed community) than on collards. *B. campestris* germinated quickly and flowered early, each plant averaging a height of 39 cm, with 12 leaves and 16 open flowers, 60 days after germination. This apparent preference of *P. cruciferae* for *B. campestris* over collards resulted in a higher concentration of flea beetles on the wild crucifer, diverting flea beetles from collards and consequently diluting their feeding on the collards. Collards grown under various levels of weed diversity exhibited significantly less leaf damage than monoculture collards grown in weed-free situations (Altieri and Gliessman 1983).

Defining periods of weed-free maintenance in crops so that numbers of pests do not surpass tolerable levels might prove to be a significant compromise between weed science and entomology, a necessary step to explore further the interactions described in this chapter.

7. THE EFFECTS OF HERBIVORES ON THE CROP–WEED BALANCE

Studies examining the effects of insect herbivory on plant competition have been almost exclusively conducted in nonagricultural systems. There is general agreement that differential herbivory on two competing plants significantly affects the population dynamics of both plant species.

Insect seed predators differentially preyed upon seeds of two-weed species, *Astragalus cibarius* Sheldon and *A. utahensis* (Torr.) Torr. and Gray, apparently responding to differences in phenology, seed and pod energy content, chemicals in the pods, and pod morphology (Green and Palmbald 1975). In a study of the effects of grazing by the chrysomelid beetle *Gastrophysa viridula* (De Geer) on the competitive interaction of two dock species *Rumex obtusifolius* and *R. crispus*, Bentley and Whittaker (1979) showed that levels of feeding that had no significant effect on either species when grown in isolation resulted in extensive damage to *R. crispus* when the two species were competing (Fig. 4). *Rumex crispus* responded by reducing its root-to-shoot dry weight ratio from 2.14 to 1.69, making more material available for consumption, whereas *R. obtusifolius* increased the ratio from 1.18 to 3.57, ensuring that material within the root system was protected from aboveground feeding. However, in further field experiments using mixed cultures subjected to normal beetle feeding pressure, total seed production, and seed weight was significantly reduced in *R. obtusifolius*, whereas in *R. crispus* the number of seeds was not reduced but seed weight was lower in one experiment. These results indicate a complex interaction in which herbivory can modify both the relative competitive fitness of the individual growing plants and their reproductive potential.

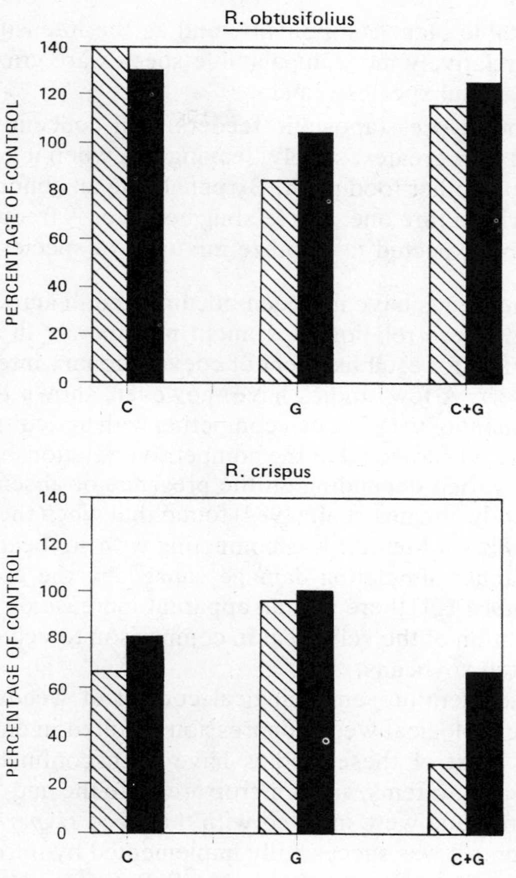

Figure 4. Effect of light grazing by the chrysomelid beetle *Gastrophysa viridula* (De Geer) on competition between two weed species, *Rumex obtusifolius* and *R. crispus*. Shaded columns represent mean leaf area and black columns represent root dry weight; C, competition; G, grazing; C + G, competition and grazing combined. (From Bentley and Whittaker 1979, by permission of Blackwell Scientific Publications.)

Based on a number of field studies, Harper (1969) developed the hypothesis that selective defoliation by insects can explain some of the vegetational diversity in certain habitats. He argued that:

1. The introduction of an herbivore freed from predator/parasite regulation can take a plant species from a dominant to a minority component position of the flora and make room for invasion by other plants.
2. Because most herbivores exhibit palatability preferences, they selectively remove plant parts or species. Undergrazing then favors the

less palatable plant components, and as the intensity of grazing increases, relatively more unpalatable species are grazed and only totally distasteful species remain.

3. Certain herbivores (apostatic feeders) will concentrate on the food plant that is in greatest supply, leaving this when it becomes rare and turning to another food plant. By penalizing an abundant food supply and favoring a rare one, this feeding behavior is frequency dependent and can be expected to stabilize mixtures of species.

The preceding trends have not been documented in agroecosystems, and possibly some of these relationships might not operate in annual cropping systems where time for establishment of coevolutionary interactions is functionally very short. A few studies have, however, shown that insects have altered development of weeds when competing with agricultural plants. Windle and Franz (1979) showed that the competitive relationships between two barley cultivars varied depending on the presence or absence of aphids. In a competition study Higgins et al. (1984) found that when the weed velvetleaf (*Abutilon theophrastii* Medic.) was competing with soybean, which was defoliated in a manner simulating damage caused by the green cloverworm (*Plathypena scabra* F.), there was an apparent increase of 13 to 55% in dry weight accumulation of the velvetleaf in comparison to velvetleaf competing with underfoliated soybeans.

Of course the literature on biological control of weeds is replete with examples where biological weed suppression resulted in enhanced yields of desired plants. Most of these studies have been confined to rangelands, plantations, aquatic systems, and noncrop areas. In the mid-1940s large areas of California rangeland were infested with the weed *Hypericum perforatum* L. Biological control was successfully implemented by introducing the phytophagous beetle *Chrysolina quadrigemina* (Suffrian), and within a few years *Hypericum* had become an uncommon plant (Huffaker and Messenger 1976).

A number of factors limit the applicability of weed biocontrol in annual crops (Altieri and Doll 1978a). In agroecosystems weeds occur in mixed species complexes and usually require rapid control in the early season. Insecticide applications, cultivation, and other disturbances can drastically affect weed herbivores. To circumvent some of these limitations, a number of management procedures are required to enhance the performance of natural enemies. Altieri and Doll (1978b) suggest the following prerequisites:

1. Predominance of a dominant weed species, normally an herbicide-resistant weed.
2. Crops tolerant to a low but continuous weed population level.
3. Maintenance of a survival level of the weed during the entire year for the herbivore to feed and self-perpetuate.
4. Minimum use of pesticides and other practices that disturb the natural balance.

Table 1. Dynamics of Pigweed (*Amaranthus dubius*) Damage Imposed by the Caterpillar (*Hymenia recurvalis*) in Corn–Bean Cropping Systems, under Various Fertilizer Regimes

			Fertilizer Effect[c]		
Cropping System	Damage Intensity (%)[a]	Damage Frequency (%)[b]	Increase in Larval Population (%)	Increase in Pigweed Damage (%)	Degree of Biocontrol at Harvest
Corn monoculture	28.8	36.4	68.0	14.2	High
Bean monoculture	23.3	32.3	99.9	61.2	Fair
Corn–bean polyculture	27.5	31.0	99.9	56.5	High

After Altieri and Doll 1978b.

[a] Percent leaves damaged, 40 days after planting.

[b] Percent plants damaged per plot, 40 days after planting.

[c] Increase in *H. recurvalis* density and pigweed damage when comparing plots with and without fertilizer (15–15–15).

Surveys of the entomofauna associated with dominant weeds in agroecosystems have revealed varying degrees of natural control. For example, in California vineyards a number of defoliating caterpillars and spider mites inflict significant damage on field bindweed (*Convolvulus arvensis* L.), a dominant weed (Rosenthal 1985). In several row and vegetable crops in California's central valley, common purslane (*Portulaca oleraceae* L.) can be severely damaged by a sawfly *Schizocerella pillicornis* (Holmgren) and a leaf-mining weevil (*Hypurus bertrandi* Perris) (Clement and Norris 1982).

In Colombia, larvae of *Hymenia recurvalis* (F.) exerted considerable damage on pigweed (*A. dubius*) in experimental mono- and polycultures of beans and corn (Altieri and Doll 1978a). Although no differences in *H. recurvalis* damage to pigweed were observed among the three cropping systems, application of nitrogen fertilizer had a stimulating effect on *H. recurvalis*, increasing its population densities and damage to *Amaranthus* in the bean monoculture and corn-bean polyculture plots (Table 1).

Very few researchers have attempted to actually manipulate weed biocontrol agents in annual crops. An isolated case in Pakistan showed that the release of *Trioza chenopodii* in wheat plots infested with the weed *Chenopodium album* L. resulted in substantial suppression of *C. album* growth, increasing wheat biomass production by 49%. The biomass of *C. album* in plots where *T. chenopodii* was released was 64% less than those in which it was not released (Mohynddin, pers. comm.).

8. CONCLUSIONS

Unquestionably, weeds stress crop plants through interference processes. However, substantial evidence suggests that weed presence in crop fields cannot be automatically judged damaging and in need of immediate control. In fact, crop–weed interactions are overwhelmingly site specific and vary according to plant species involved, environmental factors, and management practices. Thus in many agroecosystems weeds are ever-present components adding to the complexity of interacting trophic levels mediating a number of crop–insect interactions with major effects on final yields. It is here argued that in weed-diversified systems we cannot understand plant–herbivore interactions without understanding the effects of plant diversity on natural enemies, nor can we understand predator–prey and parasite–host interactions without understanding the role of the plants involved in the system.

By considering the potential role of herbivores present in mediating crop–weed interactions in weed diversified crops, we extend the theory of crop–weed interference and broaden its basis. Encouraging herbivores capable of tipping the balance of crop–weed competitive interaction bears considerable practical importance for weed management in agroecosystems.

An increasing awareness of these ecological relationships should place emphasis on weed management, as opposed to weed control, so that herbicides can be considered merely a component part of a total system for managing weeds and season-long, weed-free monocultures cannot always be assumed to be the best crop production strategy (Aldrich 1984).

ACKNOWLEDGMENTS

Critical comments by Matt Liebman and Linda L. Schmidt are greatly appreciated.

REFERENCES

Adams, J. B., and M. E. Drew. 1965. Grain aphids in New Brunswick. III. Aphid populations in herbicide-treated oatfields. *Can. J. Zool.* 43: 789–794.

Akobundu, I. O. 1980. Weed control strategies for multiple cropping systems of the humid and subhumid tropics. *In* Akobundu, I. O. (ed.), *Weeds and their Control in the Humid and Subhumid Tropics*. International Institute of Tropical Agriculture, Ibadan, Nigeria.

Aldrich, R. J. 1984. *Weed–Crop Ecology—Principles in Weed Management*. Breton, N. Scituate, MA.

Altieri, M. A., and J. D. Doll. 1978a. The potential of allelopathy as a tool for weed management in crop fields. *PANS* 24: 495–502.

Altieri, M. A., and J. D. Doll. 1978b. Some limitations of weed biocontrol in tropical crop ecosystems in Colombia. *In* Freemen, T. E. (ed.), *Proc. 4th Intl. Symp. Biological Control of Weeds*. University of Florida, Gainesville, pp. 74–82.

Altieri, M. A., and S. R. Gliessman. 1983. Effects of plant diversity on the density and herbivory of the flea beetle, *Phyllotreta cruciferae* Goeze, in California collard (*Brassica oleracea*) cropping systems. *Crop Prot.* 2: 497–501.

Altieri, M. A., and D. K. Letourneau. 1982 Vegetation management and biological control in agroecosystems. *Crop Prot.* 1: 405–430.

Altieri, M. A., and D. K. Letourneau. 1984. Vegetation diversity and insect pest outbreaks. *CRC Critical Reviews in Plant Science* 2: 131–169.

Altieri, M. A., and M. Liebman. 1986. Insect, weed and plant disease management in multiple cropping systems. *In* Francis, C. A. (ed.), *Multiple Cropping: Principles, Practices and Potentials*. Macmillan, New York.

Altieri, M. A., and W. H. Whitcomb. 1979. The potential use of weeds in the manipulation of beneficial insects. *HortScience* 14: 12–18.

Altieri, M. A., and W. H. Whitcomb. 1980. Weed manipulation for insect pest management in corn. *Environ. Mgmt.* 4: 483–484.

Altieri, M. A., A. van Schoonhoven, and J. D. Doll. 1977. The ecological role of weeds in insect pest management systems: a review illustrated with bean (*Phaseolus vulgaris* L.) cropping systems. *PANS* 23: 195–205.

Altieri, M. A., W. J. Lewis, D. A. Nordlund, R. C. Gueldner, and J. W. Todd. 1981a. Chemical interactions between plants and *Trichogramma* wasps in Georgia soybean fields. *Prot. Ecol.* 3: 259–263.

Altieri, M. A., J. W. Todd, E. W. Hauser, M. Patterson, G. A. Buchanan, and R. H. Walker. 1981b. Some effects of weed management and row spacing on insect abundance in soybean fields. *Prot. Ecol.* 3: 339–343.

Andow, D. A. 1983. *Plant Diversity and Insect Populations: Interactions among Beans, Weeds and Insects*. Ph.D. dissertation. Cornell University. Ithaca, NY. 201 p.

Bach, C. G. 1980. Effects of plant density and diversity on the population dynamics of a specialist herbivore, the striped cucumber beetle, *Acalymma vittatta* (Fab). *Ecology* 61: 1515–1530.

Bantilan, R. T., R. R. Harwood, and M. C. Palada. 1974. Integrated weed management. I. Key factors affecting crop–weed balance. *Philippine Weed Science Bull.* 1: 14–36.

Barney, R. J., W. O. Lamp, E. J. Armbrust, and G. Kapusta. 1984. Insect predator community and its response to weed management in spring-planted alfalfa. *Prot. Ecol.* 6: 23–33.

Bendixen, L. E., and D. J. Horn. 1981. *An Annotated Bibliography of Weeds as Reservoirs for Organisms Affecting Crops. III. Insects*. Ohio Agric. Res. and Development Center, Wooster.

Bentley, S., and J. B. Whittaker. 1979. Effects of grazing by a chrysomelid beetle, *Gastrophysa viridula*, on competition between *Rumex obtusifolius* and *Rumex crispus*. *J. Ecol.* 67: 79–90.

Bhandari, D. C., and D. N. Sen. 1979. Agroecosystem analysis of the Indian arid zone. I. *Indigofera cordifolia* as a weed. *Agro-Ecosystems* 5: 257–262.

Chacon, J. C., and S. R. Gliessman 1982. Use of the ''non-weed'' concept in traditional tropical agroecosystems of southeastern Mexico. *Agro-Ecosystems* 8: 1–11.

Clement, S. L., and R. F. Norris. 1982. Two insects offer potential biological control of common purslane. *Calif. Agric.* 36: 16–18.

Dempster, J. P. 1969. Some effects of weed control on the numbers of the small cabbage white (*Pieris rapae* L.) on brussel sprouts. *J. Appl. Ecol.* 6: 339–405.

Doutt, R. L., and J. Nakata. 1973. The *Rubus* leafhopper and its egg parasitoid: an endemic biotic system useful in grape pest management. *Environ. Entomol.* 2: 381–386.

Flaherty, D. 1969. Ecosystem trophic complexity and densities of the Willamette mite, *Eotetranychus willamettei* (Acarina: Tetranychidae) densities. *Ecology* 50: 911–916.

Gliessman, S. R. 1983. Allelopathic interactions in crop–weed mixtures: applications for weed management. *J. Chem. Ecol.* 9: 981–999.

Green, T. W., and I. G. Palmbald. 1975. Effects of insect seed predators on *Astragalus cibarius* and *Astragalus utahensis* (Leguminosae). *Ecology* 56: 1435–1440.

Harlan, J. R. 1975. Our vanishing genetic resources. *Science* 188: 618–622.

Harper, J. L. 1969. The role of predation in vegetational diversity. *In Diversity and Stability in Ecological Systems. Brookhaven Symposium in Biology* 22: 48–62.

Higgins, R. A., D. W. Staniforth, and L. P. Pedigo. 1984. Effects of weed density and defoliated or undefoliated soybeans (*Glycine max*) on velvetleaf (*Abutilon theophrasti*) development. *Weed Sci.* 32: 511–519.

Hodek, I. 1973. *Biology of Coccinellidae*. Academic Pub., Prague.

Huffaker, C. B., and P. S. Messenger. 1976. *Theory and Practice of Biological Control*. Academic Press, New York.

Kareiva, P. 1983. Influence of vegetation structure on herbivore population: resource concentration and herbivore movement. *In* Denno, R. F., and M. S. McClure (eds.), *Variable Plants and Herbivores in Natural and Managed Systems*. Academic Press, New York, pp. 259–289.

Leius, K. 1967. Influence of wild flowers on parasitism of tent caterpillar and codling moth. *Can. Entomol.* 99: 444–446.

Letourneau, D. K., and M. A. Altieri. 1983. Abundance patterns of a predator, *Orius tristicolor* (Hemiptera: Anthocoridae), and its prey, *Frankliniella occidentalis* (Thysanoptera: Thripidae): habitat attraction in polycultures versus monocultures. *Environ. Entomol.* 12: 1464–1469.

Mayse, M. A. 1983. Cultural control in crop fields: a habitat management technique. *Environ. Mgmt.* 7: 15–22.

Mayse, M. A., and P. W. Price. 1978. Seasonal development of soybean arthropod communities in east central Illinois. *Agro-Ecosystems* 4: 387–405.

McClure, M. S. 1982. Factors affecting colonization of an orchard by leafhopper (Homoptera: Cicadellidae) vectors of peach X-disease. *Environ. Entomol.* 11: 695–699.

Monteith, L. G. 1960. Influence of plants other than the food plants of their host on finding by tachinid parasites. *Can. Entomol.* 92: 641–652.

Moody, K. 1977. Weed control in multiple cropping. *In Proceedings of a Symposium on Cropping Systems Research and Development for the Asian Rice Farmer*. Intl. Rice Res. Inst., Los Banos, Philippines.

Nettles, W. C. 1979. *Encelatoria* sp. females: factors influencing response to cotton and okra plants. *Environ. Entomol.* 8: 619–623.

Norris, R. F. 1982. Interactions between weeds and other pests in the agroecosystem. *In* Hatfield, J. L., and G. J. Thomason (eds.), *Biometeorology in Integrated Pest Management*. Academic Press, New York.

Perrin, R. M. 1975. The role of the perennial stinging nettle *Urtica dioica* as a reservoir of beneficial natural enemies. *Ann. Appl. Biol.* 81: 289–297.

Pierce, W. D., R. A. Cushman, and C. E. Hood. 1912. The insect enemies of the cotton boll weevil. *Bureau Entomol. Bull.* No. 100, Washington, DC.

Pimentel, D. 1961. Species diversity and insect population outbreaks. *Ann. Entomol. Soc. Am.* 54: 76–86.

Price, P. W., C. E. Bouton, P. Gross, B. A. McPheron, J. N. Thompson, and A. E. Weise. 1980. Interactions among three trophic levels: influence of plants on interaction between insect herbivores and natural enemies. *Annu. Rev. Ecol. Syst.* 11: 41–65.

Putnam, A. R., and W. B. Duke. 1982. Allelopathy in agroecosystems. *Annu. Rev. Phytopathol.* 16: 431–451.

Read, D. P., R. P. Feeny, and R. B. Root. 1970. Habitat selection by the aphid parasite *Dia-*

eretiella rapae (Hymenoptera: Braconidae) and hyperparasite *Charips brassicae* (Hymenoptera: Cynipidae). *Can. Entomol.* 102: 1567–1578.

Rice, E. L. 1984. *Allelopathy*. 2d ed. Academic Press, Orlando, FL.

Risch, S. J. 1981. Insect herbivore abundance in tropical monocultures and polycultures: an experimental test of two hynotheses. *Ecology* 62: 1325–1340.

Risch, S. J., D. Andow, and M. A. Altieri. 1983. Agroecosystem diversity and pest control: data, tentative conclusions, and new research directions. *Environ. Entomol.* 12: 625–629.

Root, R. B. 1973. Organization of a plant-arthropod association in simple and diverse habitats: the fauna of collards (*Brassica olearcea*). *Ecol. Monographs* 43: 95–124.

Rosenthal, S. S. 1985. Potential for biological control of field bindweed in California's coastal vineyards. *Agric. Ecosyst. Environ.* 13: 43–58.

Schoonhoven, A. van, C. Cardona, J. Garcia, and F. Garzon. 1981. Effect of weed covers on *Empoasca kraemeri* Ross and Moore populations and dry bean yields. *Environ. Entomol.* 10: 901–907.

Shetty, S. V. R., and M. R. Rao. 1981. Weed management studies in sorghum/pigeon pea and pearl millet/groundnut intercrop systems—some observations. *In Proc. Intl. Workshop on Intercropping*. International Center for Research in the Semi-Arid Tropics, Patancheru, Andhra Pradesh, India, pp. 238–248.

Smith, J. G. 1969. Some effects of crop background on populations of aphids and their natural enemies on brussels sprouts. *Ann. Appl. Biol.* 63: 326–330.

Speight, H. R., and J. H. Lawton. 1976. The influence of weed cover on the mortality imposed on artificial prey by predatory ground beetle in cereal fields. *Oecologia* 23: 211–233.

Syme, P. D. 1975. The effects of flowers on the longevity and fecundity of two native parasites of the European pine shoot moth in Ontario. *Environ. Entomol.* 4: 337–346.

Telenga, N. A. 1958. Biological methods of pest control in crops and forest plants in the USSR. *In Report of the Soviet Delegation, 9th Int. Conf. on Quarantine and Plant Prot.*, Moscow, pp. 1–15.

Thresh, J. M. (ed.). 1981. *Pest, Pathogens and Vegetation: The Role of Weeds and Wild Plants in the Ecology of Crop Pests and Diseases*. Pitman, Concord, MA.

van den Bosch, R., and A. D. Telford. 1964. Environmental modification and biological control. *In* DeBach, P. (ed.). *Biological Control of Insect Pests and Weeds*. Chapman and Hall, London, pp. 459–488.

van Emden, H. F. 1962. Observations on the effects of flowers on the activity of parasitic Hymenoptera. *Entomologists' Monthly Magazine* 98: 225–236.

van Emden, H. F. 1965. The role of uncultivated land in the biology of crop pests and beneficial insects. *Sci. Hortic.* 17: 121–136.

Wainhouse, D., and T. H. Coaker. 1981. The distribution of carrot fly (*Psila rosae*) in relation to the flora of field boundaries. *In* Thresh, J. M. (ed.), *Pests, Pathogens and Vegetation: The Role of Weeds and Wild Plants in the Ecology of Crop Pests and Diseases*. Pitman, Concord, MA. pp. 263–272.

Walker, R. H., and G. A. Buchanan. 1982. Crop manipulation in integrated weed management systems. *Weed Sci.* 30: 17–24.

Walker, R. H., M. G. Patterson, E. Hauser, D. Isenhour, J. Todd, and G. A. Buchanan. 1984. Effects of insecticide, weed-free period, and row spacing on soybean (*Glycine max*) and sicklepod (*Cassia obtusifolia*) growth. *Weed Sci.* 32: 702–706.

William, R. D. 1981. Complementary interactions between weeds, weed control practices, and pests in horticultural cropping systems. *HortScience* 16: 508–513.

Windle, P. N., and E. A. Franz. 1979. The effects of insect parasitism on plant competition: greenbugs and barley. *Ecology* 60: 521–529.

Zimdahl, R. L. 1980. *Weed–Crop Competition—A Review*. International Plant Protection Center, Corvallis, OR.

AUTHOR INDEX

SUBJECT INDEX